OCEAN ACOUSTIC TOMOGRAPHY

CAMBRIDGE MONOGRAPHS ON MECHANICS

OCEAN ACOUSTIC TOMOGRAPHY

WALTER MUNK
Scripps Institution of Oceanography
University of California, San Diego

PETER WORCESTER
Scripps Institution of Oceanography
University of California, San Diego

CARL WUNSCH
Department of Earth, Atmosphere and Planetary Sciences
Massachusetts Institute of Technology

CAMBRIDGE UNIVERSITY PRESS
Cambridge, New York, Melbourne, Madrid, Cape Town, Singapore, São Paulo, Delhi

Cambridge University Press
The Edinburgh Building, Cambridge CB2 8RU, UK

Published in the United States of America by Cambridge University Press, New York

www.cambridge.org
Information on this title: www.cambridge.org/9780521115360

First published 1995
This digitally printed version 2009

A catalogue record for this publication is available from the British Library

Library of Congress Cataloguing in Publication data

Munk, Walter H. (Walter Heinrich), 1917–

Ocean acoustic tomography / Walter Munk, Peter Worcester, Carl Wunsch.

p. cm. – (Cambridge monographs on mechanics)

Includes bibliographical references and index.

ISBN 0-521-47095-1

1. Ocean tomography. I. Worcester, Peter. II. Wunsch, Carl.
III. Title. IV. Series.
QC242.5.023M86 1995
620.2′5–dc20 94-38469
CIP

ISBN 978-0-521-47095-7 hardback
ISBN 978-0-521-11536-0 paperback

Additional resources for this publication at www.cambridge.org/9780521115360

To our partners in the original ocean acoustic tomography group

David Behringer
Theodore Birdsall
Michael Brown
Bruce Cornuelle
Robert Heinmiller
Robert Knox
Kurt Metzger
John Spiesberger
Robert Spindel
Doug Webb

CONTENTS

6. THE INVERSE PROBLEM: DATA-ORIENTED 222

7. THE INVERSE PROBLEM: MODEL-ORIENTED 303

8. THE BASIN SCALE 323

EPILOGUE. THE SCIENCE OF OCEAN ACOUSTIC TOMOGRAPHY 346

APPENDIX 355

PREFACE

Over drinks in the Cosmos Club in 1979, Athelstan Spilhaus, who had perfected the bathythermograph for measuring temperature profiles to predict the ranges at which submarines could be detected acoustically, held forth that it should be done the other way around: the measured sonar transmission should serve to determine the ocean temperature field. Unknown to Spilhaus, we were in Washington to persuade the Office of Naval Research and the National Science Foundation to fund an experiment to do just that.

In seismology, the inversion of travel times to map the interior of the Earth has been the time-honored procedure, since the Earth is not readily accessible to direct intrusive measurements. In medicine, intrusive methods are viewed with some reluctance (at least on the part of the patient), and this has led to the development of computed tomographic inverse methods using X-rays. In contrast, the oceans are accessible to direct intrusive measurements; the limits are set by the availability of costly platforms for adequate sampling. Unlike the seismological and medical applications, ocean time variability is an essential component, and the requirements for sampling in space and time are severe. With only a few research vessels plying the world's oceans, it is not surprising that the first century of oceanography had a strong climatological flavor.

It came as a great shock in the 1960s that the oceans, like the atmosphere, had an active *weather* at all depths. The storms within the sea are called eddies. Typical spatial scales are 100 km; time scales are 100 days. Ocean eddies are far more compact and long-lived than their atmospheric counterparts. The intensity of the eddies is such that they contain the predominant fraction of kinetic energy in midocean regions. There is a great difference between an ocean with currents of 10 ± 1 cm/s and an ocean with currents of 1 ± 10 cm/s. With the appreciation of this intensive *mesoscale* field (Russian-speaking scientists refer to it as the *synoptic* scale), it became evident how inadequate the existing observational sampling strategy was, a strategy that had permitted 99% of the kinetic energy to slip through the grid. The proposal for *Ocean Acoustic Tomography* was a direct consequence of that realization.[1]

The Office of Naval Research of the United States Navy supported our initial research to develop acoustic techniques for ocean monitoring and has strongly supported our work ever since, without attempting to influence the direction of the research; we are particularly indebted to Gordon Hamilton and Hugo Bezdeck for their early encouragement. The National Science Foundation also began supporting us in 1981, and we have continued to enjoy support from both agencies. The work was an informal collaborative effort by scientists from several institutions. This book is dedicated to our partners in the first three-dimensional test of ocean acoustic tomography: D. Behringer, T. Birdsall, M. Brown, B. Cornuelle, R. Heinmiller, R. Knox, K. Metzger, J. Spiesberger, R. Spindel, and D. Webb. In addition, we have worked with many other scientists over the years, all of whom have made significant contributions to the field: S. Flatté, J. Guoliang, B. Howe, J. Lynch, P. Malanotte-Rizzoli, J. Mercer, J. Miller, J. Romm, F. Zachariasen, and B. Zetler. The following have read the manuscript and made many helpful suggestions: B. Cornuelle, J. Colosi, B. Dushaw, M. Dzieciuch, B. Howe, D. Menemenlis, and U. Send. Last, but by no means least, recognition is due the engineers, programmers, and technicians who were responsible for developing the instrumentation and performing the experiments required to test the concepts of *Ocean Acoustic Tomography*.

Elaine Blackmore and Breck Betts have worked with us for many years in preparing this volume; we are deeply indebted to them. K. Rolt has greatly contributed to the final preparation of the manuscript.

We have worked together on acoustic tomography by telemail and with shared enthusiasm, without a professional coordinator. At one stage a reviewer termed our organizational structure a disaster, but gave the proposal his reluctant support when we pointed to forty published papers. It had been intended that the tomography group would disband after a few years, but we are still working together. It is only fitting that this book is dedicated to our partners.

The reader will find a multitude of errors. Please inform Walter Munk, Scripps Institution of Oceanography, UCSD, 9500 Gilman Drive, La Jolla, CA 92093-0225. FAX 619/534-6251 or wmunk@igpp.ucsd.edu

[1] Application of tomography has subsequently broadened to include shorter- and longer-scale ocean processes.

NOTATION

Only the symbols that appear throughout the book are here identified; a notation that is used in one section only is defined locally. Several symbols have different meanings in different chapters. This duplication is impossible to avoid in a subject covering oceanography, acoustics, and inverse methods. When a choice had to be made between an established convention and some degree of ambiguity, our decision was with convention. The notation f, $\mathbf{f}$, $\mathbf{F}$, $\mathbf{F}^T$ refers to scalar, vector, matrix, and matrix transpose representations, respectively, of any quantity f.

$\mathbf{r} = (x, y, z)$	coordinates, z is upward from the sea surface
z_B, z_A, z_S	pertaining to bottom, sound axis, surface
$t, \tau^{\pm}$	clock time, travel time in direction $\pm x$
$R^{\pm}, T^{\pm}, A^{\pm}$	range, travel time, action for upper/lower ray loop
$r, t, a = n\,(R, T, A)$	total range, travel time, action for n double loops
$\delta R, \delta T, \delta A$	fractional ray loops
$R = R^+ + R^-$, etc.	range of double loop
$\theta, \ \theta = \arctan(m/k)$	inclination of ray, of modal wavenumber
$\Gamma, \Gamma(-)$	ray path, unperturbed ray path
RR, RSR, RBR	refracted refracted, refracted surface-reflected, refracted bottom-reflected
SLR, BLR	surface/bottom-limited ray
$\Delta\tau, \ \delta\tau$	perturbation in τ, error in τ
$D\tau = \tau_{n+1} - \tau_n$	interval between ray arrivals
$s, d = \frac{1}{2}\,(\Delta\tau^+ \pm \Delta\tau^-)$	sum and difference in travel-time perturbations
$C, \ S = 1/C$	sound-speed, sound-slowness

$\widetilde{C} = C(\widetilde{z}^{\pm})$	sound-speed at upper/lower turning depth, etc.
$\Delta C,\ \Delta S$	perturbations of sound-speed, sound-slowness
$\widehat{\Delta C},\ \widehat{\Delta S}$	estimated perturbations
$\widehat{\Delta C}\,(t_0, -)$	estimate prior to measurements at t_0
$c_p = \omega/k,\ c_g = d\omega/dk$	phase and group speeds
$s_p = k/\omega,\ s_g = dk/d\omega$	phase and group slowness
$\sigma^2 = (S_0^2 - S^2)/S_0^2$	dimensionless sound-slowness
$\phi^2 = \sigma^2/(\gamma_a h)$	normalized σ^2
$i\ (m, n)$	acoustic mode number, ray number
j	dynamic mode number, layer number
u, v	components of particle velocity
$T,\ T_p,\ Sa$	temperature, potential temperature, salinity
N	buoyancy frequency
$\gamma_a = 0.0113\ \text{km}^{-1}$	adiabatic gradient $-C^{-1}\,dC/dz$
k, ℓ, m	modal wavenumbers
$f,\ \omega = 2\pi f$	frequency in cyclical and circular units
a, b, c	coefficient in canonical profile (2.5.8)
RI, RD, RA, LA	range-independent, range-dependent, range-averaged, loop-averaged
$\mathbf{E}$	observation or design matrix
$\mathbf{P},\ \mathbf{P}_n$	solution uncertainty, solution variance
$\mathbf{x},\ \mathbf{y}$	statevector, observation vector
BT, XBT, AXBT	bathythermograph, expendable BT, airborne XBT
CTD	instrument measuring conductivity, temperature, and depth
$\langle\ \rangle$, rms	averaging operator, root-mean-square
$\equiv,\ \approx$	equals by definition, approximately equals
$O(\ldots)$	order of magnitude ...
i	$\sqrt{-1}$

CHAPTER 1

THE TOMOGRAPHY PROBLEM

The problem of ocean acoustic tomography is to infer from precise measurements of travel time, or of other properties of acoustic propagation, the state of the ocean traversed by the sound field. The tomographic method[1] was introduced by Munk and Wunsch (1979) in direct response to the demonstration in the 1970s that about 99% of the kinetic energy of the ocean circulation is associated with features that are only about 100 km in diameter, called the mesoscale.[2] Measuring and understanding the behaviors of both the mesoscale and the larger-scale features associated with the general circulation present a formidable sampling task. Not only are the flow elements very compact spatially, but also they have long time scales (order 100 days). To produce statistically significant measurements of the fluid behavior, even in an area as compact as 1 Mm × 1 Mm (1 megameter = 1000 km), about 1% of an ocean basin, requires several full-time vessels or several hundred fixed moorings. One is accordingly led to the technology of sound propagation to measure the properties of the fluid *between* moorings.

Ocean acoustic tomography takes advantage of the facts that (i) travel time and other measurable acoustic parameters are functions of temperature, water velocity, and other parameters of oceanographic interest and can be interpreted to provide information about the intervening ocean using inverse methods, and (ii) the ocean is nearly transparent to low-frequency sound, so that signals can be transmitted over distances of many thousands of kilometers. There is some analogy with classical seismology, in which the properties of the Earth's interior are inferred from travel times of earthquake waves. (However, in ocean tomography the emphasis is not on the *mean* field of sound-speed, but on its space-time *variability*.) One advantage of the oceanographic problem over the seismological and medical ones is that the interior of the ocean is generally accessible. In the long run, that access may have been a liability, for it delayed

[1] See Appendix A for an account of early ocean acoustic tomography.

[2] Figs. 1.5 and 1.6 show realizations of mesoscale-dominated oceans.

the development of indirect methods, which, in our prejudiced view, are made inevitable by the magnitude of the ocean sampling task.

There are numerous attractive features to ocean acoustic tomography. In common with other methods of remote sensing, it permits the monitoring of regions that are difficult to observe directly. Meanders of the Gulf Stream have been monitored with sources and receivers moored in the relatively sluggish waters to both sides of the stream. Another advantage has to do with the speed of sound (3000 knots) exceeding that of research vessels and permitting the construction of synoptic fields. The geometry of measuring the oceans *between* moorings can be exploited; with M moorings, conventional techniques yield M "spot" measurements. But M moorings consisting of S sources and R receivers yield $S \times R$ pieces of information, rather than $S + R = M$. This quadratic information growth is an attractive feature (though the quadratic loss with instrument failures is not).

A key attribute of tomographic measurements is that they are spatially *integrating*. The potential for forming horizontal and vertical averages over large ranges, up to global scales, is an attractive (but unfamiliar) tool; for example, the heat content in a vertical section across an ocean basin could be rapidly and repeatedly measured using tomographic techniques. The integrals suppress unwanted small scales that contaminate the conventional spot measurements, leading to aliasing. Transmissions over a few hundred kilometers subdue the internal wave "noise," and transmissions over a few thousand kilometers subdue the mesoscale noise. Such integral data test the skill of dynamic models and provide powerful model constraints.

Construction of a practical system, deployable at sea, accompanied by the mathematical apparatus to handle the resulting data, requires a working knowledge of elements of oceanography, acoustics, and mathematics, which are normally discussed in isolation. None of these elements is by itself particularly difficult, but in combination they present a formidable challenge. To sustain the reader's interest through much of the necessarily technical material that follows, this chapter is devoted to a heuristic summary account of ocean acoustic tomography. Most of the details are swept aside, to be taken up again in later chapters.

1.1. Ocean Acoustics

The sound-speed profile. Over much of the world ocean, the speed of sound, C, is characterized by a distinct minimum, at depths between 800 and 1200 m, called the sound channel axis. The minimum owes its existence to the dependence of the sound-speed on temperature and pressure. Sound-speed increases upward from the axis with increasing temperature, and increases downward from the axis (where the temperature gradients are small) with increasing pressure.

At high latitudes (north and south), the surface waters are increasingly colder, and the sound-channel axis shoals. During winter, convective overturning and the resulting mixing lead to near-adiabatic[3] conditions. Sound-speed then increases downward (negative z) at approximately the adiabatic rate: $-(1/C)\, dC/dz = \gamma_a = 0.0113\,\mathrm{km}^{-1}$.

We have found it useful to define two idealized model profiles for which most of the propagation characteristics can be derived analytically: the "temperate" (or canonical) sound-speed profile, and the polar winter (adiabatic) profile (fig. 1.1). Formulae are given in sections 2.17 and 2.18. In general, real profiles differ significantly from the idealized models, and change from place to place (see propagation atlas, Appendix B).

For many important dynamic purposes, the ocean is characterized by the vertical gradient of the potential density field, written as

$$N^2 = \frac{-g}{\rho}\frac{\partial \rho_p}{\partial z}, \tag{1.1.1}$$

where z is the vertical coordinate (upward from the surface), g is the force due to gravity, and ρ_p is the potential density (*i.e.*, the density corrected for adiabatic vertical displacements). N has the units of a frequency and is called the "buoyancy" or (less desirably) the Brunt-Väisälä frequency. The implications of N for ocean dynamics have been discussed by Turner (1973) and Gill (1982). In chapter 2, this quantity is related explicitly to the sound-speed gradient: under quite general conditions,

$$\frac{dC/dz}{C_A} = \gamma_a \frac{N^2 - N_A^2}{N_A^2}, \tag{1.1.2}$$

where C_A and N_A are the axial sound-speed and buoyancy frequency. For the two idealized models previously introduced,

$$\text{polar:} \quad N = 0, \qquad \text{temperate:} \quad N = N_0\, e^{z/h},$$

respectively. In polar winter, $N \to 0$ and $dC/dz \to -\gamma_a C_A$, the adiabatic profile. For the temperate profile at great depth, $N \to 0$ as in the polar profile. At the axis, $N(z) = N_A$ and $dC/dz = 0$. This depth of minimum sound-speed C_A (about 1 km in temperate latitudes) is where the acoustic waveguide is centered. Above the axis, $N(z)$ and hence dC/dz increase sharply up to a surface-mixed layer.

[3]Here *adiabatic* refers to vertical gradients in temperature, density, and sound-speed associated with the hydrostatic pressure gradient. The term is also used to refer to horizontal gradients so gradual that there is no significant acoustic scattering.

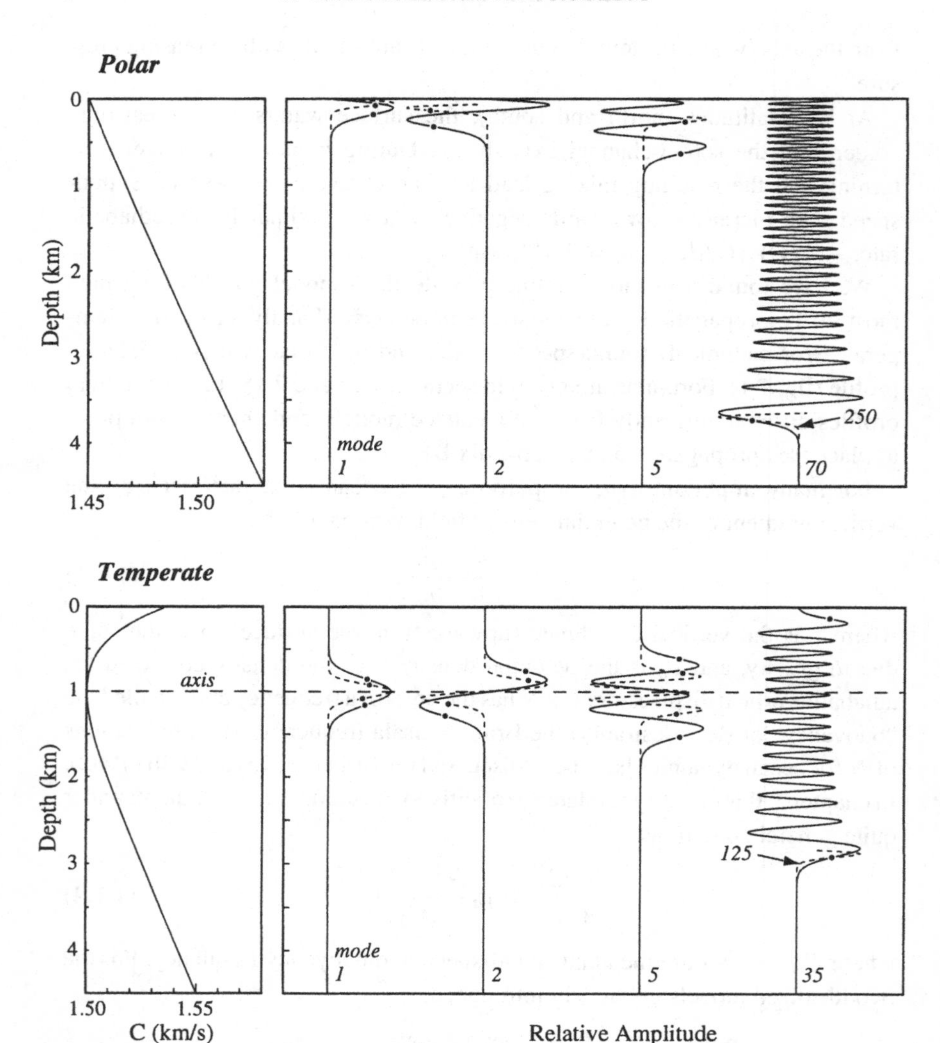

Fig. 1.1. Panels on left show the sound-speed profiles for the polar and temperate models. Polar rays are drawn for an (unrealistic) surface source and surface receiver. All rays are refracted surface-reflected (RSR); ray identifiers −7, −9, ... designate the number of (upper plus lower) turning points, including surface reflection, with minus signs indicating downward launch angles. Temperate rays +8, +9, −9, +10 are for an axial source and receiver. The vertical exaggeration in ray diagrams (right) is 25 to 1; a "small-inclination" approximation is generally valid. Selected modes $m = 1, 2, \ldots$ are shown at two frequencies, 70 Hz (solid) and 250 Hz (dashed), with WKBJ turning points indicated. Polar profile layers $j = 1$ to 6 serve as an example for the application of inverse theory.

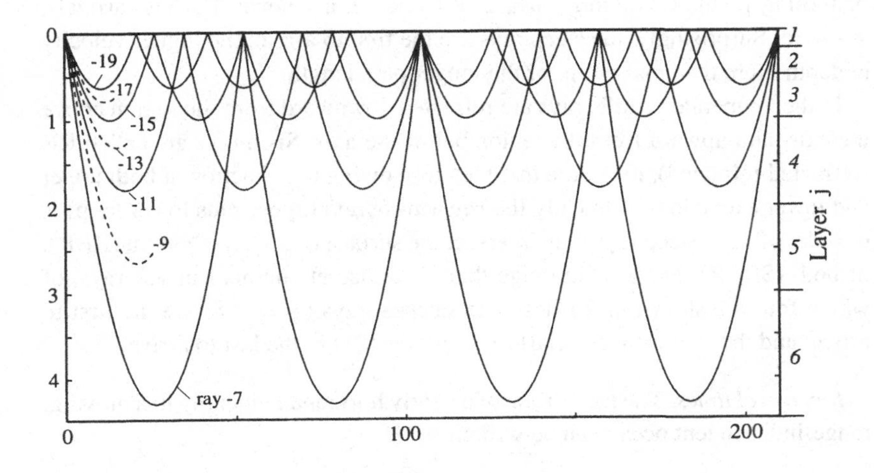

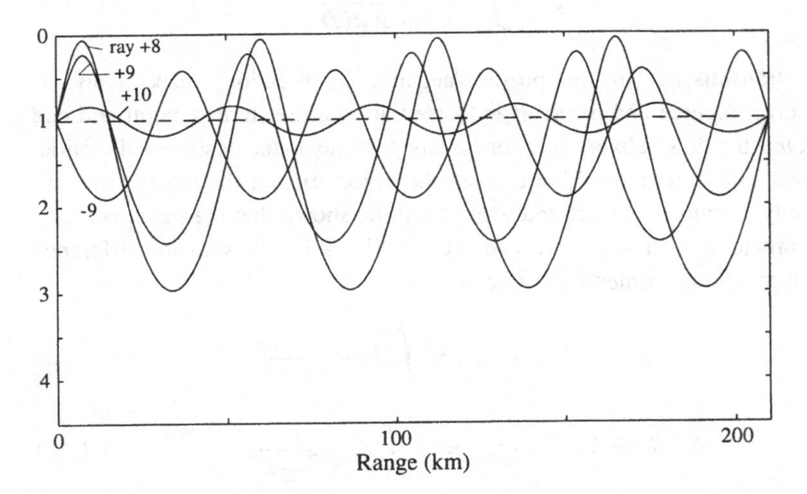

Fig. 1.1 (*cont.*)

Acoustic rays. A useful way of understanding the propagation of sound waves in the ocean is afforded by the geometric optics approach. Acoustic energy tends to propagate along arcs called "rays." In the presence of gradients in the sound-speed, Snell's law shows that the rays bend (refract) away from, or are "repelled" by, regions of higher speeds; for the polar profile, all rays are then refracted upward and subsequently reflected at the surface (fig. 1.1). They are called RSR (for refracted surface-reflected). At a specified range, rays can be designated $\pm p$ by the upward/downward launch direction and the number of turning points (including surface reflections), as shown. For late arrivals, $p \to \infty$. Surprisingly, the steep rays are the first to arrive; the higher velocity at depth more than makes up for the longer path length.

In the temperate profile, rays are refracted downward from the region above the axis, and upward from the region below the axis. Such rays are called RR (refracted refracted), to denote that they turn owing to refraction at both upper and lower excursions. Typically the *ray wavelength* (upper plus lower loop) is of order 50 km. Steep rays may intersect the surface (RSR), the bottom (RBR), or both (SRBR). For a fixed range there is a discrete number of RR rays, of which four are shown in fig. 1.1. The steepest rays ($p = \pm 8$) are the first to arrive, and the flatter (near-axial) rays ($p = \pm 10$) are the last to arrive.

Ray travel times. The travel time of a nearly horizontal eigenray in a moving, range-independent ocean can be written

$$\tau_n^{\pm} = \int_{\Gamma_n^{\pm}} \frac{ds}{C(z) \pm u(z)} \tag{1.1.3}$$

for a transmission in the positive/negative x direction, respectively. A transceiver (source and receiver) is located at both the starting point and end point; u is the flow velocity component along the ray in the positive x direction. The paths of integration $\Gamma_n^{\pm}$ are along the trajectories of the nth ray and are generally functions of $C(z)$ and $u(z)$. It will be shown that the path geometry is reciprocal to order $u/C \ll 1$, hence $\Gamma^{+} \approx \Gamma^{-} \equiv \Gamma$. The sum and difference of reciprocal travel times are defined by

$$s_n = \tfrac{1}{2}\left(\tau_n^{+} + \tau_n^{-}\right) = \int_{\Gamma} ds \, \frac{C}{C^2 - u^2}, \tag{1.1.4a}$$

$$d_n = \tfrac{1}{2}\left(\tau_n^{+} - \tau_n^{-}\right) = -\int_{\Gamma} ds \, \frac{u}{C^2 - u^2}. \tag{1.1.4b}$$

C and u are of order 10^3 and 10^{-1} m/s, respectively, so u^2 can be neglected in the denominator. The difference travel time is a small fraction of the one-way travel time τ. Hence C is well determined by one-way travel times in either direction.

Here we are more interested in the *perturbation* $\Delta\tau$ from a previous measurement, or from that inferred for the climatic ocean mean. Linearizing (1.1.4) yields

$$s_n = \int_{\Gamma} ds \, \frac{1}{C}, \quad \Delta s_n = -\int_{\Gamma} ds \, \frac{\Delta C}{C^2}, \quad d_n = -\int_{\Gamma} ds \, \frac{u}{C^2}. \tag{1.1.5a, b, c}$$

ΔC is of order 10 m/s and is still large compared with u. Hence, the perturbation ΔC is well determined by one-way travel time in either direction; measurements of u, however, require travel-time differences.[4]

Useful measurements of current profiles, $u(z)$, have been made by difference tomography (chapter 3). The method is particularly useful for separating barotropic (depth-independent) from baroclinic (depth-variable) tidal currents. Ocean current meter records are found to have comparable contributions from the barotropic and baroclinic components. Steep rays produce good estimates of vertically averaged horizontal velocities and are thus responsive predominantly to the barotropic component.

The determination of the sound-speed $C(x, y, z)$ is not of particular interest to oceanographers, except for its relation to temperature T and fluid density ρ. Both C and ρ are functions of T and salinity Sa (for fixed pressure p). We show that the effect of salinity is relatively small. An equation for C in terms of T and Sa is a complicated function that can be linearized to

$$\Delta C/C = \alpha\, \Delta T + \beta\, \Delta Sa\,,$$

with $\alpha \sim 3 \times 10^{-3}/°\mathrm{C}$, $\beta = 1 \times 10^{-3}/‰$. For a locally linear temperature–salinity relation

$$Sa = Sa(T_0) + \mu\, \Delta T, \quad \Delta T = T - T_0,$$

we have

$$\Delta C/C = \alpha\, \Delta T(1 + \mu\beta/\alpha)\,. \tag{1.1.6}$$

A typical value is $\mu\beta/\alpha = 0.03$, and a determination of ΔC is, to first order, a determination of the temperature field.

Acoustic modes. An alternative sound-field representation is in terms of acoustic modes. Application of modal theory is a straightforward way of solving the propagation problem in the range-independent case. For some analytical profiles, including the polar and temperate models, exact solutions exist.

Modes are designated by $m = 1, 2, \ldots$, having $0, 1, \ldots, m-1$ zero crossings of the vertical wave function, as shown in fig. 1.1. The scale of the mode function depends on frequency, f. Higher frequencies are more concentrated near the axis. (For the polar ocean, the axis is at the surface.) The inflection points farthest from the axis (the WKBJ "turning points") are measures of the

[4] We can regard u as a perturbation from zero flow, and write Δd_n for d_n.

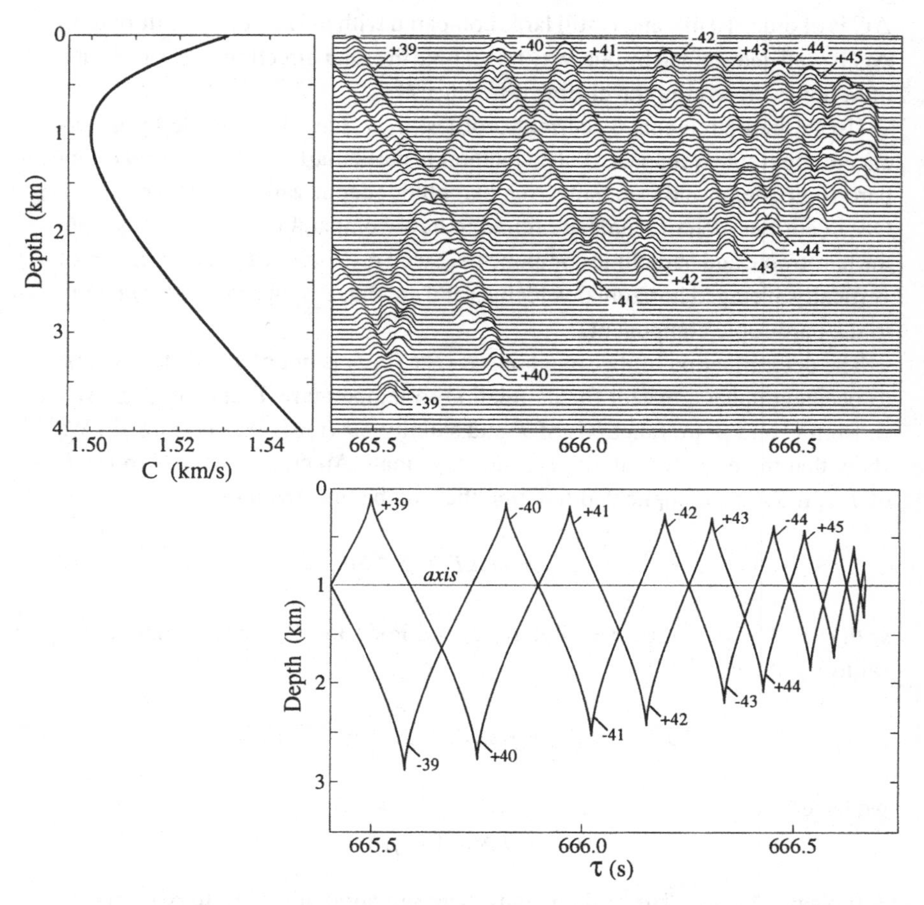

Fig. 1.2. Computed arrival pattern in a temperate ocean at 1 Mm range according to ray theory (bottom) and mode theory (top). The ray axial arrival pattern has a peak whenever one of the ray fronts crosses the line $z = z_{\text{axis}}$; the off-axis pattern is similarly derived using z_{receiver}. The modal pattern is derived by summing modes 1 to 250 at 250 Hz with bandwidth 100 Hz. It is seen that all but the latest interference ridges are ray-like. (The splitting of the ± 39 ridges is a surface effect.)

penetration of the modes into the ocean away from the axis. The figure shows only modes with turning points well above the bottom boundary.

An acoustic point source generates all these modes, with amplitudes proportional to the vertical wave function at the depth of the source. Each mode propagates with group velocity c_g (group slowness s_g), which is a known function of mode number m and frequency f. An important parameter is the "action,"

$$A = (m - \tfrac{1}{2})/f, \quad A = (m - \tfrac{1}{4})/f, \tag{1.1.7}$$

for non-surface-interacting and surface-interacting modes, respectively. Modes with the same value of A have the same turning point. For example, in a temperate-ocean mode, $m = 125$ at frequency $f = 250$ Hz and $m = 35$ at $f = 70$ Hz produce nearly the same $A \approx \frac{1}{2}$ s.

The composite arrival pattern at a receiver is obtained by summing over individual mode arrivals. Fig. 1.2 shows the result in a temperate ocean at a 1-Mm range for a 250-Hz source with 100-Hz bandwidth. Such a broadband source generates many modes with the same A and accordingly the same turning depth; these modes interfere constructively to produce the prominent accordion-like feature in fig. 1.2. For comparison, the "ray fronts" obtained from the ray theoretical approximation are shown. The patterns are in close agreement;[5] upper and lower ray turning depths occur at the two modal WKBJ turning depths. For both rays and modes, travel time at range r is conveniently written

$$\tau = s_g r \, , \tag{1.1.8}$$

but the expressions for phase and group slowness s_g are different (see section 2.11).

Under normal operating conditions, the ray-like constructive interference pattern is the most prominent and robust feature of the recorded signal. Nearly all tomography to date has been done using this feature, which is more easily interpreted by ray theory than by summing over many modes. Accordingly, in chapter 2 we first discuss the forward problem in terms of ray theory, and subsequently derive the modal interpretation, leading to some duplication.

1.2. The Forward and Inverse Problems

The "forward" or "direct" problem can be stated as follows: given $C(x, y, z)$ and $u(x, y, z)$, together with the characteristics of the sound source, compute the detailed structure of the signal at the receiver. This is the classical problem of finding solutions to the wave equation. The "inverse" problem demands calculation of the ocean properties, $C(x, y, z)$ and/or $u(x, y, z)$, given the properties of the transmitted and received signals. The inverse problem is of considerable interest in oceanography.

Application of an inverse method requires an understanding of the forward problem. When Munk and Wunsch (1979) proposed ocean acoustic tomography, they thought that the forward problem had been "solved." That notion

[5] Ray theory fails for the final, axial phase. A slight discrepancy in the early arrivals evidently has to do with the fact that RR rays do not feel the surface at all, no matter how close they are, whereas modes reaching near to the surface have some slight surface interaction.

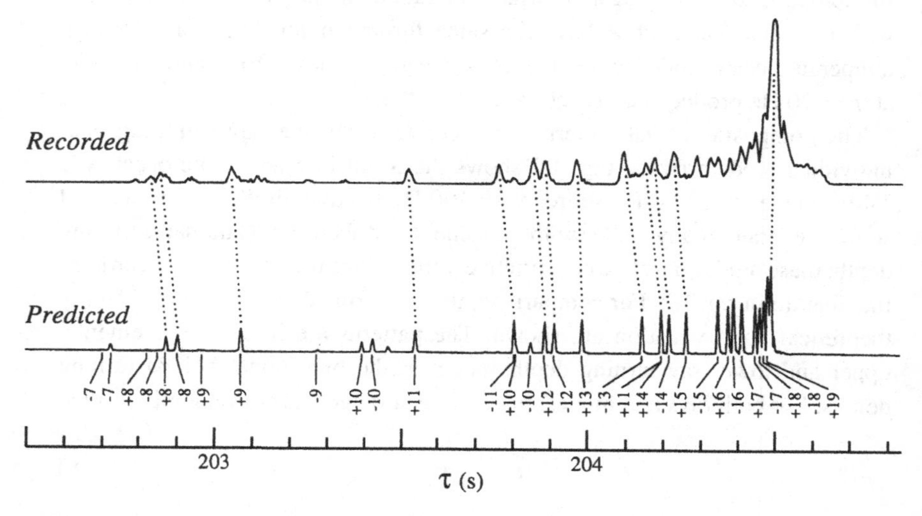

Fig. 1.3. Recorded and predicted (from historical data) axial arrival patterns for 1983, day 217, over a 300-km path west of Bermuda (Howe *et al.*, 1987). The predictions are based on ray theory, modified by Brown (1981). Recorded arrivals are generally earlier, suggesting that temperatures were above their climatological mean.

turned out to be false. Accordingly, much effort in the first decade following 1979 was devoted to the acoustic forward problem, leading to, among other things, a revision of the equation for sound-speed in sea water. Not until 1989 was the ocean adequately sampled to produce a binding comparison between prediction and measurement (Worcester *et al.*, 1994). Some problems remain, of course. For example, referring to fig. 1.2, the wedge, determined by the turning points converging onto the axis, is observed to be broader than computed, and the preceding ray-like pattern is more diffuse.[6]

Returning now to the tomography problem, fig. 1.3 compares the measured arrival pattern to that computed for the climatological-mean ocean at a temperate latitude site. Early arriving rays, which integrate over the entire ocean column, arrive slightly earlier than predicted, indicating that instantaneous temperatures were above their climatological mean. The late arrival, which "sees" only the axial ocean, is close to the prediction. Differences between the observed and predicted patterns thus contain information about the entire sound-speed profile, even though source and receiver are both located at only one depth (here they are both at the sound-channel axis). Inverse methods seek to exploit this information in a systematic way. For a successful inversion, one wishes the

[6] A possible explanation is offered by internal-wave-induced acoustic scattering (section 4.4).

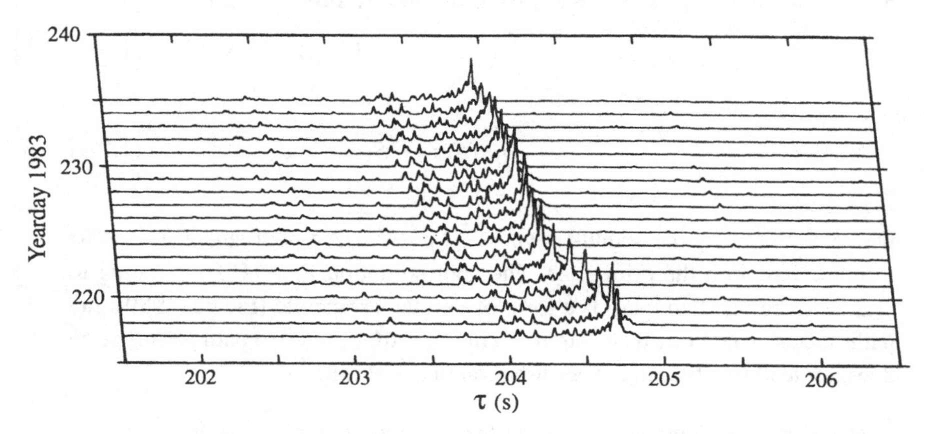

Fig. 1.4. Continuation of the arrival pattern seen in fig. 1.3, showing temporal stability. (From Worcester *et al.*, 1985*b*.)

arrivals to be resolved, identified, and stable. Measured arrival patterns following those shown in fig. 1.3 evidently fulfill these three conditions (fig. 1.4).

So far we have talked about travel time as the acoustic observable, but there are other possible parameters. Vertical receiver arrays can measure the direction of the incoming ray; the phase of a narrowband signal is another possible observable. The most ambitious procedure is to use the entire recorded pressure history $p(t)$ to infer the properties of the intervening ocean.

Prior estimates. For illustration, consider ray travel times in a motionless ocean. One can now define two related, but distinct, tomographic problems: (i) given the positions of source and receiver and a measured set of travel times $\{\tau_i\}$, find best estimates $\widehat{C}(z)$ of $C(z)$; (ii) given the best estimate, $\widehat{C}(z, -)$, of the "*prior* profile," a set of prior travel times $\{\tau_i(-)\}$ computed for the prior profile, and given the set of *measured* travel times $\{\tau_i\}$, calculate improved estimates for $\widehat{C}(z)$. The argument $(-)$ is introduced into the definitions to denote their dependence only on the prior model profile, independent of the measurements.

There are several potential data sets for the prior estimates $\widehat{C}(z, -)$. One is pure climatology – based on the historical hydrographic record. Another possibility is the use of hydrographic measurements taken during the deployment of the tomographic array, or an initialization based on a preceding tomographic inversion.

The difference between problems (i) and (ii) is in the introduction of a prior estimate of the sound-speed. Because we normally know a good deal about the ocean in any region where an experiment is to be conducted, it is the second problem leading to (1.1.5*b*), that is of most interest. From (1.1.3),

$$\Delta\tau = \tau_i - \tau_i(-) = \int_{\Gamma_i} S(z)\, ds - \int_{\Gamma_i(-)} S(z,-)\, ds \tag{1.2.1}$$

$$\approx \int_{\Gamma_i(-)} (S(z) - S(z,-))\, ds\,, \tag{1.2.2}$$

where $S = C^{-1}$ is the sound-slowness,[7] Γ_i is the true path, and $\Gamma_i(-)$ is its estimate based on the prior state. The approximation $\Gamma_i \approx \Gamma_i(-)$ leading to (1.2.2), to be justified later, is a form of Fermat's principle (Pierce, 1989): ray paths are associated with an extremum in travel time, and so a path perturbation does not lead to a first-order perturbation in travel time.[8]

Uncertainty estimates. We now write (1.2.2) in the form

$$\Delta\tau_i = \int_{\Gamma_i(-)} \Delta S(z)\, ds + \delta\tau_i\,, \tag{1.2.3}$$

where $\Delta S(z) = S(z) - S(z,-)$ is the perturbation from the prior estimate, and similarly for τ. The quantity $\delta\tau_i$ has been introduced to represent the noise contribution to $\Delta\tau_i$ that is inevitably present and that is a central element of any realistic discussion of acoustic tomography. $\delta\tau_i$ includes many sources of error: observation errors, modeling errors associated with the mathematical representation for $\Delta S(z)$, and nonlinearity errors in the approximation (1.2.3). Equations (1.2.3) for rays $i = 1$ to M represent the simplest interesting statement of a tomographic measurement.

The observational error is usually dominated by travel-time fluctuations associated with internal wave activity, so that the ultimate limit to tomographic inversions is imposed by the ocean itself. The subject has been intensively studied since the 1970s; see, for example, Flatté *et al.* (1979). The principal effect is due to temperature fluctuations at a fixed depth arising from the vertical particle displacements associated with internal waves. Internal wave "noise" increases with range roughly as $r^{\frac{1}{2}}$ and is relatively small for steep rays. A typical value is 10 ms rms at 1 Mm. For opposite transmissions along nearly reciprocal paths, much of the internal wave noise cancels, a most fortunate situation for velocity tomography.

[7] This simplifies the notation for $\Delta\tau$ from $-C^{-2}\int \Delta C\, ds$ to $+\int \Delta S\, ds$, at the expense of replacing the sound-speed C by the less familiar sound-slowness S; however, S is in close analogy with the index of refraction used in the optical literature. We use both C and S throughout this volume.

[8] In other words, $\Delta\tau$ is given to first order by integrating ΔS *along the unperturbed path*. This approximation leads to great simplification in ray travel-time tomography, as will be shown.

The inverse problem. Two simple perturbation models are (i) uniform perturbations ΔS_j within layers j and (ii) perturbations $\Delta S_j = a_j\ F_j(z)$ associated with hydrodynamic mode j or empirical orthogonal function j. The layer case is useful for illustration purposes and is physically realistic if many thin layers are used. Equation (1.2.3) can now be written

$$\Delta\tau_i = \sum_{j=1}^{N} E_{ij}\,\Delta S_j + \delta\tau_i, \quad 1 \le i \le M, \tag{1.2.4}$$

where $E_{ij}\,\Delta S_j$ is the contribution to the travel-time perturbation $\Delta\tau_i$ of ray i by the sound-slowness perturbation ΔS_j in layer $j = 1$ to N. The elements E_{ij} can be evaluated from prior information alone, according to

$$E_{ij} = \int_{\Gamma_i(-)} \delta_j\, ds\,,$$

where $\delta_j = 1$ in layer j and zero elsewhere, and so E_{ij} is the distance traveled by ray i in layer j over the entire range r. Equation (1.2.4) is a set of M equations for the N unknowns ΔS_j (and M unknown noise elements $\delta\tau_i$). We write (1.2.4) in the compact matrix notation

$$\mathbf{y} = \mathbf{E}\,\mathbf{x} + \mathbf{n}, \tag{1.2.5}$$

with

$$\mathbf{y} = [\Delta\tau_i], \quad \mathbf{E} = \{E_{ij}\}, \quad \mathbf{x} = [\Delta S_j], \quad \mathbf{n} = [\delta\tau_i]. \tag{1.2.6}$$

The inverse problem is to solve for $\mathbf{x}$ given the observations $\mathbf{y}$ in the presence of noise $\mathbf{n}$. Chapters 6 and 7 are devoted to a discussion of methods for accomplishing this task; it consists of making an *estimate*, $\widehat{\mathbf{x}}$, which will differ from the *true* $\mathbf{x}$ by $\delta\mathbf{x} = \widehat{\mathbf{x}} - \mathbf{x}$. Here we note three important features: (i) It will often be the case that there are fewer equations than unknowns ($M < N$). Indeed, this "underdetermined" case ought to be regarded as the normal situation, with "overdetermination" being exceptional.[9] (ii) The measured $\Delta\tau_i$ always contain errors $\delta\tau_i$. To allow for realistic errors is so important for attaining physically meaningful results that we regard discussions of the "noise-free" case as nearly irrelevant. The determination of the $\delta\tau_i$, $i = 1$ to M, is then an essential part of the solution. In this view there are $M + N$ unknowns and only M equations,

[9] Fluid systems have an infinite number of degrees of freedom, and oceanographers live chronically with underdetermined data sets.

and the problem is *always* underdetermined. (iii) An estimate of the uncertainty $\delta(\widehat{\Delta S_j})$ is as important as the estimate of the perturbation $\widehat{\Delta S_j}$ itself.

The estimate of $\widehat{\mathbf{x}}$ is written as a weighted linear sum of the observations,

$$\widehat{\mathbf{x}} = \mathbf{B}\,\mathbf{y} = \mathbf{B}(\mathbf{E}\,\mathbf{x} + \mathbf{n}), \tag{1.2.7}$$

using (1.2.5). With $\langle n \rangle = 0$, the expected value is

$$\langle \widehat{\mathbf{x}} \rangle = \mathbf{B}\,\mathbf{E}\,\langle \mathbf{x} \rangle \tag{1.2.8}$$

with $\langle\ \rangle$ denoting the averaging operation. The uncertainty is given by

$$\mathbf{P} = \left\langle (\widehat{\mathbf{x}} - \mathbf{x})(\widehat{\mathbf{x}} - \mathbf{x})^T \right\rangle = \left\langle (\mathbf{B}\mathbf{y} - \mathbf{x})(\mathbf{B}\mathbf{y} - \mathbf{x})^T \right\rangle. \tag{1.2.9}$$

$\mathbf{B}\,\mathbf{E}$ is called the *solution resolution matrix*. It gives the particular solution as a weighted average of the true solution in the absence of noise. If the resolution matrix is the identity $\mathbf{I}$, then the particular solution is the true solution in the absence of noise. If the row vectors of $\mathbf{B}\,\mathbf{E}$ are peaked along the diagonal, with low values elsewhere, the particular solution is a smoothed version of the true solution in the absence of noise. There are several ways of deriving $\mathbf{B}$. For illustration, consider one familiar approach.

Least-squares. Problems that are formally overdetermined, $M > N$, are often solved by classical least-squares, selecting the solution that makes the noise as small as possible through use of an "objective function,"

$$J = \mathbf{n}^T\mathbf{n} = (\mathbf{y} - \mathbf{E}\mathbf{x})^T(\mathbf{y} - \mathbf{E}\mathbf{x})\,. \tag{1.2.10}$$

Setting

$$\partial J/\partial \mathbf{x} = 0 \tag{1.2.11}$$

yields the expected values, $\langle \widehat{\mathbf{x}} \rangle$, and their uncertainties, $\mathbf{P}$, as given by (1.2.8) and (1.2.9), with

$$\mathbf{B} = (\mathbf{E}^T\mathbf{E})^{-1}\mathbf{E}^T. \tag{1.2.12}$$

The solution based on (1.2.12) has several potentially serious shortcomings: (i) the solution may have a magnitude inconsistent with what is acceptable; (ii) the smallest possible noise size may be inconsistent with what is known of its true value; (iii) if the equations are formally underdetermined, the matrix inverse does not exist, and the smallest possible noise is zero. There are several inverse methods that meet all of these objections. Consider one simple approach.

Tapered least-squares. A minor generalization of least-squares considers the redefined objective function

$$J = \alpha^2 \mathbf{x}^T \mathbf{x} + \mathbf{n}^T \mathbf{n}. \tag{1.2.13}$$

α^2 can be chosen to control the trade-off between minimizing the solution size, $\mathbf{x}^T\mathbf{x}$, and the error size, $\mathbf{n}^T\mathbf{n}$. Again setting $\partial J/\partial \mathbf{x} = 0$ yields $\widehat{\mathbf{x}} = \mathbf{B}\mathbf{y}$, with

$$\mathbf{B} = (\mathbf{I} + \alpha^{-2}\, \mathbf{E}^T\mathbf{E})^{-1}\alpha^{-2}\, \mathbf{E}^T, \tag{1.2.14}$$

which reduces properly to the plain least-squares solution (1.2.12) for $\alpha^2 \to 0$. This solution exists irrespective of the relative sizes of M and N, and it treats $\mathbf{n}$ on an equal footing with $\mathbf{x}$ as an essential part of the solution.

Constraints. Known ocean physics provides relationships among elements of $\mathbf{x}$ and can be used to reduce the solution uncertainty. As an illustration of the use of dynamic constraints, consider the determination of both sound-slowness and fluid velocity by measuring sum and difference travel times. Let $\mathbf{E}_S$ and $\mathbf{E}_u$ designate the appropriate matrices. Then the combined problem representing both fields simultaneously can again be written $\mathbf{y} = \mathbf{E}\mathbf{x} + \mathbf{n}$, provided

$$\mathbf{y} = \begin{bmatrix} \Delta s \\ \Delta d \end{bmatrix}, \quad \mathbf{E} = \begin{Bmatrix} \mathbf{E}_S & \mathbf{0} \\ \mathbf{0} & \mathbf{E}_u \end{Bmatrix}, \quad \mathbf{x} = \begin{bmatrix} \Delta \mathbf{S} \\ \mathbf{u} \end{bmatrix}, \quad \mathbf{n} = \begin{bmatrix} \mathbf{n}_s \\ \mathbf{n}_d \end{bmatrix} \tag{1.2.15}$$

[see (1.1.5)]. The zero matrices, $\mathbf{0}$, are dimensioned so as to make $\mathbf{E}$ and $\mathbf{x}$ conformable; the system is block-diagonal. Combining the $\Delta\mathbf{S}$ and $\mathbf{u}$ systems makes sense only if there are some additional equations linking velocity and sound-speed (as a surrogate for density) that can be exploited to provide improved solutions; otherwise (1.2.15) represents two disjoint problems better solved separately. This problem was discussed by Munk and Wunsch (1982*b*), who wrote

$$\mathbf{u} = \mathbf{A}\,\Delta\mathbf{S}$$

where $\mathbf{A}$ is a matrix of constants, asserting that the vertical derivative of horizontal velocity is proportional to the horizontal derivative of density in the plane normal to the velocity component (thermal-wind relation). This is a simple illustration. A more general constraint is provided by the equations of motion, that is, by combining the data analysis with numerical circulation models into a single inversion process. Parts of chapters 6 and 7 are devoted to this theme.

Table 1.1. *Observation matrix ρ_{ij} giving the fractional distances traveled by ray i in layer j of a polar ocean*[a]

		Ray i						
		1	2	3	4	5	6	7
Layer j		4.3	2.6	1.8	1.3	1.0	0.8	0.6 km
	0 km							
1		.02	.04	.06	.08	.11	.14	.17
	0.2							
2		.04	.06	.09	.13	.19	.26	.37
	0.5							
3		.06	.11	.18	.30	.70	.60	.46
	1.0							
4		.15	.29	.67	.49	0	0	0
	2.0							
5		.18	.50	0	0	0	0	0
	3.0							
6		.55	0	0	0	0	0	0
	4.5							

[a] Depth boundaries for each layer are given to the left. Turning depths for each ray are given at top. Note that no rays turn in layers 1 and 2.

Marked improvements can be achieved by allowing for what is known about the noise elements $\mathbf{n}$. For example, a clock error[10] affecting source timing leads to faulty travel times. We may not know this error, but we do know that all receptions from a single source are subject to the same clock value. Similarly, a faulty receiver timing leads to the same error from all sources. We can also exploit what is known regarding positioning errors and mooring motion (subject to mooring dynamics). In all these cases the noise elements are no longer uncorrelated. An elegant way to exploit this information is to remove clock and mooring elements from the noise vector, $\mathbf{n}$, and to add the essential clock and position parameters to the vector of unknowns, $\mathbf{x}$. In this sense, one is using some of the degrees of freedom to establish the source coordinates x, y, z, t, as in earthquake seismology (*e.g.*, Spencer and Gubbins, 1980). Tomographic experiments have in fact been done without any navigational controls, but it usually is worth the effort to keep accurate time and position, so that the observational effort can be fully devoted to the oceanographic estimates.

[10] Clock errors cancel in sum travel times; position errors cancel in difference travel times.

1.3. Vertical Slice: A Numerical Example

Consider the following overdetermined simple example of an inverse problem: seven rays in a polar ocean,[11] with six perturbation layers increasing in thickness from the surface downward (fig. 1.1, table 1.1). This example is for a range $r = 210$ km; at 2100 km there are 10 times as many available rays [but then the assumption of a range-independent mean state, $\widehat{C}(z)$, might lead to difficulties]. Let

$$\Delta S_j = x_j = -2,\ -1,\ -.5,\ -.25,\ 0,\ 0 \text{ ms/km} \tag{1.3.1}$$

for $j = 1$ to 6 be the "true" layer perturbations. For orientation, the corresponding perturbations in sound-speed and temperature are

$$\begin{aligned} \Delta C_j &= +4.5, \quad +2.2, \quad +1.1, \quad +0.6, \quad 0, \quad 0 \quad \text{m/s}, \\ \Delta T_j &= +1, \quad +\tfrac{1}{2}, \quad +\tfrac{1}{4}, \quad +\tfrac{1}{8}, \quad 0, \quad 0 \quad {}^\circ\text{C}. \end{aligned}$$

Using the values of ρ_{ij} in table 1.1 and setting $E_{ij} \approx r\rho_{ij}$ we obtain the travel-time perturbations

$$\begin{aligned} \Delta\tau_i = \mathbf{y}_i = (\mathbf{Ex} + \mathbf{n})_i &= -33{+}1.7,\ -58{-}1.9,\ -100{-}2.6, \\ &- 120{+}2.9,\ -160{+}1.3,\ -177{-}1.6,\ -200{+}0.2 \text{ ms} \end{aligned} \tag{1.3.2}$$

for $i = 1$ to 7. The values $-33,\ -58, \ldots$, for the "correct" travel-time perturbations have been modified by a "noise" of about 2 ms rms, which is a realistic estimate for the travel-time perturbation due to internal waves at 210 km range.

For illustration we shall now apply several inverse methods to the "data" given by (1.3.2).

Least-squares. It is readily confirmed that using (1.2.12) in (1.2.8) with noise-free data, $\Delta\tau_i = -33$ ms, -58 ms, ... , yields the "true" solution (1.3.1) exactly. This ideal situation is not very interesting. The central problem of oceanography is the interpretation of imperfect and inadequate data.

For the noise-corrupted travel times (1.3.2), the estimated perturbations ΔS_j and their uncertainties can be computed from (1.2.8) and (1.2.9),

$$\widehat{x}_j \pm \delta\widehat{x}_j = -1{\pm}2.4,\ -1.4{\pm}1.0,\ -0.5{\pm}0.5,\ -0.3{\pm}0.4,\ 0{\pm}0.2,\ 0{\pm}0.02 \text{ ms/km},$$

as compared with the true values $-2,\ -1,\ -0.5,\ -0.25,\ 0,\ 0$. The solution is correct within the standard errors, but the uncertainties near the surface are very

[11] Rays $i = 1$ to 7 are designated $p = -7, -9, -11, -13, -15, -17, -19$ in fig. 1.1. For geometric simplicity, source and receiver have been placed at the surface. The following results apply to source and receiver at *any* (equal) depth z^* provided z^* is above the turning depth of the flattest ray (here 0.6 km).

large. These can be attributed (as will be shown in chapter 2) to the vertical sampling properties of the rays. Ray travel-time perturbations are heavily weighted by the sound-speed perturbations near the ray turning points, and there are no turning points in the upper two layers.

The estimated travel-time errors are

$$\widehat{n}_i = y_i - (\mathbf{E}\widehat{\mathbf{x}})_i = -0.0,\ -0.0,\ -1.9,\ +2.7,\ +0.9,\ -2.3,\ +0.7 \text{ ms}.$$

It is not obvious from the formalism why the estimated noise in the two steepest rays is so small.

This overdetermined situation is not typical of the tomographic problem. Consider the situation $M < N$, by dropping the fourth and sixth arriving rays, using information from the five remaining rays to estimate the perturbations in six layers. The tapered least-squares solution with $\alpha^2 = 10^6$ produces a solution

$$\widehat{x}_j = -0.01,\ -0.2,\ -0.05,\ -0.18,\ -0.07,\ -0.03 \text{ ms/km},$$

with all uncertainties now less than 0.004 ms/km. The very small uncertainties do not encompass the correct values of $\widehat{x}_j$. The associated noise estimates

$$\widehat{n}_i = -29,\ -56,\ -97,\ -150,\ -192 \text{ ms}$$

are far larger than acceptable. On the other hand, choosing instead $\alpha^2 = 1$ produces

$$\widehat{x}_j = -0.7,\ -1.6,\ -0.5,\ -0.3,\ 0.0,\ 0.0 \text{ ms/km},$$

with all uncertainties now less than 0.04 ms/km. Evidently the solution is acceptable except very near the surface. However, the noise estimates

$$\widehat{n}_j = 1.3 \times 10^{-4},\ -2.1 \times 10^{-4},\ -2.0 \times 10^{-3},\ 1.4 \times 10^{-2},\ -2.6 \times 10^{-2} \text{ ms}$$

are now far too small. Apparently, somewhere between the values $1 \leq \alpha^2 \leq 10^6$ there is a choice of α^2 that will produce an appropriate noise norm.[12] The choice of α^2 is one aspect of the proper use of tapered least-squares as an inverse method (known as "ridge regression"). In general, only a rather narrow range of α-values produces simultaneously acceptable values of both $\widehat{x}_j$ and $\widehat{n}_j$. Least-squares provides an example of a powerful, but somewhat opaque, inverse method. The very small uncertainties in the estimates $\widehat{x}_j$ (between 0.004 and 0.04 ms/km) require explanation and interpretation.

[12]It can be shown that the proper α^2 is associated with accurate solutions except in the upper two layers.

Singular-value decomposition (SVD). Least-squares has the advantage of familiarity and ease of use. One of its disadvantages, in either the simple or the tapered form, is the difficulty in understanding the relationship between the individual measurements and the best-estimate solution. The so-called SVD is a form of least-squares that provides a complete, specific, and quantitative statement of the relationships between orthonormal structures in the data and corresponding orthonormal structures in the solution, along with a full estimate of their reliability.

The following very abbreviated treatment is to give the reader an insight into the application of SVD to the numerical example under consideration. (We refer to section 6.4 for an adequate discussion.) SVD operates with two complete sets of orthonormal vectors. One set, $\mathbf{u}_k$, lies in the M-dimensional data space, and the other, $\mathbf{v}_k$, lies in the N-dimensional solution, or model, space. The relationship between the vectors (known as "singular vectors") is given by

$$\mathbf{E}\mathbf{v}_k = \lambda_k \mathbf{u}_k \,, \qquad \mathbf{E}^T \mathbf{u}_k = \lambda_k \mathbf{v}_k \,, \qquad k = 1, \ldots, K, \tag{1.3.3}$$

where λ_k are the "singular values" (by convention, in decreasing numerical order); for the present underdetermined six layer case, $\lambda_k = 220,\ 161,\ 119,\ 89,\ 43$. The fact that none of the λ_k are very small compared with the others signifies that all of the five rays give independent information.

It can be shown that certain weighted averages of the solution, $\overline{x}_k = \mathbf{v}_k^T \mathbf{x}$, are determined by specific weighted averages of the observations, $\overline{y}_k = \mathbf{u}_k^T \mathbf{y}$, according to

$$\overline{x}_k = \overline{y}_k / \lambda_k \,. \tag{1.3.4}$$

For the present case, the $\mathbf{v}_k$ singular vectors are the columns $k = 1, \ldots, 6$ of

$$\begin{Bmatrix} .19 & .08 & .01 & -.00 & -.33 & -.92 \\ .36 & .19 & .03 & -.01 & -.83 & .39 \\ .77 & .45 & .06 & -.01 & .45 & .03 \\ .43 & -.69 & -.50 & .30 & .02 & .02 \\ .21 & -.41 & .24 & -.85 & .02 & .01 \\ .13 & -.33 & .83 & .42 & .01 & .00 \end{Bmatrix} \begin{matrix} \text{shallow layer} \\ \\ \\ \\ \\ \text{deep layer,} \end{matrix}$$

and the $\mathbf{u}_k$ singular vectors are the columns $k = 1, \ldots, 5$ of

$$\begin{Bmatrix} .24 & -.44 & .81 & .31 & .00 \\ .34 & -.46 & -.04 & -.82 & .00 \\ .46 & -.47 & -.58 & .48 & .00 \\ .61 & .48 & .08 & -.02 & .63 \\ .50 & .38 & .07 & -.02 & -.78 \end{Bmatrix} \begin{matrix} \text{steep ray} \\ \\ \\ \\ \text{flat ray.} \end{matrix}$$

From the first two columns, $\mathbf{v}_1$ and $\mathbf{u}_1$, we learn (not surprisingly) that a weighted *average* of layer slowness perturbations (with greater weight given to intermediate-depth layers[13]) is estimated from the weighted *average* of ray travel-time perturbations (with somewhat greater weight given to flatter trajectories). $\overline{x}_1$ and $\overline{y}_1$ correspond to the largest singular value, and accordingly $\overline{x}_1$ carries the most robust available information. Similar examination of the second pair $\overline{x}_2$ and $\overline{y}_2$ shows that a weighted *difference* of the perturbations in the upper ocean minus those in the lower ocean is estimated from the weighted *difference* of flatter minus steeper ray perturbations.

The fifth pair relates the perturbations in the upper three layers to only the flattest two rays; because it corresponds to the smallest singular value, the presence of noise renders uncertain its contribution to (1.3.4). Finally, because the dimension of the solution $\mathbf{x}$ (six) exceeds the dimension of the data $\mathbf{y}$ (five), there is an extra singular vector, called the "nullspace" of $\mathbf{E}$, for which there is no available information. Not surprisingly, this vector, $\mathbf{v}_6$, is largest in the two surface layers where the solution was so poor.

The SVD is a form of least-squares that can be used to understand the details of the solutions. In the present example the results are interpreted in terms of the sampling properties of ray geometry. The reason for the very small uncertainties in the tapered least-squares estimates, $\widehat{\mathbf{x}}_j$, is that they do not account for the solution uncertainty associated with the underdeterminance. For problems of large dimensionality, the amount of information provided by the SVD can become burdensome, and other methods may prove easier to use.

Gauss-Markov estimate. A quite different approach to estimation (although it is often confused with least-squares) is based on minimizing the mean square difference between each element of $\widehat{\mathbf{x}}$ and its true value, that is, minimizing the diagonal elements of

$$\mathbf{P} = \langle(\widehat{\mathbf{x}} - \mathbf{x})(\widehat{\mathbf{x}} - \mathbf{x})^T\rangle \tag{1.3.5}$$

(not the sum of the diagonals). The solution is again written $\widehat{\mathbf{x}} = \mathbf{B}\mathbf{y}$, but now

$$\mathbf{B} = \langle\mathbf{x}\,\mathbf{y}^T\rangle\langle\mathbf{y}\,\mathbf{y}^T\rangle^{-1}, \tag{1.3.6}$$

which depends on the covariance between the solution and the data and the covariance of the data with themselves (hence the name "stochastic inverse"). The matrix can be written

$$\mathbf{B} = \langle\mathbf{x}\,\mathbf{x}^T\rangle\,\mathbf{E}^T\,(\mathbf{E}\,\langle\mathbf{x}\,\mathbf{x}^T\rangle\,\mathbf{E}^T + \langle\mathbf{n}\,\mathbf{n}^T\rangle)^{-1} \tag{1.3.7}$$

[13] The weighting of the upper three layers is loosely connected to their thicknesses, 0.2, 0.3, and 0.5 km.

in terms of the solution and noise covariances. This method depends on one's ability to prescribe the covariances; in practice, adequate knowledge is usually available. The polar profile with $M = 5$, $N = 6$, can again serve as an example. For the solution, take the diagonal elements $\langle x_j^2 \rangle = 1, 1, 1, 0.1, 0.01, 0.01$ $(\text{ms/km})^2$ and zero for the off-diagonal elements. [The correct values would be 4, 1, 0.25, 0.0625, 0, 0 $(\text{ms/km})^2$ if x_j were regarded as deterministic, with $\langle x_j^2 \rangle = x_j^2$.] The noise covariances are taken as $\langle n_i n_j \rangle = 4\delta_{ij}\,\text{ms}^2$ ("white noise" of rms 2 ms). The Gauss-Markov estimate is then

$$\widehat{x}_j = -0.73 \pm 0.9, \ -1.55 \pm 0.4, \\ -0.54 \pm 0.05, \ -0.29 \pm 0.02, \ -0.02 \pm 0.02, \ +0.02 \pm 0.02$$

as compared with the correct values $-2, -1, -0.5, -0.25, 0, 0$ ms/km; the associated noise estimates are $\widehat{n}_i = +0.6, -.09, -.06, +.07, -.10$ ms. The small noise values provide a sensitive indicator that the imposed solution covariance was not completely consistent and might well lead to exploration of more appropriate values.[14] The perturbation in the uppermost layer is too small, and a comparatively large upper-ocean uncertainty remains in the solution, a result of poor physical sampling of the upper ocean. This is a basic shortcoming of our data set, and it was previously encountered in the application of the least-squares method. Had we permitted two additional rays turning in each of the upper two layers, the results would have been good indeed. Another important consideration is that one can obtain accurate estimates even for the (poorly sampled) top two layers if allowance is made for the expected strong correlation between adjoining layers, that is, $\langle x_i x_j \rangle \neq 0$ for $i \neq j$.

Many extensions and variations of these methods can be exploited to suit the needs of the investigator and the particular geometries of the experiment. Consider, for example, the intense current interest in the possible upper-ocean warming associated with the accumulation of greenhouse gases in the atmosphere. For the purpose of using tomographic observations to determine if a warming has occurred, one might regard the attempt to separate the changes into several thin upper-ocean layers as unnecessarily demanding of detail, seeking instead only the mean of the top kilometer (occupied by the first three layers). Linear programming methods permit one to compute upper and lower bounds (section 6.8). Alternatively, the Gauss-Markov solution permits the estimation of average properties of the solution, such as the *difference* between the mean warming of the top three layers and that of the bottom three layers (all weighted

[14] $\widehat{n}_i = y_i - (\mathbf{E}\widehat{\mathbf{x}})_i$ is a small difference of two large numbers, and much more sensitive than $\widehat{x}_j$ to slight misspecifications of prior statistics.

by layer thickness). The "true" values are $\overline{\Delta S} = -0.95,\ -0.07$ ms/km for the upper and lower ocean, respectively, with a difference of -0.88 ms/km, corresponding to a differential warming by $+0.44$°C. The Gauss-Markov estimate gives -0.80 ± 0.05 ms/km and $+0.40 \pm 0.03$°C for the vertical difference. We have thus derived a useful measure of upper-ocean warming, even though none of the available rays turned in the important upper two layers. This important result shows the effectiveness of tomography in obtaining *integrated* values.

It should be apparent that inversion of tomographic data is a problem in statistical-estimation theory. Linear inverse techniques are well known; the crucial problem in the application to ocean acoustic tomography is the parametrization of ocean variability and the specification of (co)variances of model and noise. As with all such problems, certain amounts of experience, skill, and insight are necessary. Statisticians without an understanding of the ocean and oceanographers without an understanding of statistical inference are both in danger of going astray. It is the reward of experience and insight that makes the problem so fascinating.

1.4. Horizontal Slice

The full oceanographic problem is, of course, three-dimensional. Transmission in a vertical slice between a single source-and-receiver pair has some range-dependent information, but is limited to ocean perturbations with scales equal to the ray loops and their harmonics (section 4.2). In practice, one has to depend on many intersecting slices for the three-dimensional information. Formally, this situation can be treated by dividing the ocean into k boxes and defining E_{ik} as the distance traveled by ray i in box k.

But insight into tomography is best gained by considering the two-dimensional horizontal problem with constant sound-speed in the vertical. Ray paths are straight lines connecting sources and receivers. This problem is not quite as artificial as it might seem at first sight. Suppose the perturbation ΔS_j in layer j has been determined from a previous vertical slice inversion, defining a pseudo-travel-time perturbation, $\Delta\tau_j^* = r\Delta S_j$, for a fictitious source-receiver pair separated by r and lying wholly within that layer. One can then perform horizontal inversions separately for each of the j layers, using $\Delta\tau^*$ as "data." The inversion formalism is exactly the same as previously discussed. Taking any one of the layers, consider a straight horizontal path i between any of the source and receiver moorings. E_{ik} is now the distance of this path inside a square k (it may be zero).

Spectral model. The idealized moving ship geometry provides a vivid example of sampling issues in mapping a field.[15] The "true" sound-speed field (fig. 1.5, top) is generated by a truncated Fourier series,

$$\Delta S(x, y) = \sum_k \sum_\ell c_{k\ell} \exp \frac{2\pi i}{L} (kx + \ell y), \quad k, \ell = 0, \pm 1, \ldots, \pm N$$

(Cornuelle *et al.*, 1989). Travel-time perturbations can be written

$$\Delta \tau_n = \sum_k \sum_\ell c_{k\ell} \int_{\Gamma_n(-)} ds \, \exp \frac{2\pi i}{L} (kx + \ell y), \tag{1.4.1}$$

and the integral in (1.4.1) evaluated *a priori* for each ray path n. The problem has again reduced to the form $\mathbf{y} = \mathbf{Ex} + \mathbf{n}$, with the solution vector $\mathbf{x}$ containing an ordered set of the complex Fourier coefficients $c_{k\ell}$.

We initially consider a scenario in which two ships start in the left and right bottom corners of a 1-Mm square and steam northward in parallel, transmitting from west to east every 71 km for a total of 15 transmissions (fig. 1.5). An inversion of the 15 travel times leads to an estimate that consists entirely of east–west contours, as all the ray paths measure only zonal averages,

$$\Delta \tau(y) = L \sum_\ell c_{0\ell} \exp \frac{2\pi i}{L} \ell y,$$

with no information on the longitudinal dependence. To interpret this result in wavenumber space, note that

$$\int_0^L dx \, \Delta S(x, y) = 0 \text{ for } k \neq 0 .$$

East–west transmissions therefore give information only on the parameters $c_{0\ell}$. All these have near-perfect prediction variances of 100% (except for the highest harmonic), while the remainder are completely unknown (see right column of fig. 1.5). The overall prediction has a skill of only 16%. Similarly, with south–north transmission, only the parameters c_{k0} are determined. Combining east–west and south–north transmissions determines both $c_{0\ell}$ and c_{k0}. Not surprisingly, this sampling is still inadequate to generate useful maps. More complex geometries that include scans at 45° give a distinct improvement.[16]

[15] Accurate mapping over a broad range of scales is the most stringent application of tomography; it is much more demanding than estimating heat content and other integral properties, or estimating the significant spectral properties.

[16] The upper two panels of fig. 1.5 each correspond to a single exposure in medical tomography, the bottom panel to exposures in four different directions.

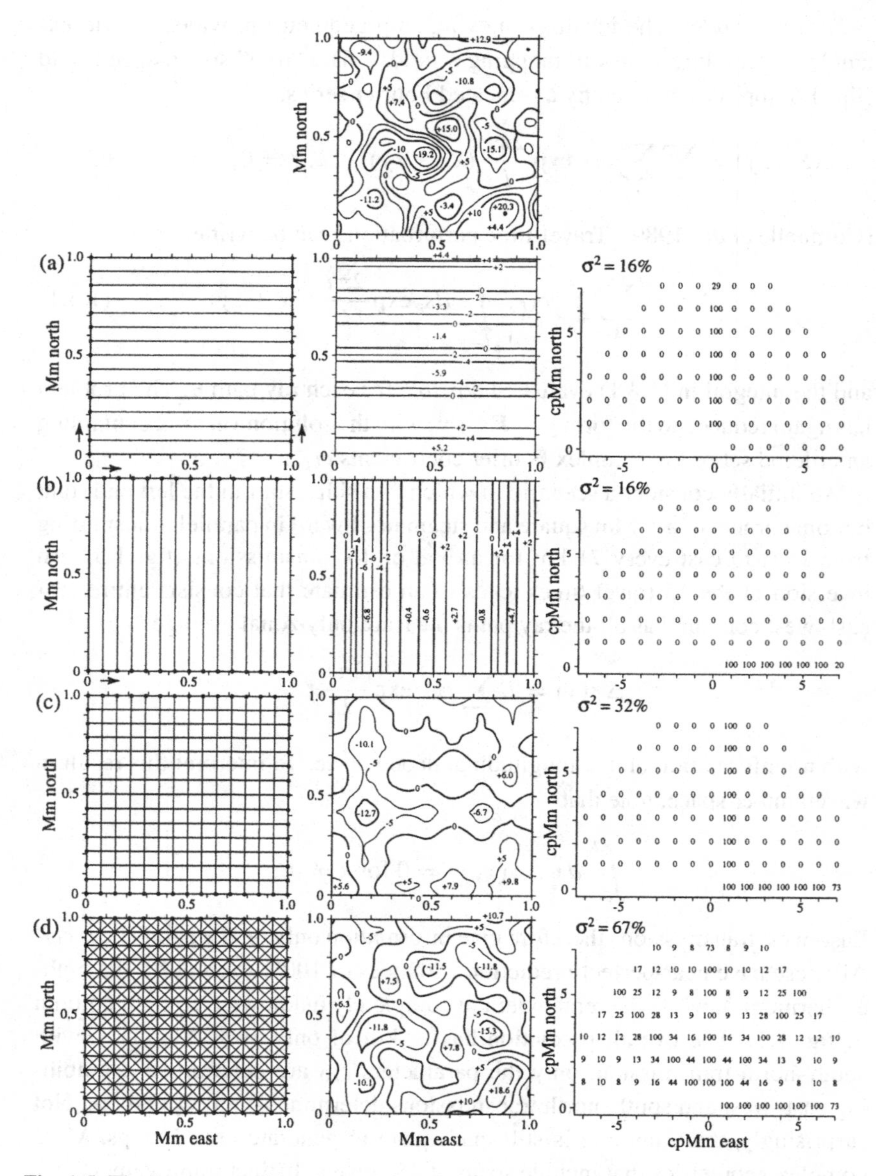

Fig. 1.5. The "true ocean" (top center) is to be mapped by tomographic inversions. The next three panels (a) display ray paths for W → E transmissions between a northward-traveling source ship on the west side of a 1-Mm square and a northward-traveling receiver ship on the east side (left), the result of inverting travel-time data from these ray paths (center), and the expected predicted variance in wavenumber space (right, 0 means no skill). (b) corresponds to S → N transmissions between two eastward-traveling ships; (c) and (d) represent increasingly ambitious sampling strategies, but only the bottom panels bear some resemblance to the "true" ocean, accounting for 67% of the variance.

Generating accurate maps requires adequate resolution in wavenumber space and hence requires ray paths at many different angles. This requirement must be independently met in any region comparable to the ocean correlation scale, a direct consequence of the projection-slice theorem (section 6.8).

Most any affordable moored array (such as the six-point array in fig. 1.6) leads to an undersampling of the mesoscale. For mapping in a 1-Mm square, one wishes to estimate at least 20×20 complex harmonics, for a total of 800 unknowns for each of two or three vertical dynamic modes. This situation has driven mesoscale applications of tomography to an augmentation of the moored arrays by moving-ship receivers. For the case shown in fig. 1.6, the receiving ship moved around a 1-Mm-diameter circle in 17 days (thus sacrificing some synopticity), for 135 receiving stops and 810 one-way paths. The number of unknowns has now increased because of the additional time dimension (see chapter 7). In practical use, ignoring variability during the finite sampling time, the tomography mapping compares favorably with an AXBT survey.[17]

Geometry. Consider the mooring configuration shown in fig. 1.6; each receiver records each source, as indicated by the white lines, corresponding to a vertical slice. The number of horizontal paths $S \times R$ increases as the *product* of the number of sources S and receivers R, whereas cost and effort are more nearly related to the sum $S + R$. For N transceivers there are $N(N-1)$ paths and $\frac{1}{2} N(N-1)$ reciprocal paths. The six-point transceiver array shown in fig. 1.6 represents a major mesoscale effort. Even so, the 15 resulting reciprocal paths are marginal for mesoscale mapping.

Vorticity. Returning now to the reciprocal transmissions within the moored transceiver array, we note that co-located sources and receivers (transceivers) at three moorings can provide a measure of the oceanographically interesting vorticity.[18] The vorticity equals the circulation $\int \mathbf{u} \cdot d\boldsymbol{\ell}$ integrated around the triangle, divided by the area. The circulation is found by a "sing-around" measurement,

$$(\tau_{A \to B} + \tau_{B \to C} + \tau_{C \to A}) - (\tau_{A \to C} + \tau_{C \to B} + \tau_{B \to A}),$$

and has given some interesting (but not unexpected) results. Vorticity in the Florida Straits was found to be of order 10^{-5} s^{-1} (see fig. 6.23), as compared

[17] AXBT is airborne expendable bathythermograph. The tomographic map in fig. 1.6 is based on a full three-dimensional inversion.

[18] Unfortunately, difference tomography is not well suited for measuring the flow divergence (see section 3.3).

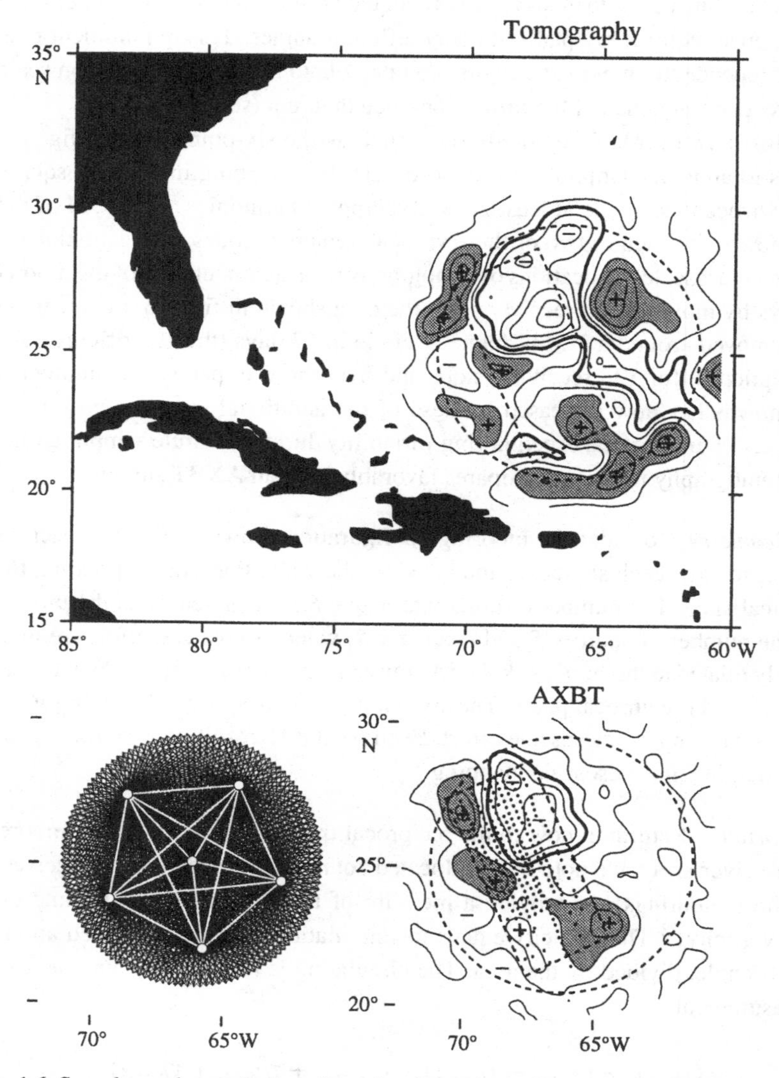

Fig. 1.6. Sound-speed perturbations from the climatological mean in 1-m/s (0.25deg C) contours at 700 m depth, as measured during the Moving Ship Tomography Experiment, July 15–30, 1991 (AMODE-MST Group, 1994). Five of the moorings are in a pentagon 700 km in diameter, with the sixth mooring in the center. The receiving ship moved around a 1-Mm circle, making 135 receiving stops. Rays from each of the moorings to the receiving stops are shown to the lower left. For comparison with the tomography map (using acoustic data only), a map using only AXBT data at 700 m is shown at the lower right. The AXBT survey was conducted 18–22 July and was confined to the rectangle. The tomography-deduced field *outside* the circle and the AXBT-deduced field *outside* the rectangle are subject to large uncertainty.

with 10^{-8} s^{-1} north of Hawaii (see fig. 6.24). It would be difficult to evaluate these quantities with traditional current meter moorings.

1.5 Estimation through Time

The ocean evolves in time, and measurements of the ocean at any time *t* carry information about its present, past, and future states. This information can be exploited if a model is available for describing the evolution, and so connecting the oceanic state and observations at differing times. One seeks to minimize objective functions similar to (1.3.5) but involving the values of **x** and **n** over some finite time span, instead of at a single instant. Such estimation problems can be solved in a number of ways. One well-known method (called a "Kalman filter") uses observations sequentially in time to make a best-possible estimate up to and including the time of observation. The results are then systematically improved by working backward in time, using the observations that subsequently became available (called a "smoothing" algorithm). An alternative approach attempts to minimize iteratively the objective function over the entire time span at once; in one formulation this leads to the so-called Pontryagin principle (section 7.3) and "adjoint modeling."

The number of model-constraint equations in time-evolving systems often exceeds by many orders of magnitude the number of observations. Chapters 6 and 7, which discuss the general problem, therefore attempt to make the blurry distinction between inverse methods in which data constraints are most numerous ("data-oriented") and those for which the model constraints are in the majority ("model-oriented").

1.6. Testing

A word needs to be said about testing[19] tomography (or any new observational method) against traditional measurements. Often this testing is like comparing apples and oranges. The fact that a new observational tool samples the environment in a distinctly new way (and therefore does not lend itself readily to a comparison) should be viewed as an asset, not a liability.

In the past, synergistic measurements (such as CTD[20] surveys and current meter records) often were withheld from the inversions so as to provide independent comparison with the acoustic results. It would be better to perform a single inversion and have the testing based on the consistency of the combined $\widehat{\mathbf{x}}$ and $\widehat{\mathbf{n}}$ with the known errors and prior statistics of all the contributing types of

[19] We prefer "testing" to the more commonly used "validating."

[20] CTD is an instrument measuring conductivity, temperature, and depth.

observations. Such inclusive tests are often regarded as too subtle to be convincing to skeptical observers, and so data have to be withheld to provide concrete proof of the new technology.

1.7. Comparisons and Comments

For comparison, medical CAT scans (for computed aided tomography) use a line array of X-ray sources and a corresponding array of receivers to measure the X-ray absorption density along a set of parallel ray paths through the patient's body. The source and receivers are then rotated, and measurements are repeated every degree or two, giving a dense set of path integrals in a straightforward geometry; these data are inverted into a two-dimensional map of the absorption densities of a section through the patient. A total of 10^4 to 10^5 path integrals is typical of modern procedures.

Geometric considerations in ocean acoustic tomography are less favorable than in CAT scans because of the cost and difficulty of installing acoustic transceivers. But it would be misleading to dwell on the relative information paucity of moored ocean acoustic tomography, for a number of reasons: (i) We normally deal with *perturbations* to a known reference state, whereas each CAT scan is conducted independently of previous knowledge. (ii) Unlike medical practice, there is no objection to using intrusive methods to establish the ocean reference state. (iii) Ocean dynamics offer a major reservoir of information that is exploited in the inverse procedure. (iv) In many ocean applications we are not striving for high-resolution mapping, but for average oceanic properties, such as heat content or volume transport. In that sense, the early tomographic emphasis on the mapping problem was unfortunate. (v) Acoustic tomography is not employed as a stand-alone procedure.[21] Current meter records, hydrographic data, and other sources of information each provide potentially unique measures of some elements of the ocean circulation. Inverse methods provide the apparatus for combining these diverse measurements with known kinematics and dynamics.

In retrospect, it may have been unfortunate that Munk and Wunsch labeled the oceanic problem "ocean acoustic tomography" in analogy with CAT scans. As we have seen, the oceanic problem differs from the medical problem in nearly all aspects: space scales, time scales, technology, and ray trajectories, not to speak of the market demand. But the basic difference is in data density.

[21] Satellite altimetry and ocean acoustic tomography supplement one another nicely, the former having good horizontal resolution, moderate time resolution, and poor depth resolution, the latter having poor horizontal resolution, good time resolution, and fair depth resolution (Munk and Wunsch, 1982*a*).

Medical tomography can afford a stand-alone procedure that makes no use of prior information. Ocean tomography depends heavily on prior information and synergistic measurements. It would be foolish, even arrogant, if our analyses of tomographic measurements did not take into account what is already known.

The reader will find that many, if not most, of the issues of ocean acoustic tomography have not been settled. It is a difficult subject, experimentally demanding and theoretically challenging. We hope that this volume will be of some help to the new investigator, who will in turn contribute to our understanding. As the late, great C.-G. Rossby once commented, "Ours is a strange business; the more we give away, the richer we become."

CHAPTER 2

THE FORWARD PROBLEM: RANGE-INDEPENDENT

The *sine qua non* of any inverse problem is an accurate treatment of the forward problem: to construct an acoustic arrival pattern given "the ocean." If the forward problem cannot be solved, then the measured data cannot be inverted to reconstruct the ocean.

When we entered this field in 1978, it was under the impression that the forward problem had been solved and that our efforts could be directed toward the oceanographically interesting inverse problem. As it turned out, the forward problem had *not* been solved; oceanographic and acoustic fields had never been simultaneously measured with accuracy adequate for a critical test. When such measurements were finally made in the late 1980s as part of tomographic work, significant discrepancies were found between the measured and computed sound fields. The most dramatic discrepancy was traced to the equations of state for sea water; some of the commonly used sound-speed equations were shown to be in error.

That era of trouble with the forward problem came to an end with the so-called SLICE89 experiment. The forward problem is now sufficiently well understood to permit oceanographically useful applications of inverse methods. Some discrepancies remain; they appear to be associated with internal waves and other range-dependent phenomena (see chapter 4).

There is a hierarchy of techniques available to predict the propagation of acoustic pulses in the ocean, given the sound-speed field. The geometric-optics approximation (*i.e.*, ray theory) is adequate to predict travel times in the majority of situations encountered in ocean acoustic tomography. Experimental measurements commonly reveal a few prominent arrivals not predicted by ray theory. These arrivals are usually associated with caustics. The WKBJ approximation accurately treats caustics and predicts the diffracted arrivals occurring in the shadow zones of geometric optics (as well as ray arrivals). Turning points, however, are not accurately described in the WKBJ approximation.

The most fundamental description of acoustic propagation is by expansion into normal modes (not an asymptotic theory). The arrival pattern computed by a brute-force summation over many hundreds of modes shows three distinct regimes: (i) an early regime of sharp arrival peaks (formed by constructive interference of many modes over a broad range of frequencies); these peaks are identified with ray-like arrivals and can accordingly be computed by simple ray theory without mode summation; (ii) A transitional interval, followed by (iii) late mode-like arrivals. In a dispersive polar sound channel, individual modes can be clearly identified in (iii), whereas in a weakly dispersive temperate channel that might not be the case. A hybrid description incorporating ray theory for the early steep-angle arrivals and mode theory for the late near-axial arrivals may be the most efficient. Normal mode theory is easily extended to the case of adiabatically varying sound-speed profiles (section 4.1); coupled mode theory is required for stronger range dependence (section 4.6).

The underlying theme of this chapter is the development of methods for the determination of the $\mathbf{E}$ matrix in $\mathbf{y} = \mathbf{E}\,\mathbf{x} + \mathbf{n}$ for both ray and mode travel-time perturbations, in preparation for chapters 6 and 7 on the inverse problem. The chapter is organized in three sections: (i) ray representation, (ii) mode representation, and (iii) comparison with observations. The mode derivations are carried out hand-in-hand with the equivalent ray representations, thus giving insight into the limits of ray optics. This leads to some duplication.

2.1. The Ocean Sound Channel

The speed of sound increases with increasing temperature, pressure, and salinity. Over most of the world's oceans, the sound-speed has a minimum (the sound axis) at about 1 km depth. Above the sound axis the effect of temperature dominates, and sound-speed increases with increasing temperature; beneath the sound axis the temperature is more uniform, and sound-speed increases with increasing pressure. The minimum (at the sound axis) forms a waveguide that permits efficient propagation of acoustic energy without lossy bottom interaction. This is the famous SOFAR channel discovered at the end of World War I. The recorded signal from a distant pulsed axial source normally consists of a series of pulses that can be interpreted in terms of rays, followed by an intense and sharp cutoff associated with axial modes (the so-called SOFAR crescendo).

At high (northern and southern) latitudes, the sound axis shoals, and eventually outcrops, giving rise to a "surface duct" between the reflecting sea surface and the upward-refracting deep water, as illustrated by an Arctic-to-Antarctic section in the mid-Pacific (figs. 2.1 and 2.2).

Propagation in temperate latitudes is very different from propagation in polar latitudes. We have found it useful to discuss propagation issues in terms of two model sound-speed profiles: a polar profile with adiabatic sound-speed

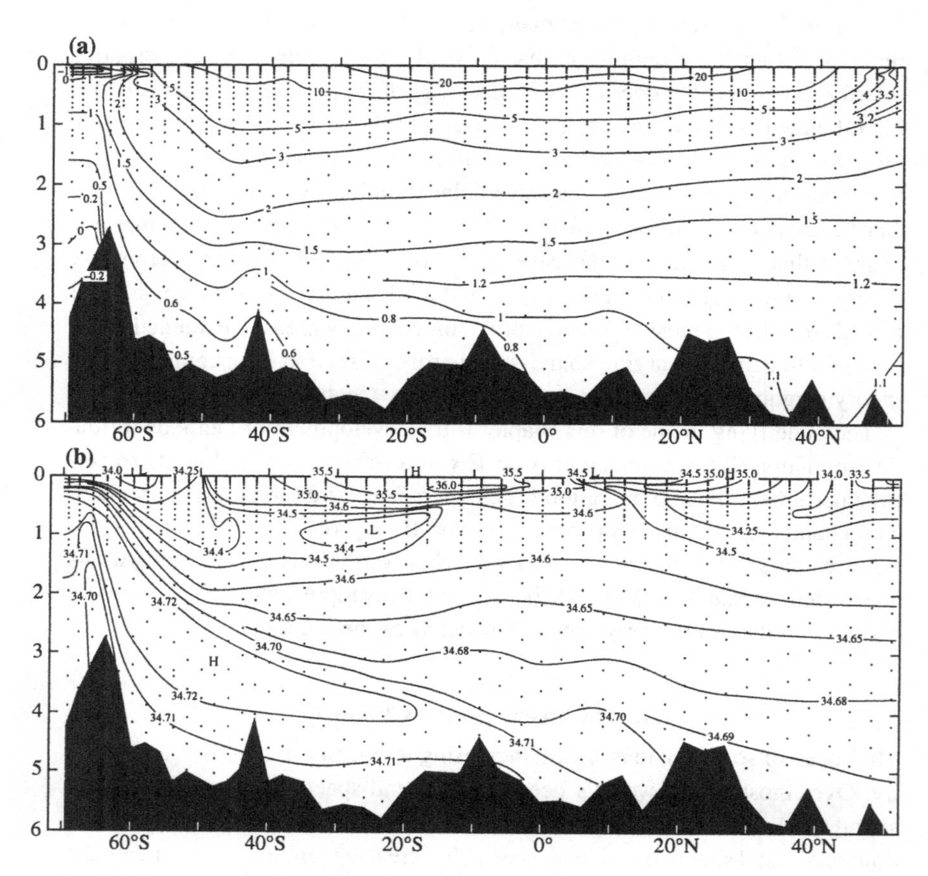

Fig. 2.1. Latitude–depth profiles (km) in the Central Pacific (170°W), from Antarctica to the Aleutians: (a) potential temperature contour (°C); (b) salinity contour (‰); (c) sound-speed contour (from 1450 to 1540 m/s; the first two digits are omitted); (d) buoyancy frequency contour (cycles per hour). Vertical exaggeration is about 1000:1; dots give measurement grid. (From Flatté *et al.*, 1979.)

gradient, and a temperate profile (sometimes referred to as the "canonical profile") that relates to an exponentially stratified ocean. In terms of the buoyancy (Brunt-Väisälä) frequency $N(z)$, the two models correspond to $N = 0$ and $N = N_0 \exp(-|z|/h)$, respectively. These models are described here and in the appendix to this chapter. But we need to stress that some of the propagation parameters are so sensitive to details in the profile[1] that the applicability of the models to any specific experiment is limited.

[1]This sensitivity is the basis of a successful inversion.

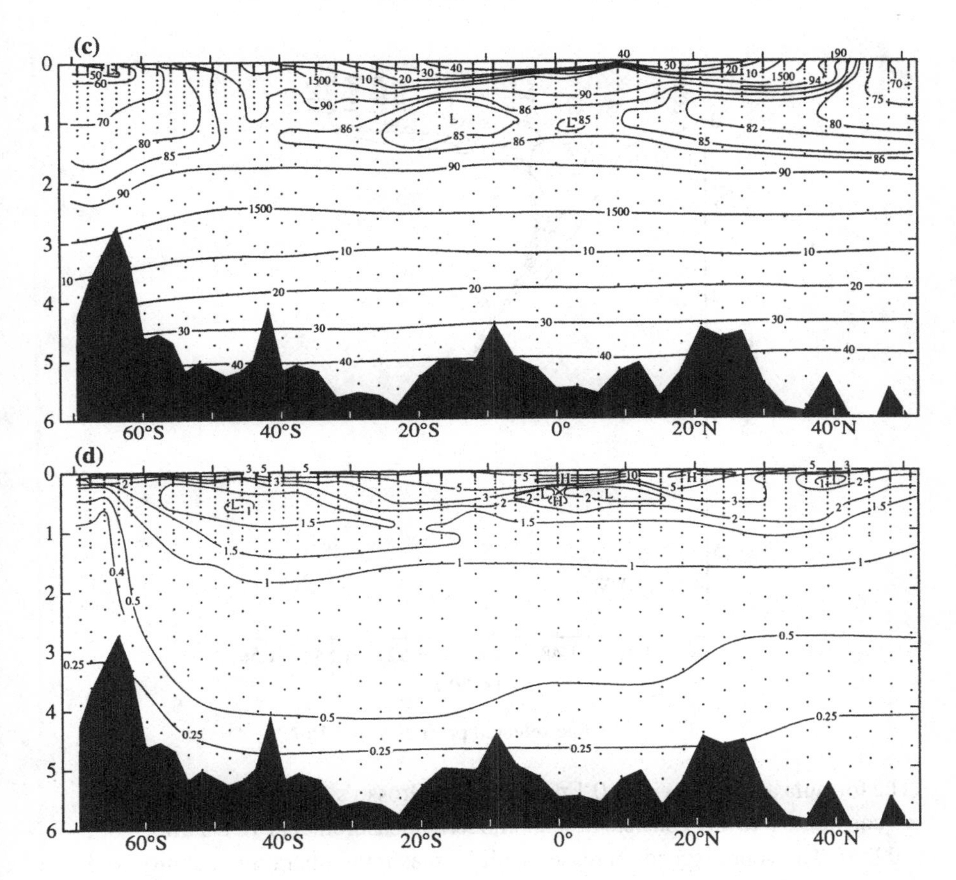

Fig 2.1 *(cont.)*

2.2. Sound-Speed

In the following discussion, we refer to standard texts for detailed information (*e.g.*, Flatté *et al.*, 1979; Urick, 1983; Brekhovskikh and Lysanov, 1991). Sound-speed in the ocean varies within narrow limits, typically between 1450 and 1550 m/s. But this small variation plays a crucial role in the character of the propagation. MacKenzie (1981) has given the following formula for sound-speed C (m/s) as a function of temperature T (°C), salinity Sa (‰), and depth D (m):

$$\begin{aligned} C(T, Sa, D) = {} & 1448.96 + 4.591T - 0.05304T^2 + 2.374 \times 10^{-4}T^3 \\ & + 1.340(Sa - 35) + 1.630 \times 10^{-2}D + 1.675 \times 10^{-7}D^2 \\ & - 1.025 \times 10^{-2}T\,(Sa - 35) - 7.139 \times 10^{-13}T\,D^3. \end{aligned} \tag{2.2.1}$$

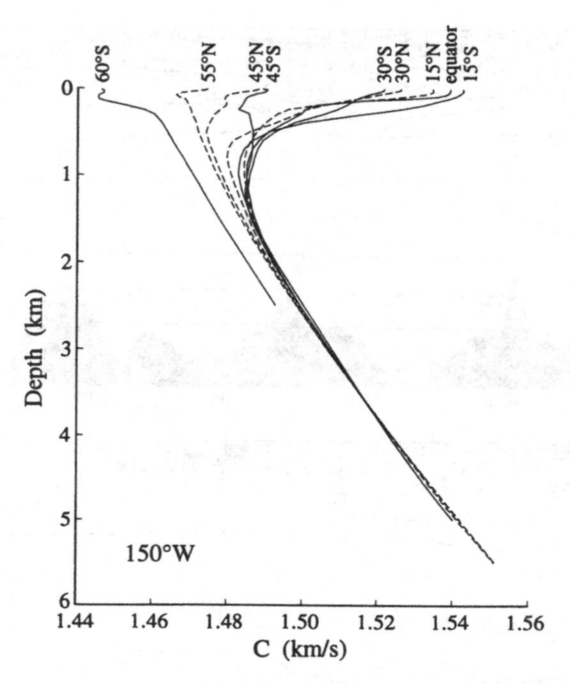

Fig. 2.2. Some selected profiles along 150°W.

The formula agrees to within 0.1 m/s with Del Grosso's (1974) equation, which is considered to be consistent with modern measurements (Dushaw *et al.*, 1993*b*). The sound-speed increases with increasing temperature, salinity, and pressure. The effect of salinity is relatively small.

It is convenient to introduce the *potential* gradient of a scalar quantity, defined as the measured gradient minus the adiabatic gradient, the latter arising from the adiabatic expansion or compression of a rising or sinking volume. Accordingly, the vertical gradient in sound-speed can be written

$$\frac{dC}{dz} = \frac{dC_p}{dz} + \frac{dC_a}{dz}, \tag{2.2.2}$$

and similarly for density and temperature. The total *in situ* density ρ appears in the horizontal momentum equations, but only the potential density gradient $d\rho_p/dz$ contributes to the stability of the water column. The total *in situ* sound-speed C determines the properties of the sound channel, but only the potential sound-speed gradient dC_p/dz contributes to the sound fluctuations associated with internal waves and other forms of vertical motion.

We write

$$\frac{1}{C}\frac{dC_p}{dz} = \alpha\frac{dT_p}{dz} + \beta\frac{dSa}{dz}, \quad \frac{1}{\rho}\frac{d\rho_p}{dz} = a\frac{dT_p}{dz} + b\frac{dSa}{dz}. \tag{2.2.3}$$

Typical numerical values are

$$\begin{aligned} \alpha &= 3.19 \times 10^{-3}(^\circ\text{C})^{-1}, & \beta &= 0.96 \times 10^{-3}(‰)^{-1}, \\ a &= 0.13 \times 10^{-3}(^\circ\text{C})^{-1}, & b &= 0.80 \times 10^{-3}(‰)^{-1}. \end{aligned} \tag{2.2.4}$$

The value of a is representative of conditions at 1 km depth, but varies considerably with temperature and pressure. The adiabatic sound-speed gradient is in part associated with the adiabatic temperature gradient, $dT_a/dz = -0.08°$ C/km, and in part with the pressure gradient. We can write

$$\frac{1}{C}\frac{dC_a}{dz} = \alpha\frac{dT_a}{dz} - \gamma\frac{dP}{dz}$$

$$= (-0.02 - 1.11)10^{-2}\ \text{km}^{-1} = -1.13 \times 10^{-2}\ \text{km}^{-1} \equiv -\gamma_a. \tag{2.2.5}$$

Note that the adiabatic sound-speed gradient is dominated by the pressure effect.

We often assume a local relationship $Sa = Sa(T)$, which can be linearized:

$$Sa = Sa(T_0) + \mu\Delta T, \quad \Delta T = T - T_0,$$

with μ in units of ‰/° C. Thus

$$\Delta C/C = \alpha\Delta T(1 + \mu\beta/\alpha). \tag{2.2.6}$$

A typical value is $\mu\beta/\alpha = 0.03$, so that a determination of ΔC is, to first order, a determination of the temperature field. The density perturbation at fixed pressure becomes

$$\frac{\Delta\rho}{\rho} = \frac{a}{\alpha}\left[1 + \left(\frac{b}{a} - \frac{\beta}{\alpha}\right)\mu\right]\frac{\Delta C}{C} \sim -0.04\,[1 - 0.65]\,\frac{\Delta C}{C}. \tag{2.2.7}$$

This relationship is sensitive to the T–Sa regression. Pond and Pickard (1983) have a good discussion of the determination of the relationships among temperature, salinity, and density. In many regions, estimation of sound-speed serves as a good estimate of density – a dynamic variable of the equations of motion. An adequate correction for salinity can be made in most regions from historical hydrographic data.[2] For some purposes the temperature field itself is of great interest, whether or not an estimate of salinity is available.

[2]There *are* exceptions: in high latitudes, salinity dominates the density field; in frontal regions, a well-defined function $Sa(T)$ does not exist.

A widely used representation of stability is the buoyancy (Brunt-Väisälä) frequency,

$$N^2(z) = -\frac{g}{\rho}\frac{d\rho_p}{dz}, \qquad \frac{d\rho_p}{dz} = \frac{d\rho}{dz} - \frac{d\rho_a}{dz} \tag{2.2.8}$$

where $d\rho_p/dz$ is the potential (true minus adiabatic) density gradient. In an isohaline (uniform salinity) ocean, or in an ocean with a linear T–Sa relation, dC_p/dz and N^2 are both proportional to dT_p/dz. One can then show (Munk, 1974) (writing C_A, N_A for values at the sound-channel axis) that

$$\frac{dC/dz}{C_A} = \gamma_a \frac{N^2 - N_A^2}{N_A^2} \tag{2.2.9}$$

is a simple way of relating the sound-speed profile to fundamental ocean conditions. Flatté *et al.* (1979, p. 5) discuss the limits to (2.2.9). At great depth, $N \to 0$ and $dC/dz \to -\gamma_a C_A$, as in the polar winter. At the axis, $N(z) = N_A$ and $dC/dz = 0$. The depth of minimum sound-speed C_A (about 1 km in temperate latitudes) is where the acoustic waveguide is centered. Above the axis, $N(z)$ and hence dC/dz increase sharply up to the surface-mixed layer.

Polar (adiabatic) profile. In polar winters, surface cooling can lead to convective instability. Thus $N \to 0$, and from (2.2.9),

$$C(z) = C_0\,(1 - \gamma_a z)\,; \tag{2.2.10}$$

the sound-speed increases linearly with depth (z is negative). For analytical convenience, we write

$$S^2(z) = S_0^2\,(1 + 2\gamma_a z) \tag{2.2.11}$$

for the squared sound-slowness $S^2 = (1/C)^2$. The two equations do not differ perceptibly, since $\gamma_a z$ is always small, less than 0.05.

Temperate (canonical) profile. The situation is different in temperate latitudes, where warming toward the surface plays a dominant role in the upper kilometer (at great depth, the adiabatic gradient is approached). To derive a representative sound-speed profile, one can take advantage of the fact that the stability of the water column diminishes smoothly from the surface downward (ignoring the surface-mixed layer), and $N(z)$ is roughly fitted by the exponential (Munk, 1974)

$$N = N_0 e^{z/h}, \quad h \approx 1\text{ km}\,. \tag{2.2.12}$$

It then follows that

$$C = C_A[1 + \tfrac{1}{2} h\gamma_a(e^{2\zeta} - 2\zeta - 1)], \quad \zeta = \frac{z - z_A}{h}, \tag{2.2.13}$$

as shown in fig. 2.3. We use the numerical values

$$C_A = 1.5\,\text{km/s}, \; \gamma_a = 0.0113\,\text{km}^{-1}, \quad z_A = -1\,\text{km}, \quad h = 1\,\text{km}.$$

The model (2.2.13) can serve as a useful guide to propagation in temperate latitudes; it is a simple way of relating the sound channel to an exponentially stratified ocean.

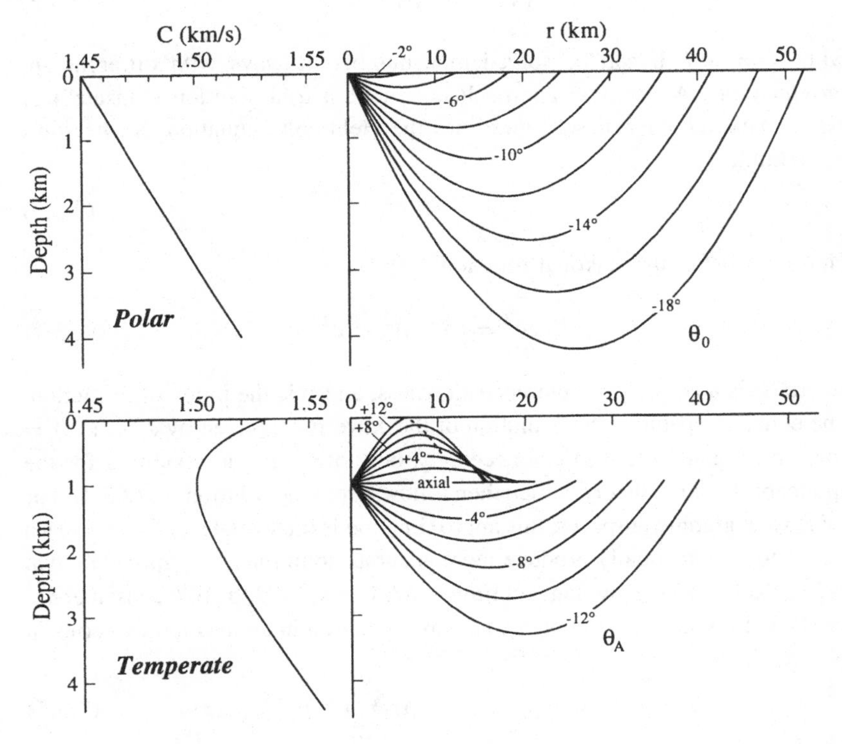

Fig. 2.3. Ray diagrams for the polar and temperate sound-speed profiles. The vertical launch angles θ_0 are indicated. Sources are at the surface and axis, respectively. For the polar profile, the loop range $\to$ 0 as $\theta_0 \to 0$. For the temperate profile, the range is finite for the axial ray $\theta_A = 0$ and *decreases* with increasing θ_A. The ray $\theta_A = +12°$ is refracted surface-reflected (RSR); the remaining rays are purely refracted (RR). Rays tend to focus at various ranges, giving rise to "convergence zones."

RAY REPRESENTATION

2.3. Ray Theory

Ray theory has proved an effective method for the study of sound propagation in the ocean and has provided the basis for most tomographic analysis thus far (*e.g.*, Boden *et al.*, 1991). The starting point is the wave equation

$$\left(\nabla^2 - S^2 \frac{\partial^2}{\partial t^2}\right) p = 0 . \tag{2.3.1}$$

With $p = p_0 e^{i\omega t}$, the resulting Helmholtz equation is

$$(\nabla^2 + k^2) p_0 = 0 , \tag{2.3.2}$$

with $k = \omega S$. It can be solved in a number of ways with different approximations. A common approach is to take a trial solution ("ansatz") of $p_0 = a \exp(ik_0\phi)$ and substitute into the Helmholtz equation. In the short wave limit,

$$\frac{\nabla^2 a}{a} \ll k_0^2 . \tag{2.3.3}$$

Phase ϕ satisfies the "eikonal equation"

$$|\nabla\phi|^2 = (S/S_0)^2 \equiv \nu^2 , \tag{2.3.4}$$

where S_0 is a convenient reference slowness, and ν is the index of refraction. The equation specifies the evolution of the wavefronts, given by $\phi(x, y, z) =$ constant. Equation (2.3.3) is a necessary, but not sufficient, condition for the applicability of ray theory, which then follows from the solutions to (2.3.4). For some tomographic purposes, this approximation is inadequate, and one seeks a means to systematically produce more accurate solutions. One approach, due to J. B. Keller (*e.g.*, Dowling and Ffowcs Williams, 1983, p. 107; Jensen *et al.*, 1994), is to write the solution to the wave equation in an asymptotic series in ω^{-1}

$$p(x, z, t) = \exp i(\omega t - k_0\phi) \sum_{n=0}^{\infty} (i\omega)^{-n} I_n(x, z) . \tag{2.3.5}$$

Substitution of (2.3.5) into (2.3.1) again produces the eikonal equation (2.3.4) at lowest order (large ω); higher-order corrections can also be found. For the moment, we examine only the lowest-order approximation. Define a "wavefront" to be the surfaces $\phi(\mathbf{r}) =$ constant, and a "ray" as the normal to the wavefront. Position along the rays is measured by the arc length s, so that $\mathbf{r} = \mathbf{r}(s)$. The unit normal to the wavefront is $\mathbf{n} = \nabla\phi/|\nabla\phi|$. But from the definition of $\mathbf{r}$, the unit normal is also given by $d\mathbf{r}/ds$, and thus

$$\frac{d\mathbf{r}}{ds} = \frac{\nabla\phi}{|\nabla\phi|} = \frac{\nabla\phi}{\mu}, \tag{2.3.6}$$

using the eikonal equation. So

$$\nu\frac{d\mathbf{r}}{ds} = \nabla\phi\,. \tag{2.3.7}$$

Differentiating, we obtain

$$\begin{aligned}\frac{d}{ds}\left(\nu\frac{d\mathbf{r}}{ds}\right) &= \frac{d}{ds}\nabla\phi = (\mathbf{n}\cdot\nabla)\nabla\phi = \left(\frac{\nabla\phi}{|\nabla\phi|}\cdot\nabla\right)\nabla\phi \\ &= \frac{1}{\nu}\nabla(|\nabla\phi|^2/2) - \frac{1}{\nu}\nabla\phi\times(\nabla\times\nabla\phi) = \nabla\nu\,,\end{aligned} \tag{2.3.8}$$

where the last step employs the eikonal equation and the result that the curl of a gradient vanishes. Substituting $\nu = S/S_0$ into

$$\frac{d}{ds}\left(\nu\frac{d\mathbf{r}}{ds}\right) = \nabla\nu\,, \tag{2.3.9}$$

with S_0 constant along the ray path, we have

$$\frac{d}{ds}\left(S\frac{d\mathbf{r}}{ds}\right) = \nabla S\,, \tag{2.3.10}$$

or, in components,

$$\frac{d}{ds}\left(S\frac{dx}{ds}\right) = \frac{\partial S}{\partial x}\,, \qquad \frac{d}{ds}\left(S\frac{dz}{ds}\right) = \frac{\partial S}{\partial z}\,. \tag{2.3.11a, b}$$

These are the ray-trajectory equations in two dimensions, whose solution is the "ray trace" $z_{\rm ray}(x)$.

Equations (2.3.11) strongly resemble the Euler equations of a variational problem. The first variation of the integral

$$\int_{s_1}^{s_2} S(z,x)\sqrt{\left(\frac{dx}{ds}\right)^2 + \left(\frac{dz}{ds}\right)^2}\,ds, \tag{2.3.12}$$

with the end points held fixed, and set to zero, leads to equations (2.3.11). We thus have the very powerful statement that the solution to the wave equation in the ray limit produces ray trajectories that are stationary values of the integral in (2.3.12). The integral reduces to $\int_{s_1}^{s_2} S\,ds$ and will be recognized as the travel time from s_1 to s_2. This result is Fermat's principle.

For the range-independent case, (2.3.11*a*) becomes

$$\frac{d}{ds}\left(S\frac{dx}{ds}\right) = \frac{d}{ds}(S\cos\theta) = 0, \tag{2.3.13}$$

where $\theta(z)$ is the ray inclination relative to the horizontal. This is Snell's law. Thus, in the limit of high frequency, the wave equations lead to Fermat's principle, and for the range-independent case this leads directly to Snell's law,

$$S(z)\cos\theta(z) = \text{constant} = \widetilde{S}, \tag{2.3.14}$$

where $\widetilde{S} = S(\widetilde{z})$ is the slowness at the ray turning points ($\widetilde{\theta} = 0$).

Geodesy. Solutions to the wave equation are here written in Cartesian coordinates. In problems of global transmissions it is necessary to use the full spheroidal coordinates (allowing for Earth flattening), and these are developed in section 8.5. For ranges up to 1 Mm, the forward problem can be treated in Cartesian coordinates, provided the Cartesian depth coordinate, z, is replaced by a "spherical" depth variable, z_s, defined by

$$e^{-z_s/a} = \rho/a \tag{2.3.15}$$

where ρ is the radial distance from the center of the Earth, with $\rho = a$ at the surface. The sound-speed profile $C(z)$ is defined by $(a/\rho)\,C_s(\rho)$. We refer to Aki and Richards (1980, pp. 463–464) for a discussion of the spherical transformation.

However, the full spheroidal equations are required for computing the range, r, even at the shortest ranges, to achieve millisecond precision. Formulae are given in books on geodesy (*e.g.*, Bomford, 1980, p. 122). These will yield better than 10-m precision, consistent with modern satellite navigational practice, up to distances of 1000 km.

2.4. Ray Diagram

For any specified source and receiver there may be many rays n, and these are identified by the *ray parameters* $\widetilde{S}_n$. The ray construction follows immediately from a solution of $dz/dx = \tan\theta(z)$ and Snell's law. Similarly, the travel time along the ray element ds is given by $ds/dt = 1/S(z)$ and $dz/ds = \sin\theta(z)$. Hence

$$x(z) = \int \frac{dz}{\tan\theta(z)}, \qquad \tau(z) = \int \frac{S(z)\,dz}{\sin\theta(z)}. \tag{2.4.1}$$

Rays and travel times can be computed numerically for any given $S(z)$, or analytically for some simple $S(z)$. In general, the ray path can be computed from the differential ray equation

$$\frac{d\theta}{ds} = \frac{1}{S}\frac{dS}{dn}, \quad \frac{dx}{ds} = \cos\theta, \quad \frac{dy}{ds} = \sin\theta\,,$$

which relates the (local) ray curvature, $d\theta/ds$, to the logarithmic gradient of sound-slowness in a direction, n, normal to the (local) ray path s. Fig. 2.3 shows rays for the polar and temperate models.

We wish to derive certain ray characteristics, such as the horizontal distance (span) R of a ray loop and the associated travel time T. Using Snell's law to express $\tan\theta(z)$ and $\sin\theta(z)$ in terms of sound-slowness,

$$R^{\pm} = \pm 2\widetilde{S}\int_{z_A}^{\widetilde{z}^{\pm}} \frac{dz}{(S^2-\widetilde{S}^2)^{\frac{1}{2}}}, \quad T^{\pm} = \pm 2\int_{z_A}^{\widetilde{z}^{\pm}} \frac{S^2 dz}{(S^2-\widetilde{S}^2)^{\frac{1}{2}}} \tag{2.4.2a, b}$$

for the upper and lower ray loops, respectively. For the polar profile, the axis is at the surface, $z_A = 0$, and there are only lower loops, $R = R^-$, $T = T^-$. For double loops in a temperate profile,

$$R = R^+ + R^- = 2\widetilde{S}\int_{\widetilde{z}^-}^{\widetilde{z}^+} \frac{dz}{(S^2-\widetilde{S}^2)^{\frac{1}{2}}}, \tag{2.4.3a}$$

$$T = T^+ + T^- = 2\int_{\widetilde{z}^-}^{\widetilde{z}^+} \frac{S^2 dz}{(S^2-\widetilde{S}^2)^{\frac{1}{2}}}. \tag{2.4.3b}$$

So far the relations are quite general. As an example, for the polar profile $S^2 = S_0^2(1+2\gamma_a z)$,

$$R = 2\gamma_a^{-1}\,\widetilde{\sigma}\,\sqrt{1-\widetilde{\sigma}^2}, \quad T = 2\gamma_a^{-1}\,S_0\,\widetilde{\sigma}\,(1-\tfrac{2}{3}\,\widetilde{\sigma}^2)\,, \tag{2.4.4}$$

where

$$\widetilde{\sigma}^2 = \sigma^2(\widetilde{S}) = \frac{S_0^2-\widetilde{S}^2}{S_0^2} = 2\gamma_a(-\widetilde{z})\,, \tag{2.4.5}$$

and so

$$T = RS_0\,(1-\tfrac{1}{6}\,\widetilde{\sigma}^2+\cdots)\,. \tag{2.4.6}$$

Steep rays arrive slightly earlier than surface-trapped rays. We defer deriving similar relations for the temperate profile until the action variable is introduced.

For simplicity, consider the case where source and receiver are at the same depth z_s in a range-independent ocean. Without loss of generality, this depth is taken as the channel axis. For the polar profile (with the axis on the surface) there are n^- lower loops, each of span R^-, so that the total range is $r = n^- R^-$ and travel time is $\tau = n^- T^-$. For the temperate profile, with $n^\pm$ upper/lower loops, $r = n^+ R^+ + n^- R^-$, $\tau = n^+ T^+ + n^- T^-$. For the case of $n = n^+ = n^-$ double loops, $r = nR$ and $\tau = nT$.

Time fronts. The arrival structure is conveniently displayed by a time-front diagram (Munk and Wunsch, 1979). Time fronts are everywhere normal to the rays, and they portray where a pulse is heard at a given instant τ in r, z-space, or, alternatively, at a fixed range r in τ, z-space. Ray arrivals correspond to the intersections of the time front $\pm p$ with the receiver depth, where $p = n^+ + n^-$ is the *total* number of upward/downward ray turning points, and $\pm$ corresponds to ray launch-angle signs. Time-front displays have received less attention than ray diagrams, but they are in many ways the more convenient representation. The geometric construction of time fronts is given in the sections 2.17 and 2.18. We shall return to this subject in section 2.12, where we show an alternate construction based on the superposition of many modes, and in section 2.16, where we show a satisfying agreement between measured and computed time fronts.

Fig. 2.4 shows the situation for the polar and temperate ocean profiles at 1 Mm range. For an axial source and axial receiver, the constituents ± 38, ± 40, ... have overlapping arrival times; see Munk and Wunsch (1979, p. 133) for a discussion of degeneracies.

Eigenrays are rays that intersect both source and receiver. In the analytical examples discussed so far, this condition could be imposed analytically. In the numerical construction, one uses trial values of launch angle θ_{source} to search for the eigenrays.[3] An equivalent procedure is to construct the ray fronts (fig. 1.2) and draw the ray from the receiver that is perpendicular to the local ray front. For a source radiating uniformly into all θ, the energy along a wave front, n, is proportional to $|dn/d\theta_{\text{source}}|^{-1}$.

The cusps where different rays are joined (*e.g.*, the lower cusp joining +37 with −37 at 3.1 km depth) represent prominent arrivals, and they cannot be neglected even if the receiver should lie too shallow or too deep for direct ray arrivals (at 4 km depth, say, for ± 37). This requires a generalized ray theory, such as the WKBJ treatment of Brown (1981, 1982).

[3]This is sometimes called the "shooting problem."

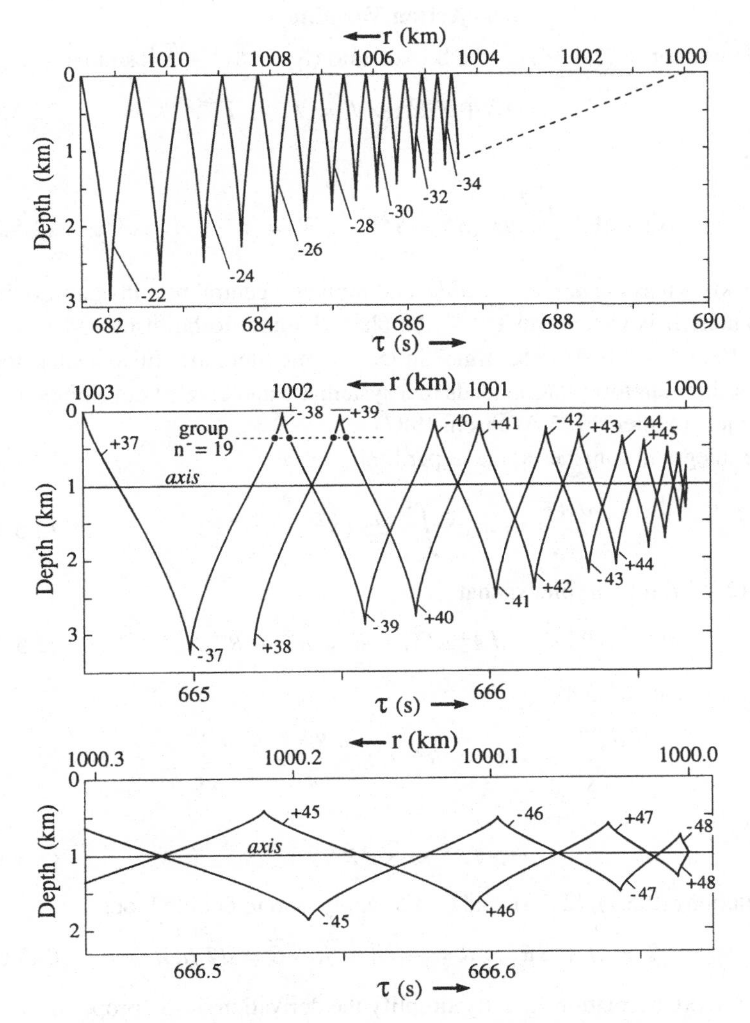

Fig. 2.4. Time fronts for the polar (top) and temperate (middle) sound-speed profiles at a 1-Mm range; source at surface and axis, respectively. The figure can be interpreted in τ, z-space for a receiver at a fixed $r = 1$ Mm; it corresponds very closely to the ray fronts in r, z-space at the moment of the final surface (axial) arrival at 1 Mm, with range extending leftward showing ray fronts beyond the receiver. For a surface receiver in a polar profile, the first refracted ray with a turning point at about 3 km depth has $n^- = 22$ lower loops, and arrives at $\tau = 681.7$, 8 s prior to the last arrival ∞ (not shown). For an axial receiver in a temperate profile, the first refracted arrival (RSRs are not shown) has $n^+ + n^- = 19 + 18 = 37$ turning points with an upward launch angle (designation $+37$) and arrives at $\tau = 664.8$ s, 1.9 s prior to the final arrivals ± 48 at 666.66 s. (Note the relatively small axial dispersion.) For a receiver well above the axis, rays arrive in groups of four (as shown for group $n^- = 19$), and the arrivals end prior to the axial arrival. The final near-axial arrivals are drawn on a larger scale at bottom.

2.5. Action Variable

The numerator of (2.4.3*b*) can be broken into $(S^2 - \widetilde{S}^2) + \widetilde{S}^2$, so that

$$T^{\pm} = A^{\pm} + \widetilde{S}R^{\pm}, \quad T = T^{+} + T^{-}, \tag{2.5.1}$$

where

$$A^{\pm}(\widetilde{S}) \equiv \pm 2 \int_{z_A}^{\widetilde{z}^{\pm}} dz\, (S^2 - \widetilde{S}^2)^{\frac{1}{2}}, \quad A = A^{+} + A^{-}. \tag{2.5.2}$$

A is known as the *action variable*, and it plays a central role in Hamiltonian mechanics. It is closely related to the phase integral to be discussed in section 2.10 and is called "delay time" in the seismic literature. In acoustics, too, the Hamiltonian formulation leads to a systematic and elegant development of the subject (Miller, 1986; Wunsch, 1987).

The integrand vanishes at the upper limit; hence

$$\frac{dA^{\pm}}{d\widetilde{S}} = \mp 2\widetilde{S} \int_{z_A}^{\widetilde{z}^{\pm}} dz\, (S^2 - \widetilde{S}^2)^{-\frac{1}{2}}. \tag{2.5.3}$$

From (2.4.3*a*) it then follows that

$$R^{\pm} = -dA^{\pm}/d\widetilde{S}, \quad R = R^{+} + R^{-}. \tag{2.5.4}$$

From (2.5.1) and (2.5.4),

$$\frac{dT^{\pm}}{d\widetilde{S}} = -R^{\pm} + R^{\pm} + \widetilde{S}\,\frac{dR^{\pm}}{d\widetilde{S}} = \widetilde{S}\,\frac{dR^{\pm}}{d\widetilde{S}},$$

so that

$$dT^{\pm} = \widetilde{S}\, dR^{\pm}. \tag{2.5.5}$$

The relations (2.5.1), (2.5.4), and (2.5.5) apply also to double loops:

$$T = A + \widetilde{S}R, \quad R = -dA/d\widetilde{S}, \quad \widetilde{S} = dT/dR. \tag{2.5.6}$$

These important relations greatly simplify the derivation of ray properties.

Polar action. For the polar profile (2.2.11),

$$A = \frac{2S_0}{3\gamma_a}\widetilde{\sigma}^3, \quad \widetilde{\sigma}^2 = \frac{S_0^2 - \widetilde{S}^2}{S_0^2}, \quad \gamma_a = 0.013\,\text{km}^{-1}, \quad S_0^{-1} = 1.45\,\text{km/s}, \tag{2.5.7}$$

and the expressions (2.4.5) for R and T are readily verified (see section 2.17). The convenience of deriving ray properties by means of these integral relations becomes more pronounced in the case of complex profiles (such as the temperate profile) and is almost mandatory for deriving second-order effects.

Temperate action. For the temperate profile, it is useful to introduce a dimensionless scale $\widetilde{\phi}$ that varies from 0 for axial rays to order 1 for steep refracted rays. We set

$$A^{\pm} = S_A h^{\frac{3}{2}} \gamma_a^{\frac{1}{2}} (a\widetilde{\phi}^2 \pm b\widetilde{\phi}^3 + c\widetilde{\phi}^4), \quad \widetilde{\phi}^2 = (\gamma_a h)^{-1} \frac{S_A^2 - \widetilde{S}^2}{S_A^2},$$

$$a = \frac{\pi}{2\sqrt{2}}, \quad b = -\tfrac{2}{9}, \quad c = \frac{\pi}{48\sqrt{2}}, \tag{2.5.8}$$

$$\gamma_a = 0.0113\,\text{km}^{-1}, \quad h = 1\,\text{km}, \quad S_A^{-1} = 1.5\,,\,\text{km/s}$$

with the upper/lower signs referring to the upper/lower ray loops.

Equation (2.5.8) is consistent with (2.2.13) to third order. The three-term expansion (2.5.8) is probably the simplest representation for the midlatitude ocean that can claim some degree of reality [but see Miller's (1982) "deep-six" sound channel].

Abel transform. The traditional procedure has been to start with a model representation of $C(z)$ and then derive the action variable and other ray parameters. We have found it an advantage to start with the simplest possible model for the action variable, $A(\widetilde{S})$, and then subsequently derive $z(S)$ and hence the profile $S(z)$. The Abel transform provides the formalism. Consider the Abel-transform pair (Sneddon, 1972; Aki and Richards, 1980)

$$f(\beta) = \int_0^\beta d\alpha \frac{dg/d\alpha}{(\beta - \alpha)^{\frac{1}{2}}}, \quad g(\alpha) = \frac{1}{\pi}\int_0^\alpha d\beta \frac{f(\beta)}{(\alpha - \beta)^{\frac{1}{2}}}. \tag{2.5.9a, b}$$

To simplify notation, z is drawn from the sound axis (z_A or z_0) upward, with $S_A = S_0$. From (2.5.4),

$$\begin{aligned} R^{\pm} &= -dA^{\pm}/d\widetilde{S} = \pm 2\widetilde{S}\int_0^{\widetilde{z}^{\pm}} dz\,(S^2 - \widetilde{S}^2)^{-\frac{1}{2}} \\ &= \pm\frac{2\widetilde{S}}{S_0}\int_0^{\widetilde{\sigma}^2} d\sigma^2 \frac{dz/d\sigma^2}{\sqrt{\widetilde{\sigma}^2 - \sigma^2}}, \end{aligned} \tag{2.5.10}$$

where

$$\sigma^2 = \frac{S_0^2 - S^2}{S_0^2}, \quad \widetilde{\sigma}^2 = \frac{S_0^2 - \widetilde{S}^2}{S_0^2}$$

are dimensionless representations of slowness as a function of z and of turning-point slowness, respectively. This is of the form (2.5.9a), provided

$$\begin{aligned} &\alpha = \sigma^2, \quad \beta = \widetilde{\sigma}^2, \quad g(\alpha) = z(\sigma)\,, \\ &f(\beta) = \tfrac{1}{2}R^{\pm}(S_0/\widetilde{S}) = \tfrac{1}{2}R^{\pm}(1 - \widetilde{\sigma}^2)^{-\frac{1}{2}}\,. \end{aligned} \tag{2.5.11}$$

From (2.5.9*b*) it follows that

$$z(\sigma) = \pm \frac{1}{2\pi} \int_0^{\sigma^2} d\widetilde{\sigma}^2 \frac{R^{\pm}(\widetilde{\sigma})}{(1-\widetilde{\sigma}^2)^{\frac{1}{2}} (\sigma^2 - \widetilde{\sigma}^2)^{\frac{1}{2}}} . \tag{2.5.12}$$

The result is quite general. For the polar profile $R = R^- = 2\gamma_a^{-1} \widetilde{\sigma} \sqrt{1-\widetilde{\sigma}^2}$ from (2.4.5), thus

$$z(\sigma) = -\frac{2}{\pi \gamma_a} \int_0^{\sigma} d\widetilde{\sigma} \frac{\widetilde{\sigma}^2}{(\sigma^2 - \widetilde{\sigma}^2)^{\frac{1}{2}}} = \frac{-\sigma^2}{2\gamma_a} \tag{2.5.13}$$

in agreement with (2.4.6). For the temperate profile, the Abel transform yields (see section 2.18)

$$\frac{z(\phi) - z_A}{h} = \pm \left(\frac{1}{\sqrt{2}} \phi \mp \frac{1}{6} \phi^2 + \frac{1}{18\sqrt{2}} \phi^3 \right), \quad \phi^2 = \frac{1}{\gamma_a h} \frac{S_A^2 - S^2}{S_A^2} \tag{2.5.14}$$

for $z \gtrless z_A$.

The Abel transform provides a direct formalism for the inverse problem (Jones *et al.*, 1990). For simplicity, we consider the case of source and receiver at the same depth, with an integral number $n = n^+ = n^-$ of double loops. We assume that the measured arrival times τ can be matched to the ray number n. Then, with travel times τ_n and range r known, we compute

$$R = r/n, \quad T = \tau_n / n,$$

and plot T/R for each value of n. We now assume (unrealistically) that the plotted points are sufficiently dense and smooth that the slope dT/dR can be meaningfully evaluated. But from (2.5.6) we have $dT/dR = \widetilde{S}$, the turning slowness. Hence we have determined $R(\widetilde{S}) \approx 2f(\beta)$, which on substitution into (2.5.9*b*) yields $z(S)$, or, equivalently, $S(z)$.[4] The Abel transform is not used in ocean tomography in practice because it does not provide a convenient linear framework for estimating errors (section 6.8).

Fractional action.[5] In analogy with (2.5.2), we define a "fractional action"

$$\delta A^{\pm}(z, \widetilde{S}) = \pm \int_0^z dz' \, [S^2(z') - \widetilde{S}^2]^{\frac{1}{2}}, \quad z \gtrless 0, \tag{2.5.15}$$

[4] See Munk and Wunsch (1983, app. A) for the determination of $R^{\pm}(\widetilde{S})$.

[5] In this section, δ refers to "fractional" rather than "error in."

in analogy with (2.5.2), so that $\pm\delta\widetilde{A}^{\pm}(\widetilde{z}^{\pm}, \widetilde{S}) = \frac{1}{2} A^{\pm}$. The factor $\frac{1}{2}$ arises because $\delta A^{\pm}(\widetilde{z}^{\pm}, \widetilde{S})$ refers to one half of the upper/lower loop, from the axis to the crest/trough. In analogy with (2.5.6), it follows that

$$\delta R(z, \widetilde{S}) = -\partial(\delta A)/\partial\widetilde{S}, \quad \delta T(z, \widetilde{S}) = \delta A + \widetilde{S}\,\delta R \tag{2.5.16}$$

are the horizontal distance and time, respectively, traveled by a ray between 0 and z. Formulae for δA, δR, and δT for polar and temperate profiles are given in appendix sections 2.17 and 2.18.

Rays $z(\delta R)$ and time fronts $z(\delta T)$ are conveniently constructed by eliminating ϕ^2 in $z(\phi^2)$, $\delta R(\phi^2)$, and $\delta T(\phi^2)$.

2.6. Structure of Ray Arrivals

In the following three sections we discuss the properties of rays and ray arrival patterns in some detail. These are illustrated for the polar and temperate oceans. We have found a familiarity with these properties to be helpful in the planning of tomographic measurements and in the interpretation of results.

The turning slowness $\widetilde{S}$ and the action $A(\widetilde{S})$ are continuous functions of launch angle θ, in accordance with Snell's law. A nondirective source radiates into all possible launch angles and therefore generates $A(\widetilde{S})$ for all possible $\widetilde{S}$. A receiver selects only those discrete values of $\widetilde{S}$, to be designated $\widetilde{S}_n$, for which the rays n intersect the receiver, thus leading to a quantization of action – with important consequences.

For simplicity, consider the case of an axial source in a range-independent ocean. Ray arrivals correspond to the intersections of time front n with z_{receiver} (fig. 2.4). For simplicity, we consider only the case of $n = n^+ = n^-$ double loops,

$$r = nR, \quad a = nA, \quad \tau = nT = a + r\widetilde{S}, \tag{2.6.1}$$

in view of (2.5.1). We now establish a set of rules for the ray arrival pattern.

The total range is fixed,

$$(n+1)R_{n+1} = nR_n = r, \tag{2.6.2}$$

so that the loop span R_n diminishes with increasing n. For any quantity x, let $Dx = x_{n+1} - x_n$ be the difference between successive (double-loop) ray arrivals. From (2.6.1),

$$D\tau = r\,D\widetilde{S} + n\,DA + A_{n+1}\,. \tag{2.6.3}$$

To first order, $DA = (\partial A/\partial\widetilde{S})D\widetilde{S}$, and from (2.5.4), $n\,DA = -n\,R\,D\widetilde{S} = -r\,D\widetilde{S}$. Thus

$$D\tau = \tau_{n+1} - \tau_n = A_{n+1} \tag{2.6.4}$$

is always positive, and independent of range r. We have the following:

Rule 1: Loop number n always increases and loop span R always decreases with arrival time.

Rule 2: The interval between successive ray arrivals is independent of range.

What does change is the total *number* of refracted ray arrivals. Let n_{AX} be the number of double loops for near-axial rays ($\widetilde{S} \equiv \widetilde{S}_{AX}$) and n_{SLR} refer to the surface-limited rays ($\widetilde{S} \equiv \widetilde{S}_{SLR}$). Thus $n_{AX}R_{AX} = n_{SLR}R_{SLR} = r$, and so $n_{AX} - n_{SLR} = (R_{AX}^{-1} - R_{SLR}^{-1})r$.

Rule 3: The number of ray arrivals and hence the total record spread (as distinct from the intervals between successive arrivals) increases linearly with range.

The results so far are quite general, and do not depend on whether the surface-limited ray comes in first and the axial ray last (the "normal sound channel") or vice versa. Equation (2.6.2) can be written

$$DR = -\frac{r}{n(n+1)} = \frac{\partial A/\partial \widetilde{S}}{n+1}, \tag{2.6.5}$$

since $r = -n\partial A/\partial \widetilde{S}$. But $DR = (\partial R/\partial \widetilde{S})D\widetilde{S} = -(\partial^2 A/\partial \widetilde{S}^2)D\widetilde{S}$, and so

$$D\widetilde{S} = \frac{R}{(n+1)\partial^2 A/\partial \widetilde{S}^2}, \tag{2.6.6}$$

Axial rays have maximum turning slowness $\widetilde{S}_0$. We define a "normal sound channel" as one where successive rays have increasing $\widetilde{S}$ (turn closer to the axis).

Rule 4: The sound channel is normal (axial rays last) if $\partial^2 A/\partial \widetilde{S}^2$ is positive, and "abnormal" if $\partial^2 A/\partial \widetilde{S}^2$ is negative.

The interval between successive ray groups equals A (2.6.4). For a normal sound channel, $D\widetilde{S}$ is positive, and $DA = (\partial A/\partial \widetilde{S})D\widetilde{S} = -RD\widetilde{S}$ is negative. Accordingly, we have the following:

Rule 5: For a normal sound channel, the group interval $D\tau$ diminishes with arrival time; for an abnormal channel, $D\tau$ increases with arrival time.

Ray groups. The discussion so far refers to ray groups. Each ray group n has four constituents. We use the ray identifier $\pm p$ to designate a ray with $\pm$ launch angle θ_{source}, having a total of $p = n^+ + n^-$ turning points (a boundary reflection is also considered a turning point). For the temperate profile, rays are clustered in groups of four, such as –37, –38, +38, +39 (fig. 2.4).

Polar profile. For the bottom-limited ray, we take (section 2.17)

$$z_{BLR} = -4\,\text{km}, \quad \widetilde{\sigma}_{BLR} = 0.30, \quad A_{BLR} = 1.2\,\text{s}. \tag{2.6.7}$$

The initial group interval $D\tau = A_{BLR} = 1.2\,\text{s}$, and the initial loop span is $2\gamma_a^{-1}\widetilde{\sigma}_{BLR} = 53\,\text{km}$; both diminish toward zero for the final arrival. The arrival spread for $r = 1$ Mm is

$$r S_0 - n_{BLR} T_{BLR} = r(S_0 - \widetilde{S}_{BLR}) - n_{BLR} A_{BLR} \approx \tfrac{1}{6} r\, S_0\, \widetilde{\sigma}^2_{BLR} = 10\,\text{s}. \tag{2.6.8}$$

Temperate profile. We use the results of section 2.18 in the appendix. For the surface-limited ray, $A^{\pm}_{SLR} = 0.28 \mp 0.11 + 0.04$ s, from (2.18.6), and

$$\widetilde{z}^{+}_{SLR} = 1\,\text{km}, \ \widetilde{z}^{-}_{SLR} = -2.18\,\text{km}, \ \widetilde{\phi}_{SLR} = 1.88,$$

$$A_{SLR} = A^{+}_{SLR} + A^{-}_{SLR} = 0.21 + 0.43 = 0.64\,\text{s},$$

where the signs refer to the upper/lower loops. The initial group interval is $A_{SLR} = 0.64$ s, and it approaches zero for the final arrivals. From (2.18.10),

$$R^{\pm} \approx R_0^{\pm}\left(1 \mp \frac{2\sqrt{2}}{3\pi}\widetilde{\phi} + \frac{1}{12}\widetilde{\phi}^2\right), \quad R_0^{\pm} \approx \frac{\pi}{\sqrt{2}}\frac{h}{\sqrt{\gamma_a h}} = 20.8\,\text{km}, \tag{2.6.9}$$

for the upper/lower-loop span, and $R = R^+ + R^-$ for the double loop. The approximation refers to $(\widetilde{S}/S_A)^2 = 1 - (\gamma_a h)\widetilde{\phi}^2 \approx 1$. The double-loop span diminishes from $R_{SLR} = R^{+}_{SLR} + R^{-}_{SLR} = 15.2\,\text{km} + 38.7\,\text{km} = 53.9$ km to an asymptotic value of $R_A = R_0^+ + R_0^- = 42$ km (not to zero as in the polar case). For the early arrivals the loops are very asymmetric, with the lower-loop spans being more than twice the upper-loop spans. For late arrivals the rays approach the form of a sine wave. Travel time relative to axial travel time over the same distance $R^{\pm}$ is given by

$$T^{\pm} - S_A R^{\pm} = A^{\pm} - R^{\pm}(S_A - \widetilde{S}) = T_0^{\pm}\left(\pm\frac{2\sqrt{2}}{9\pi}\widetilde{\phi}^3 - \frac{1}{24}\widetilde{\phi}^4\right),$$

$$T_0^{\pm} = \tfrac{1}{2}(\gamma_a h) S_A R_0^{\pm} = 0.079\,\text{s}. \tag{2.6.10}$$

For the double loop, the travel time

$$T = T^{+} + T^{-} = S_A R - \tfrac{1}{12}\, T_0^{\pm} \widetilde{\phi}^4 \tag{2.6.11}$$

is always less than the axial travel time (normal), but only very slightly less for flat rays (small $\widetilde{\phi}$). The arrival spread between surface-limited and axial rays is

$$r S_A - n_{SLR} T_{SLR} = \tfrac{1}{48}\, \gamma_a h\, r\, S_A\, \widetilde{\phi}^4_{SLR} / (1 + \tfrac{1}{12}\, \widetilde{\phi}^2_{SLR}) = 1.5\,\text{s}, \tag{2.6.12}$$

as compared with 10 s for the polar profile.

The difference in polar and temperate spreads has important consequences. For a depth $|z_B| = 4$ km and $C_0 = 1.5$ km/s,

$$\text{polar: } C_B = 1.568\,\text{km/s},\ \widetilde{\sigma}^2_{BLR} = 2\gamma_a |z_B| = 0.090,$$

$$\text{temperate: } C_B = 1.548\,\text{km/s},\ \widetilde{\sigma}^2_{BLR} = 0.065,\ \widetilde{\phi}_{BLR} = 2.40,$$

$$C_S = 1.531\,\text{km/s},\ \widetilde{\sigma}^2_{SLR} = 0.040,\ \widetilde{\phi}_{SLR} = 1.88.$$

The fractional dispersal (relative to total axial travel time) from (2.6.7) and (2.6.11) equals

$$\text{FD} = \tfrac{1}{3}\, \gamma_a h_B = 1.51 \times 10^{-2} \quad \text{for polar RSR},$$

$$\text{FD} = \tfrac{1}{48}\, \gamma_a h \frac{\widetilde{\phi}^4_{BLR}}{1 + \tfrac{1}{12}\, \widetilde{\phi}^2_{BLR}} = 0.53 \times 10^{-2} \quad \text{for temperate RSR},$$

$$\text{FD} = \tfrac{1}{48}\, \gamma_a h \frac{\widetilde{\phi}^4_{SLR}}{1 + \tfrac{1}{12}\, \widetilde{\phi}_{SLR}} = 0.23 \times 10^{-2} \quad \text{for temperate RR}.$$

Other sound-speed profiles. For a parabolic profile, $b = c = 0$ in (2.5.8). In accordance with rule 4 the parabolic channel is abnormal. The Slichter profile (Slichter, 1932),

$$C(z) = C_A \cosh[\gamma (z - z_A)], \quad A = 2\pi \gamma^{-1} (S_A - \widetilde{S}), \tag{2.6.13}$$

is focusing: all rays come in at the same time at the same location. The equatorial ocean is nearly focusing in this sense. It is evident that minor changes in the sound channel can greatly alter the arrival pattern, a potential tomographic asset.

Caustics. The condition for a caustic is

$$\frac{\partial^2 A}{\partial \widetilde{S}^2} = -\frac{\partial R}{\partial \widetilde{S}} = 0. \tag{2.6.14}$$

Eliminating $\widetilde{S}$ between (2.6.14) and $z_{\text{ray}}(x, \widetilde{S})$ yields $z_{\text{caustic}}(x)$. There are different kinds of caustics, and they can be treated as special cases of catastrophe theory (Brown and Tappert, 1987). We refer to Brekhovskikh and Lysanov (1991, sect. 4.5) for a general discussion.

The treatment of caustics goes beyond the geometric-optics approximation and requires expanding (2.3.5) to a second term. Brown's (1981, 1982) WKBJ treatment allows for caustics in a practical way and has been widely applied to the forward problem. The delta-functions in the geometric ray approximation now spread in space and time, and the caustic arrivals can be prominent even if the receiver does not lie on the caustic surface (Worcester, 1981).

2.7. Ray Weighting

For a given ray, travel time is determined by the weighted sound-speed profile, the weighting being in accordance with the distance traveled by the ray within a given depth interval. This weighting is closely related to the **E** matrix in the inverse problem and is an important consideration in determining the vertical resolution of ocean acoustic tomography. The situation is quite different for the polar ocean versus the temperate ocean. In the polar ocean, for any one arrival the weighting function increases monotonically downward toward the turning depth; successive arrivals place increasing weight onto the sound-speed at shallower depths. For the temperate sound channel the weighting function is peaked at both the upper and lower turning depths, and this leads to a troublesome ambiguity between the two *conjugate* layers (with equal sound-speeds above and beneath the sound axis). The ambiguity can be reduced (but not eliminated) by including rays with an extra upper (or lower) loop (Munk and Wunsch, 1982*b*). The problem is further considered in chapters 5 and 6.

The differential travel time is given by $dt = S\,ds = S dz \sin\theta$, with $S\cos\theta = \widetilde{S}$ by Snell's law. Hence for a temperate profile,

$$T^{\pm} = \int_{\text{upper loop}} ds\, S(z) = \pm 2 \int_{z_A}^{\widetilde{z}^{\pm}} dz\,(w^0)^{\pm}(z)\, S(z)\,, \tag{2.7.1}$$

where

$$w^0(z) = \frac{ds}{dz} = \frac{1}{\sin\theta(z)} = \frac{S(z)}{(S^2(z) - \widetilde{S}^2)^{\frac{1}{2}}}\,. \tag{2.7.2}$$

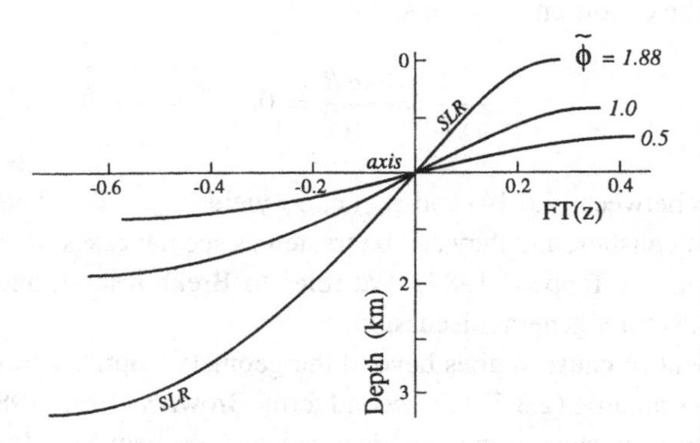

Fig. 2.5. Fractional ray travel times FT(z) in a temperate profile between the sound axis and depth z for a flat ray ($\widetilde{\phi} = 0.5$), a steep ray ($\widetilde{\phi} = 1$), and the surface-limited ray ($\widetilde{\phi} = 1.88$). Thus FT($\widetilde{z}^{+}$)+ FT($\widetilde{z}^{-}$) = 1. The departure from linearity is a measure of weighting at the turning depth.

Accordingly, $w^0(z)$ is a weighting function with integrable singularities at the turning depths; $2w^0(z)\,\delta z$ is the distance traveled by a loop within the layer $z \pm \frac{1}{2}\,\delta z$, and $2S(z)\,w^0(z)\,\delta z$ is the ray travel time in this layer. Fig. 2.5 shows the cumulative travel time.

2.8. Ray Perturbations

Applications of the tomographic process to ocean problems have featured a perturbation treatment: the determination of small departures from a known (or estimated) reference field. In this section we shall derive, or make plausible, certain key features in the perturbation treatment.

The travel time for a pulse traveling along an acoustic ray path Γ_n is

$$\tau_n = \int_{\Gamma_n} ds\, S(\mathbf{x})\,. \tag{2.8.1}$$

Travel-time tomography uses τ_n to determine $S(\mathbf{x})$. The exact travel-time perturbation is

$$\Delta\tau_n = \int_{\Gamma_n} ds\, S(\mathbf{x}) - \int_{\Gamma_n(-)} ds\, S(\mathbf{x}, -) \tag{2.8.2}$$

where $\Gamma_n(-)$ is the ray path in the unperturbed field $S(\mathbf{x}, -)$. This problem is nonlinear, since the ray path Γ_n depends on $S(\mathbf{x})$. Equation (2.8.2) can be linearized using perturbation theory, so that standard linear inverse techniques are applicable. The ocean sound-speed differs from the mean state by only a few percent. Accordingly,

$$S(\mathbf{x}) = S(\mathbf{x}, -) + \Delta S(\mathbf{x}), \quad \Delta S(\mathbf{x}) << S(\mathbf{x}, -)\,. \tag{2.8.3}$$

"Frozen-ray" approximation. The perturbation in travel time is given by the perturbation in sound-slowness along the unperturbed path:

$$\Delta\tau_n = \int_{\Gamma_n(-)} ds\ \Delta S(z) \tag{2.8.4}$$

to first order. The effect of the path perturbation on travel time requires a second-order treatment, given at the end of this section. We first derive the results from the basic definitions, using the polar profile for illustration. This gives a straightforward interpretation of the so-called frozen-ray approximation. A more general derivation using perturbed action follows.

To proceed, we suppose $S(\mathbf{x})$ is a function of a finite number of variables, $\boldsymbol{\ell}$. For the case when $S(z)$ does not depend on range, we can write, to first order,

$$\Delta\tau_n = \Delta \int_{\text{source}}^{\text{receiver}} ds\ S = \int_{\Gamma_n(-)} ds \cdot \Delta S + \int_{\Gamma_n(-)} \Delta(ds) \cdot S(-)$$

$$= \underset{\mathbf{A}}{\int_{\Gamma_n(-)} ds\ \frac{\partial S(-)}{\partial \boldsymbol{\ell}} \cdot \Delta\boldsymbol{\ell}} + \underset{\mathbf{B}}{\int_{\Gamma_n(-)} ds\ \frac{\partial S(-)}{\partial z} \cdot \Delta z} + \underset{\mathbf{C}}{\int_{\Gamma_n(-)} d\,(\Delta s) \cdot S(-)}, \tag{2.8.5}$$

where the interpretation of **A** is the traditional one for the leading approximation, namely, of integrating the perturbation in sound-slowness along the unperturbed path. Term **B** is the change in travel time due to the change in the (unperturbed) sound-slowness associated with the vertical path displacement, and **C** is the result of the perturbation in path lengths; **B** and **C** arise from the perturbations in the ray path that are consequences of the perturbation of the ocean model.

The three terms in (2.8.5) are explicitly evaluated for the special case in which the gradient of the polar profile is perturbed. We identify the gradient with a single model parameter $\boldsymbol{\ell} = \gamma_a$, which is perturbed from γ_a to $\gamma_a + \Delta\gamma$ (fig. 2.6). Using the relations (2.4.1) and (2.4.2),

$$dz = \frac{dS^2}{2\gamma_a S_0^2} = \sin\theta\, ds, \quad \cos\theta = \frac{\widetilde{S}}{S},$$

the three terms become

$$\begin{aligned} \mathbf{A} &= -\tfrac{1}{3}\, r\, S_0\, \widetilde{\sigma}^2\, (1-\widetilde{\sigma}^2)^{-\frac{1}{2}}\, (\Delta\gamma/\gamma_a), \\ \mathbf{B} &= -\mathbf{C} = 3\,(1-\widetilde{\sigma}^2)\,(1-2\widetilde{\sigma}^2)^{-1}\,\mathbf{A}, \end{aligned} \tag{2.8.6}$$

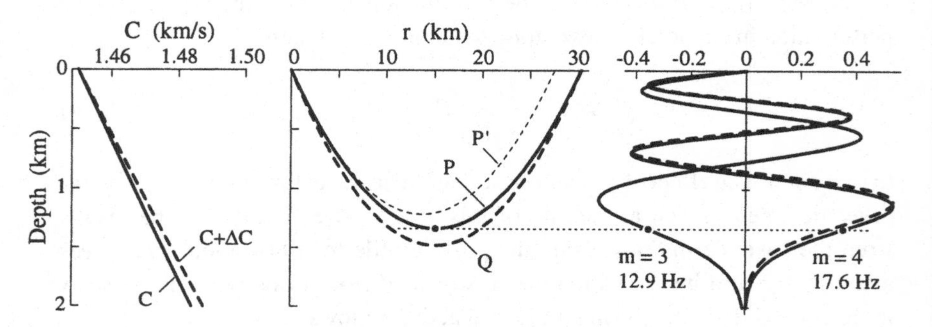

Fig. 2.6. The solid curves correspond to an unperturbed polar profile, with $\gamma_a = 0.0113$ km^{-1}. A ray P with a launch angle $\theta_0 = -10°$ intercepts a receiver at 30 km. When the gradient is perturbed from γ_a to $1.1\gamma_a$ (dashed), the $-10°$ ray P' falls short of the receiver, but a ray Q with a $-11.1°$ launch angle reaches the receiver. The perturbed path Q is longer and deeper (faster) than the unperturbed path P. In accordance with Fermat's principle, these two effects of path perturbation cancel. The travel-time perturbation is computed to first order by allowing for the sound-speed perturbation ΔC along the unperturbed path P. For later reference, unperturbed and perturbed mode functions are shown to the right.

where $\widetilde{\sigma}^2 = (S_0^2 - \widetilde{S}^2)/S_0^2$. The perturbed ray is deeper (faster) and longer than the original ray, and the two effects just cancel. The travel-time perturbation is given to first order by the expression **A** (the sound-speed perturbation along the unperturbed path), not because **B** and **C** are individually smaller than **A** (they are here larger), but because they cancel owing to the stationarity condition. This is sometimes referred to as the frozen-ray approximation. The wording is unfortunate, as no such assumption has been made.

The stationarity is a special case of applying Fermat's principle. This calculation has to be done with care, because Fermat's principle has nothing to do with perturbing the medium; rather, it applies to a *virtual* perturbation of a ray path in a *fixed* ocean model. But because we are free to choose any virtual path perturbation, we select that particular perturbation that would occur if the ocean model changed from ℓ to $\ell + \Delta\ell$, and we assume that the virtual perturbation takes place in the as-yet-undisturbed ocean ℓ (prior to $\ell + \Delta\ell$). It follows that **B** and **C** must generally cancel, in accordance with Fermat's principle governing stationarity in travel time with path perturbation in the unperturbed ocean.[6]

[6]No corresponding principle exists in mode tomography, where the perturbation of the modal wave function enters to first order.

Perturbed action. A more convenient derivation follows from the perturbed actions. Travel time can be written in terms of the group slowness s_g:

$$\tau = r s_g, \quad s_g = T/R = \widetilde{S} + A/R\,. \tag{2.8.7}$$

For the *polar* profile, using (2.17.8),

$$s_g(\widetilde{\sigma}) = \frac{T(\widetilde{\sigma})}{R(\widetilde{\sigma})} = S_0 \frac{1 - \frac{2}{3}\,\widetilde{\sigma}^2}{\sqrt{1-\widetilde{\sigma}^2}} \approx S_0(1 - \tfrac{1}{6}\,\widetilde{\sigma}^2)\,, \tag{2.8.8}$$

with $\widetilde{\sigma}^2 = 2\,\gamma_a(-\widetilde{z})$. For rays with n double loops at a range $r = n R(\widetilde{\sigma}_n)$, the dimensionless slowness parameter at the turning point is

$$\widetilde{\sigma}^2 = \widetilde{\sigma}_n^2 = \tfrac{1}{2}\left[1 - \sqrt{1 - (\gamma_a r/n)^2}\,\right] \approx (\tfrac{1}{2}\,\gamma_a r/n)^2\,. \tag{2.8.9}$$

The group slowness perturbation is given by

$$\Delta s_g = \Delta \widetilde{S} + R^{-1}\,\Delta A + A\,\Delta(R^{-1})\,. \tag{2.8.10}$$

For any ray path with n double loops within a constant range r, $R = r/n$ is constant, and thus

$$\Delta R = 0 = \frac{\partial R}{\partial \boldsymbol{\ell}} \cdot \Delta \boldsymbol{\ell} + \frac{\partial R}{\partial \widetilde{S}}\,\Delta \widetilde{S}\,. \tag{2.8.11}$$

(The discussion is easily extended to the case of an extra upper or an extra lower loop.) Equation (2.8.11) relates, to first order, the ray perturbation $\Delta\widetilde{S}$ to the ocean perturbation $\Delta\boldsymbol{\ell}$. Further

$$\Delta A = \frac{\partial A}{\partial \boldsymbol{\ell}} \cdot \Delta \boldsymbol{\ell} + \frac{\partial A}{\partial \widetilde{S}}\,\Delta \widetilde{S} \tag{2.8.12}$$

for the first-order action perturbation. From (2.5.4), R^{-1} times the second term in (2.8.12) equals $-\Delta\widetilde{S}$, and this cancels the first term in (2.8.10). We are left with the simple expression

$$(\Delta s_g)_{\rm ray} = \frac{1}{R}\frac{\partial A}{\partial \boldsymbol{\ell}} \cdot \Delta \boldsymbol{\ell}\,. \tag{2.8.13}$$

For the polar ocean, $A = \frac{2}{3}\,\gamma_a^{-1}\,S_0\,\widetilde{\sigma}^3$, and it follows at once from (2.8.9) that

$$(\Delta s_g)_{\rm ray} = -\tfrac{1}{3}\,S_0\,\widetilde{\sigma}^2\,(1 - \widetilde{\sigma}^2)^{-\frac{1}{2}}\,(\Delta\gamma/\gamma_a)\,, \tag{2.8.14}$$

in agreement with (2.8.6): the perturbation in travel time $\Delta\tau = r\,\Delta s_g$ is equivalent, to first order, to integration along the unperturbed path.

Second order. The second-order ray problem has been considered by Mercer and Booker (1983), Spiesberger and Worcester (1983), Spiesberger (1985*a,b*), Munk and Wunsch (1985, 1987), and Wunsch (1987). In general, $A(\boldsymbol{\ell}, \widetilde{S})$ is a function of the ocean parameter vector $\boldsymbol{\ell}$ and the ray parameter $\widetilde{S}$. For the unperturbed state, $\Delta\boldsymbol{\ell} = 0$; ocean perturbations $\Delta\boldsymbol{\ell}$ give rise to ray perturbations $\Delta\tau,\ \Delta A,\ \Delta\widetilde{S}$.

The expansion to second order requires an expression for the path perturbation $\Delta\widetilde{S}$ to first order. This is obtained from (2.8.11) by requiring $\Delta R = 0$. Details are tedious and uninteresting. The result is (Munk and Wunsch, 1985, 1987)

$$\Delta T = L\,\Delta\ell + M\,(\Delta\ell)^2\,,$$
$$L = \frac{\partial A}{\partial \ell}, \qquad M = \tfrac{1}{2}\left[\frac{\partial^2 A}{\partial \ell^2} - \frac{(\partial^2 A/\partial\ell\,\partial\widetilde{S})^2}{\partial^2 A/\partial\widetilde{S}^2}\right]. \tag{2.8.15}$$

A different derivation of (2.8.15) by Hamiltonian methods is given by Wunsch (1987).

For rays in a polar profile, substituting $A = \frac{2}{3}\,\ell\,S_0\,\widetilde{\sigma}_n^3$ into (2.8.15) yields

$$\Delta\tau = \tfrac{2}{3}\,n\,\widetilde{\sigma}^3\,S_0\,\Delta\ell\ -\ n\,\widetilde{\sigma}^3\,S_0\,\ell^{-1}(\Delta\ell)^2\,. \tag{2.8.16}$$

The ratio of the second-order term to the first-order term is $-\frac{3}{2}\,(\Delta\ell/\ell)$. The negative sign in the second-order term (2.8.16) corresponds to a *warm bias*; that is, the path perturbation leads to an earlier arrival than predicted by the linear term; this could be misinterpreted as a warmer ocean than is in fact the case.

An equivalent calculation by Munk and Wunsch (1987) for perturbations by dynamic (Rossby) modes in a temperate ocean leads to qualitatively similar results.

2.9. Parametric and Functional Perturbation

So far we have considered only cases where the sound profile is specified by a few parameters $\boldsymbol{\ell}$ and their perturbations by $\Delta\boldsymbol{\ell}$. For the polar ocean we have set $\Delta\boldsymbol{\ell} = \Delta\gamma$, and for the temperate ocean we can set $\Delta\boldsymbol{\ell} = (\Delta a,\ \Delta b,\ \Delta c)$. This formulation in terms of parameter perturbations is useful for illustrative purposes, but it is hardly useful for any actual application. In general, we wish to treat the case of arbitrary $S(z) + \Delta S(z)$, with the goal of deriving a perturbation weighting function $w(z)$ such that

$$\Delta s_g = \int_{\widetilde{z}^-}^{\widetilde{z}^+} dz\, w(z)\,\Delta S(z)$$

for a given $S(z)$. But even if the reference state is well represented by an analytical model with a few parameters, the expected sound-slowness *perturbation* may not be adequately described by perturbations in the reference-state parameters.[7] In these cases we set

$$\Delta S(z) = \sum_j \ell_j F_j(z) \tag{2.9.1}$$

where $F_j(z)$ are functions that are selected *a priori* to represent the expected slowness perturbations. Many choices of $F_j(z)$ are possible, including layers, dynamic ocean modes (Rossby-wave modes), and empirical orthogonal functions constructed from historical data.

To treat the general case, we interpret the perturbation parameters in terms of the slowness perturbation ΔS_j in a layer j extending from z_j to $z_j + \delta z_j$. But with

$$A = 2\int_{\tilde{z}^-}^{\tilde{z}^+} dz\,(S^2 - \tilde{S}^2)^{\frac{1}{2}}, \quad \frac{\partial A}{\partial S} = 2\int_{\tilde{z}^-}^{\tilde{z}^+} dz\,\frac{S}{(S^2 - \tilde{S}^2)^{\frac{1}{2}}},$$

we have

$$\frac{\partial A}{\partial \boldsymbol{\ell}} \cdot \Delta\boldsymbol{\ell} = \sum_j \frac{\partial A}{\partial S_j} \Delta S_j = \sum_j 2\int_{z_j}^{z_j+\delta z_j} dz\, S(S^2 - \tilde{S}^2)^{-\frac{1}{2}}\, \Delta S_j,$$

and in the limit of many thin layers,

$$\frac{\partial A}{\partial \boldsymbol{\ell}} \cdot \Delta\boldsymbol{\ell} = 2\int_{\tilde{z}^-}^{\tilde{z}^+} dz\, w^0(z)\, \Delta S(z), \quad w^0 = \frac{S(z)}{(S^2 - \tilde{S}^2)^{\frac{1}{2}}}, \tag{2.9.2}$$

where $w^0(z)$ is identical with the unperturbed weighting function (2.7.2). Thus

$$\begin{aligned} \Delta s_g &= \frac{1}{R}\frac{\partial A}{\partial \boldsymbol{\ell}} \cdot \Delta\boldsymbol{\ell} \\ &= \frac{2}{R}\int_{\tilde{z}^-}^{\tilde{z}^+} dz\, w^0 \Delta S = \frac{2}{R}\int_{\Gamma_n(-)} ds\, \Delta S(z) \end{aligned} \tag{2.9.3}$$

[7]The mean in the sound-speed profile is determined by large-scale, long-term oceanographic processes, *e.g.*, the global-scale thermohaline circulation, while the perturbations are largely, although not completely, due to processes with shorter time and space scales, such as the mesoscale eddy field.

since $w^0 = ds/dz$. To summarize,

$$s_g = \frac{2}{R}\int_{\widetilde{z}^-}^{\widetilde{z}^+} dz\, w^0(z)\, S(z), \qquad \Delta s_g = \frac{2}{R}\int_{\widetilde{z}^-}^{\widetilde{z}^+} dz\, w^0(z)\, \Delta S(z)\,,$$

with the important result that the ray weighting function is the same for s_g and Δs_g. This is a consequence of Fermat's principle, which leads to the cancellation of terms in deriving (2.8.13). Fermat's principle is equivalent to $w(z) = w(-)$ and $\widetilde{z}^{\pm} = \widetilde{z}^{\pm}(-)$.

For illustration, the polar profile with the gradient perturbation $\Delta\gamma/\gamma_a$ and

$$\Delta S(z) = S_0\, z\, \Delta\gamma = -\tfrac{1}{2}\, S_0\, \widetilde{\sigma}^2\, (\Delta\gamma/\gamma_a)\, \zeta, \qquad \zeta = z/\widetilde{z},$$

yields

$$(\Delta s_g)_{\text{ray}} = -\tfrac{1}{4}\, S_0\, \widetilde{\sigma}_n^2 (1 - \widetilde{\sigma}_n^2)^{-\frac{1}{2}} (\Delta\gamma/\gamma_a) \int_0^1 d\zeta\, (1 - \zeta)^{-\frac{1}{2}},$$

in agreement with (2.8.14).

In section 2.15 we present some examples of perturbation models for both rays and modes. Chapters 5 and 6 discuss the implications and relative merits of various choices in the context of the inverse problem. The formalism to be used is

$$\Delta\tau = r\, \Delta s_g, \qquad \Delta S = \sum_j \ell_j\, F_j(z),$$

so that the forward problem can be written in the matrix form

$$\Delta\boldsymbol{\tau} = \mathbf{E}\, \boldsymbol{\ell}, \tag{2.9.4}$$

where

$$E_{nj} = 2n \int dz\, w_n^0(z)\, F_j(z) \tag{2.9.5}$$

relates the travel-time perturbation $\Delta\tau_n$ of ray n to the ocean perturbation $F_j(z)$, with $n = r/R$ designating the number of ray double loops.

MODE REPRESENTATION

2.10. Modes

In the ray approximation, the response to an impulsive source is a series of delta-functions of variable amplitude. This description is not a bad one for the observed early arrivals. The usual inverse problem is to infer information about the intervening ocean from the arrival times alone, although this clearly does not use all available information. But the measured arrival pattern is more complex than just a series of delta-functions, particularly near the final arrivals.

Ideally, one would like to compute the entire arrival pattern from an exact solution of the wave equation and then infer the intervening ocean from a comparison of the computed and measured $p(t)$. In the seismic literature the computed pattern is referred to as the "synthetic seismogram." So we would like the forward problem to yield a synthetic sonogram for comparison with the complete measured record. This is called "matched-field processing."

In a range-independent case, the procedure is to expand the source function in terms of a complete set of orthogonal normal modes. Each mode is propagated with the appropriate group velocity, and the synthetic sonogram is constructed by a linear superposition of all modes at the receiver. We shall think of this procedure as yielding an exact solution to a horizontally homogeneous ocean (although of course it does not).[8]

Given these conditions, the question must arise: Why even bother with the ray solutions? The answer is that they are the most prominent features in the record (except for the arrival *finale*), they are robust, they yield an astonishingly good approximation over most of the record (compare fig. 2.4 with fig. 2.7), they are easily visualized, and they are not demanding computationally. Further, they can be computed for the case of small-scale range-dependent structures. [In principle, (2.3.11) can be integrated across discontinuities.] As yet, practical range-dependent modal solutions are limited to the very restrictive case of adiabatic range dependence (see chapter 4).

It is useful to understand the relation between ray and mode representations and the advantages and disadvantages of each. In choosing a representation in terms of a sum of rays or a sum of modes, one normally chooses the one that requires the lesser number of terms. Kamel and Felsen (1982) have proposed a hybrid analysis, using rays to represent the outgoing radiation for the early, steep source angles, and modes for the late, near-axial arrivals.

[8]The equations have been linearized, and the waviness of the surface boundary and the geologic structure of the bottom boundary have been ignored, among other approximations.

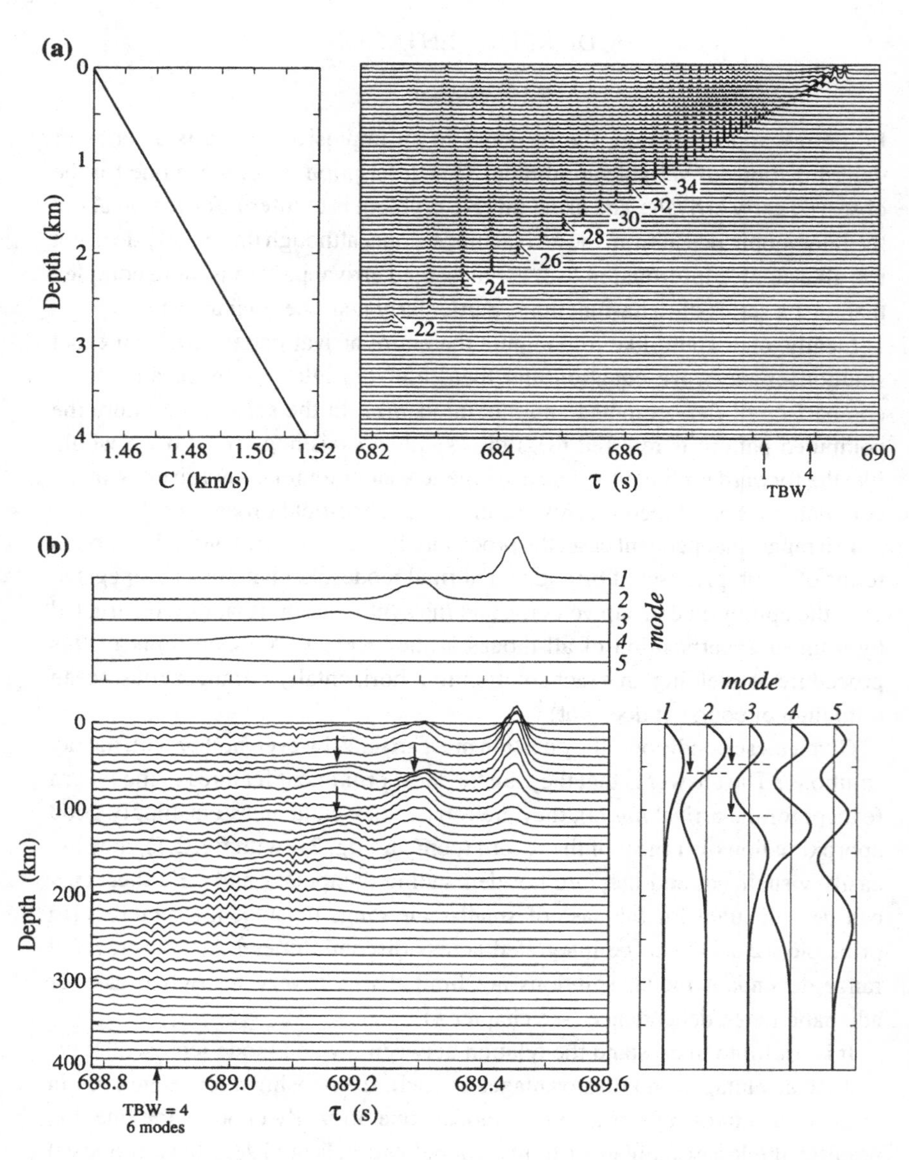

Fig. 2. 7. (a) Computed arrival pattern in a polar-ocean profile from superposition of modes 1 to 150 (non bottom-interacting): range 1 Mm, center frequency 250 Hz, bandwidth 100 Hz, source depth 30 m. The ray designators −22, −24, ... are obtained by comparison with the ray solution (fig. 2.4, top) and refer to the number of downward ray loops. (b) The final 0.8 s in more detail, with the time functions and depth function for modes 1 to 5 shown separately. Arrows indicate depths of modal zero crossings.

There is an extensive literature concerning rays and modes. (The ray–mode duality has been known since antiquity.) Some of the relationships have been reinvented (seemingly) independently in various fields, but application to the ocean channel has introduced some aspects that appear novel (Officer, 1958; Tolstoy and Clay, 1966; Brekhovskikh and Lysanov, 1991). We follow the discussion by Munk and Wunsch (1983). The emphasis is on developing an understanding of the ray–mode relationship, which is a necessary preliminary to all of the numerical details that are encountered with the real ocean and real data.

The starting point is again the wave equation (2.3.1),

$$\left(\nabla^2 - S^2(z)\frac{\partial^2}{\partial t^2}\right) p = 0,$$

which has a separable solution in cylindrical coordinates,

$$p(r, z, t) = Q(r) P(z) e^{i\omega t} . \tag{2.10.1}$$

The radial wave function $Q(r)$ must satisfy

$$\frac{1}{r}\frac{d}{dr}\left[r\frac{dQ}{dr}\right] + k_H^2 r = 0 .$$

The solution appropriate to outgoing energy only is

$$Q(r) = H_0^1(k_H r) \rightarrow \sqrt{\frac{2}{\pi k_H r}}\, e^{i(k_H r - \frac{1}{4}\pi)} \quad \text{for} \quad r \rightarrow \infty ,$$

where H_0^1 is a Hankel function. The asymptotic solution at large $k_H r$ identifies the separation constant k_H as the horizontal wavenumber; note that k_H is not a function of z.

The vertical wave function $P(z)$ must satisfy

$$\frac{d^2 P}{dz^2} + (\omega^2 S^2(z) - k_H^2)\, P = 0 . \tag{2.10.2}$$

We recognize $\omega S(z) = k(z)$ as the scalar wavenumber and

$$k_V(z) = (\omega^2 S^2(z) - k_H^2)^{\frac{1}{2}} \tag{2.10.3}$$

as the local vertical wavenumber: k_V changes from real to imaginary at depth $\widetilde{z}$ (to be called the modal turning depth), where

$$k_H/\omega = S(\widetilde{z}) \equiv \widetilde{S}, \tag{2.10.4}$$

so that

$$k_V(z) = \omega(S^2(z) - \widetilde{S}^2)^{\frac{1}{2}}\,. \tag{2.10.5}$$

$P(z)$ is oscillatory near the axis and exponential far from the axis, with its behavior changing at the modal turning points.

Equation (2.9.2) is now written (using primes for d/dz)

$$P'' + \omega^2\,(S^2 - \widetilde{S}^2)\,P = 0.$$

We multiply both terms by $P(z)$ and integrate:

$$\int_{-\infty}^{\infty} dz\,P_m\,P_m'' + \omega_m^2 \int_{-\infty}^{\infty} dz\,(S_m^2(z) - \widetilde{S}^2)\,P_m^2 = 0\,.$$

We write $PP'' = (PP')' - (P')^2$. For a temperate sound channel with $P(z)$ exponentially small at both boundaries, PP' vanishes at the limits. Thus

$$-\int_{-\infty}^{\infty} dz\,(P_m')^2 + \omega_m^2 \int_{-\infty}^{\infty} dz\,(S^2(z) - \widetilde{S}_m^2)\,P_m^2 = 0\,. \tag{2.10.6}$$

This is Rayleigh's principle; ω_m is stationary with $S(z)$ for a normal mode $P_m(z)$ (Morse and Feshbach, 1953, p. 1112).

Phase and group slownesses are defined by

$$s_p = k_H/\omega, \quad s_g = dk_H/d\omega = s_p + \omega \cdot ds_p/d\omega\,. \tag{2.10.7}$$

From (2.10.4), $s_p = \widetilde{S}$: the phase slowness equals the sound-slowness at the turning depth.

To obtain group slowness, write $\omega\widetilde{S} = k_H$ in the second integral of (2.10.6) and differentiate with respect to ω, remembering that $P_m(z)$ is stationary. The result is

$$s_p\,s_g = \overline{S^2} \equiv \frac{\int_{-\infty}^{\infty} dz\,S^2(z)\,P^2(z)}{\int_{-\infty}^{\infty} dz\,P^2(z)}\,, \tag{2.10.8}$$

where $\overline{S^2}$ is defined as the P^2-weighted mean of $S^2(z)$.

The exact solutions to (2.10.2) for both the polar and the temperate profiles can be written in terms of Bessel functions (section 2.18). For most purposes the WKBJ approximation is adequate (Brekhovskikh and Lysanov, 1991, sect. 6.7). This implies the substitution

$$P^2(z) \rightarrow 2\,(\widetilde{S}/R)\,[S^2(z) - \widetilde{S}^2]^{-\frac{1}{2}}, \tag{2.10.9}$$

Table 2.1. *Comparison of WKBJ approximation with exact modal solution (2.9.8) for a polar ocean*[a]

	s_g (s/km)		Δs_g	
m	Modal	WKBJ	Modal	WKBJ
1	0.666078	0.666083	-3.96×10^{-6}	-3.88×10^{-6}
2	0.665637	0.665642	-6.75×10^{-6}	-6.82×10^{-6}
5	0.66467	0.66468	-13.68×10^{-6}	-13.20×10^{-6}
10	0.66344	0.66346	-21.07×10^{-6}	-21.18×10^{-6}

[a]Parameters are $\gamma_a = 0.0113\,\text{km}^{-1}$, $S_0 = (1/1.5)$ s/km, $f_0 = 250$-Hz. The left two columns refer to the group slowness, the right two columns (for later reference) to the perturbation Δs_g in group slowness associated with $\Delta\gamma/\gamma_a = 0.01$.

Source: We are indebted to E. C. Shang for providing the modal values.

with the normalizations

$$\int_{-\infty}^{\infty} dz\, P^2 = 1, \quad 2\,(\tilde{S}/R) \int_{\tilde{z}^-}^{\tilde{z}^+} dz\,(S^2 - \tilde{S}^2)^{-\frac{1}{2}} = 1\,. \tag{2.10.10}$$

Here $R(\tilde{S}_m)$ has no immediate interpretation (as it does in ray theory) but is simply defined by the integral (2.4.4a) between the WKBJ turning points $\tilde{z}_m^{\pm}$ of mode m. The result is again $s_p\, s_g = \overline{S^2}$, with $\overline{S^2}$ now defined by

$$\overline{S^2} = \frac{\int_{\tilde{z}^-}^{\tilde{z}^+} dz\, S^2 (S^2 - \tilde{S}^2)^{-\frac{1}{2}}}{\int_{\tilde{z}^-}^{\tilde{z}^+} dz\,(S^2 - \tilde{S}^2)^{-\frac{1}{2}}}, \tag{2.10.11}$$

in place of (2.10.8). The approximation is excellent (table 2.1). With this in mind, it is simpler to derive s_g by the action principle, with the added advantage that this leads to a uniform treatment of rays and modes and sheds light on the important ray–mode duality.

2.11. WKBJ Approximation: Ray/Mode Equivalence

In the WKBJ approximation, the "phase integral"

$$\int_{\tilde{z}^-}^{\tilde{z}^+} dz\, k_V(z) = \omega \int_{\tilde{z}^-}^{\tilde{z}^+} dz\,(S^2 - \tilde{S}^2)^{\frac{1}{2}} = \tfrac{1}{2}\,\omega A \tag{2.11.1}$$

gives the phase difference between the modal turning points. Identification of the phase integrals with $\frac{1}{2}\omega A$ follows from the definition (2.5.2). The solution is oscillatory within the turning points $\widetilde{z}^{\pm}$, and exponential outside. In the exponentially decaying regions, the WKBJ solution to (2.10.2) consists of two terms containing $(\exp \pm i \int_0^z k_V \, dz)$. For finite solutions, the growing term must vanish. The condition for this to be the case is

$$\omega A_m = 2\pi (m - \tfrac{1}{2}), \quad \omega A_m = 2\pi (m - \tfrac{1}{4}), \quad m = 1, 2, \dots, \tag{2.11.2}$$

for a temperate and polar sound channel, respectively (Brekhovskikh and Lysanov, 1991, sect. 6.7). Mode numbers $m = 1, 2, \dots$, equal the number of extrema in $P(z)$ (fig. 2.9).

From (2.11.2) we have

$$\omega_m = 2\pi A_m^{-1} (m - \tfrac{1}{2}), \quad \omega_m = 2\pi A_m^{-1} (m - \tfrac{1}{4}). \tag{2.11.3}$$

Fig. 2.6 gives some examples of two modes of different frequencies, with the same turning points. Substituting for ω_m in (2.10.7),

$$\begin{aligned} s_g &= \widetilde{S} - A_m / (\partial A_m / \partial \widetilde{S}) \\ &= T/R = \overline{S^2}/s_p, \end{aligned} \tag{2.11.4}$$

using the definitions (2.5.1), (2.5.4), and (2.10.11). This is in agreement with the WKBJ approximation to the modal solution derived in the previous section, but the present derivation is much more compact. It is also in agreement with ray derivation (2.8.7).

To recapitulate,

$$s_p = \widetilde{S}, \quad s_g = T/R = \widetilde{S} + A/R, \tag{2.11.5}$$

are the phase and group slownesses for both rays and modes, but the interpretation is different. For a ray with n double loops, $R(A) = r/n$ determines A and hence $T(A)$ and $s_g(A) = T(A)/R(A)$. For a mode with mode number m and frequency f, $A = (m - \frac{1}{2})/f$ or $(m - \frac{1}{4})/f$ determines A and hence $s_g(A)$.

Polar profile. References to specific profiles will clarify the results. From (2.17.6), (2.17.7), and (2.17.8),

$$s_g(\widetilde{\sigma}) = \frac{T(\widetilde{\sigma})}{R(\widetilde{\sigma})} = S_0 \frac{1 - \frac{2}{3}\widetilde{\sigma}^2}{\sqrt{1 - \widetilde{\sigma}^2}} \approx S_0 (1 - \tfrac{1}{6}\widetilde{\sigma}^2) \tag{2.11.6}$$

for both rays and modes, with $\widetilde{\sigma}^2 = 2\,\gamma_a(-\widetilde{z})$. For rays with n double loops at a range $r = n\,R(\widetilde{\sigma}_n)$,

$$\widetilde{\sigma}^2 = \widetilde{\sigma}_n^2 = \tfrac{1}{2}\left[1 - \sqrt{1 - (\gamma_a r/n)^2}\right] \approx (\tfrac{1}{2}\,\gamma_a r/n)^2\,, \qquad (2.11.7n)$$

whereas for modes with $A(\widetilde{\sigma}_m) = (m - \frac{1}{4})/f$,

$$\widetilde{\sigma}^3 = \widetilde{\sigma}_m^3 = \frac{A_m}{\frac{2}{3}\,\gamma_a^{-1}\,S_0} = \frac{m - \frac{1}{4}}{f/F_P}, \qquad F_P = \frac{3\gamma_a}{2S_0} = 0.0254\,\text{Hz}. \qquad (2.11.7m)$$

Temperate profile. From (2.18.6), (2.18.10), and (2.18. 13),

$$s_g(\widetilde{\phi}) = \frac{T(\widetilde{\phi})}{R(\widetilde{\phi})} = S_A\left[1 - \tfrac{1}{48}\,\gamma_a h\,\frac{\widetilde{\phi}^4}{1 + \frac{1}{12}\,\widetilde{\phi}^2}\right]$$

for both rays and modes, but

$$\widetilde{\phi}^2 = \widetilde{\phi}_n^2 = 12\left[\frac{\rho}{n} - 1\right], \qquad \rho = \frac{r\gamma_a}{\pi\,\sqrt{2\gamma_a h}}, \qquad (2.11.8n)$$

$$\widetilde{\phi}^2 = \widetilde{\phi}_m^2 = 12\left[\sqrt{1 + \frac{m - \frac{1}{2}}{f/F_T}} - 1\right], \qquad F_T = \frac{2c/a^2}{S_A h\sqrt{\gamma_a h}} = 1.059\,\text{Hz}, \qquad (2.11.8m)$$

for rays and modes, respectively. This is easily extended to an extra upper or an extra lower loop [see appendix equation (2.18.17)].

The foregoing equations permit the computation of group slowness (and hence travel time) for any specified ray n, or alternatively for any specified mode number m and frequency f. Some numerical values for modes are given in table 2.2. These follow from (2.11.7m) and (2.11.8m), which are conveniently written

$$\frac{c_g - C_0}{C_0} = \left(\frac{m - \frac{1}{4}}{f/f_1}\right)^{\frac{2}{3}}, \qquad \frac{c_p - C_0}{C_0} = \left(\frac{m - \frac{1}{4}}{f/f_2}\right)^{\frac{2}{3}}, \qquad (2.11.9P)$$

$$\frac{c_g - C_A}{C_A} = \left(\frac{m - \frac{1}{2}}{f/f_3}\right)^2, \qquad \frac{c_p - C_A}{C_A} = \frac{m - \frac{1}{2}}{f/f_4}, \qquad (2.11.9T)$$

Table 2.2. *Modal group and phase velocity excesses over surface sound speed C_0 in a polar profile, and over axial sound-speed C_A in a temperature profile, at stated frequencies f (in Hz) and mode numbers m*

	f					
m	50 Hz	60 Hz	100 Hz	50 Hz	60 Hz	100 Hz
Polar sound channel						
	$c_g - C_0$ (m/s)			$c_p - C_0$ (m/s)		
1	1.01	0.90	0.64	3.03	2.69	1.91
2	2.10	1.86	1.32	6.31	5.59	3.97
5	4.38	3.87	2.76	13.1	11.6	8.30
10	7.20	6.38	4.54	21.6	19.1	13.6
Temperate sound channel						
	$c_g - C_A$ (m/s)			$c_p - C_A$ (m/s)		
1	0.0013	0.0009	0.0004	0.54	0.45	0.27
2	0.012	0.008	0.003	1.62	1.35	0.81
5	0.107	0.075	0.027	4.86	4.05	2.43
10	0.53	0.33	0.12	10.3	8.6	5.1

for the polar and temperate sound channels, respectively, with

$$f_1 = \frac{\gamma_a}{4\sqrt{6}\,S_0} = 1.75 \times 10^{-3}\ \text{Hz}, \qquad f_2 = \frac{3\gamma_a}{4\sqrt{2}\,S_0} = 9.09 \times 10^{-3}\ \text{Hz},$$

$$f_3 = \frac{\sqrt{1 - 6\gamma_a h}}{2\sqrt{6}\pi\, h\, S_A} = 9.4 \times 10^{-2}\ \text{Hz}, \qquad f_4 = \frac{\sqrt{2\gamma_a h}}{2\pi\, h\, S_A} = 3.6 \times 10^{-2}\ \text{Hz}.$$

The foregoing equations provide the starting point for the perturbation treatment in section 2.14, which is at the heart of the forward problem: Given a perturbation in the sound-speed profile, what is the perturbation in travel times? But first we wish to leave the reader with some illustrations of what the ray–mode duality is all about under representative ocean conditions.

A ray with a ray turning depth $\tilde{z}_n$ and associated $\tilde{\sigma}_n$ is the result of constructive interference of all modes having mode numbers m and frequency f such that $\tilde{\sigma}_m = \tilde{\sigma}_n$, and hence WKBJ turning depth $\tilde{z}_m$ equals ray turning depth $\tilde{z}_n$. They all have the same group slowness $s_g(\tilde{\sigma})$ and the same phase slowness $s_p = S(\tilde{z})$. Rays can be constructed by adding the modal solutions; modes can be constructed by adding ray solutions. We refer to Brekhovskikh and Lysanov (1991, sect. 6.7) and to Munk and Wunsch (1983).

2.12. Modal τ, z-display

Figs. 2.7–2.9 show the arrival pattern in time-and-depth space. We have found this display to be the most telling résumé of the forward problem. In section 2.16 this display will serve as a stringent test for comparison with observations.

Fig. 2.7 shows the result for the *polar* profile. The figure was constructed by Phil Sutton (personal communication) by a brute-force summation of the lowest 150 modal solutions. The figure displays the signal in τ, z-space from an impulsive source 1 Mm distant at times 682 to 690 s after transmission, and for 1 s just prior to termination. At the trailing end there is a strong near-surface disturbance. The figure gives a very close approximation to the disturbance in x, z-space at a fixed instant, with the source being 1 Mm to the right.

The most prominent feature is the accordion-like intensification pattern that can be traced past 688 s. The upper termination is at the surface; the lower termination results in a sharp peak, with a steep cutoff beneath. This is the signature of a caustic. The accordion pattern turns out to be in exact coincidence with the wave fronts constructed from ray theory (fig. 2.4, top) and has been labeled accordingly. Yet there has not been any assumption of ray-like propagation. Clearly the pattern resulting from constructive mode interference dominates most of the record, and any analysis ignoring this feature would be ill-advised. One can imagine the discovery of modal interference patterns prior to the development of ray theory, leading to a subsequent effort to account for the pattern directly without the need for summing many modes.

The last second of the arrival structure has an entirely different appearance (fig. 2.7b). The pattern is now vertical, not slanting. The bottom termination is more gradual than at earlier times (fig. 2.8). The arrival peaks coincide with the computed modal arrivals, and the vertical modulation coincides with the computed modal wave functions. We can identify and resolve modes 1 to 3, perhaps more. This mode-like finale can be interpreted as the result of constructive interference of the many rays that have turning points very near the surface (just as the ray-like overture was interpreted in terms of constructive interference of the many high order modes with deep turning points). The finale is the Cartesian analogue of the "whispering-gallery" effect (Budden, 1961, p. 203).

A similar construction for the *temperate* profile is shown in fig. 2.9. The non-boundary interacting arrival lasts 1.4 s and so is much less dispersed than that for the polar profile (note the different time scales in figs. 2.7 and 2.9). Once again the interference pattern is in excellent accord with the ray fronts derived from ray theory[9] (fig. 2.4, middle). An enlarged display of the final

[9]The construction prior to the arrival of ray 41 suffers from an inadequate number of interfering modes.

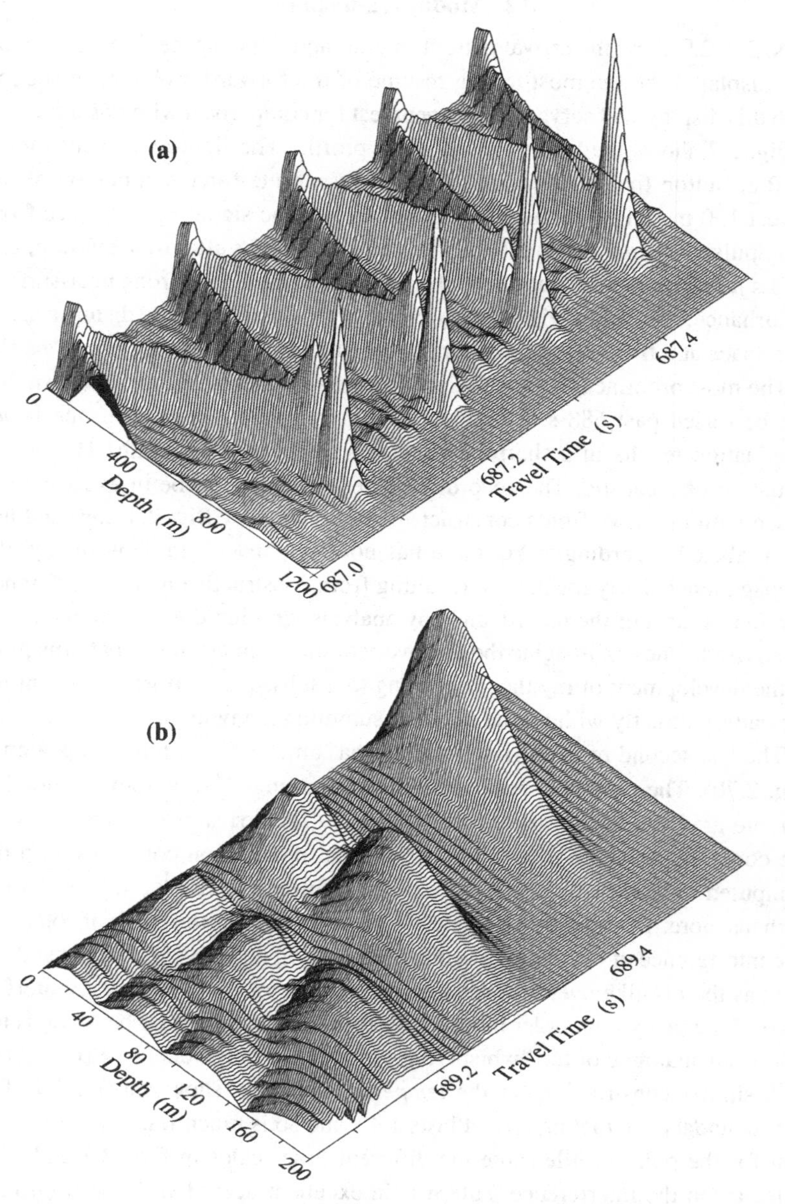

Fig. 2.8. The lower termination (a) during the ray-like regime and (b) during the mode-like regime.

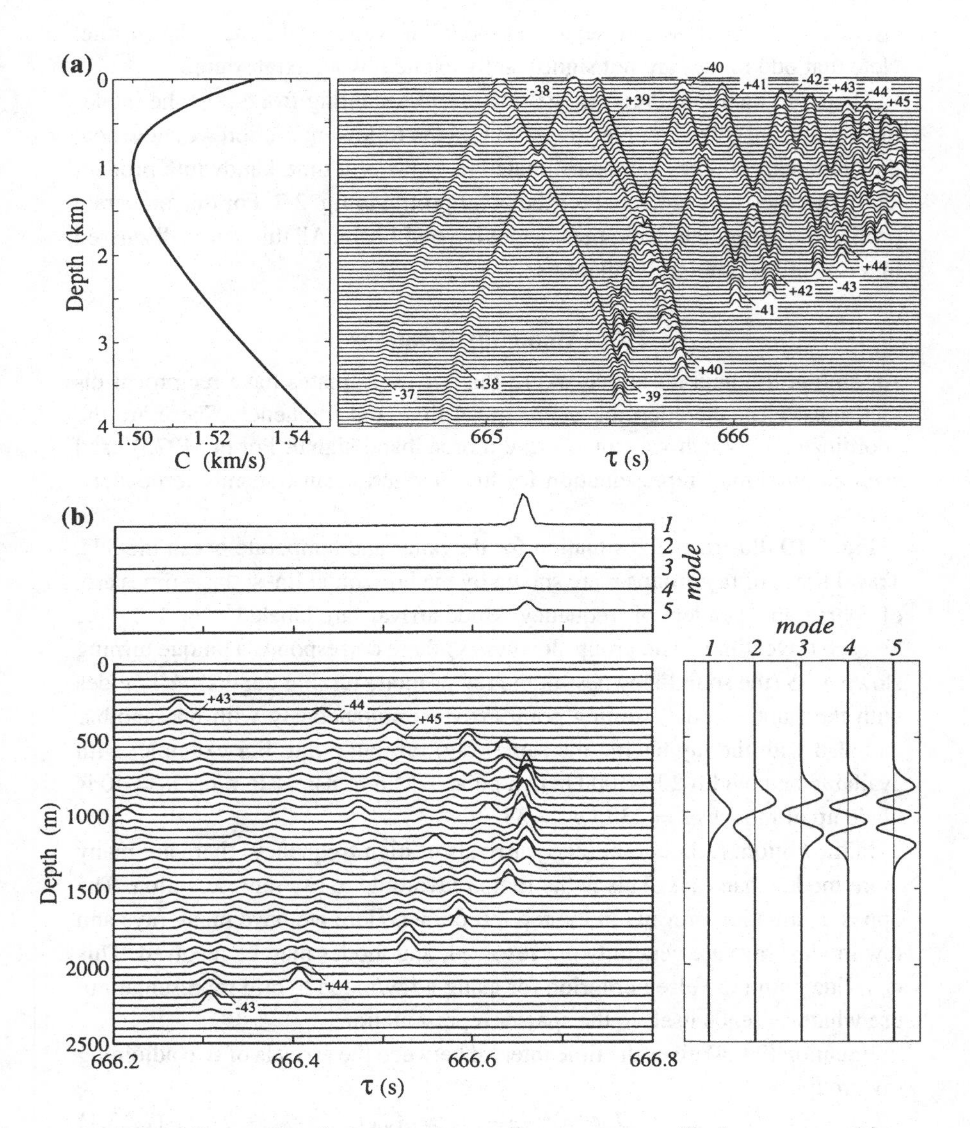

Fig. 2.9. (a) Computed arrival pattern in a temperate ocean profile from superposition of modes 1 to 250 (non-boundary-interacting): range 1 Mm, center frequency 250 Hz, bandwidth 100 Hz, source depth 1000 m (on axis). (b) The final 0.5 s in more detail, with the time functions and depth functions for modes 1 to 5 shown separately. The labels ±39, ±40, ... have been obtained by comparison with the ray solution (fig. 2.4, middle) and refer to the ± launch angle and the sum of upward and downward loops.

arrivals does not show any separated modal arrivals, unlike the polar profile. Note that odd modes are not significantly excited by an axial source.

There are then three regimes: (i) the ray-like slanting fronts, (ii) the mode-like finale, and (iii) a complex transition. In the following section we show how the demarcation times can be estimated in terms of a time-bandwidth product TBW of order 1, as indicated for the polar profile in fig. 2.7. For the temperate profile, mode separation requires ranges beyond 1 Mm. All this can be discussed in terms of the ambiguity relations.

2.13. Ambiguity Relations

An ambiguity diagram is one where the two coordinates have reciprocal dimensions. Here the coordinates are travel time and frequency. These are the coordinates in which we can observe a broadband signal. Porter (1973) used such an ambiguity representation for his Mediterranean transmission experiments.

Fig. 2.10 illustrates the situation for the polar and temperate ocean models. Travel times of ray groups n are shown by the horizontal lines; these times are, of course, independent of frequency. Mode arrivals are labeled $m = 1, 2, \ldots$. To each travel time τ and group slowness s_g there corresponds a unique turning slowness $\widetilde{S}$ (the sound-slowness at the ray or mode turning depths). All modes with the same (modal) turning point arrive simultaneously with one another and also with the ray having the same (ray) turning point. For example, with available bandwidth 200–300 Hz in a polar 1-Mm transmission, ray $n = 40$ is made up of modes $m = 23$ to $m = 35$.

In the bottom right corner (early arrivals of high frequency) there are many more modes than rays. This is the region favorable to ray representation. The opposite situation prevails in the top left corner. Here we have many rays and few modes; rays are generally not resolved, and modes may be resolved. This is in line with the Felsen criterion for using a ray/mode hybrid representation: use whatever lends itself to the sparser representation.

Equation (2.6.4) gives the time interval between the arrivals of two adjoining ray groups:

$$D_n\tau = \tau_{n+1} - \tau_n = A_{n+1}\,. \tag{2.13.1}$$

Equation (2.11.3) gives the frequency interval between two adjoining modes at some fixed time:

$$D_m f = f_{m+1} - f_m = A^{-1}\,. \tag{2.13.2}$$

It can be verified on the ambiguity diagrams that

$$D_n\tau \cdot D_m f = 1 \tag{2.13.3}$$

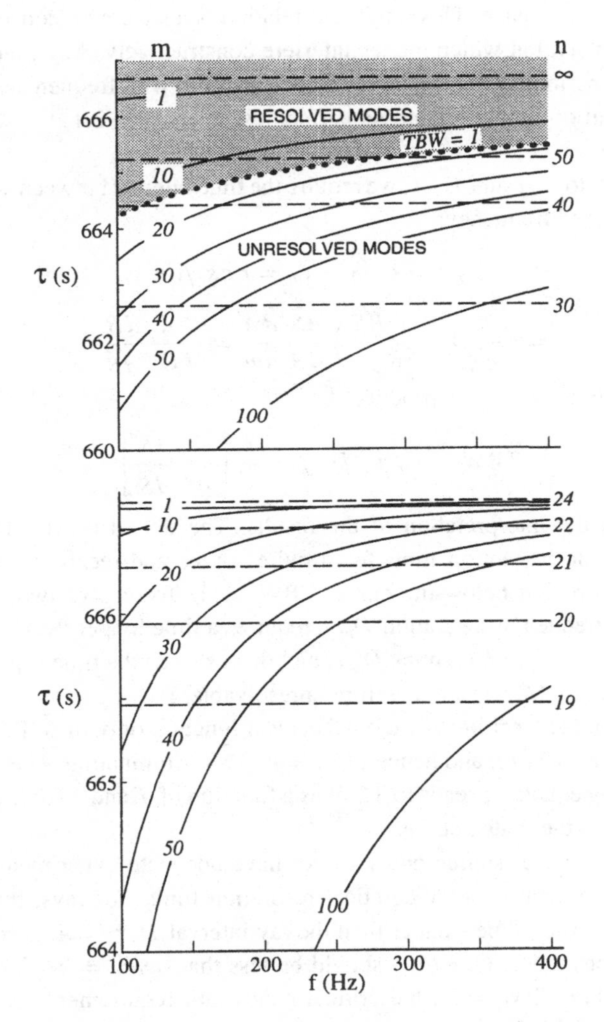

Fig. 2.10. Ambiguity diagrams for 1-Mm polar (top) and temperate (bottom) ocean transmissions. The coordinate frequency f and travel time τ have reciprocal dimensions, and this permits the representation of "ambiguity relations." Travel times for ray groups n are shown by the horizontal dashed lines. For the polar profile, the final axial (surface) arrival has an infinite number of ray loops of zero length; for the temperate profile, the final axial arrival has 24 double loops of length 41.6 km. Mode arrivals are labeled $m = 1, 10, \ldots$ For a bandwidth of 200–300 Hz in a polar transmission, modes 23 to 35 interfere constructively to form ray $n = 40$, all arriving simultaneously at 664.4 s. The shaded area designates conditions under which late, low-order modes can be resolved in the time domain, given adequate bandwidth. The dotted boundary designates the time-bandwidth product TBW = 1. For the temperate low-dispersive ocean, mode resolution requires ranges in excess of 1 Mm.

anywhere in f, τ-space. This important relation is a statement concerning the frequency interval at which modes interfere constructively. Any function that occurs with periodicity $D\tau = A$ has a line spectrum with frequencies equal to the reciprocal of the period and its harmonics, $f = A^{-1}, 2A^{-1}, \ldots$, and so $Df = A^{-1}$.

In addition to $D_n\tau$ and $D_m f$, we require the time interval between successive modes at a fixed frequency:

$$D_m\tau = \tau_{m+1} - \tau_m = r\, ds_g/dm$$

$$= r\frac{d}{d\widetilde{S}}\left(\frac{A+R\widetilde{S}}{R}\right)\frac{d\widetilde{S}}{dA}\frac{dA}{dm} = \frac{rA}{fR^3}\frac{dR}{d\widetilde{S}}\,. \tag{2.13.4}$$

The modal time-bandwidth product

$$\mathrm{TBW} = D_m\tau \cdot D_m f = \frac{r}{f}\left[\frac{1}{R^3}\frac{dR}{d\widetilde{S}}\right] \tag{2.13.5}$$

is central to the interpretation of the results. The dotted curve TBW $= 1$ gives the mode resolution limit for infinite source-and-receiver bandwidth. Consider the region below the curve, TBW < 1. To resolve two adjoining modes with frequency separation $D_m f$ requires a time longer than $(D_m f)^{-1}$. But $(D_m f)^{-1} = D_m\tau/\,\mathrm{TBW} > D_m\tau$, and this exceeds the time separation of adjoining modes. Modes are therefore unresolvable.

In (2.13.5), the term in square brackets is a function only of $\widetilde{S}$. For a given TBW, this yields $\widetilde{S}(f)$, and hence $\tau(\widetilde{S}) = rs_g(\widetilde{S})$. Eliminating $\widetilde{S}$ between the two foregoing equations leads to TBW as a function of f and τ for fixed range r, as shown by the dotted curve.

It is assumed that source and receiver have adequate experimental bandwidth $D_E f$ to achieve the theoretical resolution limit. For rays, this means that $(D_E f)^{-1}$ should be smaller than the ray interval $D_n\tau = A$, according to (2.10.13). For modes, $(D_E f)^{-1}$ should be less than $D_m\tau = \mathrm{TBW}/D_m f = \mathrm{TBW} \times A$. For TBW $= 1$, the critical bandwidth requirement can thus be written

$$D_E f = (D_n\tau)^{-1} = (D_m\tau)^{-1} = D_m f = A^{-1}\,. \tag{2.13.6}$$

We need to illustrate the results with specific examples. For the *polar* profile,

$$A = \tfrac{2}{3}\,\gamma_a^{-1}\,S_0\,\widetilde{\sigma}^3 \tag{2.13.7}$$

for rays and modes. Given A, we can evaluate (2.13.5):

$$\mathrm{TBW} = \tfrac{1}{4}\,\frac{r\,\gamma_a^2}{S_0 f\,(\widetilde{\sigma}^*)^4}\,. \tag{2.13.8}$$

The asterisk implies the value critical for mode resolution. Substituting into (2.13.4) yields

$$\tau^* - \tau_A = -\tfrac{1}{6}\, r\, S_0\, (\widetilde{\sigma}^*)^2 = -\tfrac{1}{12}\, r^{\frac{3}{2}}\, S_0^{\frac{1}{2}}\, \gamma_a\, f^{-\frac{1}{2}}\, \mathrm{TBW}^{-\frac{1}{2}} \tag{2.13.9}$$

for the travel time relative to the axial arrival. From (2.11.2),

$$m^* - \tfrac{1}{4} = f A^* = \frac{\sqrt{2}}{6}\, \gamma_a^{\frac{1}{2}}\, S_0^{\frac{1}{4}}\, f^{\frac{1}{4}}\, r^{\frac{3}{4}}\, \mathrm{TBW}^{-\frac{3}{4}}, \tag{2.13.10}$$

using (2.13.7) and (2.13.8). At 1 Mm and for 250 Hz, the result is that modes equal to or less than

$$m^* = \tfrac{1}{4} + 16.0\, (\mathrm{TBW})^{-\frac{3}{4}}$$

can be resolved, and these arrive at or later than

$$\tau^* = \tau_A - 1.54\, (\mathrm{TBW})^{-\frac{1}{2}}\ \mathrm{s}.$$

The required experimental bandwidth is

$$D_E f = (A^*)^{-1} = (\tfrac{3}{2})\, 4^{\frac{3}{4}}\, \gamma_a^{-\frac{1}{2}}\, S_0^{-\frac{1}{4}}\, f^{\frac{3}{4}}\, r^{-\frac{3}{4}} = 156\,\mathrm{Hz}. \tag{2.13.11}$$

This is a challenging, but attainable, requirement.

Demarcation times for TBW $= 1$ and 4 corresponding to modal cutoffs $m^* = 16$ and 6 are shown in fig. 2.7. There is general agreement, though we fail to achieve modal resolution to the theoretical limit of TBW $= 1$. This may have something to do with the fact that the bandwidth in the mode construction was taken at 100 Hz and falls short of the "required" 156 Hz.

The situation is very different for the *temperate* profile. Substituting into (2.13.5),

$$\mathrm{TBW} = K^3 \left(1 + \tfrac{1}{12}\, (\widetilde{\phi}^*)^2\right)^{-3}, \quad K^3 = \frac{r}{12\pi^2\, f\, S_A\, h^2}. \tag{2.13.12}$$

For the assumed values of $r = 1$ Mm and $f = 250$ Hz, this gives $K^3 = 0.051$, far short of $K^3 = 1$ required for TBW $= 1$. The conclusion is that modes cannot be resolved at any time! This is a consequence of the relatively weak dispersion of the temperate ocean.

To pursue this case, consider parameters consistent with some planned experiments, $r = 6$ Mm, $f = 70$ Hz, giving $K^3 = 1.086$, and

$$(\widetilde{\phi}^*)^2 = 12\left(K (\mathrm{TBW})^{-\frac{1}{3}} - 1\right),$$

or $(\widetilde{\phi}^*)^2 = 0.33$ for TBW = 1. Then, from (2.13.1),

$$\tau^* - \tau_A = -\frac{r\, S_A\, \gamma_a h}{48} (\widetilde{\phi}^*)^4 = -0.10\,\text{s}, \tag{2.13.13}$$

$$m^* - \tfrac{1}{2} = f A^* = f \frac{\pi}{\sqrt{2}} S_A\, h\, (\gamma_a h)^{\frac{1}{2}}\, \widetilde{\phi}^{*2} \left(1 + \tfrac{1}{24} (\widetilde{\phi}^*)^2\right) = 3.6. \tag{2.13.14}$$

So even under these rather extreme circumstances, the opportunity for mode resolution is limited, though the required experimental bandwidth of $(A^*)^{-1} =$ 19 Hz is achievable. All of this discussion refers to mode separation in the time domain. It does not preclude the separation of modes in the vertical wavenumber domain, using vertical arrays in an actual experiment.

2.14. Modal Perturbations

Fermat's principle does not apply to modal perturbation. One might have thought that to first order, $(\Delta\tau)_{\text{mode}}$ would be given by a modal propagation of the unperturbed model wave function through the perturbed ocean, and references to such a procedure appear in the literature (*e.g.*, Munk and Wunsch, 1983). But that is not the case: there is no "frozen-mode" approximation. Analysis leads to three terms of the same magnitude (such as **A**, **B**, and **C** in section 2.8), but no cancellation.

But clearly the perturbed modal field must be able to sustain a ray representation by an appropriate constructive interference. The answer is that if f_m are the frequencies of the constructively interfering mode generating a given ray in the unperturbed ocean, then f'_m are the corresponding frequencies in the perturbed ocean, and $f_m \neq f'_m$. Nor are the m constructively interfering mode functions the same, $P_m(z) \neq P'_m(z)$. Because modal group slowness is a function of frequency, the perturbation $(\Delta s_g)_{\text{ray}}$ will not be the same as that for one of the constituent modes if the modal frequency is kept unchanged.

We first give the perturbation for the modal representation. This is the form in which tomographic modal inversions have been accomplished by numerical treatment (Romm, 1987; Lynch *et al.*, 1991). The solutions are greatly simplified in the WKBJ approximation, particularly for ray perturbations, and the reader may wish to omit the tedious algebra of the modal perturbation.

The starting point is (2.10.2):

$$P'' + (\omega^2\, S^2(z) - k_H^2)\; P = 0, \tag{2.14.1}$$

from which we derived Rayleigh's principle (2.10.6), here written in the slightly altered form

$$\omega^2 \int dz\, S^2\, P^2 - k_H^2 \int dz\, P^2 - \int dz\, (P')^2 = 0, \tag{2.14.2}$$

with $P' = dP/dz$, and all integrals from $-\infty$ to $+\infty$ for the temperate profile. We have previously derived the group slowness by perturbing (2.10.6) with $S(z)$ fixed. The terms multiplying δP add to zero, by Rayleigh's variational principle, and

$$\omega\,\delta\omega \int dz\, S^2 P^2 - k_H\,\delta k_H \int dz\, P^2 = 0 \tag{2.14.3}$$

leads at once to the expression (2.10.8) for group slowness,

$$s_g = \delta k_H/\delta\omega = \widetilde{S}^{-1} \int dz\, S^2 P^2 \Big/ \int dz\, P^2\,, \tag{2.14.4}$$

allowing for $s_p = k_H/\omega = \widetilde{S}$.

In this section we derive the perturbation Δs_g associated with any ocean perturbation $\Delta S(z)$. The resulting perturbation in the dispersion relations $f_m(\omega, k_H) = 0$ can be computed by either keeping ω_m fixed and deriving Δk_H or keeping k_H fixed and deriving $\Delta\omega_m$. We keep ω_m fixed, in accordance with the experimental practice of maintaining the frequency spectrum of the source even as the ocean changes by $\Delta S(z)$. The perturbed group slowness follows from

$$s_g = \frac{dk_H}{d\omega}, \qquad \Delta s_g = \frac{d\,\Delta k_H}{d\omega}\,. \tag{2.14.5}$$

The derivation is again greatly simplified by Rayleigh's variational principle. From (2.14.2), with $\int dz\, P^2 = 1$,

$$\omega^2 \int dz\, S\,\Delta S\, P^2 - k_H\,\Delta k_H = 0\,. \tag{2.14.6}$$

Using the notation $(P^2)_\omega$ for $dP^2/d\omega$, $s_g = (k_H)_\omega$, $\Delta s_g = (\Delta k_H)_\omega$,

$$2\omega \int dz\, S\,\Delta S\, P^2 + \omega^2 \int dz\, S\,\Delta S\,(P^2)_\omega = s_g\,\Delta k_H + k_H\,\Delta s_g\,, \tag{2.14.7}$$

$$\Delta s_g = \widetilde{S}^{-1} \int dz\, S\,\Delta S\, P^2 \Big[2 - s_g/\widetilde{S}\Big] + \widetilde{S}^{-1}\,\omega \int dz\, S\,\Delta S\,(P^2)_\omega, \tag{2.14.8}$$

in agreement with Shang (1989), Shang and Wang (1991, 1992), and Lynch *et al.* (1991).[10]

To evaluate $(P^2)_\omega = 2P P_\omega$, differentiate (2.14.7),

$$P''_\omega + (\omega^2 S^2 - k^2)\, P_\omega = -(2\omega S^2 - 2k_H s_g)\, P,$$

[10] After changing the signs of the first and third terms in their equation 18 (J. F. Lynch, private communication).

and solve for P_ω. Rodi *et al.* (1975) discuss efficient numerical methods. This is the formalism followed by Lynch *et al.* (1991) and Romm (1987) in their studies of modal perturbations.

To obtain the WKBJ approximation to Δs_g, we proceed as previously, using (2.10.9) and (2.10.10). Note that with $\widetilde{S} = k_H/\omega$,

$$\frac{\partial}{\partial \omega} = \frac{\partial}{\partial \widetilde{S}} \frac{\partial \widetilde{S}}{\partial \omega} + \frac{\partial}{\partial k_H} \frac{\partial k_H}{\partial \omega} = \frac{s_g - \widetilde{S}}{\omega} \frac{\partial}{\partial \widetilde{S}} .$$

From (2.11.5), $s_g - \widetilde{S} = A/R$, and from (2.10.9), $P^2 = (2\widetilde{S}/R)\,(S^2 - \widetilde{S}^2)^{-\frac{1}{2}}$. It follows that

$$(P^2)_\omega = \frac{2A}{R^2 \omega}\left[1 - \frac{\widetilde{S}}{R}\frac{\partial R}{\partial \widetilde{S}} + \widetilde{S}\frac{\partial}{\partial \widetilde{S}}\right](S^2 - \widetilde{S}^2)^{-\frac{1}{2}} .$$

Substitution of P^2 and $(P^2)_\omega$ into (2.14.8) leads to

$$\Delta s_g = \frac{2}{R}\left[1 - \frac{A}{R^2}\frac{\partial R}{\partial \widetilde{S}} + \frac{A}{R}\frac{\partial}{\partial \widetilde{S}}\right] \int dz \frac{S}{(S^2 - \widetilde{S}^2)^{\frac{1}{2}}}\, \Delta S, \qquad (2.14.9)$$

in agreement with the WKBJ result (to be derived next).

Perturbed action. The foregoing equation can be derived directly starting with the action

$$A^{\pm} = 2 \int_{z_A}^{\widetilde{z}^{\pm}} dz \left[S^2(z) - \widetilde{S}^2\right]^{\frac{1}{2}} . \qquad (2.14.10)$$

This has several advantages. The results are more easily visualized, and the WKBJ approach gives parallel treatments for rays and modes. The starting point is (2.5.6)

$$T = A + R\widetilde{S}, \qquad R = -\,\partial A/\partial \widetilde{S}, \qquad (2.14.11)$$

where T and R are defined by the integrals (2.4.4*a,b*). These are readily interpreted for rays as loop travel time and range; for modes, only the ratio $s_g = T/R$ has a simple physical interpretation.

Suppose the sound-slowness is perturbed from $S(z, -)$ to $S(z, -) + \Delta S(z)$. As a result, the ray path is perturbed so that the turning slowness changes from $\widetilde{S}(-)$ to $\widetilde{S}(-) + \Delta \widetilde{S}$. The group slowness and group slowness perturbation are then given by

$$s_g = T\,R^{-1} = \widetilde{S} + A\,R^{-1} , \qquad (2.14.12)$$

$$\Delta s_g = \Delta \widetilde{S} + R^{-1}\,\Delta A + A\,\Delta(R^{-1}) , \qquad (2.14.13)$$

where

$$\Delta A = \frac{\partial A}{\partial \ell} \cdot \Delta \ell + \frac{\partial A}{\partial \widetilde{S}} \Delta \widetilde{S}. \tag{2.14.14}$$

From here on, the treatments of ray and mode perturbations diverge. We have shown, (2.8.11)–(2.8.13), that the condition $\Delta R = 0$ for constant range $r = nR$ leads to a cancellation of terms, with the simple result

$$(\Delta s_g)_{\text{ray}} = \frac{1}{R} \frac{\partial A}{\partial \ell} \cdot \Delta \ell, \tag{2.14.15}$$

which is equivalent to integration along the unperturbed path. For modes, there is no simple constraint such as $\Delta R = 0$, and we are left with the three terms (2.14.13) of comparable magnitudes.

We take some special cases. For long polar ranges, the final record consists of a few well-separated modes $m = 1, 2, 3$ (fig. 2.7*b*). If we can identify modes, and assume that their frequency spectrum is not appreciably perturbed, so that m and f, and hence (for the temperate and polar oceans)

$$A = \frac{m - \frac{1}{2}}{f}, \quad A = \frac{m - \frac{1}{4}}{f},$$

are unperturbed, then it follows that

$$\Delta A = 0 = \frac{\partial A}{\partial \ell} \cdot \Delta \ell + \frac{\partial A}{\partial \widetilde{S}} \Delta \widetilde{S}, \tag{2.14.16}$$

so that $\Delta \widetilde{S} = R^{-1}(\partial A/\partial \ell) \cdot \Delta \ell$. Equation (2.14.13) now reduces to

$$(\Delta s_g)_{\text{mode}} = R^{-1} \frac{\partial A}{\partial \ell} \cdot \Delta \ell + A \, \Delta(R^{-1}). \tag{2.14.17}$$

The first term will be recognized as $(\Delta s_g)_{\text{ray}}$. Writing

$$\Delta(R^{-1}) = -R^{-2} \Delta R = -R^{-2} \left(\frac{\partial R}{\partial \ell} \cdot \Delta \ell + \frac{\partial R}{\partial \widetilde{S}} \Delta \widetilde{S} \right) \tag{2.14.18}$$

and substituting for $\Delta \widetilde{S}$ in accordance with (2.14.16),

$$(\Delta s_g)_{\text{mode}} = \frac{1}{R} \left(1 - \frac{A}{R^2} \frac{\partial R}{\partial \widetilde{S}} + \frac{A}{R} \frac{\partial}{\partial \widetilde{S}} \right) \frac{\partial A}{\partial \ell} \cdot \Delta \ell. \tag{2.14.19}$$

With the substitution $R = -\,\partial A/\partial \widetilde{S}$, (2.11.19) expresses the mode perturbation as a function of $A(\ell, \widetilde{S})$ and its derivatives. The substitution (2.9.2) for $(\partial A/\partial \ell)\cdot$

$\Delta\ell$ leads to the previous result (2.14.9). The foregoing expression can also be written in the form

$$(\Delta s_g)_{\text{mode}} = (\Delta s_g)_{\text{ray}} + \frac{A}{R}\,\frac{\partial\,(\Delta s_g)_{\text{ray}}}{\partial \widetilde{S}}\,. \tag{2.14.20}$$

We shall find that none of the terms in (2.14.19) can be neglected. So the ray and mode perturbations are numerically quite different. The difference concerns not so much a distinction between rays and modes, but rather the difference in the observational constraints;[11] in the former case $\Delta R = 0$, in the latter case $\Delta A = 0$.

The constraint $\Delta A = 0$ is not the only one that can be envisioned. For example, with adequate vertical arrays it may be possible to select that particular wave function in the incoming signal that most resembles the unperturbed wave function (a "frozen-mode" approximation). This will be illustrated in the next section by considering four possible constraints. None of them is quite the same as picking the times of maximum modal amplitudes, which is the procedure followed by Lynch *et al.* (1991) and by Worcester *et al.* (1993). A more robust procedure is that of matched-field processing, that is, of matching the full time history of the computed modal arrivals with the observations, taking fully into account the configuration of source and receiver arrays, the source frequency spectrum, and any subsequent modification of the spectrum by attenuation and analysis. We have little experience as to the potential pitfalls.

Parameter perturbation. For illustration, we again take the case of a polar ocean gradient perturbation: $\Delta\ell = \Delta\gamma$. For the polar ocean, $A = \frac{2}{3}\,\gamma_a^{-1}\,S_0\,\widetilde{\sigma}^3$, and it follows at once from (2.14.15) and (2.14.19) that

$$(\Delta s_g)_{\text{ray}} = -\tfrac{1}{3}\,S_0\,\widetilde{\sigma}^2\,(1-\widetilde{\sigma}^2)^{-\frac{1}{2}}\,(\Delta\gamma/\gamma_a)\,, \tag{2.14.21}$$

$$(\Delta s_g)_{\text{mode}} = -\tfrac{1}{9}\,S_0\,\widetilde{\sigma}^2\,(1-\widetilde{\sigma}^2)^{-\frac{3}{2}}\,(1-2\widetilde{\sigma}^2)\;\Delta\gamma/\gamma_a\,. \tag{2.14.22}$$

Note that the modal perturbations are only about one-third the ray perturbations! To first order, $(\Delta s_g)_{\text{mode}} = [1 + \frac{1}{3} - 1]\,(\Delta s_g)_{\text{ray}}$, with the three numbers in square brackets corresponding to the three terms in (2.14.19).

The results can be easily confirmed by starting with $s_g = S_0\,(1 - \frac{1}{6}\,\widetilde{\sigma}^2 + \cdots)$ for both rays and modes and computing $\Delta s_g = (\partial s_g/\partial\widetilde{\sigma})\,\Delta\widetilde{\sigma}$, there being no explicit dependence on $\Delta\ell$. Conservation of R or A, respectively, will lead to

$$\Delta\widetilde{\sigma}_{\text{ray}} = \widetilde{\sigma}\,\Delta\gamma/\gamma_a\,, \qquad \Delta\widetilde{\sigma}_{\text{mode}} = \tfrac{1}{3}\,\widetilde{\sigma}\,\Delta\gamma/\gamma_a\,.$$

[11] The constraint $\Delta A = 0$ has nothing to do with the preservation of action with range in an adiabatically range-dependent environment. We are dealing with observations conducted at different times in different range-independent environments.

The 3:1 ratio in the group slowness perturbations is surprising. To gain some insight, consider the unperturbed rays and unperturbed wave functions (fig. 2.6) (see section 2.17 for construction details). Unperturbed rays and modes have the same turning depth, $\tilde{z} = -1.33$ km. For modes $m = 3, 4$, this requires frequencies of 12.9 Hz and 17.6 Hz, respectively. The unperturbed ray can be considered the result of the superposition of many such modes m and frequencies f_m all having the same turning depth. The perturbed wave function for $m = 4$ is drawn under the constraint that the frequency remains 17.6 Hz. With this constraint, the perturbation displaces the wave function *upward*, opposite to the ray displacement. It follows that the perturbed mode no longer contributes constructively to the perturbed ray. This is the result of having imposed the constraint $\Delta A = 0$.

We consider other constraints, again using the polar gradient perturbation for illustration. Neglecting terms of order $\tilde{\sigma}^4$,

$$s_g = S_0 \left(1 - \tfrac{1}{6}\tilde{\sigma}^2\right), \quad \tilde{\sigma}^2 = 2\gamma(-\tilde{z}), \tag{2.14.23}$$

with $\tilde{z}$ negative. For a polar gradient perturbation γ_a to $\gamma_a + \Delta\gamma$,

$$\frac{\Delta\tilde{z}}{\tilde{z}} = 2\,\frac{\Delta\tilde{\sigma}}{\tilde{\sigma}} - \frac{\Delta\gamma}{\gamma}, \quad \Delta s_g = -\tfrac{1}{3}\,S_0\,\tilde{\sigma}^2\,\frac{\Delta\gamma}{\gamma}. \tag{2.14.24}$$

For a ray n, $r = 2n\gamma^{-1}$, $\tilde{\sigma}_n$ and hence

$$\frac{\Delta\tilde{\sigma}_n}{\tilde{\sigma}_n} = \frac{\Delta\gamma}{\gamma}, \quad \frac{\Delta\tilde{z}_n}{\tilde{z}_n} = \frac{\Delta\gamma}{\gamma}, \quad (\Delta s_g)_n = -\tfrac{1}{3}\,S_0\,\tilde{\sigma}_n^2\,\frac{\Delta\gamma}{\gamma}. \tag{2.14.25}$$

For modes, the starting relation is $\sigma_m^3 = 3\,(m - \frac{1}{4})\gamma/(2fS_0)$, so

$$\frac{\Delta\tilde{\sigma}_m}{\tilde{\sigma}_m} = \tfrac{1}{3}\,\frac{\Delta\gamma}{\gamma} - \tfrac{1}{3}\,\frac{\Delta f}{f} \quad \frac{\Delta\tilde{z}_m}{\tilde{z}_m} = -\tfrac{1}{3}\,\frac{\Delta\gamma}{\gamma} - \tfrac{2}{3}\,\frac{\Delta f}{f},$$
$$(\Delta s_g)_m = -\tfrac{1}{9}\,S_0\,\tilde{\sigma}_m^2\left(\frac{\Delta\gamma}{\gamma} - \frac{\Delta f}{f}\right). \tag{2.14.26}$$

There are many ways to go:

(A) For a continuous-wave (CW) transmission, or a sharply frequency filtered CW record,

$$\Delta f = 0, \quad \frac{\Delta\tilde{\sigma}_m}{\tilde{\sigma}_m} = \tfrac{1}{3}\,\frac{\Delta\gamma}{\gamma}, \quad \frac{\Delta\tilde{z}_m}{\tilde{z}_m} = -\tfrac{1}{3}\,\frac{\Delta\gamma}{\gamma},$$
$$(\Delta s_g)_m = -\tfrac{1}{9}\,S_0\,\tilde{\sigma}_m^2\,\frac{\Delta\gamma}{\gamma}. \tag{2.14.27}$$

Note that the modal turning depth is slightly diminished, whereas there was an increase for the ray turning depth. This is the case corresponding to $\Delta A = 0$.

(**B**) A fixed vertical array favors the unperturbed wave function, and accordingly the peak arrival in the perturbed ocean is shifted in frequency,

$$\Delta\tilde{z}_m = 0, \quad \frac{\Delta\tilde{\sigma}_m}{\tilde{\sigma}_m} = \tfrac{1}{2}\,\frac{\Delta\gamma}{\gamma}, \quad \frac{\Delta f}{f} = -\tfrac{1}{2}\,\frac{\Delta\gamma}{\gamma},$$
$$(\Delta s_g)_m = -\tfrac{1}{6}\,S_0\,\tilde{\sigma}_m^2\,\frac{\Delta\gamma}{\gamma}. \tag{2.14.28}$$

This may be considered the modal equivalent to a frozen-mode approximation. It can be treated formally by setting $\Delta\tilde{S} = 0$, which leads to

$$(\Delta s_g)_{\text{mode}} = \frac{1}{R}\left(1 + \frac{A}{R}\frac{\partial}{\partial\tilde{S}}\right)\frac{\partial A}{\partial\ell}\cdot\Delta\ell,$$

in place of (2.14.19).

(**C**) By appropriate frequency filtering, the travel time can be kept unchanged:

$$(\Delta s_g)_m = 0, \quad \frac{\Delta f}{f} = \frac{\Delta\gamma}{\gamma}, \quad \frac{\Delta\tilde{\sigma}_m}{\tilde{\sigma}_m} = 0, \quad \frac{\Delta\tilde{z}_m}{\tilde{z}_m} = -\frac{\Delta\gamma}{\gamma}.$$

(**D**) Finally, for modes to interfere constructively to form rays,

$$\frac{\Delta\tilde{\sigma}_m}{\tilde{\sigma}_m} = \frac{\Delta\tilde{\sigma}_n}{\tilde{\sigma}_n} = \frac{\Delta\gamma}{\gamma}, \quad \frac{\Delta f}{f} = -2\,\frac{\Delta\gamma}{\gamma},$$
$$\frac{\Delta\tilde{z}_m}{\tilde{z}_m} = \frac{\Delta\gamma}{\gamma}, \quad (\Delta s_g)_m = -\tfrac{1}{3}\,S_0\,\tilde{\sigma}_m^2\,\frac{\Delta\gamma}{\gamma}, \tag{2.14.29}$$

so that $\Delta z_m = \Delta z_n$ and $(\Delta s_g)_m = (\Delta s_g)_n$, as it has to be for constructive interference.

The situation is neatly summarized in an ambiguity perturbation diagram (fig. 2.11). For the unperturbed polar situation (top), modes 28 and 29 (among others) contribute constructively to ray 40 (at $f = 243$ and 251.5 Hz, respectively). Fig. 2.11 (bottom) shows the unperturbed ($n = 40$, $m = 29$) and perturbed ($40'$, $29'$) locations of ray 40 and mode 29. The situation for case **A** is illustrated by the arrow OA; there is no change in frequency, $\Delta f = 0$, and hence $\Delta A = 0$. But at point A the perturbed mode $29'$ does not contribute constructively to the perturbed ray $40'$; it does so at point D, but with a frequency shifted from 251.5 to 246.5 Hz. We note that the travel-time perturbation for OA is about one-third of that for OD.

The arrow OB (for case **B**) corresponds to no change in the modal turning depth $\tilde{z}_m$. For arrows clockwise from OB, the ocean perturbation deepens the modal turning point; for the case OD it is deepened by the same amount as the ray turning depth $\tilde{z}_n$. For the case OA, the perturbation raises the modal

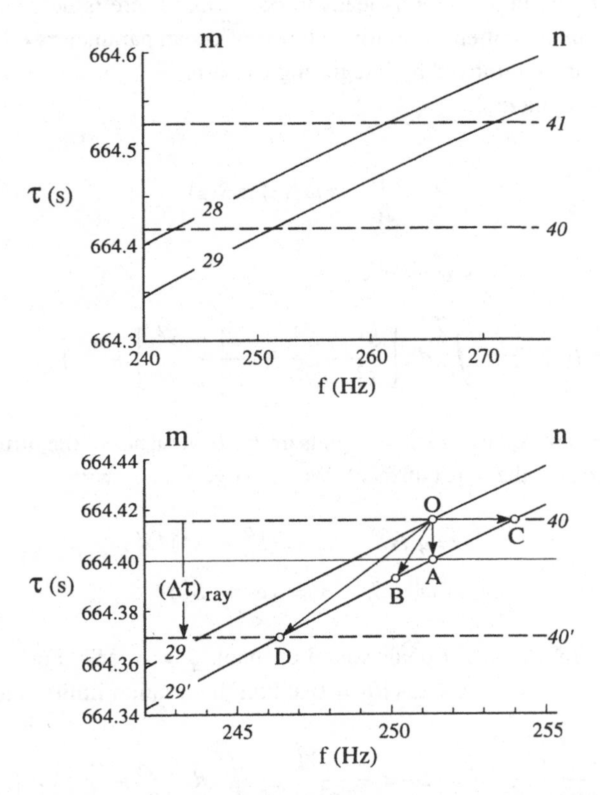

Fig. 2.11. Ambiguity perturbation diagram. Top: Magnified section of the unperturbed ambiguity diagram (fig. 2.10, top). Bottom: Ray 40 displacement from an unperturbed state 40 to a perturbed state 40′, and similarly for mode 29, as a result of a perturbation of the sound-speed gradient from γ_a to $1.1\gamma_a$. Perturbations in travel time are shown by the arrows OA, OB, OC, and OD and depend on the conditions of recording and analysis.

turning depth. For case **C** the perturbation is taken up by an increase in the center frequency.

Functional perturbation. The discussion so far has dealt with highly artificial simple models for purpose of illustration. For any application we need to consider any perturbation $\Delta S(z)$, as we did for rays in section 2.9. The goal is to derive modal perturbation weighting functions $w(z)$. The starting point is (2.14.19), with $(\partial A/\partial \ell) \cdot \Delta \ell$ given by (2.9.2). We note that $\partial w^0/\partial \widetilde{S} \sim (S^2 - \widetilde{S}^2)^{-\frac{3}{2}}$ leads to a nonintegral singularity and that the integrand is infinite at the limits $\widetilde{z}^{\pm} = z(\widetilde{S})$. Application of Leibnitz's rule for differentiating definite integrals leads to $\infty - \infty$. [There is no such problem when $A(\boldsymbol{\ell})$ can be written explicitly in terms of ocean parameters $\boldsymbol{\ell}$.] The indeterminance can be removed by integrating by parts.

Consider the integral

$$I = \int_0^{\widetilde{z}} dz\, w^0(z)\, \Delta S(z)\,.$$

This can be written quite generally,

$$I = I_1 + I_2 = \int_0^{\widetilde{z}} dz \left[\frac{S^2 - \widetilde{S}^2}{S_0^2 - \widetilde{S}^2} + \frac{S_0^2 - S^2}{S_0^2 - \widetilde{S}^2} \right] w^0(z)\, \Delta S(z)$$

where the term in square brackets equals unity. I_1 vanishes at the turning point, and I_2 vanishes at the axis (surface). We set $I_2 = \int u\, dv$, with

$$u = q\, \Delta S, \quad q = (S_0^2 - S^2)/S'\,,$$
$$dv = w^0\, dS, \quad v = (S^2 - \widetilde{S}^2)^{\frac{1}{2}}\,,$$

where $S' = dS/dz$. For a polar sound channel, $q = -2Sz$. For a parabolic sound channel, $q = -Sz$. Thus $uv = 0$ at both integration limits, and

$$I_2 = \frac{1}{S_0^2 - \widetilde{S}^2} \int u\, dv = -\frac{1}{S_0^2 - \widetilde{S}^2} \int_0^{\widetilde{z}} dz\, (S^2 - \widetilde{S}^2)^{\frac{1}{2}}\, u'(z)\,,$$
$$I = \frac{1}{S_0^2 - \widetilde{S}^2} \int_0^{\widetilde{z}} dz\, (S^2 - \widetilde{S}^2)^{\frac{1}{2}}\, [S\, \Delta S - u']\,,$$

$$\frac{\partial I}{\partial \widetilde{S}} = \frac{\widetilde{S}}{S_0^2 - \widetilde{S}^2} \int_0^{\widetilde{z}} dz \frac{1}{(S^2 - \widetilde{S}^2)^{\frac{1}{2}}} \left(-1 + 2\frac{S^2 - \widetilde{S}^2}{S_0^2 - \widetilde{S}^2} \right) \left[(S \Delta S - u' \right].$$

Writing $u' = q'\, \Delta S + q\, \Delta S'$, the term in square brackets equals $3S\, \Delta S + 2z\, S\, \Delta S'$ for the polar profile, and $2S\, \Delta S + zS\, \Delta S'$ for the temperate profile. The final result can be written

$$(\Delta s_g)_{\text{ray}} = \frac{2}{R} \int dz\, w^0\, \Delta S, \quad (\Delta s_g)_{\text{mode}} = \frac{2}{R} \int dz \left[w^I\, \Delta S + w^{II}\, \Delta S' \right],$$

$$w^0(z) = S(z)\, (S^2 - \widetilde{S}^2)^{-\frac{1}{2}}\,,$$

$$w^I(z) = \left[1 - \frac{A}{R^2}\frac{\partial R}{\partial \widetilde{S}} + \frac{A}{R}\frac{\widetilde{S}}{S_0^2 - \widetilde{S}^2}\left(-1 + 2\frac{S^2 - \widetilde{S}^2}{S_0^2 - \widetilde{S}^2}\right)\left(1 - \frac{q'}{S}\right)\right] w^0, \tag{2.14.30}$$

$$w^{II}(z) = \frac{A}{R}\frac{\widetilde{S}}{S_0^2 - \widetilde{S}^2}\left(-1 + 2\frac{S^2 - \widetilde{S}^2}{S_0^2 - \widetilde{S}^2}\right)\left(-\frac{q}{S}\right) w^0. \tag{2.14.31}$$

The integration is in the positive z direction, from $\widetilde{z}^-$ to 0 for the polar ocean and from $\widetilde{z}^-$ to $\widetilde{z}^+$ for the temperate ocean.

The important new feature is the dependence on $\Delta S'(z) = d\,\Delta S/dz$ as well as on $\Delta S(z)$, with both functions being heavily weighted by the w^0 singularity at the turning points. For the polar and temperate profiles, the weights can be reduced to rather simple analytical functions (sections 2.17 and 2.18). In the general case the weights are derived numerically from a measured $S(z)$ and hence known $A(z)$.

In (2.14.31), w^0 and w^I are given in dimensionless units (w^{II} has the dimension of a length). It is now convenient to incorporate the factor $2/R$ in the weights by writing $W(z) = (2/R)w(z)$, so that W^0 and W^I are per kilometer, W^{II} is dimensionless, and

$$(\Delta s_g)_{\text{ray}} = \int dz\, W^0\, \Delta S, \quad (\Delta s_g)_{\text{mode}} = \int dz\,[W^I\, \Delta S + W^{II}\, \Delta S']. \tag{2.14.32}$$

All integrations are in the positive (upward) direction.

Weights $W^0(z)$, $W^I(z)$, and $W^{II}(z)$ are plotted in fig. 2.12. The figure illustrates the difference between ray and mode weighting, the dependence of mode weighting on frequency (the two frequencies are characteristic of short-range tomography and very long range tomography), and the major distinction between polar and temperate tomography. For both the polar and temperate cases, and for each frequency, the upper panels represent flat rays and low-order modes, and the lower panels steep rays and high modes. The contribution of W^{II} is relatively small for surface-trapped rays and modes.

2.15. Perturbation Models

From here on, rays and modes are treated side-by-side. The forward problem is written in the matrix form

$$\mathbf{\Delta\tau} = \mathbf{E}\, \boldsymbol{\ell} \tag{2.15.1}$$

for both rays and modes. The inverse procedure evaluates ℓ_j from the measured $\Delta\tau_i$, given E_{ij}. We refer to a dimensionless ℓ_j, so that $\mathbf{E}$ has dimensions of time.

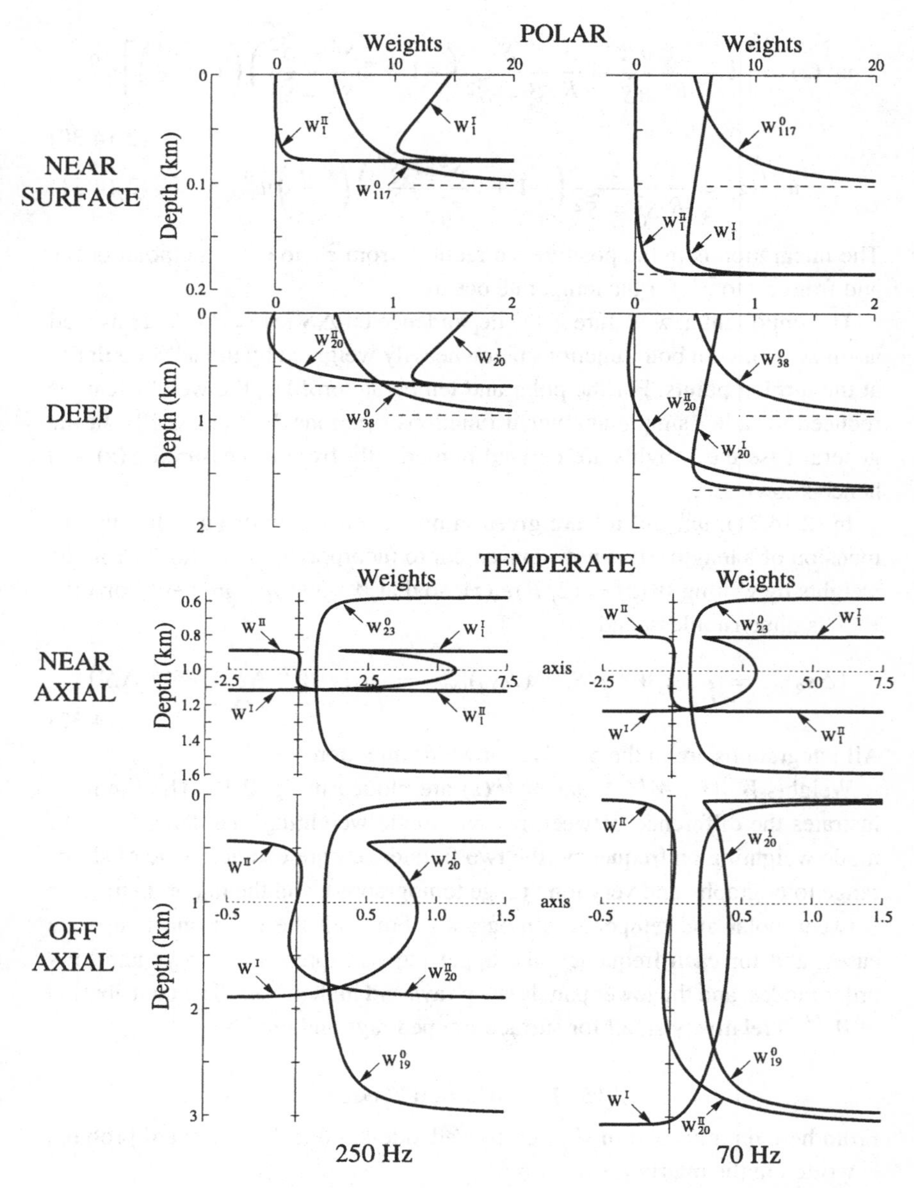

Fig. 2.12. Perturbation weighting functions. The travel-time perturbation is written as a weighted mean of the sound-slowness perturbation: $\Delta\tau = \int dz\, W(z)\, \Delta S(z)$. W_n^0 is the ray weighting for ray n; W_m^I and W_m^{II} are weights associated with mode m. Left panels correspond to 250 Hz, right panels to 70 Hz [ray weights $W_n^0(z)$ are the same at the two frequencies]. For both the polar and temperate profiles, the upper panels correspond to flat rays and low-order modes, the lower panels to steep rays and high modes.

The following examples will serve as illustrations for the inverse method in chapters 6 and 7. We confine ourselves to functional perturbations; parameter perturbations have been found useful for illustrating the forward problem, but they do not make good examples for the inverse problem. Accordingly,

$$\Delta S(z) = \sum_j \ell_j \, F_j(z) \,, \tag{2.15.2}$$

$$\begin{aligned} (E_{ij})_{\text{ray}} &= \frac{2r}{R_i} \int dz \, w_i^0(z) \, F_j(z) \,, \\ (E_{ij})_{\text{mode}} &= \frac{2r}{R_i} \int dz \, [w_i^I(z) \, F_j(z) + w_i^{II}(z) \, F_j'(z)] \,, \end{aligned} \tag{2.15.3}$$

where $F' = dF/dz$, and the weights w are given by (2.14.30) and (2.14,31). Dimensions of $(w^0, \, w^I, \, w^{II})$ are (0, 0, L).

Layer perturbation, rays. We assume a constant perturbation $\Delta S = \ell_j$ in layer j with upper boundary z_j and lower boundary z_{j+1} (layers $1, 2, \ldots$ are numbered from the surface downward). $F_j(z) = 1$ within layer j, and 0 outside. For $n = r/R_n$ ray loops in a polar ocean,

$$E_{nj} = 2n \int_{z_{j+1}}^{z_j} dz \, w_n^0(z), \tag{2.15.4}$$

with E in kilometers. The *polar* weighting function is

$$w_n^0 \approx \tfrac{1}{2} \, \gamma_a^{-\frac{1}{2}} \, (z - \tilde{z}_n)^{-\frac{1}{2}}, \quad \tilde{z}_n = -r^2 \, \gamma_a/(8n^2), \tag{2.15.5}$$

and the integral is readily performed. There are three cases, depending on whether the ray n turns above, within, or beneath the layer j:

$$\begin{aligned} (i) \quad & E_{nj} = 0 \quad \text{for } \tilde{z}_n > z_j \,, \\ (ii) \quad & E_{nj} = 2n \, \gamma_a^{-\frac{1}{2}} \sqrt{z_j - \tilde{z}_n} \quad \text{for } z_j \geq \tilde{z}_n \geq z_{j+1} \,, \\ (iii) \quad & E_{nj} = 2n \, \gamma_a^{-\frac{1}{2}} \left[\sqrt{z_j - \tilde{z}_n} - \sqrt{z_{j+1} - \tilde{z}_n} \right] \quad \text{for } z_{j+1} \geq \tilde{z}_n \,. \end{aligned} \tag{2.15.6}$$

Fig. 2.13 is drawn with source and receiver at $z_s = -0.1$ km. For $n = 119$ to ∞, sources and receivers are beneath the turning points, and there are no rays.

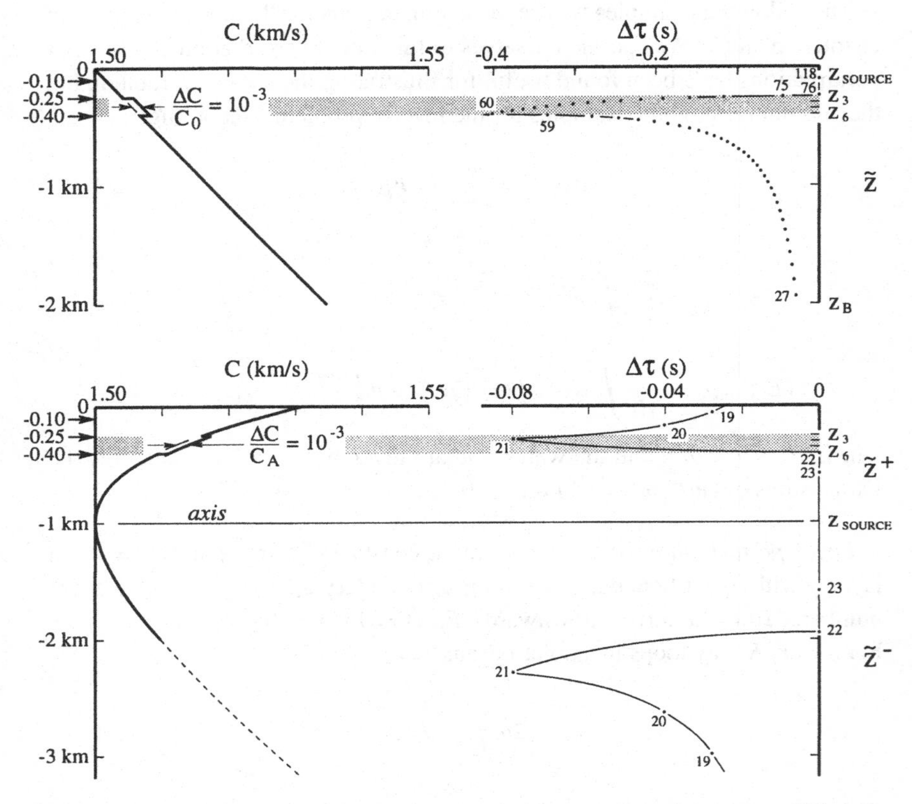

Fig. 2.13. Ray travel-time perturbations by layers. The upper panels (polar profile) and bottom panels (temperate profile) refer to a layer perturbation between 250 m and 400 m of depth in which the sound-speed is increased by 1.5 m/s. The profiles and their perturbation are shown at the left. The right panels give the travel-time perturbation $\Delta\tau$ as a function of the upper and lower ray turning depths $\tilde{z}^{\pm}$. Ray numbers n (we consider only double loops) are indicated. For the polar profile, the source and receiver are taken at $z_{\text{source}} = -100$ m, and there are no rays with turning points above 100 m. For the temperate profile, the source and receiver are at the axis. The range is 1 Mm.

We have taken three layers, $j = 3, 4, 5$, at depths of 250–300, 300–350, and 350–400 m, with equal perturbations $\ell_j = -10^{-3}\, S_0$, hence $\Delta C_j = 1.5$ m/s (+ 0.3° C); $\ell_j = 0$ for all other layers. Perturbations reach a maximum of $\Delta\tau = -400$ ms for the ray turning just above the bottom of the lowest perturbed layers.

A similar treatment for the *temperate* sound channel leads to the problem of up/down ambiguity (Munk and Wunsch, 1982*b*); it is difficult to distinguish perturbations in "conjugate" layers (layers with equal sound-speeds above and beneath the sound axis). The ambiguity is an inverse problem; the forward problem causes no such difficulty.

Again layer j has an upper boundary z_j and a lower boundary z_{j+1}. It is convenient to introduce the notation z_j^A as the boundary of layer j nearer to the axis (whether above or below the axis), and z_j^B as the boundary farther from the axis, so that

$$z_j^A = z_{j+1}, \quad z_j^B = z_j, \qquad z > z^A,$$

$$z_j^A = z_j, \quad z_j^B = z_{j+1}, \quad z < z^A.$$

The associated values of ϕ_j^A, ϕ_j^B are obtained from a numerical solution of (2.18.8).

Then for n double loops,

$$\widetilde{\phi}_n^2 = 12\left[\frac{r}{2n\, R_0^{\pm}} - 1\right], \tag{2.15.7}$$

$$E_{nj} = 2n\, R_0^{\pm}\, K_{nj}\left[F_4(\upsilon_{nj}) \mp F_5(\upsilon_{nj})\,\widetilde{\phi}_n + F_6(\upsilon_{nj})\,\widetilde{\phi}_n^2\right]_{\phi^A}^{\phi^B}, \tag{2.15.8}$$

$$0 \le \widetilde{\phi}_n \le \phi_j^A : \quad K_{nj} = 0,$$

$$\phi_j^A \le \widetilde{\phi}_n \le \phi_j^B : \quad K_{nj} = 1, \quad \sin\upsilon_{nj}^A = \phi_j^A/\widetilde{\phi}_n, \quad \sin\upsilon_{nj}^B = \widetilde{\phi}_n/\widetilde{\phi}_n,$$

$$\phi_j^B \le \widetilde{\phi}_n : \quad K_{nj} = 1, \quad \sin\upsilon_{nj}^A = \phi_j^A/\widetilde{\phi}_n, \quad \sin\upsilon_{nj}^B = \phi_j^B/\widetilde{\phi}_n.$$

The $F(\upsilon_{nj})$ functions are defined in (2.18). Results are shown in fig. 2.13, bottom.

A maximum perturbation of $\Delta\tau = -80$ ms is associated with the ray turning near the top of the perturbed three layers. Note that $\Delta\tau$ is much smaller than for the polar case. The reason is that only upper loops pass through the perturbation layer, and further that the upper loops have generally a shorter span than the lower loops. An equal sound-speed perturbation in the conjugate layer beneath the axis produces a larger signal, but is otherwise undistinguishable. In addition to the four perturbed ray groups $n = 19, 20, 21, 22$, there are groups 23 and 24 that turn beneath the layer and thus have $\Delta\tau = 0$. Since there are only six ray groups, one will want to take advantage of rays with extra upper or lower loops, giving a total of 24 rays at 1 Mm. The extra loops weight the upper and lower oceans differently and help to reduce the up/down ambiguity, especially for shorter ranges. Nonetheless, the existence of a relatively sparse ray set and the tendency toward up/down ambiguity make the inversion problem more difficult for the temperate ocean than for the polar ocean (chapter 6).

For modes, the discontinuity between layers gives infinite values of $\Delta S'$ at the layer boundaries. There are no solutions. This is related to the "soloton effect" in the seismic literature (Lapwood, 1975). Triangular layers yield finite solutions.

Dynamic mode perturbations in a temperate profile (rays and modes). The representation of ocean perturbations by discrete layers is in many ways objectionable; it is a statement of minimum *a priori* knowledge of ocean dynamics.[12] A more satisfying treatment expands the perturbation in terms of ocean normal modes (dynamic, not acoustic). An equivalent statement is that the perturbation is represented by internal planetary (Rossby) waves. The general experience is that the lowest few modes contain most of the energy. For long ranges, the present range-independent treatment is, of course, invalid.

Planetary waves cannot be sustained in an adiabatic profile $N = 0$, and so we shall treat only the case of a temperate profile $N(z) \sim e^{z/h}$. The solution to the hydrodynamic equation in a temperate ocean for dynamic mode j is given by (2.18.31):

$$\eta_j(z) = \ell_j \, B_j(\xi), \quad B_j(\xi) = (-1)^{j-1}\left[J_0(\xi_j) - \frac{J_0(\xi_B)}{Y_0(\xi_B)}\, Y_0(\xi_j) \right], \tag{2.15.9}$$

where $\eta(z)$ is the vertical displacement, with $\eta(0) = 0$ and $\eta(z_B) = 0$, and ℓ_j is a vertical amplitude. To meet the boundary conditions, (2.18.23)–(2.18.29),

$$\xi_j(z) = (\xi_0)_j \, e^{z/h}, \quad (\xi_0)_j = 2.87, 6.12, \ldots \ \text{for } j = 1, 2, \ldots . \tag{2.15.10}$$

The term $(-1)^{j-1}$ specifies that the displacement $\eta_j(z)$ is upward near the sea surface for both even and odd mode numbers.

Recall that it is the *potential* sound-speed C_p that is vertically advected, (2.2.2):

$$\frac{\Delta C_j}{C} = -\frac{1}{C}\frac{dC_p}{dz}\,\eta.$$

In general, dC_p/dz is positive, and so an upward displacement η produces a negative sound-speed perturbation. In a canonical profile, $C^{-1} dC_p/dz = \gamma_a N^2(z)/N_A^2$, and so

$$\Delta C_j/C = -\gamma_a \ell_j \, B_j(z)\, N^2(z)/N_A^2,$$

with $N(z)/N_A = e^{(z-z_A)/h}$. Using $\Delta S_j(z) = \ell_j \, F_j(z)$,

$$F_j(z) = +\gamma_a \, S_A \, B_j(z)\, N^2(z)/N_A^2, \tag{2.15.11}$$

[12] Specifying a realistic vertical covariance between layers can make the layer representation quite rewarding.

so F has the dimension T/L^2, and ℓ_j the dimension of a length.[13]

It is convenient to switch to ν coordinates, with $z^{\pm}(\widetilde{\phi},\ \nu)$ given by (2.18.35). Hence

$$(E_{ij})_{\text{ray}} = r \int_0^{\pi/2} d\nu\, V_i^0\, F_j, \quad (E_{ij})_{\text{mode}} = r \int_0^{\pi/2} d\nu \left[V_i^I\, F_j + V_i^{II}\, F_j' \right], \tag{2.15.12}$$

with $V(\widetilde{\phi},\ \nu) = V^+ + V^-$ as given by (2.18.32), and $F_j' = d\, F_j/dz$. Here E has the dimension $T\, L^{-1}$.

For axial rays,

$$\widetilde{\phi}_n = 12 \left[\frac{r}{2n\, R_0^{\pm}} - 1 \right] \to 0, \quad E_{nj} \to r\, F_j(0), \quad \Delta\tau_{nj} \to r\, \Delta S_j(0),$$

as expected. The situation is illustrated in fig. 2.14. We have taken a maximum perturbation $\Delta C_{\text{max}} = 10^{-3} C$ for both modes $j = 1$ and $j = 2$. The associated amplitudes of vertical displacement are $\ell_1 = -0.0508$ km, $\ell_2 = -0.0456$ km. The corresponding axial perturbations are $\Delta C_A/C = -\Delta S_A/S = +0.43 \times 10^{-3}, -0.24 \times 10^{-3}$. For near-axial rays, the travel-time perturbations are approximately $r \Delta S_A = -0.29$ s, $+0.16$ s. The perturbation $|\Delta\tau_n|$ diminishes with increasing ray steepness (decreasing n) as the lower loop (where ΔC is relatively small) lengthens relative to the upper loop.

The dynamic mode–acoustic mode interaction matrix E_{ij} summarizes the situation (lower panels in fig. 2.15). Contours are in units of seconds per kilometer for a 1-Mm transmission, in accordance with (2.15.12). The upper right panel is drawn to assist with the interpretation of the E plots. Note that for dynamic mode $j = 1$, the axial value of the wave function $F(z)$ is near 0.006 s/km^2, and that $F'(z)$ is nearly constant. Accordingly, the near-axial averages of $F(z)$ between the low-order modal turning points to both sides of the axis remain nearly constant at 0.006 s/km^2, and accordingly the $E \approx 0.006\, r = 6$ s/km is insensitive to the acoustic mode and ray numbers. The associated travel-time perturbation is $l_1\ E$ s, where l_j is the vertical amplitude of mode j (fig. 2.14). The upper left panel is drawn for comparison with ray–mode interaction. Note that the E values (hence group velocity perturbations) are similar for rays and modes, unlike the 3:1 ratio for gradient perturbations in a polar profile.

[13] We use the unnormalized vertical coefficient ℓ_j. Since B_j approaches $N^{-\frac{1}{2}}(z)$ for large z, $F(z) \sim N^{\frac{3}{2}}$. One may prefer a depth-normalized amplitude, which makes $\overline{\eta_j^2} = 1$. A traditional approach is to normalize the mode energy: $\overline{\eta_j^2\, N^2} = 1$. Or one may wish to normalize ΔC^2, which leads to $\overline{\eta_j^2\, N^4} = 1$ in an isohaline ocean. For a discussion, see Cornuelle *et al.* (1989, app. A).

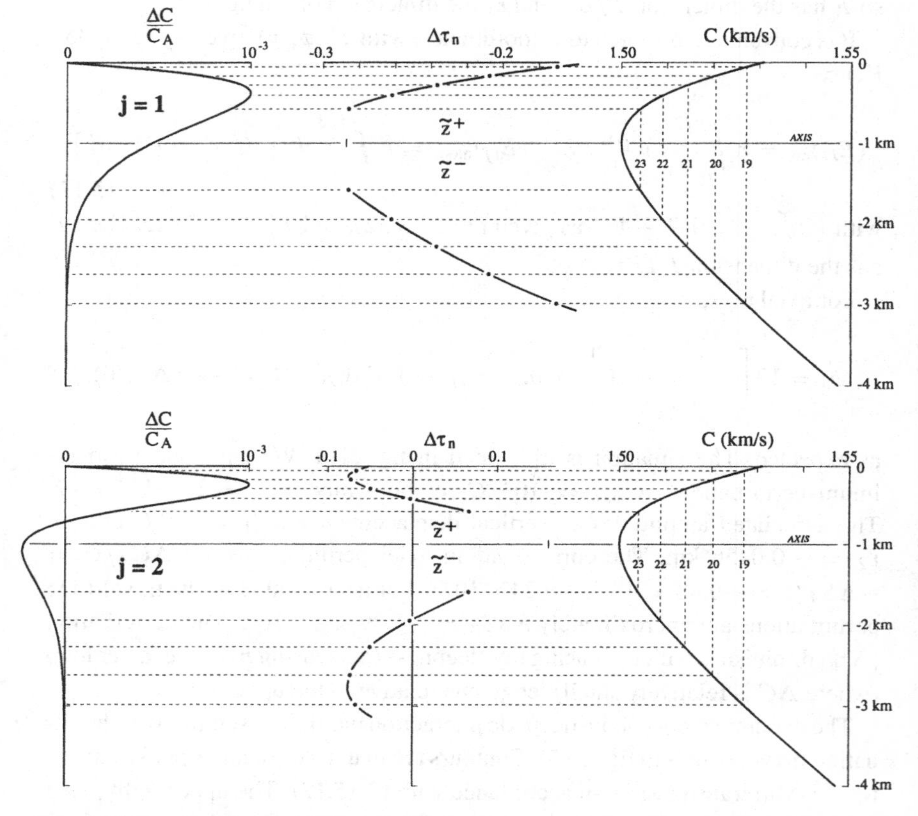

Fig. 2.14. Ray travel-time perturbations by dynamic modes in a temperate profile $C(z)$ (right). The velocity perturbation $\Delta C_j(z)$ for modes $j = 1$ and $j = 2$ (left) has been normalized for a maximum $\Delta C = 10^{-3}C$. Upper and lower turning depths for rays $n = 19$ to 23 are shown by the horizontal dashed lines. The resulting travel-time perturbations $\Delta\tau_{jn}$ (center) are plotted as functions of upper and lower turning depths. The tick on the axis corresponds to $\Delta\tau = r(\Delta S_j)_{\text{axis}}$.

For dynamic mode 2, it happens that $F(z)$ has a minimum near the axis, and so E is negative for axial modes and rays, but decreases in magnitude with increasing m and decreasing n values.

The important consideration is that dynamic modes, which are trapped near the ocean surface by the large surface buoyancy $N(z)$, nonetheless reach to axial depths and interact strongly with acoustic modes and rays. On the other hand, seasonal variability and other near-surface phenomena are not accessible to low modes and flat rays in the temperate ocean.

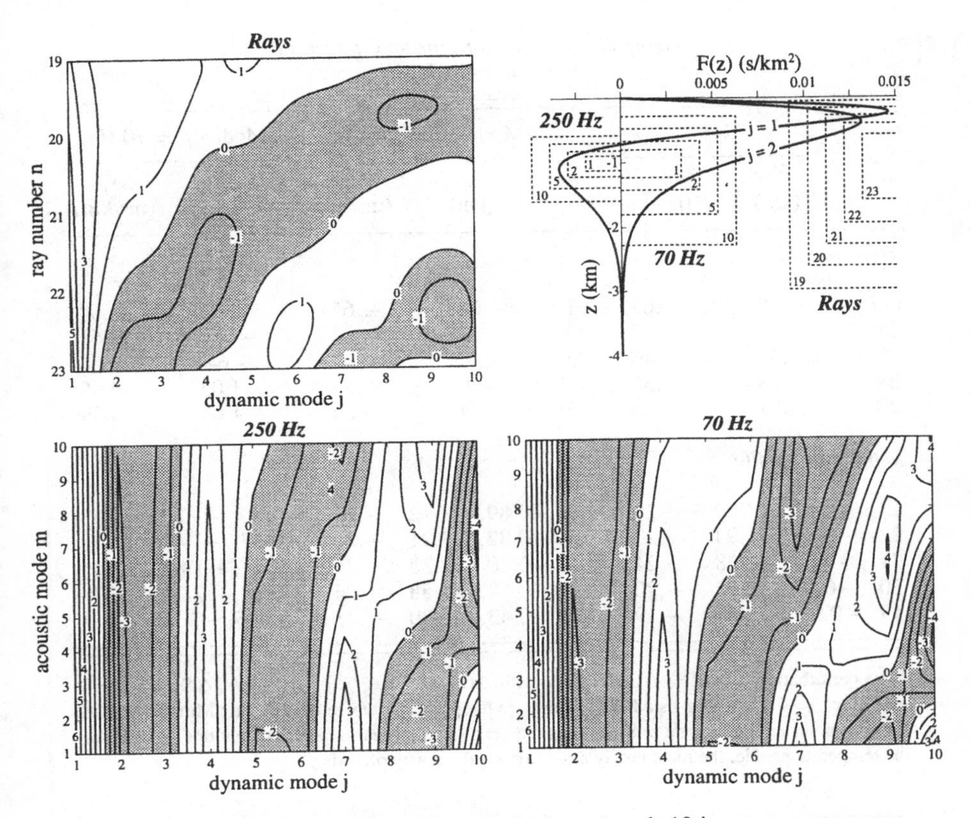

Fig. 2.15. The E_{ij} matrix for dynamic perturbations j = 1–10 in a temperate ocean. Upper left panel corresponds to the previous figure and gives the perturbations of acoustic rays $i = n = 19$ (steep) to 23 (near axial). The two lower panels correspond to perturbations of the acoustic modes $i = m = 1$ (axial) to $m = 10$. Interactions are strongest for low dynamic modes and low (axial) acoustic modes and axial rays. The upper right panel shows the wave functions of the two lowest dynamic modes, and the turning depths of the acoustic modes and rays. The E matrix is for a range of 1 Mm, contour interval 1 s/km (negative E is shaded). Travel time perturbations are given by multiplying by the vertical amplitudes l_j of the dynamic modes.

An example of surface warming. All this can be summarized with a simple example of an exponential surface warming, represented by

$$\Delta C(z) = \Delta C_0 \ \exp(-z/z^*), \quad \Delta C_0 = 10^{-3} C_0, \quad z^* = -1 \text{ km}.$$

The surface warming by 0.3° C (1.5 m/s) might be of the order expected over a decade, due to greenhouse warming. The perturbation in group slowness is in units of milliseconds per kilometer (table 2.3), or seconds per 1 Mm. Thus,

Table 2.3 *Perturbations for an exponential surface heating*[a]

n	Ray $\tilde{z}$ (km)	Ray Δs_g (ms/km)	m	Mode ($f = 250$ Hz) $\tilde{z}$ (km)	Mode ($f = 250$ Hz) Δs_g (ms/km)	Mode ($f = 70$ Hz) $\tilde{z}$ (km)	Mode ($f = 70$ Hz) Δs_g (ms/km)
Polar profile							
117	−.10	−.62	1	−.080	−.65	−.18	−0.64
75	−.25	−.57	2	−.14	−.64	−.33	−0.61
50	−.50	−.46	5	−.27	−.62	−.64	−0.56
38	−.98	−.36	10	−.44	−.59	−1.03	−0.49
26	−2.1	−.20	20	−.70	−.54	−1.64	−0.39
Temperate profile							
23	−0.58, −1.59	−.23	1	−0.89, −1.12	−.25	−0.80, −1.23	−.25
22	−0.41, −1.94	−.22	2	−0.82, −1.21	−.24	−0.67, −1.42	−.24
21	−0.28, −2.28	−.21	5	−0.70, −1.38	−.24	−0.47, −1.80	−.24
20	−0.17, −2.62	−.20	10	−0.58, −1.58	−.24	−0.28, −2.28	−.23
19	−0.055, −2.98	−.19	20	−0.43, −1.90	−.23	−0.025, −3.08	−.21

[a] The perturbation is $\Delta C = \Delta C_0\, e^{-z/z^*}$, with $\Delta C_0 = +10^{-3} C_0 = 1.5$ m/s, $\Delta S_0 = -.667$ ms/km, $z^* = -1$ km. The listed Δs_g values are the perturbations in ms/km, or s/Mm. The table, for both polar and temperate profiles, lists the near-surface (near-axial) propagations on top. For the temperate profile, the most nearly axial ray is quite a way off axis.

for a 10,000-km range, the warming leads to a reduction in travel time by a few seconds.

Table 2.3 is arranged so that the near-surface (near-axial) trapped rays and modes are on top. For these flat rays or low-order modes, $\Delta s_g \approx \Delta S_0 = -S_0(\Delta C_0/C_0) \approx -0.66$ ms/km. The effect diminishes to -0.36 ms/km for rays turning at 1 km and to -0.49 for modes with a WKBJ turning point at 1 km.

Weights $W^0(z), W^I(z), W^{II}(z)$ have been plotted in fig. 2.12. Because $\Delta S = S_0\, e^{-z/z^*}$ and $\Delta S' = -(1/z^*)\, S_0$ are numerically equal in the selected units for $z^* = -1$ km, the relative contributions can be directly compared. (The weights are, of course, independent of the assumed perturbation.) The contribution of W^{II} is relatively small for the surface-trapped rays and modes (upper panels), but not for the deeply penetrating rays and modes (lower panels). For equivalent turning points, modes put greater weights on the shallow water than do rays.

The polar ocean, with its surface-trapped acoustic radiation, is a favorable environment for the detection of surface warming. The situation is quite different for the temperate ocean, where the sound is axially trapped. For the near-axial rays and modes (top of the temperate profile) we have the expected result that the perturbation represents axial warming,

$$\Delta s_g \approx \Delta S_0 \, \exp(z_A/z^*) = \Delta S_0/e = -0.245 \text{ ms/km}$$

(there are no rays *very* near the axis for the present geometry). The surprising result is that the steeper rays and higher modes that extend into the surface-warmed layers actually suffer a *smaller* perturbation! This is because steeper rays have shorter upper loops, and equivalently higher modes put less weight in the upper ocean. It can be shown that for a more concentrated surface warming ($z^* = 0.5$ km) the situation is reversed, and the steeper rays and higher modes experience a *larger* perturbation.

OBSERVATIONS

2.16. Observations

In the preceding sections we have studied range-independent acoustic propagation for idealized temperate and polar sound-speed profiles. How useful are results derived for these idealized cases in understanding propagation in the real ocean? Two issues need to be addressed: (i) How representative of measured sound-speed profiles are the idealized profiles? (ii) Is the range-independent approximation adequate to make qualitatively useful propagation predictions?

Propagation atlas. For any given location and geometry, climatological data can be used to predict the climatological acoustic arrival pattern. Sound-speed perturbations in the ocean are generally sufficiently small that climatological data are adequate to obtain meaningful qualitative descriptions of measured acoustic propagation. Although detailed calculations for the proposed geometry of any experiment should be done during the planning process, a worldwide atlas of acoustic propagations for a standard 500-km range gives some feeling for the variability to be expected.

Appendix B at the end of this book gives the sound-speed profiles and predicted acoustic arrival patterns for locations distributed throughout the world's oceans. Annual average climatological temperature and salinity data (Levitus, 1982) have been used to obtain annual average sound-speed profiles for use in propagation calculations. The Levitus climatology gives a horizontally

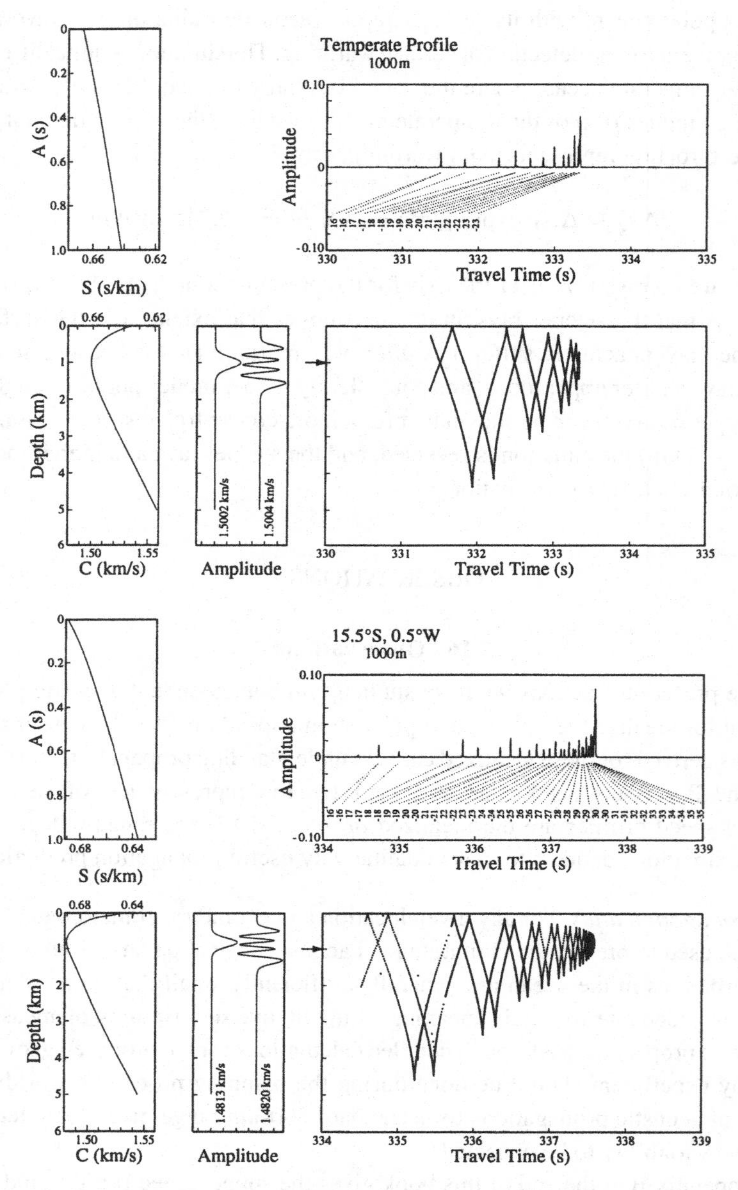

Fig. 2.16. Sound-speed profile, action as a function of turning-point slowness, ray wavefront diagram for an axial source, WKBJ arrival pattern for axial source and receiver, and acoustic normal mode functions 1 and 7 for the temperate profile (top), and for climatological data at 15.5°S, 0.5°W (bottom). Additional parameters used in the calculations are given in the text.

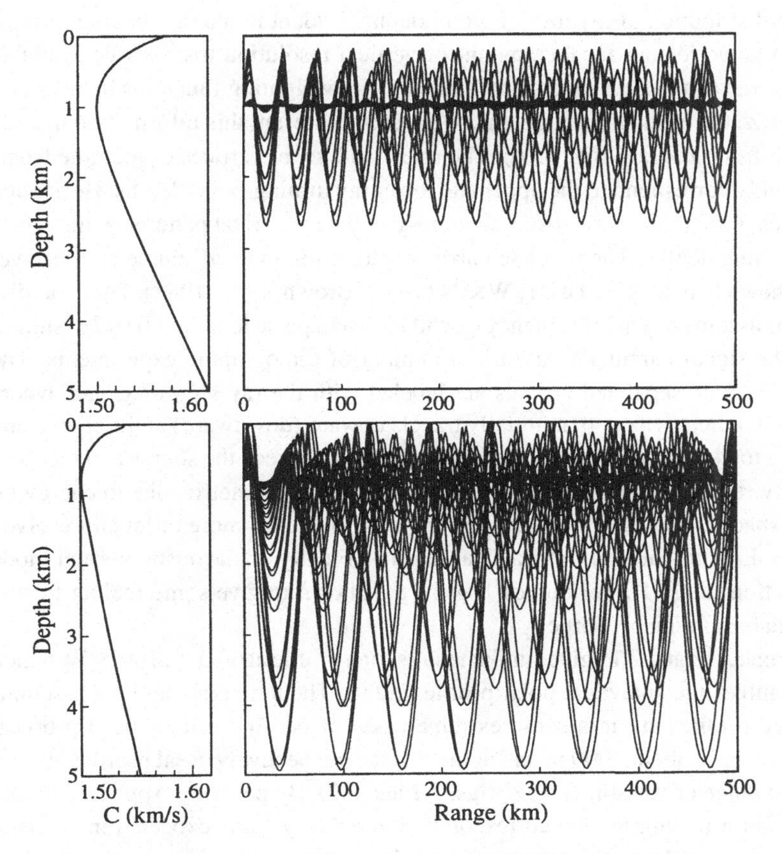

Fig. 2.17. Eigenrays between an axial source and receiver 500 km apart for the temperate profile (top) and for the climatological data at 15.5°S, 0.5°W (bottom). Only rays starting upward at the source are shown.

smoothed picture of the ocean, so the results do not properly represent the behavior in frontal regions. Nonetheless, over most of the oceans the propagation calculations can be expected to give qualitative agreement with measurements.

Propagation calculations for the South Atlantic Ocean at 15.5°S, 0.5°W roughly correspond to the analytical temperate case (fig. 2.16). The annual average sound-speed profile is similar to the analytical temperate profile, except for slightly shallower axial depth. (For simplicity, the axial sound-speed for the analytical profile was assumed to be 1500 m/s, rather faster than is typical of real profiles.) The eigenrays between an axial source and receiver 500 km apart are qualitatively similar to those in the analytical temperate ocean (fig. 2.17).

The distribution of ray turning-point depths evident in the ray diagram gives a qualitative feeling for the amount of vertical resolution that will be available from ray travel-time inversions, since the ray weighting functions heavily emphasize the turning-point depths. We prefer to present this information in a ray time-front diagram (fig. 2.16). The upper and lower vertices of the time fronts roughly correspond to the upper and lower ray turning-point depths. Horizontal slices through the wave-front diagram give the ray arrival patterns at the corresponding depths. The predicted arrival pattern for an axial source and receiver is shown in more detail using WKBJ theory (Brown, 1981, 1982). These predictions assume a center frequency of 250 Hz and a pulse length of 0.012 s, similar to the signal parameters used in a number of tomographic experiments. The earlier, well-separated arrivals are labeled with the ray identifier, $\pm p$, where $+(-)$ indicates a ray that initially travels upward (downward) at the source and has a total of p upper plus lower turning points between the source and receiver. Early resolved arrivals without identifiers are nongeometric, due to energy on the shadow-zone side of a caustic. The final cluster of more or less unresolved arrivals is not labeled for practical reasons. Finally, the acoustic normal mode functions at 70 Hz for modes 1 and 7 are shown to give some feeling for the modal sampling properties.

Figs. 2.18 and 2.19 give similar results for a profile at 60.5°S, 100.5°W, which resembles the analytical polar profile (2.2.9). The temperate and polar sound-speed profiles are in a sense extreme cases of oceanic sound-speed profiles, with axes at about 1000 m and at the surface, respectively. Real profiles show a wide range of variability (*e.g.*, figs. 2.1 and 2.2). By perusing Appendix B, one can get a feeling for the corresponding variability to be expected in acoustic propagation. Acoustic ray trajectories depend on the vertical gradients in the sound-speed profile and are therefore sensitive to the detailed structure. There is no guarantee that steep rays will arrive first, and axial rays last, as was found for the idealized temperate and polar profiles. In some cases, the axial paths arrive first and the steep paths last. Equatorial sound-speed profiles tend to be focusing, with all paths having nearly the same travel times. In other cases, complex sound-speed profiles give complicated arrival patterns, with no simple relation between ray angle and arrival time. In the eastern North Atlantic, for example, the Mediterranean outflow gives a warm, salty layer at about 1000 m depth, resulting in sound-speed minima above and below the Mediterranean layer. Propagation in this region is complex.

Measured arrival patterns. The validity of the range-independent approximation (using the range-average profile) depends strongly on location. Using a range-independent profile to predict propagation through strongly

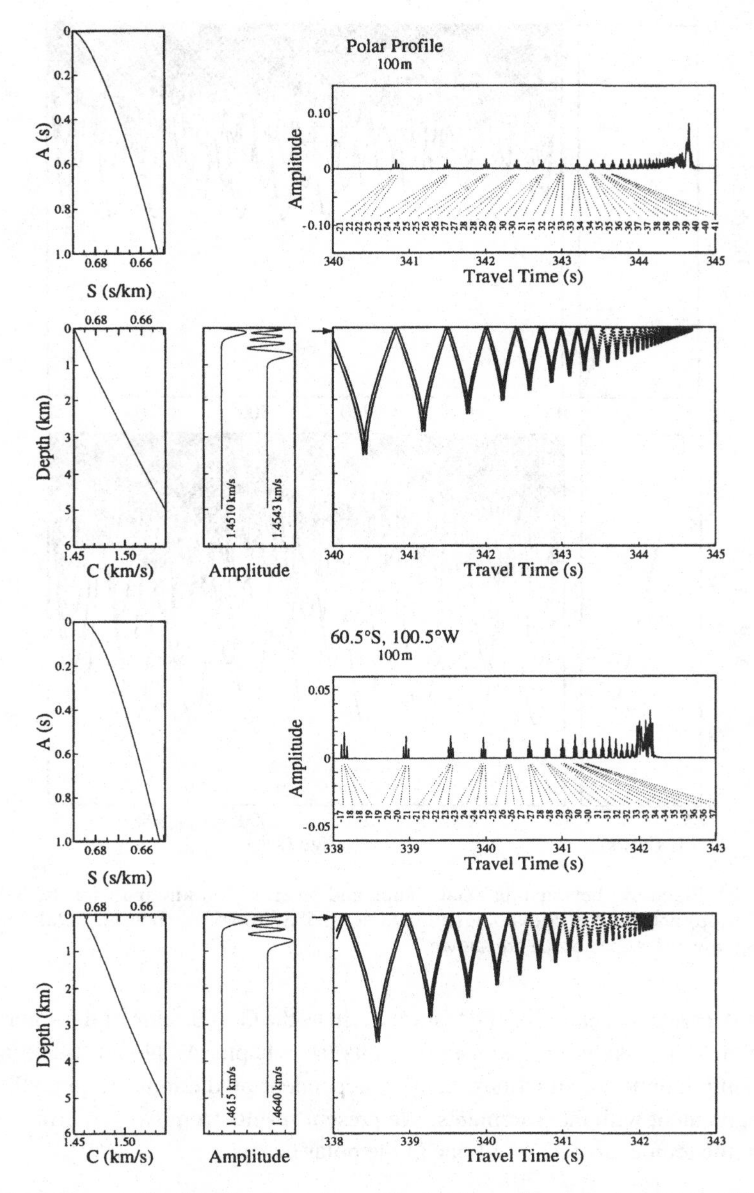

Fig. 2.18. Same as fig. 2.16 for the polar profile (top) and climatological data at 60.5°S, 100.5°W (bottom).

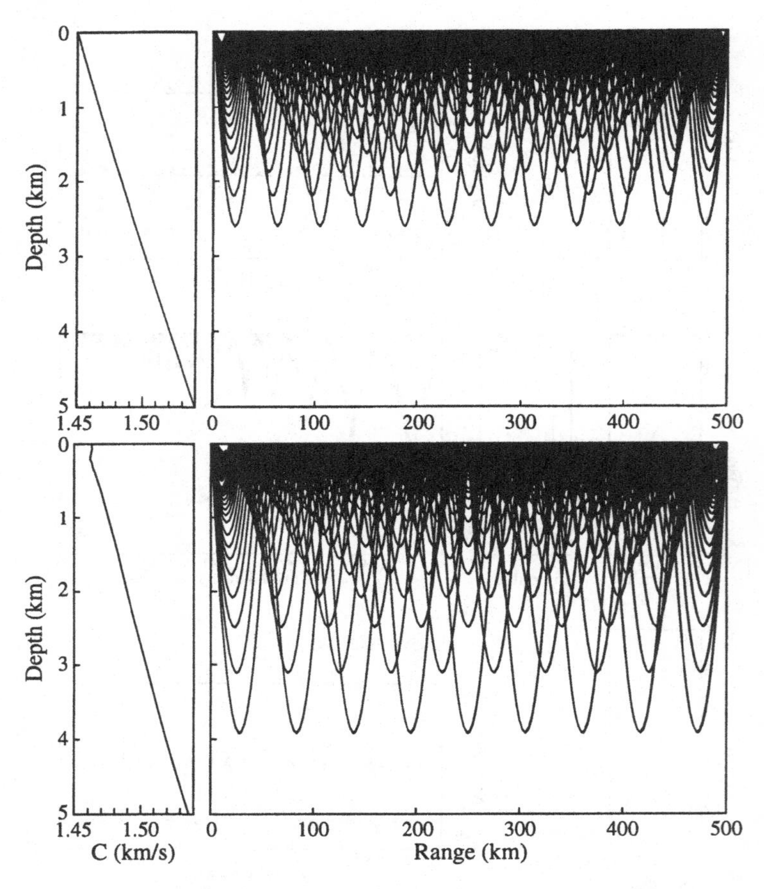

Fig. 2.19. Eigenrays between an axial source and receiver 500 km apart for the polar profile (top) and for the climatological data at 60.5°S, 100.5°W (bottom). Only rays starting upward at the source are shown.

range-dependent oceanographic features such as the Gulf Stream or the Antarctic Front is not likely to yield useful results (see chapter 4). Nonetheless, in a surprising number of situations, range-independent predictions are in qualitative agreement with measurements. We present results from two experiments, one in the temperate ocean and one in the polar ocean.

Temperate ocean. During July 1989, broadband acoustic signals were transmitted from a 250-Hz source to a 3-km-long vertical array of hydrophones 1 Mm distant in the north central Pacific Ocean (Howe *et al.*, 1991; Duda *et al.*, 1992;

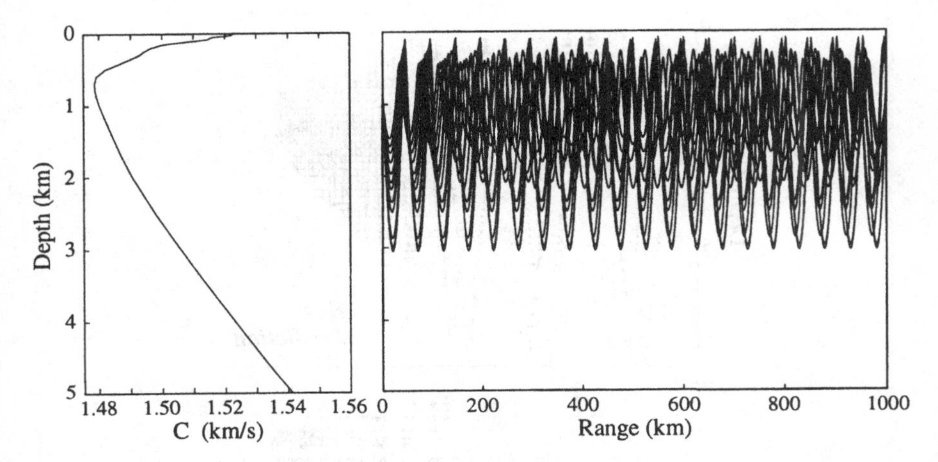

Fig. 2.20. Eigenrays between the source, at 32°00′N, 150°26′W, and the 733-m-deep receiver, at 34°00′N, 140°00′W, for the range-averaged sound-speed profile constructed by combining the CTD, XBT, and AXBT data obtained during the 1989 Vertical Slice Tomography Experiment in the north central Pacific Ocean. Only rays starting down at the source are shown.

Cornuelle *et al.*, 1992, 1993; Worcester *et al.*, 1994). The acoustic source was moored at 804 m depth, near the sound channel axis. The receiving array was suspended from the research platform (R/P) *FLIP*. The temperature and salinity fields along the great circle-path connecting the source and receiver were measured with a ship-lowered conductivity-temperature-depth (CTD) sensor, expendable bathythermographs (XBTs), and airborne expendable bathythermographs (AXBTs). The range-averaged sound-speed profile (fig. 2.20) constructed by combining the CTD, XBT, and AXBT data is qualitatively similar to the analytical temperate profile (2.2.11), except that the axis is shallower, 750 m compared with 1000 m.

Plotting the locations of the measured arrival peaks for an individual reception as a function of travel time and hydrophone depth reveals acoustic wave fronts sweeping across the array (fig. 2.21). The wave fronts seen early in the reception are in one-to-one correspondence with the wave fronts predicted using either ray theory or broadband normal mode theory, making identification of the measured wave fronts with particular ray paths straightforward. The measured and predicted wave front patterns are not identical, however. Most of this difference can be eliminated using range-dependent propagation codes, as will be discussed in chapter 4. The most striking discrepancy, however, is the

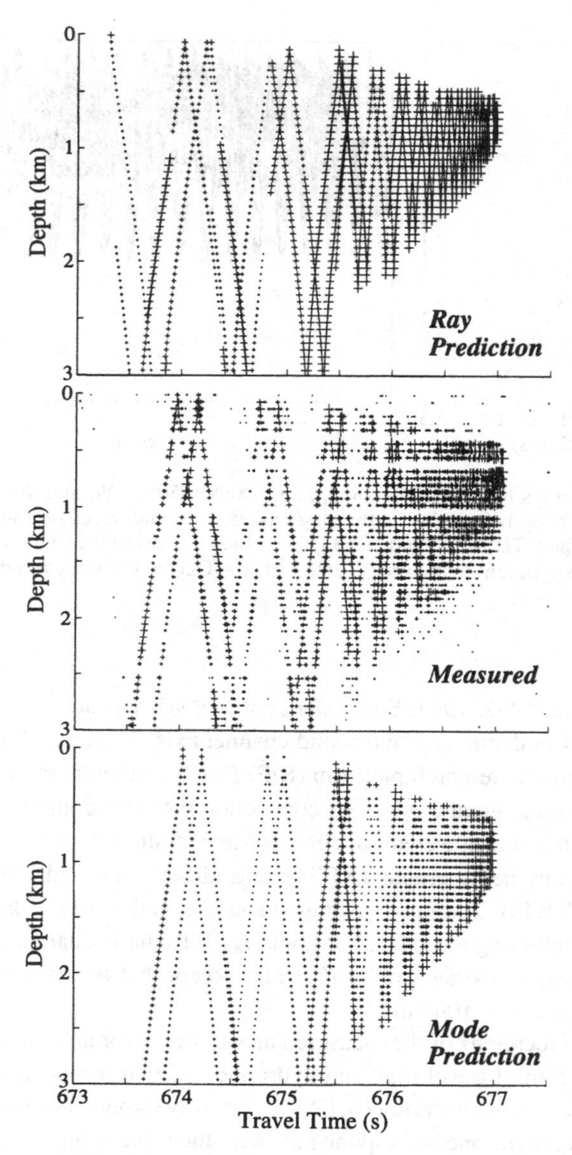

Fig. 2.21. Top: Predicted travel times vs. hydrophone depth computed using a range-independent ray-tracing algorithm. Middle: Measured pulse travel times for day 192 of 1989 at 19:20 UTC. The four prominent horizontal data gaps are the result of hydrophone failures. Bottom: Same as the top panel, except that a range-independent, broadband acoustic normal algorithm was used to make the predictions. Symbol size is proportional to intensity (in dB).

difference between the predicted cutoffs of the acoustic receptions and the times at which the final cutoffs actually occur. Whereas the near-axial hydrophones, at 700–800-m depths, have final cutoff times roughly consistent with ray theoretical predictions, off-axis hydrophones have final cutoff times up to several hundred milliseconds later than predicted using ray theory. The difference is especially dramatic for hydrophones below the sound-channel axis (fig. 2.21). In other words, the measured cutoff wedge is *blunter* than predicted by ray theory.

One possible explanation for this discrepancy is that geometric ray theory does not treat turning points and caustics correctly. Rather than thinking of the discrepancy as being energy arriving later than predicted for the off-axis arrivals, one can think of it as energy leaking vertically away from the slow axial rays. Worcester (1981), for example, observed a diffracted arrival containing significant energy associated with a caustic roughly 400 m above the receiver depth (Brown, 1981). For range-independent profiles, acoustic normal modes give complete and accurate solutions for both turning points and caustics. In fact, broadband normal mode predictions show significant broadening of the final cutoff. However, the predicted broadening does not appear adequate to account for all of the observed broadening, particularly above the sound-channel axis. Colosi *et al.* (1994) have shown that scattering from internal-wave-induced sound-speed fluctuations also significantly broadens the final cutoff, as will be discussed in chapter 4.

Polar ocean. During 1988 and 1989, an array of six 250-Hz acoustic transceivers was deployed in the Greenland Sea for one year (Worcester *et al.*, 1993; Sutton *et al.*, 1993, 1994). Five of the instruments were in a pentagon, centered on the sixth instrument at 75°04′N, 02°58′W. The radius of the pentagon was about 105 km; the range between the most distant instruments was about 200 km. The sources were moored 95 m below the surface, with small vertical receiving arrays immediately below (six hydrophones maximum). During February and March 1989, extensive Sea Soar[14] and CTD measurements were made in the vicinity of the array (SIZEX Group, 1989; Greenland Sea Project Group, 1990). The range-averaged sound-speed profile between the central and northern moorings is nearly adiabatic (fig. 2.22), similar to the analytical profile (2.2.9). All of the predicted ray paths for this source-receiver pair reflect off the surface. The agreement between the measured and predicted ray arrival *patterns*

[14]Prescribes a vertical zigzag course when towed behind a vessel.

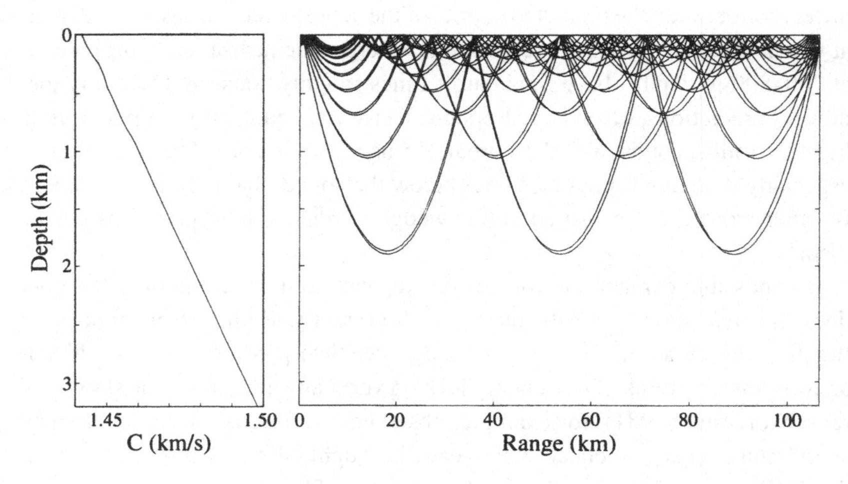

Fig. 2.22. Eigenrays between a source, at 75°04′N, 2°58′W, and a receiver, at 75°58′N, 1°50′W, for the range-averaged sound-speed profile constructed by combining Sea Soar and CTD data obtained on the line between the instruments during the March 1989 SIZEX experiment in the Greenland Sea. Only rays starting upward at the source are shown.

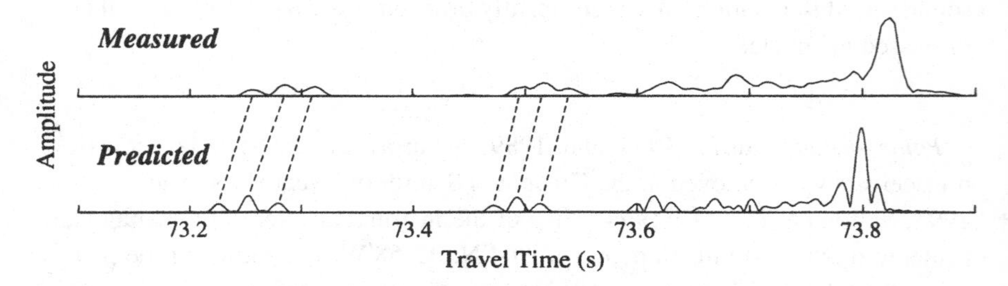

Fig. 2.23. Measured and predicted arrival amplitudes for the source-receiver pair of fig. 2.22. The measured arrival pattern is a daily average constructed from receptions in March 1989, with the initial arrivals aligned prior to averaging. The predicted arrival pattern was constructed using a range-independent, broadband acoustic normal mode algorithm. Dotted lines connect corresponding measured and predicted ray arrivals.

is excellent for the early arrivals,[15] but deteriorates for the later, unresolved arrivals (fig. 2.23). This behavior is typical of other arrival patterns previously discussed.

[15]The *displacement* between measured and computed arrivals is largely the result of a range error in the computed arrival pattern.

APPENDIX

2.17. Polar (Adiabatic) Profile

In polar winter, the sound-speed profile approaches adiabatic conditions, $S = S_0(1 + \gamma_a z)$ with $\gamma_a = 0.0113\ \text{km}^{-1}$ (2.2.5). Thus slowness deceases (sound-speed increases) with increasing depth $-z$. For analytical simplicity, we take as our starting point

$$S^2(z) = S_0^2(1 + 2\gamma_a z), \tag{2.17.1}$$

which does not differ appreciably from adiabatic conditions.

Define

$$\sigma^2(S) = \frac{S_0^2 - S^2}{S_0^2} = 2\gamma_a(-z), \tag{2.17.2}$$

which varies from $\sigma_0 \equiv \sigma(S_0) = 0$ to

$$\tilde{\sigma}^2 = \sigma^2(\tilde{S}) = \frac{S_0^2 - \tilde{S}^2}{S_0^2}. \tag{2.17.3}$$

We recognize

$$\tilde{\sigma} = \sin\theta_0, \tag{2.17.4}$$

where θ_0 is the ray inclination at the surface.

Define the "loop action"

$$A(\tilde{S}) = -2\int_0^{\tilde{z}} dz\,(S^2 - \tilde{S}^2)^{\frac{1}{2}} = \tfrac{2}{3}\,\gamma_a^{-1}\,S_0\,\tilde{\sigma}^3. \tag{2.17.5}$$

The awkward factor 2 is called for because we subsequently deal mostly with complete ray loops. Accordingly,

$$R(\tilde{\sigma}) = -\partial A/\partial\tilde{S} = 2\gamma_a^{-1}\tilde{\sigma}\sqrt{1-\tilde{\sigma}^2}, \tag{2.17.6}$$

$$\begin{aligned} T(\tilde{\sigma}) = A + R\tilde{S} &= 2\gamma_a^{-1}S_0\,\tilde{\sigma}\,(1 - \tfrac{2}{3}\,\tilde{\sigma}^2) \\ &\approx RS_0(1 - \tfrac{1}{6}\,\tilde{\sigma}^2 + \cdots). \end{aligned} \tag{2.17.7}$$

Thus steep rays arrive only slightly ahead of the surface-trapped rays.

Group and phase slownesses follow from (2.9.12):

$$s_g = \frac{T}{R} = \frac{S_0\,(1 - \frac{2}{3}\,\tilde{\sigma}^2)}{\sqrt{1-\tilde{\sigma}^2}} \approx S_0\,(1 - \tfrac{1}{6}\,\tilde{\sigma}^2), \tag{2.17.8}$$

$$s_p = \frac{dT}{dR} = \frac{dT/d\tilde{\sigma}}{dR/d\tilde{\sigma}} = \tilde{S} = S_0\sqrt{1-\tilde{\sigma}^2}. \tag{2.17.9}$$

The equations (2.17.6)–(2.17.9) apply equally to a modal representation, provided $\widetilde{\sigma}$ is appropriately defined.

Rays. For a fixed range with n complete ray loops $r = nR$, it follows from (2.17.6) that

$$\widetilde{\sigma}^2 = \widetilde{\sigma}_n^2 = \tfrac{1}{2}\left[1 - \sqrt{1 - (\gamma_a r/n)^2}\,\right] \approx (\tfrac{1}{2}\,\gamma_a r/n)^2 . \qquad (2.17.10)$$

To deal with fractional ray loops, we introduce a "fractional action"

$$\delta A(S, \widetilde{S}) = -\int_0^z dz\,(S^2 - \widetilde{S}^2)^{\frac{1}{2}} = -S_0 \int_0^z dz\,(\widetilde{\sigma}^2 - \sigma^2)^{\frac{1}{2}}, \qquad (2.17.11)$$

with $z > \widetilde{z}$. For the polar profile (2.17.1),

$$dz = \frac{1}{2\gamma_a}\, d\Big(\frac{S^2}{S_0^2}\Big) = -\frac{1}{2\gamma_a}\, d\sigma^2 ,$$

hence

$$\delta A(\sigma, \widetilde{\sigma}) = \frac{S_0}{2\gamma_a} \int_0^\sigma d\sigma^2\,(\widetilde{\sigma}^2 - \sigma^2)^{\frac{1}{2}} = \frac{S_0}{3\gamma_a}\big[\widetilde{\sigma}^3 - (\widetilde{\sigma}^2 - \sigma^2)^{\frac{3}{2}}\big]. \qquad (2.17.12)$$

The fractional action varies from 0 at $z = 0$ to $\frac{1}{2}\, A(\widetilde{S})$ at the turning point. Fractional range is given by

$$\delta R\,(\sigma, \widetilde{\sigma}) = -\frac{\partial(\delta A)}{\partial \widetilde{S}} = \frac{1}{\gamma_a}\big[\widetilde{\sigma} - \sqrt{\widetilde{\sigma}^2 - \sigma^2}\,\big]\sqrt{1 - \widetilde{\sigma}^2} .$$

A convenient notation is

$$\sin^2 \upsilon = \frac{\sigma^2}{\widetilde{\sigma}^2} = \frac{S_0^2 - S^2}{S_0^2 - \widetilde{S}^2}, \qquad 0 \le \upsilon \le \tfrac{1}{2}\pi , \qquad (2.17.13)$$

giving

$$\delta A(\upsilon,\, \widetilde{\sigma}) = \tfrac{1}{3}\,\gamma_a^{-1} S_0 \widetilde{\sigma}^3 (1 - \cos^3 \upsilon) , \qquad (2.17.14)$$

$$\delta R(\upsilon,\, \widetilde{\sigma}) = \gamma_a^{-1} (1 - \cos \upsilon)\, \widetilde{\sigma}\, \sqrt{1 - \widetilde{\sigma}^2} . \qquad (2.17.15)$$

The loop range is accordingly $R(\widetilde{\sigma}) = 2\,\delta R\,(\frac{1}{2}\pi, \widetilde{\sigma})$, in agreement with (2.17.6).

Travel time is given by

$$\delta T(S, \widetilde{S}) = \delta A + \widetilde{S}\, \delta R = S_0\, \delta R - (S_0 - \widetilde{S})\, \delta R + \delta A .$$

Thus

$$\delta T(\nu, \widetilde{\sigma}) = \gamma_a^{-1} S_0 (1 - \cos \upsilon)\, \widetilde{\sigma}\,\big[1 - \tfrac{1}{2}\widetilde{\sigma}^2 (2 - \cos \upsilon - \cos^2 \upsilon)\big]. \quad (2.17.16)$$

The loop travel time is $T(\widetilde{\sigma}) = 2\,\delta T(\frac{1}{2}\,\pi,\, \widetilde{\sigma})$, in agreement with (2.17.7).

Time fronts. $z(\tau)$ are constructed by solving the quadratic equation

$$r = nR(\widetilde{\sigma}) + \delta R(\upsilon, \widetilde{\sigma})$$

for $\widetilde{\sigma}(\upsilon, n)$, and then computing

$$z(\upsilon, n) = -(\widetilde{\sigma} \sin \upsilon)^2 / 2\gamma_a, \quad \tau(\upsilon, n) = nT(\widetilde{\sigma}) + \delta T(\upsilon, \widetilde{\sigma}),$$

for $0 \le \upsilon \le \pi$, given n (fig. 2.4).

Modes. For solutions $Q(r)\, P(z) e^{i\omega t}$ to the wave equation (2.9.2), the vertical wave function $P(z)$ must satisfy

$$\frac{d^2 P}{dz^2} + k_\upsilon^2(z)\, P = 0, \quad k_\upsilon(z) = \omega (S^2(z) - \widetilde{S}^2)^{\frac{1}{2}} \qquad (2.17.17a, b)$$

(see section 2.9 for further discussion). For the polar profile (2.17.1), equation (2.17.17) can be transformed into the Airy equation

$$\frac{d^2 P}{d\xi^2} - \xi P = 0,$$

with the solution

$$P(\xi) = a \operatorname{Ai}(\xi) + b \operatorname{Bi}(\xi), \qquad (2.17.18)$$

provided

$$\xi(z) = -\left[\tfrac{3}{2}\right]^{\frac{2}{3}} \frac{z - \widetilde{z}}{H_{\text{Ai}}}, \quad H_{\text{Ai}} = \left[\frac{9}{32\pi^2} \frac{1}{\gamma_a f^2 S_0^2}\right]^{\frac{1}{3}}. \qquad (2.17.19)$$

Note that ξ is positive *downward* from the turning depth $\widetilde{z}$. For $\gamma_a = 0.0113$, $f = 2\pi/\omega = 250$ Hz, and $C_0 = 1/S_0 = 1.45$ km/s, we have an Airy scale $H_{\text{Ai}} = 0.044$ km: the wave function extends essentially 44 m beneath the turning depth. For the case that the bottom is much deeper, we can set $b = 0$. At the surface, $z = 0$, $\xi = \xi_0$, hence the surface boundary condition $P' = 0$ yields $\operatorname{Ai}'(\xi_0) = 0$, with the roots $(\xi_0)_m = -1.019, -3.248, -4.820, \ldots$ for modes $m = 1, 2, 3, \ldots$. The asymptotic formula

$$(\xi_0)_m = -\left[(3\pi/2)\,(m - \tfrac{3}{4})\right]^{\frac{2}{3}} = -1.115,\ -3.262,\ -4.826, \ldots$$

is in close agreement. Equating these roots to $\xi(0)$ in (2.17.19) gives the frequencies f_m for constructive interference at turning point $\widetilde{z}$.

Group and phase velocities again follow from (2.17.8) and (2.17.9), provided $\widetilde{\sigma}$ is interpreted as follows:

$$\widetilde{\sigma}^3 = \widetilde{\sigma}_m^3 = \frac{A_m}{\frac{2}{3}\gamma_a^{-1} S_0} = \frac{m - \frac{1}{4}}{f/F_P}, \quad F_P = \frac{3\gamma_a}{2S_0} = 0.0254\,\text{Hz}.$$

Perturbations. Using the dimensionless coordinate $\zeta = z/\widetilde{z}$, with $-\widetilde{z} = \widetilde{\sigma}^2/2\,\gamma_a$, the weights W in (2.14.32) for the polar profile are given by

$$W^0(z) = \tfrac{1}{2}\,(1-\zeta)^{-\frac{1}{2}}/(-\widetilde{z})\,,$$
$$W^I(z) = \tfrac{1}{2}\,(1-\zeta)^{-\frac{1}{2}}\,(\tfrac{7}{3} - 2\zeta)/(-\widetilde{z})\,, \qquad (2.17.20)$$
$$W^{II}(z) = -\tfrac{1}{3}\;\zeta\;(1-\zeta)^{-\frac{1}{2}}\,(1-2\zeta)\,.$$

Note that $\int dz\, W^0 = 1$ and $\int dz\, W^I = 1$, so that $(\Delta s_g)_{\text{ray}} = (\Delta s_g)_{\text{mode}} = \Delta S$ for $\Delta S' = 0$. Because W^I and $W^{II}/(-\widetilde{z})$ are of comparable magnitudes, it follows that the second term in the expression for $(\Delta s_g)_{\text{mode}}$ dominates if the turning depth $-\widetilde{z}$ exceeds the perturbation scale $\Delta S/\Delta S'$. For a polar gradient perturbation $\Delta\gamma$,

$$\Delta S = -\tfrac{1}{2}\,\widetilde{\sigma}^2\,S_0\,(\Delta\gamma/\gamma)\,\zeta, \qquad \Delta S' = S_0\,\Delta\gamma\,,$$

and

$$(\Delta s_g)_{\text{ray}} = -\tfrac{1}{3}\;S_0\,\widetilde{\sigma}^2\,\Delta\gamma/\gamma\,,$$

$$(\Delta s_g)_{\text{mode}} = S_0\,\widetilde{\sigma}^2\,\frac{\delta\gamma}{\gamma}\,(-\tfrac{11}{45} + \tfrac{2}{15}) = -\tfrac{1}{9}\;S_0\,\widetilde{\sigma}^2\,\Delta\gamma/\gamma\,, \qquad (2.17.21)$$

to order $\widetilde{\sigma}^2$, in agreement with previous results. The contribution of ΔS is about twice that of $\Delta S'$.

The $(1-\zeta)^{-\frac{1}{2}}$ singularity at the turning point can be avoided by writing $z = \widetilde{z}\,\sin^2\nu$, with $0 \le \nu \le \frac{1}{2}\pi$ as the independent variable. Thus with $\Delta S(\widetilde{z},\,\nu)$ and $\Delta S'(\widetilde{z},\,\nu) = (d/dz)\,\Delta S$ given,

$$(\Delta s_g)_{\text{ray}} = \int_0^{\pi/2} d\nu\, V^0\,\Delta S, \quad (\Delta s_g)_{\text{mode}} = \int_0^{\pi/2} d\nu \left[V^I\,\Delta S + V^{II}\,\Delta S'\right],$$

$$V^0 = \sin\nu,\;\; V^I = \sin\nu(\tfrac{1}{3} + 2\cos^2\nu),\;\; V^{II} = -\tfrac{2}{3}\,\sin^3\nu\,(1 - 2\sin^2\nu)(-\widetilde{z})\,. \qquad (2.17.22)$$

2.18. Temperate (Canonical) Profile

A fundamental property of most of the world's oceans is that the upper ocean is more stably stratified than the deep ocean. An *exponential* decrease with depth of the vertical gradient in potential density does not give a bad fit to some average profiles (Munk, 1974). Stratification is usually expressed in terms of the Brunt-Väisälä (or buoyancy) frequency N, with $N^2 = -(g/\rho)d\rho_p/dz$; z is upward from the surface. We set $N(z) = N_0 e^{z/h}$, or

$$N(z) = N_A e^{\zeta}, \quad \zeta = (z - z_A)/h, \tag{2.18.1}$$

referred to the situation at some depth z_A that we identify with the sound axis.

In an isohaline ocean, the gradients of potential temperature, potential density, and potential sound-speed are all proportional, and so $dC_p/dz \sim N^2$. The potential gradient dC_p/dz equals the absolute gradient dC/dz minus the adiabatic gradient dC_a/dz, and we can write

$$\frac{1}{C}\frac{dC}{dz} = \gamma_a \frac{N^2 - N_A^2}{N_A^2}, \quad \gamma_a = -\frac{1}{C}\frac{dC_a}{dz}. \tag{2.18.2}$$

Note that $dC/dz = 0$ at the axis (sound-speed minimum), and approaches dC_a/dz at great depth ($N \ll N_A$). For the polar profile, $N = 0$ and $dC/dz = -\gamma_a C$.

All this can be put together by setting

$$\phi^2 \equiv \frac{1}{\gamma_a h}(S_A^2 - S^2)/S_A^2 = e^{2\zeta} - 2\zeta - 1, \tag{2.18.3}$$

where $S = 1/C$ is the sound-slowness. The temperate profile $S(z)$ follows from (2.18.3). The adiabatic gradient is taken at $\gamma_a = 1.13 \times 10^{-2}\,\text{km}^{-1}$. We set $h = 1$ km, and $z_A = -1$ km.

The temperate profile $S(z)$ has a reassuring appearance of up/down asymmetry (fig. 2.3) and the "normal" dispersive properties (high modes first, axial modes last, 0.2% spread at 1 Mm range). But it does not readily lend itself to analytical treatment. An expansion of (2.18.3) yields

$$\phi^2 = 2\zeta^2 + \frac{4}{3}\zeta^3 + \frac{2}{3}\zeta^4 + \cdots, \tag{2.18.4}$$

which *does* lend itself to analytical treatment, but the truncated series does not adequately describe the sound channel below 2 km and is, in fact, abnormally dispersive (axial modes first).

For some years, we have sought a tractable model in rough agreement with the properties of the real sound channel (Munk and Wunsch, 1979, 1985, 1987). The usual procedure is to start with a model profile $S(z)$ and then derive the action

$$\begin{aligned} A^{\pm} &= \pm 2\int_{z_A}^{\tilde{z}^{\pm}} dz\,(S^2 - \tilde{S}^2)^{\frac{1}{2}}, \\ A &= A^+ + A^- = 2\int_{\tilde{z}^-}^{\tilde{z}^+} dz\,(S^2 - \tilde{S}^2)^{\frac{1}{2}}, \end{aligned} \tag{2.18.5}$$

where $\widetilde{S} = S(\widetilde{z}^+) = S(\widetilde{z}^-)$ is the slowness at the upper and lower turning depths. We have found that a preferable procedure is to start with a model action $A(\widetilde{z})$ and *then* (if necessary) to compute the sound profile via an Abel transform.

In particular, we choose

$$A^{\pm} = S_A h\,(\gamma_a h)^{\frac{1}{2}}\,(a\,\widetilde{\phi}^2 \pm b\,\widetilde{\phi}^3 + c\,\widetilde{\phi}^4) \tag{2.18.6}$$

exactly, with

$$\widetilde{\phi}^2 = (\gamma_a h)^{-1}\,\widetilde{\sigma}^2 = (\gamma_a h)^{-1}\,(S_A^2 - \widetilde{S}^2)/S_A^2, \tag{2.18.7}$$

in analogy to (2.18.3). We recognize $\widetilde{\sigma} = \sin\theta_A$. Most of the interesting parameters (such as group slowness s_g) follow directly from A. To evaluate the constants a, b, and c, we perform the Abel transform (2.5.12). Set

$$\alpha = \phi^2 = \frac{\sigma^2}{\gamma_a h}, \qquad \beta = \widetilde{\phi}^2 = \frac{\widetilde{\sigma}^2}{\gamma_a h},$$

$$f(\beta) = \frac{R^{\pm} S_A}{2\widetilde{S}} = \frac{h}{\sqrt{\gamma_a h}}\left(a \pm \tfrac{3}{2}\,b\widetilde{\phi} + 2c\,\widetilde{\phi}^2\right),$$

$$\zeta = \pm\frac{1}{\pi}\int_0^{\phi} 2\widetilde{\phi}\,d\widetilde{\phi}\,\frac{a \pm \frac{3}{2}\,b\widetilde{\phi} + 2c\,\widetilde{\phi}^2}{(\phi^2 - \widetilde{\phi}^2)^{\frac{1}{2}}}$$

$$= \pm\left[\frac{2a}{\pi}\phi \pm \frac{3b}{4}\phi^2 + \frac{8c}{3\pi}\phi^3\right], \tag{2.18.8}$$

exactly. To be consistent with the temperate ocean (2.18.3) to order ϕ^3 requires that

$$a = \frac{\pi}{2\sqrt{2}}, \quad b = -\frac{2}{9}, \quad c = \frac{\pi}{48\sqrt{2}}. \tag{2.18.9}$$

The expression (2.18.8) for $\zeta(\phi)$ carried to three terms gives a good fit to the canonical ocean and yields normal dispersion; in contrast, the three terms in (2.18.4) for $\phi(\zeta)$ do not. Most of the analytical results in this volume follow from the foregoing three-term expansion, to be referred to as the *polynomial temperate* sound channel. Results differ slightly from those for the *exponential temperate* profile (2.18.3).

From (2.5.4),

$$R^{\pm} = -\frac{\partial A^{\pm}}{\partial \widetilde{S}} = R_0^{\pm}\left(1 \mp \frac{2\sqrt{2}}{3\pi}\widetilde{\phi} + \frac{1}{12}\widetilde{\phi}^2\right)\frac{\widetilde{S}}{S_A}, \tag{2.18.10}$$

$$R_0^\pm = \frac{\pi}{\sqrt{2}} \frac{h}{\sqrt{\gamma_a h}} = 20.8\,\mathrm{km}\,,$$

for the upper/lower-loop span, and $R = R^+ + R^-$ for the double loop. We can set $\widetilde{S}/S_A = \sqrt{1 - \gamma_a h\,\widetilde{\phi}^2} \approx 1$. Similarly, from (2.5.1),

$$\begin{aligned} T^\pm &= A^\pm + \widetilde{S}R^\pm = S_A R^\pm + A^\pm - R^\pm(S_A - \widetilde{S}) \\ &= S_A R^\pm - T_0^\pm \left(\mp \frac{2\sqrt{2}}{9\pi}\widetilde{\phi}^3 + \frac{1}{24}\widetilde{\phi}^4\right), \end{aligned} \tag{2.18.11}$$

$$T_0^\pm = \tfrac{1}{2}\,(\gamma_a h) S_A R_0^\pm = 0.079\,\mathrm{s}.$$

For the double loop, to first order in $\gamma_a h$,

$$R = R^+ + R^- = 2R_0^\pm \left(1 + \tfrac{1}{12}\,\widetilde{\phi}^2\right)(\widetilde{S}/S_A)\,, \tag{2.18.12}$$

$$T = T^+ + T^- = R S_A \left(1 - \gamma_a h \frac{\widetilde{\phi}^4/48}{1 + \widetilde{\phi}^2/12}\right). \tag{2.18.13}$$

Group and phase slownesses are given by

$$s_g = \frac{T}{R} = S_A \left[1 - \gamma_a h \frac{\widetilde{\phi}^4/48}{1 + \widetilde{\phi}^2/12}\right], \tag{2.18.14}$$

$$s_p = \frac{dT}{dR} = \frac{dT/d\widetilde{\sigma}^2}{dR/d\widetilde{\sigma}^2} = S_A \left[1 - \gamma_a h\,\widetilde{\phi}^2\right]^{\frac{1}{2}} = \widetilde{S}\,. \tag{2.18.15}$$

The equations (2.18.10)–(2.18.15) apply equally to ray and mode representations provided $\widetilde{\sigma}$ and $\widetilde{\phi}$ are appropriately defined (2.11.8).

Extra ray loops. For a fixed range with n complete ray loops, $r = nR$, it follows from (2.18.12) that

$$\widetilde{\phi}^2 = \widetilde{\phi}_n^2 = 12\left[\frac{r}{2n\,R_0^\pm} - 1\right]. \tag{2.18.16}$$

For the case of an extra upper or an extra lower loop,

$$r = n^+R^+ + n^-R^- = \tfrac{1}{2}\,(n^+ + n^-)(R^+ + R^-) + \tfrac{1}{2}\,(n^+ - n^-)(R^+ - R^-)\,,$$

$$\rho = \tfrac{1}{2}\,(n^+ + n^-)\,(1 + \tfrac{1}{12}\,\widetilde{\phi}^2) - \tfrac{1}{2}\,(n^+ - n^-)\,\frac{2\sqrt{2}}{3\pi}\,\widetilde{\phi}\,,$$

$$\widetilde{\phi}_n = \lambda + \sqrt{\lambda^2 + 12(\frac{\rho}{n} - 1)}, \quad \lambda = \frac{4\sqrt{2}}{3\pi}\,\frac{n^+ - n^-}{n^+ + n^-}\,. \tag{2.18.17}$$

Modes. The relation $A = (m - \frac{1}{2})/f$, together with (2.18.6) for $A = A^+ + A^-$ yields the solution (2.11.8) for $\widetilde{\phi}_m(m, f)$. Analytical expressions for $\widetilde{z}(m, f)$ readily follow from (2.18.20).

Partial loops. The discussion has dealt with complete loops, that is, complete upper loops, complete lower loops, and complete double loops. To deal with fractional loops it is convenient to use "fractional action," as defined in (2.7.1). In analogy with the development in the preceding section, we write

$$\phi^2(S) = \frac{1}{\gamma_a h}\frac{S_A^2 - S^2}{S_A^2}, \quad \widetilde{\phi}^2(\widetilde{S}) = \frac{1}{\gamma_a h}\frac{S_A^2 - \widetilde{S}^2}{S_A^2},$$
$$\sin^2\nu = \frac{\phi^2}{\widetilde{\phi}^2} = \frac{S_A^2 - S^2}{S_A^2 - \widetilde{S}^2}, \quad 0 \le \nu \le \tfrac{1}{2}\pi, \tag{2.18.18}$$

giving

$$\begin{aligned}\delta A^{\pm} &= \pm\int_{z_A}^{z} dz' \, [\, S^2(z) - \widetilde{S}^2\,]^{\frac{1}{2}}, \qquad z \gtrless z_A \\ &= \pm\sqrt{\gamma_a h}\; S_A \int_{z_A}^{z} dz' \, (\widetilde{\phi}^2 - \phi^2)^{\frac{1}{2}} \\ &= \pm\sqrt{\gamma_a h}\; S_A \,\widetilde{\phi} \int_0^{\nu} d\nu' (dz/d\nu') \cos\nu' .\end{aligned} \tag{2.18.19}$$

The Abel transform of $A(\widetilde{S})$ yields

$$\frac{z - z_A}{h} = \pm\left(\frac{1}{\sqrt{2}}\phi \mp \frac{1}{6}\phi^2 + \frac{1}{18\sqrt{2}}\phi^3\right) \tag{2.18.20}$$

and hence $z(\nu)$. The result is

$$\delta A^{\pm}(\nu, \widetilde{\phi}) = T_0^{\pm}\,[F_1(\nu)\,\widetilde{\phi}^2 \mp F_2(\nu)\,\widetilde{\phi}^3 + F_3(\nu)\,\widetilde{\phi}^4], \tag{2.18.21}$$

where

$$\begin{aligned}F_1(\nu) &= \pi^{-1}(\nu + \cos\nu\ \sin\nu), \\ F_2(\nu) &= (2\sqrt{2}/9\pi)(1 - \cos^3\nu), \\ F_3(\nu) &= (1/24\pi)(\nu - \cos\nu\ \sin\nu + 2\cos\nu\ \sin^3\nu).\end{aligned}$$

With δA given, δR and δT are computed according to (2.7.2). Because ν and $\widetilde{\phi}$ are both functions of $\widetilde{S}$, we use the notation

$$\frac{\partial}{\partial\widetilde{S}} = -\frac{\widetilde{S}}{\gamma_a h\,\widetilde{\phi}\,S_A^2}\left(\frac{\partial}{\partial\widetilde{\phi}} - \frac{\tan\nu}{\widetilde{\phi}}\frac{\partial}{\partial\nu}\right). \tag{2.18.22}$$

The result is

$$\delta R^{\pm}(\nu, \widetilde{\phi}) = R_0^{\pm} \, [F_4(\nu) \mp F_5(\nu) \, \widetilde{\phi} + F_6(\nu) \, \widetilde{\phi}^2] , \tag{2.18.23}$$

where

$$\begin{aligned} F_4(\nu) &= \pi^{-1} \nu , \\ F_5(\nu) &= (\sqrt{2}/3\pi)(1 - \cos \nu) , \\ F_6(\nu) &= (1/12\pi)(\nu - \cos \nu \, \sin \nu) ; \end{aligned}$$

and

$$\begin{aligned} \delta T^{\pm} &= \delta A^{\pm} + \widetilde{S} \, \delta R^{\pm} = \delta A^{\pm} - (S_A - \widetilde{S}) \, \delta R^{\pm} + S_A \, \delta R^{\pm} \\ &= S_A \, \delta R^{\pm} + T_0^{\pm} \, [F_7 \, \widetilde{\phi}^2 \mp F_8 \, \widetilde{\phi}^3 + F_9 \, \widetilde{\phi}^4] , \end{aligned} \tag{2.18.24}$$

where

$$\begin{aligned} F_7 &= F_1 - F_4 = (1/\pi) \cos \nu \, \sin \nu , \\ F_8 &= F_2 - F_5 = (\sqrt{2} / 9\pi) \, (1 - 3 \cos \nu + 2 \cos^3 \nu) , \\ F_9 &= F_3 - F_6 = (1/24\pi) \, (-\nu + \cos \nu \, \sin \nu + 2 \cos \nu \, \sin^3 \nu) . \end{aligned}$$

Time fronts. Time fronts (fig. 2.4) are constructed by solving the quadratic equation

$$r(\nu, \widetilde{\phi}) = n^+ R^+(\widetilde{\phi}) + n^- R^-(\widetilde{\phi}) - \operatorname{sgn} \delta R^+(\nu, \widetilde{\phi}) \tag{2.18.25}$$

for $\widetilde{\phi}(\nu)$, with $\sin \nu = \phi/\widetilde{\phi}$, $-\pi/2 \le \nu \le \pi/2$, and sgn = +/− for upward/downward launch angles. The plot $\tau(z)$ then follows from eliminating ν between two equations: equation (2.18.20) for $z(\phi) = z(\widetilde{\phi} \sin \nu)$, and $\tau(\widetilde{\phi}, \nu)$, as given by

$$\begin{aligned} \frac{\tau(\nu, \widetilde{\phi}) - \tau_{AX}}{T_0^{\pm}} &= \frac{2\sqrt{2}}{9\pi} \, (n^+ - n^-) \widetilde{\phi}^3 - \frac{1}{24} \, (n^+ + n^-) \widetilde{\phi}^4 \\ &\quad - \operatorname{sgn} [F_7 \widetilde{\phi}^2 - F_8 \widetilde{\phi}^3 + F_9 \widetilde{\phi}^4] . \end{aligned} \tag{2.18.26}$$

Dynamic modes. For a canonical ocean, the dynamic modes (Rossby waves) can be written in terms of Bessel functions. The starting point is the geostrophic balance on a β-plane, with a Coriolis frequency $f + \beta y$ (*e.g.*, Pedlosky, 1987, chap. 6). Over a flat bottom with a rigid lid (the barotropic component is therefore suppressed), the northward velocity component can be written

$$v\,(x, y, z, t) = \exp\left[i\,(\kappa_x x + \kappa_y y - \omega t)\right] V\,(z)\,,$$

with

$$\frac{d^2V}{dz^2} - \frac{dN^2/dz}{N^2 - \omega^2}\,\frac{dV}{dz} + \frac{N^2 - \omega^2}{f^2 - \omega^2}\,\lambda^2 V = 0\,, \tag{2.18.27}$$

where

$$\lambda^2 = -\left(\frac{\beta\kappa_x}{\omega} + \kappa_x^2 + \kappa_y^2\right). \tag{2.18.28}$$

For the case $N = N_0\,e^{z/h}$, to first order in the small parameter σ/N, (2.18.27) becomes

$$\xi^2\,\frac{d^2V}{d\xi^2} - \xi\,\frac{dV}{d\,\xi} + \xi^2 V = 0, \quad \xi(z) = h\,\lambda\,\frac{N(z)}{(f^2 - \omega^2)^{\frac{1}{2}}}, \tag{2.18.29}$$

with the solution $V = \xi\,C_1(\xi)$, where $C_1(\xi)$ denotes any linear combination of the Bessel functions J_1, Y_1. The vertical displacement η is proportional to $N^{-2}\,dV/dz \sim \xi^{-1}\,dV/d\xi$. Using $dV/d\xi = \xi\,C_0(\xi)$,

$$\eta = a\,J_0(\xi) + b\,Y_0(\xi)\,. \tag{2.18.30}$$

Let $\xi_0 = \xi(0)$ and $\xi_B = \xi(-H)$ designate surface and bottom conditions, so that $\xi_B = \xi_0\,e^{-H/h}$ for the exponential $N(z)$. The bottom boundary condition $\eta_B = 0$ is satisified by choosing a, b such that

$$\eta = c\left[J_0(\xi) - \frac{J_0\,(\xi_B)}{Y_0\,(\xi_B)}\,Y_0(\xi)\right]. \tag{2.18.31}$$

The surface boundary condition $\eta_0 = 0$ requires that

$$J_0(\xi_0)\;Y_0(\xi_B) - Y_0(\xi_0)\;J_0(\xi_B) = 0, \quad \xi_B = \xi_0\,e^{-H/h}\,. \tag{2.18.32}$$

For the case $H = 4$ km, $h = 1$ km, the first roots of (2.18.32) are

$$\xi_0(j) = 2.872,\ 6.117,\ 9.351,\ 12,576, \quad j = 1, 2, 3, 4. \tag{2.18.33}$$

The mode functions $\eta_j(z)$ are now obtained from (2.18.31) by setting

$\xi_j(z) = \xi_0(j)\, e^{z/h}$. The dispersion is found from (2.18.28) with $\lambda_j = \xi_0(j)\, h^{-1} \sqrt{f^2 - \omega^2}/N_0$. The asymptotic solution for large arguments is

$$\eta_j(z) = c_j\, N^{-\frac{1}{2}} \sin\left(j\pi \, \frac{N_S - N(z)}{N_S - N_B} \right). \qquad (2.18.34)$$

Perturbations. From (2.18.20) we have

$$\frac{z - z_A}{h} = \pm \frac{\phi\, f_c^{\pm}(\phi)}{\sqrt{2}}, \quad f_c^{\pm}(\phi) = 1 \mp \frac{\sqrt{2}}{6}\, \phi + \frac{1}{18}\, \phi^2, \qquad (2.18.35)$$

for $z \gtrless z_A$, with

$$\phi = \widetilde{\phi} \sin \nu, \quad 0 \le \nu \le \tfrac{1}{2}\pi,$$

thus yielding $z^{\pm}(\widetilde{\phi}, \nu)$, and so the perturbation can be written $\Delta S(\widetilde{\phi}, \nu)$. The general weights $W(z)$ in (2.14.32) can be evaluated for the temperate profile:

$$W^0(z) = \frac{\sqrt{2}}{\pi} \, \frac{1}{h\widetilde{\phi}\left(1 + \frac{1}{12}\, \widetilde{\phi}^2\right)} \, \frac{1}{\cos \nu},$$

$$W^I(z) = W^0 \left[1 + \tfrac{1}{12}\, \widetilde{\phi}^2\, F(\widetilde{\phi}) \right.$$

$$\left. + \frac{\sqrt{2}\, f_b^{\pm}(\phi)}{\pi\, h\, \widetilde{\phi}\, f_a^{\pm}(\phi)} \quad F(\widetilde{\phi}) \quad \frac{1 - 2\sin^2 \nu}{\cos \nu} \right], \qquad (2.18.36)$$

$$W^{II}(z) = \pm \frac{f_a^{\pm}(\phi)}{2\pi} \quad F(\widetilde{\phi}) \quad \tan \nu \; (1 - 2\sin^2 \nu),$$

where

$$f_a^{\pm}(\phi) = 1 \mp \tfrac{\sqrt{2}}{3}\, \phi + \tfrac{1}{6}\, \phi^2, \quad f_b^{\pm}(\phi) = 1 \mp \tfrac{\sqrt{2}}{2}\, \phi + \tfrac{1}{3}\, \phi^2,$$

$$f_c^{\pm}(\phi) = 1 \mp \tfrac{\sqrt{2}}{6}\, \phi + \tfrac{1}{18}\, \phi^2, \quad F(\widetilde{\phi}) = \frac{1 + \frac{1}{24}\, \widetilde{\phi}^2}{(1 + \frac{1}{12}\, \widetilde{\phi}^2)^2}. \qquad (2.18.37)$$

Then, with

$$dz = 2^{-\frac{1}{2}}\, h\, \widetilde{\phi}\, f_a^{\pm}(\phi) \cos \nu \, d\nu,$$

it can be demonstrated that $\int_{\widetilde{z}^-}^{\widetilde{z}^+} dz\, W^0 = 1$ and $\int_{\widetilde{z}^-}^{\widetilde{z}^+} dz\, W^I = 1$, using the appropriate signs for $f^{\pm}(\phi)$ according to $z \gtrless z_A$. So again $(\Delta s_g)_{\text{ray}} = (\Delta s_g)_{\text{mode}} = \Delta S$ for $\Delta S' = 0$. Using the definitions (2.18.37),

$$(V^0)^{\pm} = \frac{1}{\pi(1 + \frac{1}{12}\,\widetilde{\phi}^2)}\; f_a^{\pm}(\phi),$$

$$(V^I)^{\pm} = (V^0)^{\pm}\left[\,1 + \tfrac{1}{12}\,\widetilde{\phi}^2\, F(\widetilde{\phi})\,\right] + \pi^{-1}\, F(\widetilde{\phi})\; f_b^{\pm}(\phi)\,(1 - 2\,\sin^2\nu),$$

$$(V^{II})^{\pm} = \pm\frac{1}{2\sqrt{2}\,\pi}\, h\,\widetilde{\phi}\, F(\widetilde{\phi})\,(f_a^{\pm})^2\; \sin\nu\,(1 - 2\,\sin^2\nu).$$

Again, $\int_0^{\pi/2} V\, d\nu = 1$ for $V^0 = (V^0)^+ + (V^0)^-$, and similarly for V^I; further, $\int_0^{\pi/2} V^{II}\, d\nu \ll 1$. The E_{ij} matrix is conveniently computed in ν-space (2.15.12), as there are no (integrable) infinities.

CHAPTER 3

CURRENTS

An acoustic pulse propagating with a current travels faster than one propagating against the current. Ocean currents are typically of order 10 cm/s rms or less, except in strong western boundary currents such as the Gulf Stream, whereas ocean sound-speed perturbations are typically of order 5 m/s rms. Travel-time perturbations due to ocean currents are correspondingly one to two orders of magnitude smaller than travel-time signals due to sound-speed perturbations. It is nonetheless possible to measure ocean currents using acoustic techniques, by *differencing* the travel times of signals traveling in opposite directions. As was briefly summarized in chapter 1, travel-time signals due to sound-speed perturbations cancel in the difference travel time, leaving only the effect of currents.

Section 3.1 describes ray theory as applied to moving media. The presence of a current introduces anisotropy. Perturbation expressions for the sum and difference of reciprocal travel times are then presented in section 3.2. When the flow is in geostrophic balance, the current and sound-speed fields are related. In section 3.3, quantitative estimates of their relative sizes are made, confirming the rough orders of magnitude cited earlier.

Using a horizontal-slice approximation, section 3.4 shows that the averaging properties of acoustic travel times make acoustic techniques uniquely suited for measuring the fluid circulation by integrating around a closed contour. By Stokes's theorem, the circulation is equivalent to the areal-average relative vorticity. This result is then generalized to show that differential travel times are sensitive to the solenoidal component of the flow, from which relative vorticity can be mapped, but are not functions of the irrotational component of the flow between the transceivers, which is needed to map the horizontal flow divergence. Even though the divergence is of great oceanographic importance, differential travel times unfortunately are not particularly well suited for measuring it.

To this point we have implicitly assumed that acoustic pulses traveling with and against a current follow the same ray path. Section 3.5 discusses the nonreciprocity of acoustic propagation in a moving medium. (This problem is closely related to the discussion of travel-time nonlinearity in chapter 2.) In general, errors due to nonreciprocity are small, although exceptions are possible.

Finally, measured receptions from oppositely traveling signals are shown in section 3.6, together with some time series of differential travel times. The differential travel times are found to contain tidal signals consistent with empirical numerical barotropic-tide models and with independent tidal measurements. The principal limitations on precise one-way (and sum) travel-time measurements are travel-time fluctuations due to temperature (sound-speed) fluctuations associated with internal waves. These travel-time fluctuations have been found to largely cancel in the differential travel times, demonstrating that the ray paths are nearly reciprocal out to 1-Mm ranges, at least for the cases for which measurements are available. This result is in accord with the analysis in section 3.5. The reduced levels of internal-wave-induced noise in differential travel times significantly improve the precision of current estimates made from them.

3.1. Ray Theory in an Inhomogeneous Moving Medium

In a motionless medium, the direction of the wave-front normal (*i.e.*, the normal to surfaces of constant phase) and the direction in which acoustic energy propagates (*i.e.*, the vector tangent to the ray path), are identical. In a moving medium, the presence of a current introduces anisotropy into the problem, so that the directions of the wave-front normal and the tangent to the ray differ. The ray velocity depends not only on the local sound-speed and water velocity but also on the local direction of the ray.

Consider a fluid moving with vector velocity $\mathbf{v}$. In a coordinate system moving with the fluid, each wave-front expands normal to itself at the local sound-speed C, in accord with Huygens's principle; that is, each point on the wave front moves with velocity $C(\mathbf{r})\widehat{\mathbf{n}}$, where $C(\mathbf{r})$ is the local sound-speed and $\widehat{\mathbf{n}}$ is the unit vector in the direction normal to the wave front. Transforming to a coordinate system at rest, the velocity of each point on the wave front becomes $C(\mathbf{r})\,\widehat{\mathbf{n}} + \mathbf{v}$ (fig. 3.1). A point $\mathbf{r}(t)$ lying on the wave front at some initial time will therefore always lie on the wave front if its velocity is (Pierce, 1989)

$$\frac{d\mathbf{r}}{dt} = \mathbf{v}(\mathbf{r}) + C(\mathbf{r})\,\widehat{\mathbf{n}}(\mathbf{r}) \equiv \mathbf{v}_{\text{ray}}\,. \tag{3.1.1}$$

The line described in space by $\mathbf{r}(t)$ as a function of time is the ray path in a moving medium and is the path followed by an acoustic pulse. This equation is the extension of Huygens's principle to a moving medium.

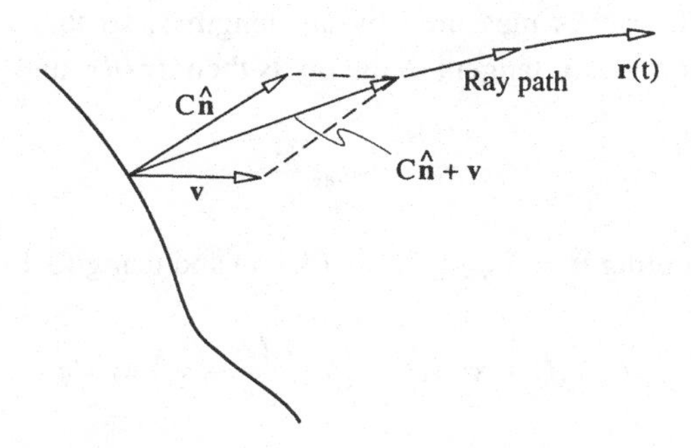

Fig. 3.1. Concept of a ray path in a moving medium. The point $\mathbf{r}(t)$ moves with velocity $C\widehat{\mathbf{n}} + \mathbf{v}$ such that it is always on the wave front and in so doing traces out a ray path.

We now proceed to derive explicit equations for the ray trajectories in a moving medium. The first step is to obtain the analogue to the eikonal equation (2.3.4) derived in chapter 2 for a motionless medium,

$$|\nabla\phi|^2 = (S/S_0)^2 = (C_0/C)^2 \ . \tag{3.1.2}$$

This equation specifies the evolution of the wave fronts, given by the phase $\phi =$ constant, in a motionless medium. In chapter 2, the eikonal equation was derived directly from the wave equation, although it can be obtained using Huygens's principle. The eikonal equation in a moving medium can also be derived directly from the wave equation (*e.g.*, Blokhintsev, 1946, 1952, 1956; Heller, 1953; Keller, 1954; Thompson, 1972), but for our purposes a simple heuristic derivation using Huygens's principle suffices (Kornhauser, 1953; Pierce, 1989). The velocity perpendicular to surfaces of constant phase (*i.e.*, in the direction of $\nabla\phi$) appears in (3.1.2). By definition, $\widehat{\mathbf{n}}$ is perpendicular to surfaces of constant phase, so the speed of the wave front normal to itself in a moving medium is the dot product of the right side of (3.1.1) with $\widehat{\mathbf{n}}$, giving, for the normal velocity,

$$C + \mathbf{v} \cdot \widehat{\mathbf{n}} \, . \tag{3.1.3}$$

When $\mathbf{v} = 0$, the velocity normal to the wave front reduces to C, as expected. The eikonal equation in a motionless medium can then be simply generalized to a moving medium by replacing C by $C + \mathbf{v} \cdot \widehat{\mathbf{n}}$ in (3.1.2) to obtain

$$|\nabla\phi|^2 = (C_0/\,(C + \mathbf{v} \cdot \widehat{\mathbf{n}}\,))^2 \ . \tag{3.1.4}$$

Explicit ray-trajectory equations for a moving medium can be derived using (3.1.1) and (3.1.4), following a procedure analogous to that used in section 2.3.

Position along the ray is measured by arc length s, so that $\mathbf{r}(s)$ is the ray trajectory. The unit vector tangent to the ray is then $d\mathbf{r}/ds$, and

$$\mathbf{v}_{\text{ray}} = v_{\text{ray}} \frac{d\mathbf{r}}{ds} \tag{3.1.5}$$

in (3.1.1). Substituting $\widehat{\mathbf{n}} = \nabla\phi / |\nabla\phi|$ in (3.1.1) and using (3.1.4) to eliminate $|\nabla\phi|$ gives

$$(C_0/C)\,(C + \mathbf{v}\cdot\widehat{\mathbf{n}})^{-1} \left(v_{\text{ray}} \frac{d\mathbf{r}}{ds} - \mathbf{v} \right) = \nabla\phi\,. \tag{3.1.6}$$

Define

$$N \equiv (C_0/C)\,(C + \mathbf{v}\cdot\widehat{\mathbf{n}})^{-1}\, v_{\text{ray}} \tag{3.1.7}$$

and

$$\mathbf{V} \equiv (C_0/C)\,(C + \mathbf{v}\cdot\widehat{\mathbf{n}})^{-1}\, \mathbf{v}\,. \tag{3.1.8}$$

Equation (3.1.6) is then

$$\nabla\phi = N\frac{d\mathbf{r}}{ds} - \mathbf{V}\,. \tag{3.1.9}$$

Differentiating with respect to arc length s,

$$\frac{d}{ds}\nabla\phi = \frac{d}{ds}\left(N\frac{d\mathbf{r}}{ds} - \mathbf{V} \right), \tag{3.1.10}$$

and, rearranging,

$$\begin{aligned} \frac{d}{ds}\left(N\mathbf{r}'\right) &= \frac{d}{ds}\left(\mathbf{V} + \nabla\phi\right) \\ &= \left(\mathbf{r}'\cdot\nabla\right)\left(\mathbf{V} + \nabla\phi\right) + \mathbf{r}''\cdot\nabla'\mathbf{V}\,, \end{aligned} \tag{3.1.11}$$

where $\mathbf{r}' = d\mathbf{r}/ds$, and the gradient operator ∇' operates on the direction components $\mathbf{r}' = \left(x_i', y_i', z_i'\right)$. The last term arises because $\mathbf{V} = \mathbf{V}(\mathbf{r}, \mathbf{r}')$ is an explicit function of both the ray path $\mathbf{r}$ and the ray direction $\mathbf{r}'$ (Ugínčius, 1965, 1972). Solving (3.1.9) for $\mathbf{r}'$ and substituting in the first term on the right-hand side gives

$$\begin{aligned} \frac{d}{ds}\left(N\mathbf{r}'\right) &= \frac{1}{N}\left[(\nabla\phi + \mathbf{V})\cdot\nabla\right](\nabla\phi + \mathbf{V}) + \mathbf{r}''\cdot\nabla'\mathbf{V} \\ &= \frac{1}{N}\left\{\frac{1}{2}\nabla\left(|\nabla\phi + \mathbf{V}|^2\right) - (\nabla\phi + \mathbf{V})\times\left[\nabla\times(\nabla\phi + \mathbf{V})\right]\right\} \\ &\quad + \mathbf{r}''\cdot\nabla'\mathbf{V}\,, \end{aligned} \tag{3.1.12}$$

where we have used the identity $\mathbf{u} \cdot \nabla \mathbf{u} = (1/2)\nabla \mathbf{u} \cdot \mathbf{u} - \mathbf{u} \times (\nabla \times \mathbf{u})$. Because $|\nabla \phi + \mathbf{V}|^2 = N^2$, and because the curl of a gradient is identically zero, (3.1.12) simplifies to

$$\frac{d}{ds}(N\mathbf{r}') - \mathbf{r}'' \cdot \nabla' \mathbf{V} + \mathbf{r}' \times (\nabla \times \mathbf{V}) = \nabla N . \tag{3.1.13}$$

Equation (3.1.13) is the generalization to a moving medium of the ray-trajectory equation (2.3.9). The Cartesian components of (3.1.13) constitute a set of three second-order differential equations that can be integrated numerically to yield the equations of the ray paths.

Ray travel times are then given by

$$\tau_i = \int_{\Gamma_i} \frac{ds}{v_{\text{ray}}} \tag{3.1.14}$$

where Γ_i is the trajectory of ray i. The results to this point are quite general. No assumptions have been made about the relative sizes of $|\mathbf{v}|$ and C, about the stratification of the medium, or about the direction of $\mathbf{v}$ with respect to the direction of propagation.

Considerable simplification is possible for propagation in the ocean, because $|\mathbf{v}| \ll C$. The term $\mathbf{r}'' \cdot \nabla' \mathbf{V}$ in (3.1.13) can be shown to be zero to first order in $|\mathbf{v}|/C$ (Uginčius, 1972). Furthermore, (3.1.7) and (3.1.8) simplify to

$$N = C_0/C, \tag{3.1.15}$$

and

$$\mathbf{V} = \frac{C_0}{C}\frac{\mathbf{v}}{C}\left[1 - \frac{\mathbf{v}}{C} \cdot \frac{d\mathbf{r}}{ds}\right] . \tag{3.1.16}$$

A number of investigators have derived solutions to the ray equations in a variety of specific cases of oceanographic interest (*e.g.*, Hayre and Tripathi, 1967; Stallworth and Jacobson, 1970, 1972*a,b*; Franchi and Jacobson, 1972, 1973a,b; Widfeldt and Jacobson, 1976; Hamilton *et al.*, 1977, 1980; Newhall *et al.*, 1977).

The expression for the travel time can also be simplified when $|\mathbf{v}|/C \ll 1$. From (3.1.1) and (3.1.5),

$$C\widehat{\mathbf{n}} = v_{\text{ray}}\mathbf{r}' - \mathbf{v} . \tag{3.1.17}$$

The magnitude of the ray velocity therefore satisfies the quadratic equation

$$v_{\text{ray}}^2 - 2v_{\text{ray}}\mathbf{v} \cdot \mathbf{r}' - \left(C^2 - v^2\right) = 0 \tag{3.1.18}$$

whose positive solution, for $C^2 > v^2$, is

$$v_{\text{ray}} = \mathbf{v} \cdot \mathbf{r}' + \left[C^2 - v^2 + (\mathbf{v} \cdot \mathbf{r}')^2\right]^{1/2} . \tag{3.1.19}$$

Substituting into (3.1.14),

$$\tau_i = \int_{\Gamma_i} \frac{ds}{\mathbf{v} \cdot \mathbf{r}' + \left[C^2 - v^2 + (\mathbf{v} \cdot \mathbf{r}')^2\right]^{1/2}} . \tag{3.1.20}$$

To first order in $|\mathbf{v}|/C$ this is

$$\tau_i = \int_{\Gamma_i} \frac{ds}{C(\mathbf{r}) + \mathbf{v}(\mathbf{r}) \cdot \mathbf{r}'} \tag{3.1.21}$$

($\mathbf{r}'$ is an implicit function of $\mathbf{r}$, through its dependence on the ray trajectory). This expression has the virtue that knowledge of the direction of the wave-front normal is not required. All that is needed is the ray trajectory. Equation (3.1.21) is convenient when deriving expressions for the travel-time perturbations due to sound-speed and current perturbations. The sound-slowness, $S = 1/C$, which was used extensively in chapter 2, is not as convenient a notation in a moving medium.

3.2. Travel-Time Perturbations

As was the case in the absence of a current, this problem is nonlinear, because the ray path Γ_i depends on $C(\mathbf{r})$ and $\mathbf{v}(\mathbf{r})$. We linearize again by setting

$$C(\mathbf{r}) = C(\mathbf{r}, -) + \Delta C(\mathbf{r}) , \tag{3.2.1}$$

$$\mathbf{v}(\mathbf{r}) = \mathbf{v}(\mathbf{r}, -) + \Delta \mathbf{v}(\mathbf{r}) , \tag{3.2.2}$$

where $C(\mathbf{r}, -)$, $\mathbf{v}(\mathbf{r}, -)$ are known reference states. The travel-time perturbation is

$$\Delta\tau_i = \int_{\Gamma_i} \frac{ds}{C(\mathbf{r}) + \mathbf{v}(\mathbf{r}) \cdot \mathbf{r}'} - \int_{\Gamma_i(-)} \frac{ds}{C(\mathbf{r}, -) + \mathbf{v}(\mathbf{r}, -) \cdot \mathbf{r}'(-)} \tag{3.2.3}$$

where $\Gamma_i(-)$, $\mathbf{r}'(-)$ are the ray path and ray tangent vector in the unperturbed fields $C(\mathbf{r}, -)$, $\mathbf{v}(\mathbf{r}, -)$. Normally, $\Delta C(\mathbf{r}) \ll C(\mathbf{r}, -)$, but $|\Delta \mathbf{v}(\mathbf{r})| > |\mathbf{v}(\mathbf{r}, -)|$, although $|\Delta \mathbf{v}(\mathbf{r})|$ is always small compared with $C(\mathbf{r}, -)$. Away from strong western boundary currents such as the Gulf Stream, most of the kinetic energy in the ocean is not in the mean, but in mesoscale fluctuations with periods of weeks to months and spatial scales of 50–150 km. We therefore set $\mathbf{v}(\mathbf{r}, -) \equiv 0$, $\Delta\mathbf{v}(\mathbf{r}) \equiv \mathbf{v}(\mathbf{r})$. Then, to first order in $\Delta C/C$ and $|\mathbf{v}|/C$,

$$\Delta\tau_i = -\int_{\Gamma_i(-)} \frac{(\Delta C(\mathbf{r}) + \mathbf{v}(\mathbf{r}) \cdot \mathbf{r}'(-))\, ds}{C^2(\mathbf{r}, -)}. \tag{3.2.4}$$

By interchanging the positions of source and receiver, the sign of $\mathbf{r}'$ can be reversed in a fixed coordinate system. The sum of the reciprocal travel-time perturbations

$$\Delta s_i = \tfrac{1}{2}\,(\Delta\tau_i^+ + \Delta\tau_i^-) = -\int_{\Gamma_i(-)} \frac{ds}{C^2(\mathbf{r}, -)}\, \Delta C(\mathbf{r}) \tag{3.2.5}$$

depends only on ΔC, whereas the difference

$$\Delta d_i = \tfrac{1}{2}\,(\Delta\tau_i^+ - \Delta\tau_i^-) = -\int_{\Gamma_i(-)} \frac{ds}{C^2(\mathbf{r}, -)}\, \mathbf{v}(\mathbf{r}) \cdot \mathbf{r}'(-) \tag{3.2.6}$$

depends only on $\mathbf{v}(\mathbf{r})$. [For $\mathbf{v}(\mathbf{r}, -) \equiv 0,\ d_i = \Delta d_i$.]

Measurements of sum and difference travel times provide powerful tomographic tools. First, they separate the effects of ΔC and $\mathbf{v}$. This separation leads to a mildly improved determination of ΔC. It is crucial for measuring $\mathbf{v}$, however, because $|\mathbf{v}|$ is generally much smaller than ΔC. Stallworth (1973) was perhaps the first to suggest using differential travel times to measure large-scale ocean currents. Equation (3.2.6) was first applied in the ocean by Worcester (1977*a,b*) on a 25 km scale. Second, when the $\mathbf{v}$ and ΔC fields are related, as in geostrophic flow (section 3.3), combined inversions provide improved estimates for both fields (section 6.5). Finally, the covariance of $s(t)$ and $d(t)$ is related to oceanographically interesting fluxes; $s(t)$ is a measure of the vertical displacement $\zeta(t)$ along the transmission path, and hence of the temperature perturbation ΔT; $d\zeta/dt$ is the vertical velocity $w(t)$. The in-phase component of the cross-spectrum of $s(t)$ and $d(t)$ is related to the heat flux and the out-of-phase component is related to the momentum flux. Attempts to evaluate these quantities have not been successful (Munk *et al.*, 1981*a*; Munk and Wunsch, 1982*a*; Munk, 1986).

3.3. Geostrophic Flow

Large-scale, steady ocean currents tend to be in geostrophic balance, with the horizontal pressure-gradient force balanced by the Coriolis effect. The flow is perpendicular to the pressure gradient, and the sound-speed and current fields are related by

$$\rho f u = \Delta p / L_H , \tag{3.3.1}$$

where ρ is density, $f = 2\Omega_e \sin\theta \approx 10^{-4}\,\mathrm{s}^{-1}$ (at midlatitudes) is the Coriolis parameter, $\Omega_e = 7.29 \times 10^{-5}\,\mathrm{s}^{-1}$ is the angular velocity of the earth, θ is latitude, $\Delta p = g\,\Delta\rho\,L_V$ is the pressure perturbation, and L_H and L_V are the horizontal and vertical scales of motion. Using (2.2.7) to relate $\Delta\rho$ to ΔC,

$$\left|\frac{u}{\Delta C}\right| = \left|\left(\frac{g}{Cf}\right)\left(\frac{\Delta\rho/\rho}{\Delta C/C}\right)\frac{L_V}{L_H}\right| \approx (1.3)\,\frac{L_V}{L_H}. \tag{3.3.2}$$

For geostrophic flow, the ratio of the current to the sound-speed perturbation is nearly the same as the ratio of the vertical and horizontal length scales of the flow. For mesoscale perturbations, $L_V = 1$ km, $L_H = 10^2$ km; for the gyre-scale, $L_V = 1$ km, $L_H = 10^3$ to 10^4 km. The effect of geostrophic currents on travel times is therefore of the order of 10^{-2} to 10^{-4} of the effect of the corresponding sound-speed perturbations.

3.4. Circulation, Vorticity, and Divergence

Rossby (1975) first suggested using reciprocal acoustic transmissions to measure the relative vorticity, $\zeta \equiv \nabla \times \mathbf{v}$. The difference in travel times (counter-clockwise minus clockwise) around an array of transceivers directly measures the circulation along the periphery, or, by Stokes's theorem, the integrated relative vorticity within the array. It is difficult to overestimate the importance of the ability to obtain direct measurements of vorticity. Almost all theoretical discussions of the large-scale oceanic circulation and its variability are formulated in terms of the evolution of the vorticity field. In a rotating fluid, angular momentum conservation, in the guise of the fluid vorticity, plays a fundamental role in understanding the flow field (Pedlosky, 1987). In practice, oceanographers who must treat a stratified, rotating fluid tend to work with a derived quantity called the "potential vorticity," for which conservation laws are more readily formulated than for the vorticity itself. For example, in a fluid represented by a sequence of nearly homogeneous density layers, the potential vorticity of layer i can be written as

$$\zeta_i^{\mathrm{pot}} = \frac{(\nabla \times \mathbf{v}_i)_z + f}{h_i} \tag{3.4.1}$$

where h_i is the thickness of the ith layer,

$$(\nabla \times \mathbf{v}_i)_z = \left(\frac{\partial v}{\partial x} - \frac{\partial u}{\partial y} \right)$$

is the vertical component of the relative vorticity, and f/h_i is the planetary vorticity. Theories of the ocean circulation suggest that the relative vorticity should be negligible over most of the ocean interior, but should become as important as the planetary vorticity in the boundary-current areas and other regions with strong currents. If one seeks observational confirmation of this deduction, and other theoretical predictions such as the sign, fluxes, and flux divergences of the potential vorticity, one must be able to directly measure the different terms in (3.4.1). Measurement of the relative vorticity by current meters or other devices has proved extremely difficult. But velocity tomography around a closed circuit produces a direct estimate of the relative vorticity as the corresponding circulation as a function of depth. One-way tomography is capable of producing estimates of the thickness (from the temperature inversions) around the same closed circuit. Thus a single tomographic triangle is capable of providing estimates of both the relative-vorticity and planetary-vorticity contributions to the potential vorticity as functions of depth.

For simplicity, we specialize to the case of propagation in a horizontal plane. (In general, one must invert the differential travel times to obtain the range-averaged current profile on each leg of the array and combine currents from the same depth on each leg.) The difference in travel times around an array of m transceivers enclosing an area A can then be written

$$\tau_{123\cdots m1} - \tau_{1m\cdots 321} = -\frac{1}{C^2} \oint \mathbf{v}(\mathbf{r}) \cdot \mathbf{r}' \, ds = -\frac{1}{C^2} \iint_A [\nabla \times \mathbf{v}(\mathbf{r})] \cdot \mathbf{n} \, da, \tag{3.4.2}$$

where the reference sound-speed $C(\mathbf{r}, -)$ has been chosen to be a constant C, and $\mathbf{n}$ is the unit normal to the surface.

An array of m transceivers contains $m(m-1)(m-2)/3!$ triangles. In addition to the circulation around the entire array, the circulation for each triangle can be calculated. Longuet-Higgins (1982) showed that only

$$(m - 1)(m - 2)/2$$

of the measurements are independent, however. With $m = 4$, for example, there are four triangles, of which three are independent. This number provides sufficient information to determine three coefficients in a Taylor expansion:

$$\zeta = a + b\frac{\partial \zeta}{\partial x} + c\frac{\partial \zeta}{\partial y}. \tag{3.4.3}$$

In general, one can estimate the vorticity and its derivatives up to order $m - 3$. Five transceivers therefore yield estimates of the vorticity ζ, its gradient $\nabla\zeta$, and its Laplacian $\nabla^2\zeta$.

For orientation, the magnitude of the gyre-scale relative vorticity can be estimated under the assumption that the flow is in Sverdrup balance (Munk and Wunsch, 1982*a*),

$$\rho\beta v h = \nabla \times \boldsymbol{\tau}, \qquad (3.4.4)$$

where $\beta = \partial f/\partial y \approx f/r_{\text{earth}}$, f is the Coriolis parameter, y is northward, h is the layer depth, and $\boldsymbol{\tau}$ is the wind stress. For zonal winds with wind stress varying from $-\tau_0$ in the trade winds to $+\tau_0$ in the midlatitude westerlies,

$$\boldsymbol{\tau} = \tau_x(y), \qquad (3.4.5)$$

and $\partial v/\partial x = 0$. Using the fact that the vertically integrated geostrophic flow is nondivergent, $\partial u/\partial x = -\partial v/\partial y$, and integrating over a distance x from the eastern boundary to obtain u will give, for the vertical component of vorticity,

$$\zeta_{\text{gyre}} = \frac{\partial v}{\partial x} - \frac{\partial u}{\partial y} = \frac{x}{\rho\beta h}\frac{\partial^2}{\partial y^2}(\nabla \times \tau_x) \approx \frac{8 x r_{\text{earth}} \tau_0}{\rho f h L^3}. \qquad (3.4.6)$$

Then

$$|\zeta_{\text{gyre}}| \approx 10^{-8}\,\text{s}^{-1} \approx 10^{-4} f, \qquad (3.4.7)$$

where we have used $x = 3\,\text{Mm}$, $\tau_0 = 1\,\text{dyn/cm}^2$, $r_{\text{earth}} = 6\,\text{Mm}$, $h = 1\,\text{km}$, and $L_{\text{gyre}} = 2500\,\text{km}$, the distance between the trade winds and the midlatitude westerlies. The integrated vorticity over a square array that is $L = 1$ Mm on a side is then $L^2\zeta_{\text{gyre}} \approx 10^4\,\text{m}^2/\text{s}$, and

$$\Delta d_i \approx -\frac{1}{C^2} L^2 \zeta_{\text{gyre}} \approx -4 \times 10^{-3}\,\text{s}.$$

For comparison, the mean vorticity of an eddy in solid-body rotation with a velocity $v = 10$ cm/s at a radius r_{eddy} of 50 km is

$$\zeta_{\text{eddy}} = 2v/r_{\text{eddy}} \approx 4 \times 10^{-6}\,\text{s}^{-1} \approx 10^{-1} f, \qquad (3.4.8)$$

as compared with the gyre-scale vorticity $10^{-4} f$. The integrated vorticity $\pi r_{\text{eddy}}^2 \zeta_{\text{eddy}} \approx 3 \times 10^4\,\text{m}^2/\text{s}$ is comparable to the integrated gyre-scale vorticity in a 1-Mm square. There will be significant fluctuations as mesoscale eddies enter and exit the array area.

Divergence. Acoustic measurements naturally produce the areal-average relative vorticity. But the large-scale horizontal divergence of the flow equals $-\partial w/\partial z$ and is of great oceanographic importance as a measure of vertical circulation (Munk and Wunsch, 1982*a*).[1] To compute the circulation around an enclosed region requires the component of flow parallel to the perimeter, while the divergence requires the components perpendicular to the perimeter. Differential travel times provide the former, but acoustic travel time is essentially unaffected by flow perpendicular to the ray path. One therefore suspects that acoustic techniques are not well suited for directly measuring divergence. [Cross-correlations of the amplitude or travel-time scintillations at two horizontally separated receivers provide the flow perpendicular to the path, under some conditions (Farmer and Clifford, 1986; Crawford *et al.*, 1990; Farmer and Crawford, 1991).]

We can show formally that acoustic travel times are not sensitive to the divergence of the flow in the interior of a transceiver array by decomposing the field $\mathbf{v}(\mathbf{r})$ into its irrotational and solenoidal components:

$$\mathbf{v}(\mathbf{r}) = \nabla\Phi(\mathbf{r}) + \nabla \times \boldsymbol{\Psi}(\mathbf{r}) \tag{3.4.9}$$

where $\Phi(\mathbf{r})$ and $\boldsymbol{\Psi}(\mathbf{r})$ are the scalar and vector potentials (Norton, 1988). Helmholtz's theorem states that any finite, continuous vector field vanishing at infinity can be written uniquely as (3.4.9). For the two-dimensional case, $\mathbf{v}(\mathbf{r}) = \mathbf{v}(x, y)$, only the z-component $\boldsymbol{\Psi}(\mathbf{r}) \equiv \Psi(x, y)\hat{\mathbf{z}}$ contributes, and the two scalar functions $\Phi(x, y)$ and $\Psi(x, y)$ determine $\mathbf{v}(x, y)$. Substituting in (3.2.6),

$$\Delta d_i = -\frac{1}{C^2}\int_{\Gamma_i} [\nabla\Phi(x, y) + \nabla \times \Psi(x, y)\hat{\mathbf{z}}] \cdot \mathbf{r}'\, ds, \tag{3.4.10}$$

where we have again set $C(\mathbf{r}, -)$ to a constant C. Because the line integral of the gradient of a scalar field depends only on the end points of the path, this becomes

$$\Delta d_i = -\frac{1}{C^2}[\Phi(\mathbf{r}_{\text{end}}) - \Phi(\mathbf{r}_{\text{start}})] - \frac{1}{C^2}\int_{\Gamma_i} [\nabla \times \Psi(x, y)\hat{\mathbf{z}}] \cdot \mathbf{r}'\, ds, \tag{3.4.11}$$

where $\mathbf{r}_{\text{start}}$, $\mathbf{r}_{\text{end}}$ are the beginning and end points of the ray path Γ_i. The implication is that differential travel-time data do not contain information on the

[1] In the geostrophic approximation on a β-plane, the horizontal divergence is $-\partial w/\partial z = -\beta v/f$.

irrotational component $\nabla\Phi(x, y)$ between the acoustic transceivers. Only the field $\nabla \times \Psi(x, y)\hat{\mathbf{z}}$ can be determined from the data Δd_i. Norton (1988) derived a vector projection-slice theorem showing that the solenoidal component $\nabla \times \Psi(x, y)\hat{\mathbf{z}}$ can be uniquely reconstructed from an infinite set of line integrals of $\mathbf{v}(x, y)$ through a bounded region, but the irrotational component $\nabla\Phi(x, y)$ cannot.

In summary, the relative vorticity

$$\nabla \times \mathbf{v}(\mathbf{r}) = \nabla \times \nabla \times \boldsymbol{\Psi}(\mathbf{r}) \tag{3.4.12}$$

can be determined from reciprocal travel times [$\nabla \times \nabla\Phi(\mathbf{r}) \equiv 0$, because the curl of a gradient is identically zero], whereas the divergence

$$\nabla \cdot \mathbf{v}(\mathbf{r}) = \nabla^2\Phi(\mathbf{r}) \tag{3.4.13}$$

cannot be directly determined [$\nabla \cdot \nabla \times \boldsymbol{\Psi}(\mathbf{r}) \equiv 0$, because the divergence of a curl is identically zero]. If there are acoustic sources and receivers in the interior of the region, however, (3.4.11) shows that one will obtain information on the scalar potential $\Phi(\mathbf{r})$ at the location of each instrument, so that discrete estimates of $\nabla\Phi$ can be made, provided that the irrotational component of the flow is slowly changing over the scale of the tomographic array.

3.5. Nonreciprocity

In a motionless medium, the ray-trajectory equations, (2.3.11), are independent of the direction of propagation. Rays traveling in opposite directions follow the same ray path, and there is a general reciprocity relation

$$\frac{p(\mathbf{r}_2; \mathbf{r}_1)}{\rho(\mathbf{r}_2)} = \frac{p(\mathbf{r}_1; \mathbf{r}_2)}{\rho(\mathbf{r}_1)}, \tag{3.5.1}$$

where $\rho(\mathbf{r})$ is density and $p(\mathbf{r}_j; \mathbf{r}_i)$ is the acoustic pressure at $\mathbf{r}_j$ due to a unit source of pressure at $\mathbf{r}_i$ (Worcester, 1977*b*). Because density in the ocean varies by a few percent at most, the acoustic pressure fields from reciprocal transmissions are essentially identical. In the presence of vertical current shear, rays traveling in opposite directions are displaced vertically in opposite directions, one upward and the other downward, and (3.5.1) is no longer valid. One can construct situations in which acoustic ray paths for oppositely traveling signals will be significantly different (Sanford, 1974; Mercer, 1988), so that oppositely traveling signals will sample different parts of the current field, and (3.2.6) will no longer apply.

Nonreciprocity in ray paths is most easily estimated for the case of nearly horizontal rays, so that C becomes $C \pm v$, where v is the horizontal component of current in the direction of propagation, and arc length s becomes horizontal position x. Then (2.3.11b) is approximately

$$\frac{d^2(z_{\text{ray}} + \Delta z_{\text{ray}})}{dx^2} = -\frac{1}{C}\left(\frac{dC}{dz} + \frac{dv}{dz}\right), \tag{3.5.2}$$

so that the displacement Δz_{ray} from the unperturbed path z_{ray} is due to the current shear. Ray-path nonreciprocity will be significant when $dv/dz \geq dC/dz$. [Sanford (1974) considered degenerate cases in which $dC/dz \approx 0$ and $dv/dz \gg dC/dz$.]

For small angles, (3.5.2) has the solution

$$\Delta z_{\text{ray}} = \frac{1}{2C}\frac{dv}{dz}x(R - x). \tag{3.5.3}$$

Consider a single upper ray loop with $R = 25$ km, similar to the geometry of Worcester's (1977a,b) experiment. A reasonable value for current shear is $dv/dz = 3 \times 10^{-4}\,\text{s}^{-1}$, and the maximum vertical displacement (at $x = \frac{1}{2}R$) is then

$$\Delta z_{\text{ray}} = \frac{1}{8C}\frac{dv}{dz}R^2 = 16\,\text{m}. \tag{3.5.4}$$

Oppositely traveling signals are therefore separated by twice this distance. Ray paths are of course not infinitely thin, but have finite widths (the Fresnel zone) that typically are of order 50–100 m. The ray separation needs to be compared to the Fresnel-zone width to determine if the oppositely traveling rays are indeed significantly different. In this case they are not.

The sound-speed perturbation associated with a 16-m displacement can be large. To see this, multiply Δz_{ray} by a representative sound-speed gradient. For the temperate profile at $z = \frac{1}{2}z_A = -\frac{1}{2}h$, the potential sound-speed gradient is $C\gamma_a e = 0.046\,\text{s}^{-1}$, from (2.18.2), and so the perturbation arising from the ray-path displacement at $x = R/2$ is

$$\Delta C_{nr} = \tfrac{1}{8}\frac{dv}{dz}R^2\gamma_a e = 0.7\text{ m/s}. \tag{3.5.5}$$

One might expect that the current, v, would have to exceed 0.7 m/s to be measured by reciprocal transmissions. Although Δz and ΔC_{nr} are indeed of first order in dv/dz, the perturbation in travel time due to ΔC_{nr} is of second order, owing to a cancellation between the increased speed and increased length of the perturbed path, as a consequence of Fermat's principle. [This relationship was derived for motionless media in section 2.8, but applies also to moving media (Pierce, 1989).]

Thinking of the current speed v as a perturbation to the sound-speed C, with $|v|/C \ll 1$, the second-order expansion for the travel-time perturbation given in section 2.8,

$$\Delta\tau = L\Delta\ell + M(\Delta\ell)^2, \tag{3.5.6}$$

is applicable, with $\Delta\ell = \pm v/C$ depending on the direction of propagation. Because the second-order terms are by definition quadratic in the perturbation, they cancel when the differential travel time

$$\Delta d_i = \tfrac{1}{2}\left(\Delta\tau_i^+ - \Delta\tau_i^-\right) \tag{3.5.7}$$

is computed. Even though ray paths in a moving medium are nonreciprocal, the linearized expression for the differential travel time is correct to second order. (One-way and sum travel times are correct only to first order.)

The dependence of Δz_{ray} on R^2 in (3.5.4) suggests a failure of reciprocity at large ranges. But for ranges exceeding the double-loop length of the ray, the solution for Δz_{ray} becomes oscillatory, with the same spatial period as the ray. (A numerical example from an actual experiment is provided in the next section.) This periodic behavior is characteristic of many properties of the sound channel and applies to both the temperate channel and the polar surface duct. The ray displacement due to current shear can be obtained in general from the condition that the overall range is fixed, $\Delta R = 0$. From (2.8.11), and using (2.5.4),

$$\Delta\tilde{S} = -\frac{\partial^2 A/\partial\tilde{S}\partial\ell}{\partial^2 A/\partial\tilde{S}^2}\Delta\ell \equiv -f(\tilde{S})\Delta\ell\,.$$

For the temperate $A(\tilde{S})$ it can be shown that $f(\tilde{S})$ is nearly $-\tilde{S}$.

For orientation, consider the case of positive dv/dz. Setting $\Delta\ell = -v_{\text{AX}}/C$,

$$\Delta\tilde{S} = -\tilde{S}\,(v_{\text{AX}}/C), \tag{3.5.8}$$

where v_{AX} is the axial sound-speed. For propagation in the direction of v, $\Delta\tilde{S}$ is negative. The ray loop above the axis is displaced upward, and the ray loop beneath the axis is displaced downward. For transmission against the current, $\Delta\tilde{S}$ is positive, and both ray loops are displaced toward the axial slowness maximum. Taking the adiabatic gradient as representative of the lower ocean, the displacement is of order

$$\gamma_a^{-1}(v/c) = 1\text{ m to }10\text{ m}, \tag{3.5.9}$$

independent of range. These values are comparable to the estimate (3.5.4) for a single ray loop. It would appear that the measurement of currents by reciprocal transmissions can be carried out at long ranges.

3.6. Reciprocal-Transmission Experiments

A series of experiments at steadily increasing ranges, from 25 km to 1275 km, have explored the feasibility of using reciprocal acoustic transmissions to measure large-scale ocean currents; see Dushaw *et al.* (1993*a*) for a review. The numbers and spacings of specific arrivals have been sufficiently similar to permit one to identify corresponding arrivals and measure their travel-time differences (fig. 3.2). (The differential travel times are only a few milliseconds and cannot be seen when the entire arrival pattern is displayed.)

Amplitudes of corresponding arrivals can be significantly different, however, suggesting some degree of nonreciprocity. That was particularly evident at 25 km, where the 0.6-ms pulse length resolved individual micromultipaths in the arrival at about 17,530 ms. The cluster of arrivals at that time corresponded to a single deterministic ray path for the average sound-speed profile (Worcester, 1977*a,b*; Worcester *et al.*, 1981). Internal waves can split the basic ray path into micromultipaths, which then interfere at the receivers, sometimes destructively (chapter 4). Daily averages for the longer ranges suppress internal-wave-related variability (fig. 3.2).

Differential travel times at a range of 300 km to the west of Bermuda (Howe *et al.*, 1987) and at ranges of 745, 995, and 1275 km in the central North Pacific Ocean (Dushaw *et al.*, 1994) are displayed in figs. 3.3 and 3.4. The order of magnitude of the currents corresponding to the differential travel times can be obtained from

$$\Delta d_i = \frac{1}{2}\left(\Delta\tau_i^{+} - \Delta\tau_i^{-}\right) \approx -\frac{\bar{u}}{C_0}\frac{R}{C_0}, \tag{3.6.1}$$

where the ray path is taken to be a straight line, $\bar{u}$ is the range-averaged current, and C_0 is a reference sound-speed (*e.g.*, Worcester, 1977*a,b*). Approximate current magnitudes computed in this way are shown on the right-hand sides of figs. 3.3 and 3.4. Low-frequency, large-scale currents are much larger in the northwest Atlantic Ocean than in the central North Pacific Ocean, as expected.

Current shear is evident in fig. 3.3 from the difference in the differential travel times of the two rays plotted, with upper turning depths at 99 m and 881 m. Similarly, in fig. 3.4 the differential travel times for the axial rays differ from those for the deep turning rays (*i.e.*, all of the other rays), indicating current shear as well. Acoustic travel times are sensitive to both the baroclinic (depth-dependent) and barotropic (depth-independent) components of the ocean current field.

Barotropic tides. Barotropic tidal currents are depth-independent and of large horizontal scale, $O(1\ \text{Mm})$, in midocean. In contrast, baroclinic tidal currents are of much shorter horizontal scale and have vertical coherence scales of order

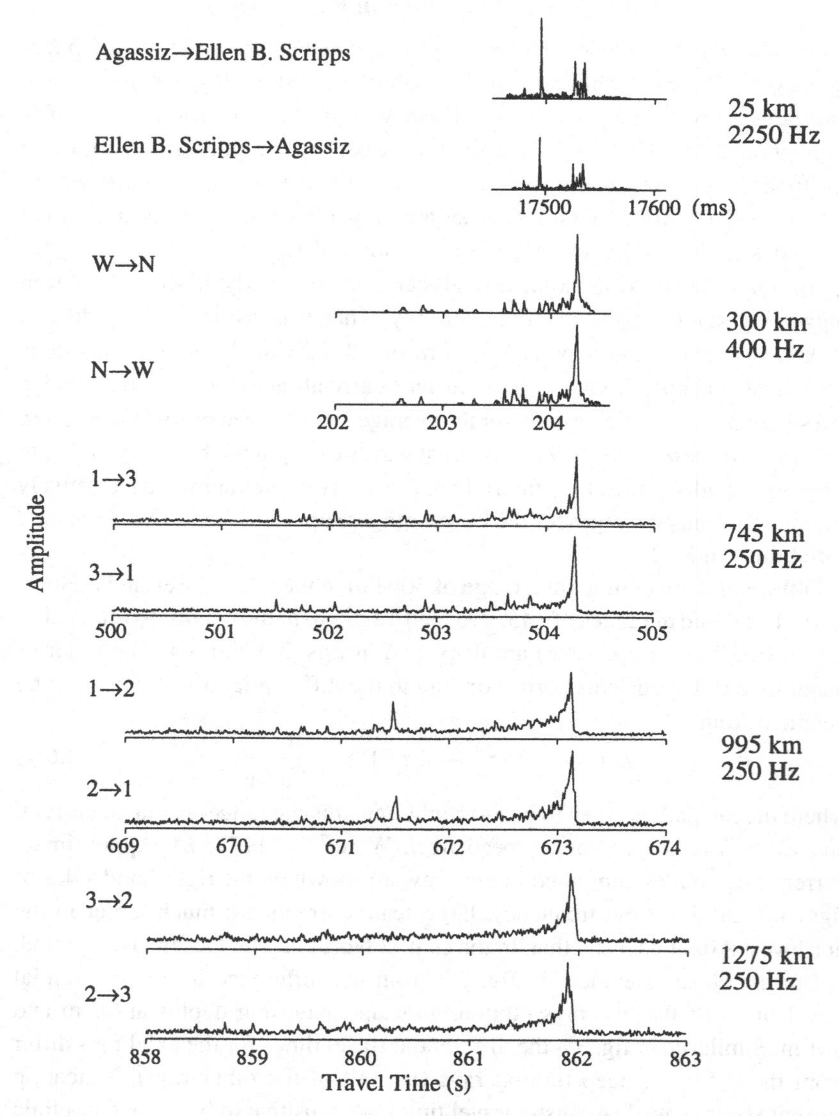

Fig. 3.2. Reciprocal arrival patterns at a variety of ranges and acoustic frequencies. The vertical axis is proportional to pressure. The arrivals at 25 km are for 0914Z, 3 April 1976. (From Worcester, 1977*b*.) The arrivals at 300 km are daily averages for year day 223 of 1983. (From Worcester *et al.*, 1985*b*.) The arrivals at 745 km and 995 km are daily averages for year day 153 of 1987; the arrivals at 1275 km are daily averages for year day 165 of 1987. (From Worcester *et al.*, 1991*b*.) Note that an expanded scale is used at the 25-km range. *Agassiz* and *Ellen B. Scripps* were the ship platforms in the 1976 experiment.

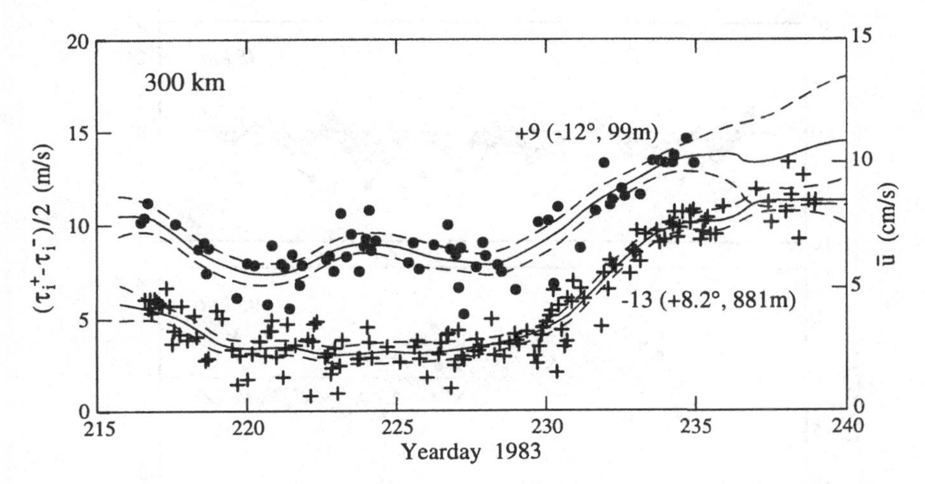

Fig. 3.3. Differential travel times for the +9 and −13 rays in the 1983 Reciprocal Acoustic Transmission Experiment. The range is 300 km. The computed arrival angle and upper turning depth are given for each ray. High-frequency noise (>0.5 cycles per day) is suppressed by low-pass filtering prior to performing inversions (solid curve). The estimated rms uncertainties for the low-pass series are given by the dashed curves. The right-hand axis gives rough estimates for the corresponding range-averaged current magnitudes, $\bar{u}$. (Adapted from Howe *et al.*, 1987.)

100 m (Wunsch, 1975). Barotropic and baroclinic tidal currents are of comparable magnitudes, $O(1\ \text{cm/s})$, and have the same frequencies. Acoustic techniques can be expected to yield excellent estimates of the barotropic tidal currents, because the vertical and horizontal averaging inherent in long-range, steep ray paths that cycle through most of the water column suppresses the baroclinic contribution to acoustic travel times.

In the 1987 Reciprocal Tomography Experiment, the tidal signal accounted for roughly 90% of the differential travel-time variance at 1 Mm range in the central North Pacific Ocean (Dushaw, 1992; Dushaw *et al.*, 1994, in press). A least-squares fit of components at eight tidal frequencies to the currents deduced from the differential travel times gave tidal amplitudes and phases in excellent agreement with barotropic tidal currents computed from a current meter mooring (Luther *et al.*, 1991), as well as with numerical models of the tides (Schwiderski, 1980; Cartwright *et al.*, 1992) (table 3.1). [An unweighted least-squares fit of the tidal constituents to the differential travel times does not differ significantly from these results (Worcester *et al.*, 1991*b*; Luther *et al.*, 1991).]

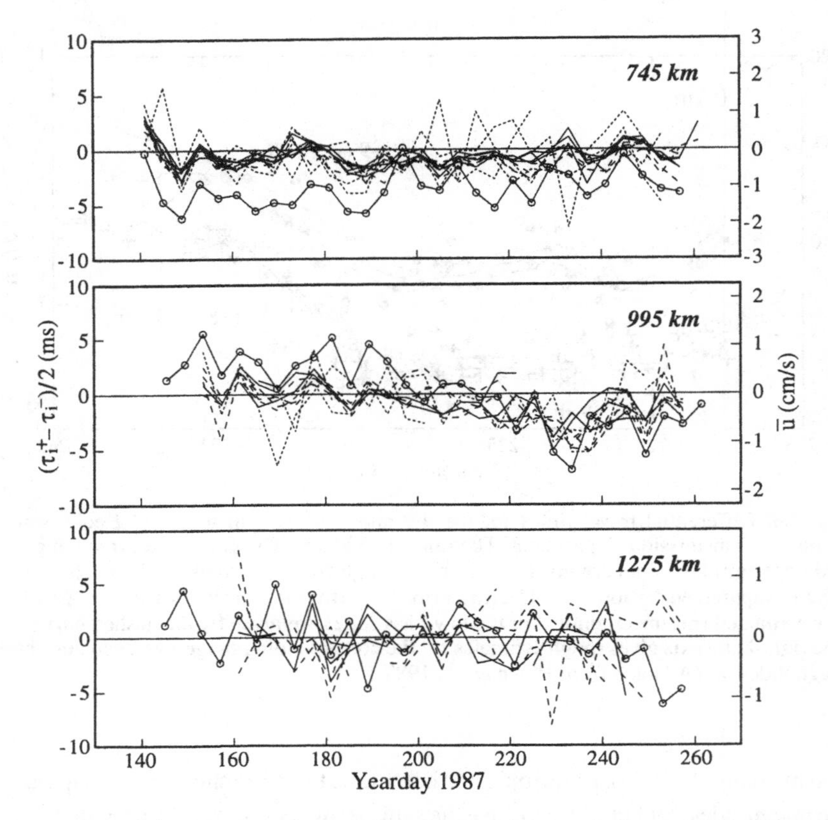

Fig. 3.4. Low-pass-filtered (<1 cycle/day) differential travel times at 745, 995, and 1275 km from the 1987 Reciprocal Tomography Experiment. Solid, dashed, and dotted curves indicate different ray groups. Axial-ray travel times are indicated by open circles. At a 1275-km range, the acoustic signal-to-noise ratio is low, giving noisy differential travel times. The right-hand axis gives rough estimates for the corresponding range-averaged current magnitudes, $\bar{u}$. (Adapted from Dushaw *et al.*, 1994.)

Under some conditions, the tidal travel-time signal can even be sufficiently great to be recognizable in one-way travel times. For example, Munk *et al.* (1981*b*) found a barotropic tidal component of 8-ms amplitude in a one-way 1978 transmission from Bermuda toward the east coast of the United States.

Reciprocity. An experimental indication of the degree to which oppositely traveling signals follow similar ray paths can be obtained from examination of high-frequency travel-time variances. Travel-time fluctuations at periods shorter than the inertial period are predominantly due to sound-speed

Table 3.1. *Amplitude and Greenwich epoch of the eastward barotropic tidal current components along the north leg of the 1987 Reciprocal Tomography Experiment*

Tidal component		Acoustic tomography	Current-meter mooring	Schwiderski model	CRS[a] model
K_2	cm/s	0.12 ± 0.04	0.10 ± 0.04	–	0.14
	°G	268 ± 13	280 ± 21	–	279
S_2	cm/s	0.53 ± 0.04	0.53 ± 0.05	0.66	0.63
	°G	272 ± 4	280 ± 6	270	276
M_2	cm/s	1.31 ± 0.03	1.32 ± 0.05	1.28	1.42
	°G	223 ± 1	218 ± 2	222	218
N_2	cm/s	0.14 ± 0.03	0.15 ± 0.05	0.16	0.15
	°G	191 ± 13	216 ± 17	184	201
K_1	cm/s	0.75 ± 0.04	0.74 ± 0.03	0.45	0.53
	°G	128 ± 2	135 ± 3	127	100
P_1	cm/s	0.18 ± 0.04	0.27 ± 0.02	–	0.10
	°G	132 ± 11	141 ± 4	–	129
O_1	cm/s	0.43 ± 0.03	0.46 ± 0.03	0.33	0.38
	°G	101 ± 4	122 ± 4	99	104
Q_1	cm/s	0.10 ± 0.03	0.07 ± 0.02	–	0.05
	°G	64 ± 16	110 ± 14	–	125

[a] The CRS model is the Cartwright, Ray, and Sanchez (1992) model.
Source: Adapted from Dushaw *et al.* (1994).

perturbations associated with internal wave displacements (Flatté *et al.*, 1979; Flatté, 1983; Flatté and Stoughton, 1986, 1988). If oppositely traveling signals follow ray paths sufficiently close in space and time that the internal-wave-induced sound-speed perturbations are correlated, the resulting one-way travel-time fluctuations will also be correlated and will cancel in the differential travel times. A reduction of the high-frequency travel-time variance for differential travel times relative to one-way travel times therefore indicates that the ray paths of the oppositely traveling signals are separated by less than the internal wave correlation scale, roughly 100 m in the vertical. The correlation will be reduced at ranges sufficiently great that the internal wave field will change significantly during the pulse travel time. (In this case the precision of the differential travel times can be improved by time averaging to suppress internal-wave-induced travel-time fluctuations.)

Table 3.2. *Nontidal, high-frequency (>1 cycle/day) travel-time variances*[a]

Range	Ray-path geometry	$\langle\tau^2\rangle$ (ms^2)	$\frac{1}{2}\langle(\tau^+ - \tau^-)^2\rangle$ (ms^2)	$\langle\tau^+\tau^-\rangle/\langle\tau^2\rangle$
300 km (RTE83)	RR	10.6	0.8	0.92
745 km (RTE87)	RSR	8	3	0.60
	RSR/RR	38	15	0.60
995 km (RTE87)	RSR	15	7	0.51
	RSR/RR	45	21	0.53
1275 km (RTE87)	RSR	17	7	0.59
	RSR/RR	39	10	0.50

[a]Ray paths denoted by RR are refracting, and those denoted by RSR reflect from the surface. $\langle\tau^2\rangle = \frac{1}{2}(\langle(\tau^+)^2\rangle + \langle(\tau^-)^2\rangle)$. The differential travel-time variance is normalized such that if $\langle\tau^+\tau^-\rangle = 0$, the differential and one-way variances will be equal.

Source: Adapted from Dushaw *et al.* (1993*a*).

High-frequency travel-time variances were found to be highly correlated at a range of 300 km in the 1983 Reciprocal Transmission Experiment conducted west of Bermuda, with one-way travel-time variance $\langle\tau^2\rangle = 10.6$ ms^2 and differential travel-time variance $\frac{1}{2}\langle(\tau^+ - \tau^-)^2\rangle = 0.8$ ms^2 averaged over all rays (table 3.2). Stoughton *et al.* (1986) have shown that measured differential travel-time fluctuations are predominantly due to internal wave currents, rather than to ray-path nonreciprocity. Data from the 1987 Reciprocal Tomography Experiment conducted in the central North Pacific Ocean indicate that out to the longest range for which differential travel-time data are available, 1275 km, the oppositely traveling signals are still significantly correlated, although less so than at shorter ranges (table 3.2). The reduced correlation is due at least in part to the nonsimultaneity of the oppositely traveling pulses. The travel time at 1275 km is almost 15 min, comparable to the shortest period of internal waves near the ocean surface. (Both the mean and rms currents are small in the central North Pacific, however, so that the correlation might be expected to be a maximum for this case.)

One can also perform a consistency check by tracing rays in opposite directions for the large-scale current field deduced from the differential travel times. For the 1983 Reciprocal Transmission Experiment, the computed vertical separation between oppositely traveling rays was less than 40 m (fig. 3.5), compared

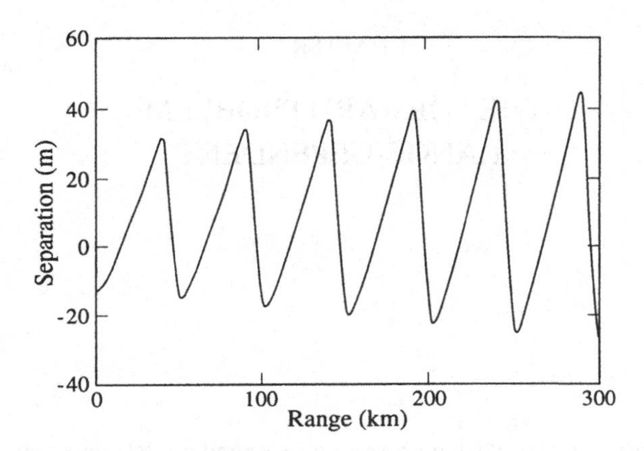

Fig. 3.5. Vertical separation between oppositely traveling rays from the 1983 Reciprocal Transmission Experiment. The separation was computed using the range-independent current profile derived from the differential travel times (Howe, 1986). The separation is due both to mean current shear, $\partial u(z)/\partial z$, and to the fact that the sources and receivers are not precisely coincident. (Adapted from Stoughton *et al.*, 1986.)

with the internal wave vertical correlation scale of roughly 100 m. This result is consistent with the high degree of correlation found from the high-frequency variances.

CHAPTER 4

THE FORWARD PROBLEM: RANGE-DEPENDENT

The discussion so far has dealt with a range-independent (RI) sound channel. But the ocean certainly varies horizontally and is always range-dependent (RD); one of the chief goals of tomography is to derive its range-averaged (RA) properties.

The RD treatment will vary depending on whether the scale of the horizontal variations is larger, comparable to, or very much smaller than the ray-loop range (typically 50 km). The term "adiabatic[1] range dependence" is defined to apply to the case of small fractional variation over a ray loop. Variations on a gyre scale can accordingly be treated by the adiabatic approximation, assuming there are no sharp frontal surfaces.[2]

The term "loop resonance" applies to ray travel-time perturbations due to ocean perturbations with horizontal scales equal to the ray-loop scale, or to a fraction of the loop scale. This includes mesoscale activity (which accidentally has a scale comparable to the loop scale) and ranges down to the longer components in the internal wave spectrum. Cornuelle and Howe (1987) have shown that measured travel-time perturbations associated with loop resonance can provide some RD information for even a single source-receiver transmission path.

Internal waves are generally included among the small-scale processes for which the forward problem yields estimates of the variance and other statistical properties of the travel time (Flatté *et al.*, 1979). These estimates are required for inversion of the measured data set. In turn, the measured variances can provide useful information about the small-scale structure (Flatté and Stoughton, 1986). It is ultimately the small-scale structure, not measurement error, that imposes the fundamental limits on acoustic tomography.

[1]This is a different use of the word than in (2.2.5), where *adiabatic* refers to vertical gradients in a mixed fluid.

[2]The wind-driven ocean circulation in all northern and southern ocean basins can be classified according to latitudinal belts. Subtropical gyres extend from the trade winds to the westerly winds (roughly from latitude 15° to 50°); subpolar gyres are poleward from the subtropical gyres. Boundaries between gyres can be quite sharp.

4.1. Adiabatic Range Dependence

It has been shown (Milder, 1969; Wunsch, 1987) that the adiabatic ray invariant is the wave action A. With A = constant in a sound channel varying adiabatically with range x, one can compute the adiabatic variation of $\widetilde{S}(x)$ and $\widetilde{z}(x)$. Fig. 4.1 illustrates the modification of the ray structure for propagation from a mid-latitude into a polar sound channel (Dashen and Munk, 1984). This transition is modeled by a sloping channel axis that outcrops at B. Assuming conservation of the action variable, the refracted rays follow the sound channel upward and change at point A from refracted (RR) to surface-reflected (RSR) propagation. Some observations of propagation across an RR to RSR transition are discussed in chapter 8.

For a simple example, consider a polar profile with a very slow latitudinal variation of the vertical sound-speed gradient (such as might arise from a salinity gradient),

$$\gamma(x) = \gamma_a + \mu(x - x_0)\,,$$

where γ_a is the adiabatic vertical sound-speed gradient in an isohaline ocean (2.2.5), and μ is a coefficient (units of km^{-2}). Then, with $A = \frac{2}{3}\,\gamma^{-1}\,S_0\,\widetilde{\sigma}^3$ a constant, we have

$$\frac{3\,\delta\widetilde{\sigma}}{\widetilde{\sigma}} = +\frac{\delta\gamma}{\gamma_a} = +\frac{\mu(x - x_0)}{\gamma_a}\,,$$

and so the latitudinal variation of $R(x)$, and similarly of T, s_g, and s_p, is readily computed according to (2.17.6)–(2.17.10). The latitude dependence of the phase slowness s_p leads to horizontal refraction.

Adiabatic range dependence has a very simple interpretation in terms of modes. Milder (1969) demonstrated that the adiabatic invariants of wave propagation are the mode numbers m. In other words, individual modes propagate independently, without mutual coupling, despite adiabatic changes in the channel profile along the path. We refer to Brekhovskikh and Lysanov (1991, sect. 7.2) for a discussion of the adiabatic approximation.

Second-order effects can be troublesome. For the RI case, we previously found

$$\Delta T = L\,\Delta\ell + M\,(\Delta\ell)^2, \quad \Delta\tau = n\,\Delta T, \tag{4.1.1}$$

for the simple case of a perturbation of a single parameter ℓ; M and N are given functions (2.8.15) of A and its derivative in the unperturbed state. For the RD adiabatic case, the result is [Munk and Wunsch, 1987, eq. (4.4); Wunsch, 1987]

$$\Delta T = L\langle\Delta\ell\rangle + M\langle\Delta\ell\rangle^2 + N\langle(\Delta\ell - \langle\Delta\ell\rangle)^2\rangle, \tag{4.1.2}$$

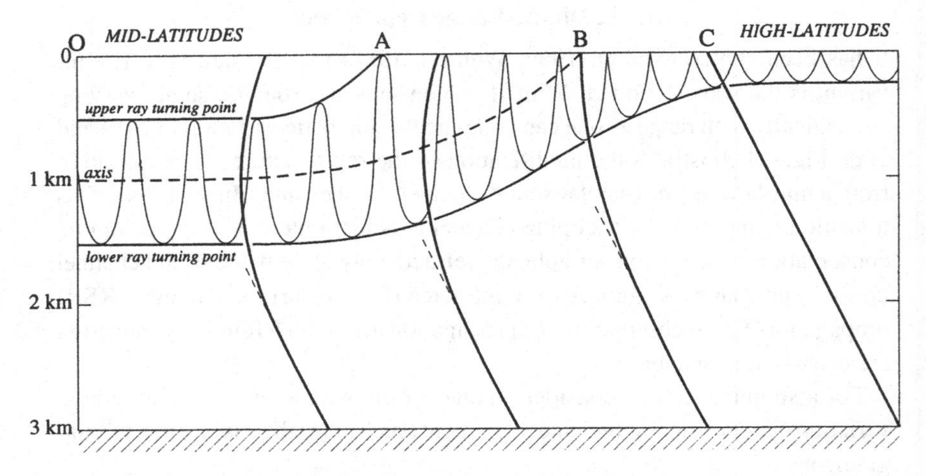

Fig. 4.1. Adiabatic transformation of ray paths with latitude. Four sound-speed profiles are plotted. Conditions are taken to be uniform from the source at 0 to the left profile. The sound axis reaches the surface at B. The sound-speed gradient becomes constant (adiabatic) at C. Rays are drawn for RR transmission at midlatitude, and RSR at high latitudes, with A the point where the upper turning point first reaches the surface. There are lower loops only (ducted transmisson) beyond B.

with $\Delta\tau = n\,\Delta T$, where

$$N = \tfrac{1}{2}\,\frac{\partial^2 A}{\partial \ell^2} - \left(1 - \tfrac{1}{2}\,\frac{(\partial A/\partial \ell)\,(\partial^2 A/\partial \widetilde{S}^2)}{(\partial A/\partial \widetilde{S})\,(\partial^2 A/\partial \ell\,\partial \widetilde{S})}\right)\frac{(\partial A/\partial \ell)\,(\partial^2 A/\partial \ell\,\partial \widetilde{S})}{(\partial A/\partial \widetilde{S})}, \tag{4.1.3}$$

and angle brackets denote range averages. For the adiabatic case, $A = \langle A \rangle \neq f(x)$, but the perturbed A differs from the initial $A(-)$.

The first term in (4.1.2) is the "frozen" RD ray approximation. The second and third terms give the nonlinear corrections. There are two important special cases: (i) for the RI case we have only the second term, whereas (ii) for the RD case with $\langle \Delta\ell \rangle = 0$ we have only the third term. An important result is that the third term depends only on the spatial variance of $\Delta\ell(x)$. For RD variations other than adiabatic, the travel time will depend on the moments of $\Delta\ell(x)$, which will vary from ray to ray.

Trouble could arise in long-range transmissions, for which the linear term associated with the mean gyre perturbation $\langle \Delta\ell \rangle$ may be comparable to, or smaller than, the quadratic $\langle (\Delta\ell - \langle \Delta\ell \rangle)^2 \rangle$ associated with the adiabatic RD perturbation $\Delta\ell(x)$ along the propagation path. Under these circumstances, the quadratic bias is comparable to the linear perturbation.

4.2. Loop Resonance

We follow the analysis by Cornuelle and Howe (1987) in their general treatment of perturbation of travel time in an RD environment. In a Fourier decomposition of the RD sound channel, each ray is most sensitive to those components of the RD perturbation that have the same wavelength as the ray double loop and its harmonics; hence the term "loop resonance." Conversely, the inverse treatment will give some information concerning the spectrum of the RD perturbation. We call this "loop harmonic analysis." This restriction to just a few discrete components of the ocean wavenumber spectrum is both good and bad. From the point of view of the forward problem, it means that the RI approximation is relatively good, given the highly selective interaction with only a small fraction of the ocean variability. The bad aspect is from the inverse point of view: information about the perturbation spectrum obtained by loop harmonic analysis from a single source-receiver pair is sparse in wavenumber space. In practice, one must depend on two-dimensional arrays for x, y-mapping. An interesting display of path perturbations in θ, z-space due to loop resonance is shown by Smith *et al.* (1992*a*, fig. 2).

We have previously written $\Delta S(z) = \sum_j \ell_j \, F_j(z)$ for the RI slowness perturbations, where $F_j(z)$ is some convenient set of modal functions (dynamic, empirical, ...). We generalize to

$$\Delta S(x, z) = \sum_k \sum_j \left[a_{kj} \cos \frac{2\pi kx}{r} + b_{kj} \sin \frac{2\pi kx}{r} \right] F_j(z) \,, \qquad (4.2.1)$$

where k is the number of cycles between source and receiver. The sinusoids can have any fundamental period; the use of range r simplifies some of the algebra.

The travel-time perturbation is

$$\Delta \tau_n = \int_{\Gamma_n} ds \, \Delta S\,(x,\, z_n(x))\,,$$

where the integration Γ_n is along the nth ray path $z_n(x)$. In the linear approximation we replace Γ_n by the unperturbed path $\Gamma_n(-)$. Replacing ds by $(ds/dx)\, dx$,

$$\Delta \tau_n = \sum_k \sum_j \int_{\Gamma_n(-)} dx \, G_{nj}(x) \left(a_{kj} \cos \frac{2\pi kx}{r} + b_{kj} \sin \frac{2\pi kx}{r} \right), \qquad (4.2.2)$$

where

$$G_{nj}(x) = (ds/dx) \;\, F_j\,[z_n(x)] \qquad (4.2.3)$$

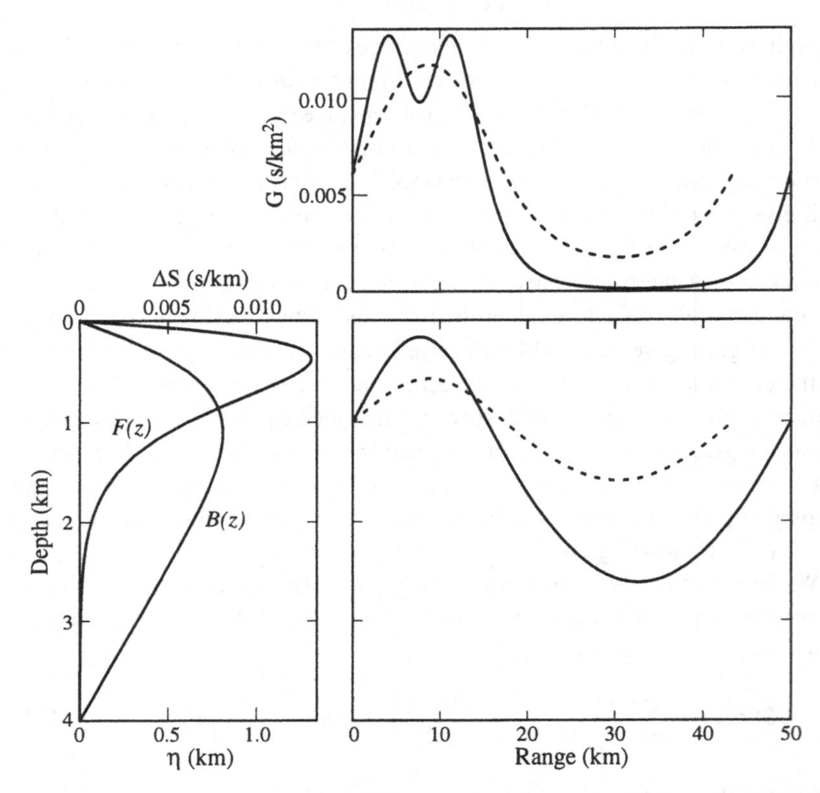

Fig. 4.2. Bottom left: Wave functions $B(z)$ of vertical displacement, and $F(z)$ of the sound slowness perturbation for Rossby waves of mode $j = 1$ in a temperate ocean. Bottom right: Ray orbits $z_n(x)$ for a flat (dashed) and a steep ray. Top: The ray weighting functions $G_n(x) \sim F(z_n(x))$ for the two rays.

is the projection of ray n on vertical mode j. Cornuelle and Howe call G the "ray weighting function." We now make the severe restriction to an RD perturbation in an otherwise RI ocean. Fig. 4.2 portrays the situation for Rossby waves of mode $j = 1$ in a temperate profile. A flat ray (dashed) stays beneath the maximum of $F(z)$, and so $G(x)$ has a peak at the ray summit. A steep ray (solid) extending above the maximum is associated with a double peak in $G(x)$. The effectiveness of loop resonance is determined by the Fourier components of $G(x)$.

As an illustration, take source and receiver both on the channel axis and consider a near-sinusoidal axial-ray pair $z_n(x) = z_A \pm \beta \sin(2\pi n x/r)$. For a single mode (suppressing the subscript j), expand $F(z)$ about the axis, $F = F_A + F'_A (z - z_A)$. The ray tier correction ds/dx is ignored. The result is

$$\Delta\tau_n = \int_0^r dx \sum_k \left[a_k \cos\frac{2\pi kx}{r} + b_k \sin\frac{2\pi kx}{r}\right]\left[F_A \pm F'_A\,\beta \sin\frac{2\pi nx}{r}\right], \tag{4.2.4}$$

$$= r a_0\, F_A \pm \tfrac{1}{2}\, r\, b_n\, F'_A\, \beta, \tag{4.2.5}$$

with the $\pm$ sign corresponding to upward/downward launch angles, leading to a "splitting" in travel times. In the RI case, for an integral number of n double loops, there is an up/down degeneracy, with upward and downward rays arriving simultaneously, provided source and receiver are at the same depth (not necessarily axial). The first term in (4.2.5) is the RI perturbation, $k = 0$, arising from the near-axial perturbation $\Delta S = a_0\, F_A$. The second term is the RD perturbation for $k = n$, with the perturbation harmonic k having the same horizontal wavelength as ray n. There is no information concerning a_k other than for $k = 0$, because of the orthogonality of sines and cosines.

The general case is complicated, for a number of reasons. (i) For multiple modes j, the first term in (4.2.5) becomes

$$\Delta\tau_n = r \sum_j a_{0j}\, (F_A)_j\,.$$

(ii) Nonaxial rays (smaller n) are increasingly more nonsinusoidal, so that one needs to consider the harmonics h of ray n:

$$z_n(x) - z_A = \sum_h \beta_{nh}\, \sin(2\pi nhx/r)\,.$$

All harmonics arrive at the same time, and hence ray n has contributions from $k = n, 2n, \ldots, hn$ perturbation cycles over range r.

With the foregoing restrictions to an RD perturbation in an otherwise RI ocean, it is convenient to refer to the matrix formalism (section 2.9):

$$\Delta\tau_n = \sum_p E_{n,p}\, \ell_p\,, \tag{4.2.6a}$$

where $p(j, h)$ is a combined index for all possible values of j and h, and where

$$\ell_p = a_{0j} \text{ for } k = h = 0, \qquad \ell_p = \tfrac{1}{2}\, b_{k,j} \text{ for } k = nh \neq 0\,. \tag{4.2.6b}$$

For the case of only one mode $j = 1$, the subscript j is suppressed, and, writing k for p,

$$\Delta\tau_n = \sum_k E_{n,k}\, \ell_k\,, \tag{4.2.7}$$

$$E_{n,0} = n \int_0^{r/n} dx\, G_n(x), \quad \ell_0 = a_0, \text{ for } k = 0,$$

$$E_{n,k} = n \int_0^{r/n} dx \sin\frac{2\pi kx}{r}\, G_n(x), \; \ell_k = \tfrac{1}{2} b_k, \quad \text{for } k = nh \neq 0. \tag{4.2.8}$$

$E_{n,k} = 0$ for all other k. For a 1-Mm range in the temperate ocean, we have only $n = 19, 20, \ldots, 23$ double loops per megameter, and so $E_{n,k} = 0$ except for $k = 0$, $k = n = 19, 20, \ldots, 23$, $k = 2n = 38, 40, \ldots, 46$, and so forth (table 4.1). The first column, $k = 0$, corresponds to the RI case for $j = 1$ and is identical with $(E_{ij})_{\text{ray}}$ given in (2.15.12) and portrayed in fig. 2.15. The matrix is sparse. The result, as shown in chapter 6, is that the information (resolution) gained by an inverse analysis is very restricted in horizontal wavenumber space.

Rossby-wave ocean. We shall discuss at some length the case of a monochromatic two-dimensional Rossby-wave spectrum. The resulting configuration (fig. 4.4), though far from realistic, will serve as an illustration of what can be learned from inverse methods applied to a simple transmission in a complex ocean. (The reader may wish to skip the remainder of this section.) Let

$$\eta(x, y, z, t) = c\,(\kappa_x, \kappa_y, \omega)\, B(z) \cos(\kappa_x x + \kappa_y y - \omega t - \phi) \tag{4.2.9}$$

designate the vertical displacement of a Rossby wave, and

$$\Delta S(x, y, z, t) = c\, F(z) \cos(\kappa_x x + \kappa_y y - \omega t - \phi) \tag{4.2.10}$$

the associated perturbation in sound-slowness, with $F(z)$ given by the term in square brackets in (2.18.31). All possible solutions lie on a circle of radius κ_r centered at κ_c (fig. 4.3),

$$\kappa_r = \sqrt{\kappa_c^2 - \lambda^2}, \quad \kappa_c = -\beta/(2\omega),$$

with $\lambda = \xi_0(1)\, h^{-1} f/N_0$ [see equations 2.18.27–2.18.34]. For any $|\kappa_y|$ there are four elementary wave components corresponding to

$$\kappa_x^{L/R} = \kappa_c \mp \sqrt{\kappa_r^2 - \kappa_y^2}, \quad \kappa^{L/R} = \sqrt{(\kappa_x^{L/R})^2 + \kappa_y^2}, \tag{4.2.11}$$

with $\kappa_y = \pm|\kappa_y|$, corresponding to northward- and southward-traveling waves, respectively; κ_x is always negative: the phase velocity is westward. The shortest and longest wavelengths are $2\pi/\kappa_x^L(0) = 10.5$ km and $2\pi/\kappa_x^R(0) = 2371$ km, respectively.

Table 4.1. *Transposed observation matrix E_n^T (s/km) for a 1-Mm range in a temperate profile*[a]

	n				
κ_y	19	20	21	22	23
0	3.39	4.12	4.69	5.18	5.64
1	0	0	0	0	0
⋮					
18	0	0	0	0	0
19	2.03	0	0	0	0
20	0	2.50	0	0	0
21	0	0	2.76	0	0
22	0	0	0	2.76	0
23	0	0	0	0	2.33
24	0	0	0	0	0
⋮					
37	0	0	0	0	0
38	1.00	0	0	0	0
39	0	0	0	0	0
40	0	1.20	0	0	0
41	0	0	0	0	0
42	0	0	1.10	0	0
43	0	0	0	0	0
44	0	0	0	0.79	0
45	0	0	0	0	0
46	0	0	0	0	0.34
47	0	0	0	0	0

[a] n is the number of ray double loops per 1 Mm, and κ_y is the number of Rossby wave perturbation cycles per 1 Mm north–south.

For a north–south transmission over a 1-Mm range, we expand the y-component of the perturbations into harmonics of range $r = 1$ Mm:

$$\kappa_y = (2\pi/1000)k, \quad k = 0, \pm 1, \pm 2, \ldots . \tag{4.2.12}$$

This simplifies the algebra. As previously stated, loop resonance occurs for $k = n = 19, 20, 21, 22, 23$, $k = 2n = 38, 40, 42, 44, 46$, and so forth. We have chosen Rossby parameters so that the circle subtends harmonics up to and

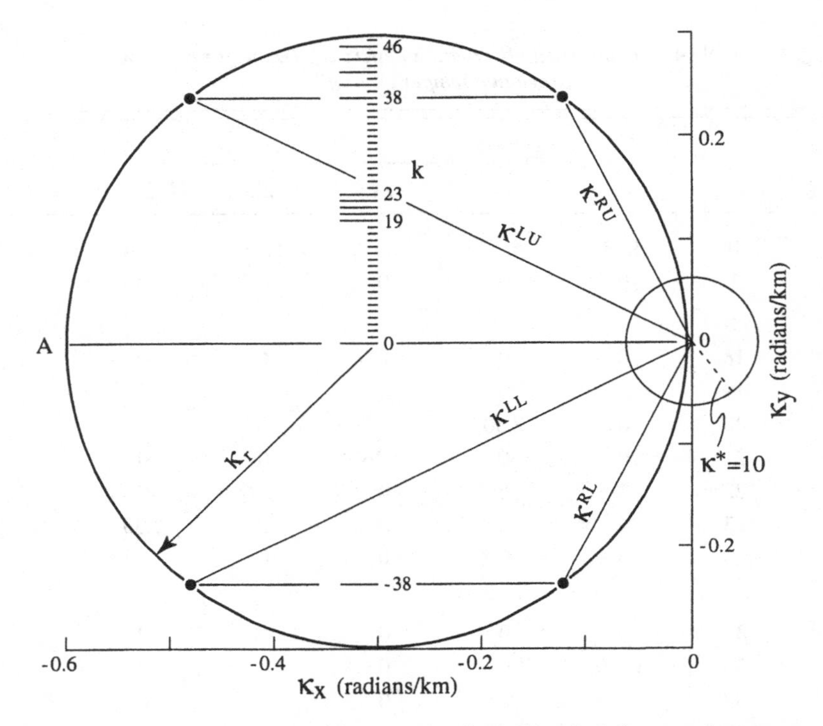

Fig. 4.3. Rossby wave dispersion diagram (the dispersion circle is not quite tangent to the y-axis). All possible wavenumbers κ are drawn from the origin to a point on the dispersion circle. The circle center is located at $\kappa_x = -\kappa_c$, $\kappa_y = 0$. The northward components $\kappa_y = (2\pi/1000)k$ radians/km for harmonics $k = 0$ to 47 cycles per 1 Mm are indicated along the circle diameter; long ticks correspond to loop resonance for the ray fundamental and first harmonics. Four elementary wave trains are associated with each harmonic, as indicated for $|k| = 38$. The small circle about the origin corresponds to $\kappa^* = 10$ (wavelength 100 km); larger wavenumbers (smaller wavelengths) diminish rapidly with increasing κ. Parameters are $\omega = 6.65 \times 10^{-8}$ radians per second (1/3 cycle per year), $f = 7.27 \times 10^{-5}$ s^{-1} (1 cycle per day), $N_0 = 5.2 \times 10^{-3}$ s^{-1} (3 cycles per hour), $h = 1$ km, $\zeta_0 = 2.872$.

including 47. This range requires an unrealistically low frequency of $\frac{1}{3}$ cycle per year. The implication is that Rossby waves are generally too long to resonate with ray harmonics.

We assume a power spectrum $c^2(\kappa)$ diminishing with increasing scalar wavenumber κ as

$$c(\kappa) = g(\kappa)c_0, \quad g^2(\kappa) = \kappa^{*2}/(\kappa^2 + \kappa^{*2}), \tag{4.2.13}$$

with $c_0 = 0.04$ km and $k^* = 0.628$ radian/km, or 10 cycles/Mm (fig. 4.3). The

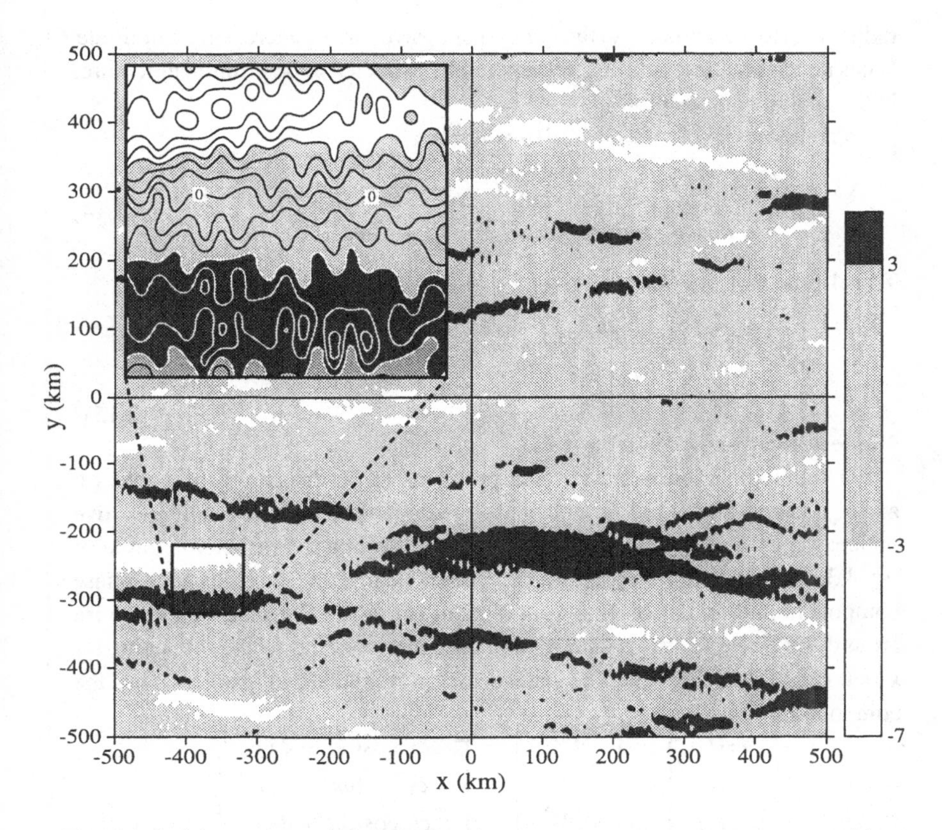

Fig. 4.4. Axial sound-speed perturbations in a temperate megameter square resulting from a Rossby-wave spectrum. Shades indicate $\Delta C > +3$ m/s and $\Delta C < -3$ m/s; the contour $\Delta C = 0$ is shown. A 100-km × 100-km square is shown on an enlarged scale. Travel-time perturbations for a 1-Mm transmission along the y-axis are given in table 4.2. This model serves as an example for the discussion on range-dependent inverse methods in chapters 6 and 7.

largest wavenumber is about $10\kappa^*$, $\frac{1}{10}$ of the amplitude of c_0. The choice of c_0 is arbitrary; the foregoing value leads to reasonable perturbation amplitudes.

For purposes of mapping $\Delta S(x, y, z, t)$ for some fixed z, the terms (4.2.10) are summed from $k = -47$ to $k = +47$ for the left and right sides of the dispersion circle, with κ_y, $\kappa_x^{L/R}$, $\kappa^{L/R}$, and $c^{L/R}$ all functions of k according to (4.2.9)–(4.2.12). Random phases $\phi_k^{L/R}$ are generated for each of the 190 terms. Fig. 4.4 shows a map over a 1-Mm × 1-Mm area. The slanting ridges and trenches can be related to the wavenumber vectors in fig. 4.3 drawn from the origin to the points where the large dispersion circle intersects the small circle of radius κ^*. Beyond this circle the component waves are of decreasing amplitude. Lines normal to these limiting vectors can be expected to be prominent features in the spatial contour plot.

To compute travel times, (4.2.10) is written

$$\begin{aligned} \Delta S(x, y, z, t) &= F(z)[a \cos(\kappa_x x + \kappa_y y) + b \sin(\kappa_x x + \kappa_y y)], \\ a &= c \cos(\omega t + \phi), \quad b = c \sin(\omega t + \phi). \end{aligned} \tag{4.2.14}$$

Travel-time perturbations are given by (4.2.7), provided

$$\begin{aligned} \ell(k) &= \ell^L(k) + \ell^L(-k) + \ell^R(k) + \ell^R(-k), \\ \ell^{L/R}(k) &= \tfrac{1}{2} a^{L/R}(k) \text{ for } k = 0, \; = \tfrac{1}{2} b^{L/R}(k) \text{ for } k = 1, \ldots, 47. \end{aligned}$$

(4.2.15)[3]

The results are shown in table 4.2.

For the application to the inverse problems to be discussed in chapters 6 and 7, it is convenient to introduce an alternate notation that uses only positive indices. There are 190 elementary wave trains around the dispersion circle of fig. 4.3, with amplitudes c_1 to c_{190} and phases ϕ_1 to ϕ_{190}. Each has an in-phase component and an out-of-phase component (in time). We proceed clockwise around the circle, starting with c_1 and ϕ_1 corresponding to the left point for $k = 0$ (point A in fig. 4.3). The statevector (to be defined in chapter 6) has the components

$$\mathbf{x}(t) = \begin{bmatrix} x_1(t) \\ x_2(t) \\ x_3(t) \\ x_4(t) \\ \vdots \\ x_{39}(t) \\ x_{40}(t) \\ \vdots \\ x_{379}(t) \\ x_{380}(t) \end{bmatrix} = \begin{bmatrix} a_0^L(t) \\ b_0^L(t) \\ a_1^L(t) \\ b_1^L(t) \\ \vdots \\ a_{19}^L(t) \\ b_{19}^L(t) \\ \vdots \\ a_{-1}^L(t) \\ b_{-1}^L(t) \end{bmatrix} = \begin{bmatrix} c_1 \cos(\omega t + \phi_1) \\ c_1 \sin(\omega t + \phi_1) \\ c_2 \cos(\omega t + \phi_2) \\ c_2 \sin(\omega t + \phi_2) \\ \vdots \\ c_{20} \cos(\omega t + \phi_{20}) \\ c_{20} \sin(\omega t + \phi_{20}) \\ \vdots \\ c_{190} \cos(\omega t + \phi_{190}) \\ c_{190} \sin(\omega t + \phi_{190}) \end{bmatrix}. \tag{4.2.16}$$

The dot in the left upper quarter of fig. 4.3 corresponds to the lowest (fundamental) resonance for ray $n = 19$ and is now designated by x_{39} and x_{40}.

[3]The factors $\frac{1}{2}$ in these equations arise for two different reasons. The summation (4.2.7) is for positive k only, and the term $\ell(0)$ appears twice. The $\frac{1}{2}b$ term arises from the fact that $\overline{\sin^2} = \frac{1}{2}$ in (4.2.8).

Table 4.2. *Travel-time perturbations $\Delta\tau_n$ in seconds*[a]

	n				
t	19	20	21	22	23
0	−.11	−.06	−.12	−.11	−.22
0.5	−.15	−.18	−.21	−.17	−.30
1	−.04	−.12	−.09	−.06	−.08
1.5	.11	.06	.12	.11	.22

[a] n is the number of ray double loops per 1 Mm, and t is calendar time in years. The first and last lines are at a time interval of exactly one-half wave period and therefore are equal and of opposite signs.

4.3. Mesoscale Variability

As an example, we refer to the 1-Mm transmission of the SLICE89 experiment (see appendix A, table A.1). The RI time fronts were shown in fig. 2.21. A similar plot for the RD time fronts, with loop resonance taken into account, does not differ visibly from fig. 2.21. Accordingly, fig. 4.5 shows the *differences* in travel times for the RD and RI constructions. The experiment took place in the North Pacific subtropical gyre, a location having relatively weak horizontal gradients. In other locations the RI and RD constructions would differ significantly.

4.4. Internal Waves

We now consider ocean variability on scales much smaller than the loop range. One would not attempt deterministic mapping of such features. Rather, the object is to relate the *statistics* of small-scale ocean variability to the travel-time statistics. The geographic distribution of small-scale variability is of great oceanographic interest. The statistics also provide information on the time limit for coherent processing, which is an important element of the tomographic recording strategy (chapter 5), and on the uncertainties in the measured travel times, which need to be known for proper inversion procedures (chapter 6). Even for mesoscale processes, though they can be tomographically mapped, it will probably prove more interesting to treat them statistically in megameter transmissions.

Perhaps the most characteristic property of acoustic transmissions is their inherent instability. Fade-outs, or scintillations, are the rule rather than the exception. Early attempts at understanding these acoustic scintillations were in terms of a homogeneous, isotropic ocean "fine structure." Although that assumption

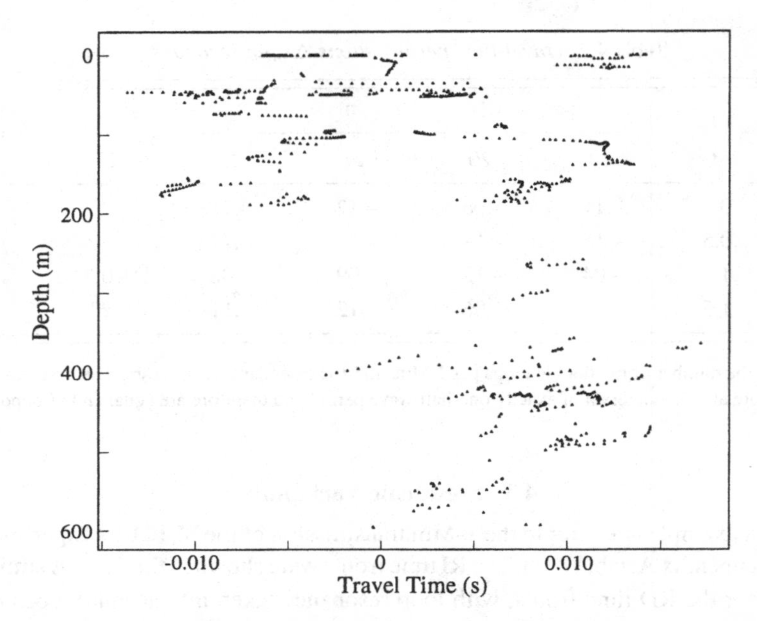

Fig. 4.5. The *difference* in travel times τ_n (RD) $-$ τ_n (RA) as a function of the upper turning depths $\tilde{z}_n^+$ for the SLICE89 experiment. The difference has been computed for each of the 50 receiver depths. The rms difference is 7.5 ms.

was mathematically convenient, it had nothing to do with reality: the fine structure of the ocean is neither homogeneous nor isotropic. As understood today, the fine structure is dominated by internal waves. The "GM model" of internal wave statistics (Garrett and Munk, 1972; Munk, 1981) was developed from a synthesis of diverse oceanographic observations.[4] Computations by Munk and Zachariasen (1976) of acoustic phase and intensity variances based on the GM model led to reasonable agreement with a transmission experiment in the Straits of Florida conducted 10 years earlier (Steinberg and Birdsall, 1966; Clark and Kronengold, 1974). The GM model has served as a convenient reference for subsequent work by Dashen, Flatté, and others on sound transmission through a fluctuating ocean (Flatté *et al.*, 1979; Flatté, 1983). Here it should be emphasized that not all ocean fine structure is due to internal waves and that any measured realization of internal waves is only crudely represented by the GM spectrum. The GM model serves for purposes of illustration, in a manner similar to the polar and temperate model representations of the sound channel.

[4]The GM spectrum does not allow for internal tides, which may have important acoustic effects (Colosi *et al.*, 1994; Dushaw *et al.*, in press).

Internal wave scattering. Travel-time fluctuations at frequencies above one cycle per day (cpd) are predominantly caused by sound-speed perturbations due to the vertical displacements associated with internal waves (Flatté *et al.*, 1979). High-frequency fluctuations in *differential* travel times from reciprocal transmissions are principally caused by internal wave currents. Flatté (1983), Flatté and Stoughton (1986, 1988), and Colosi *et al.* (1994) have given detailed discussions and presented numerical predictions at 1-Mm range in the North Pacific.

The rms travel-time fluctuation due to internal waves is given by

$$\tau_{\mathrm{rms}} = \Phi/\omega, \tag{4.4.1}$$

where ω is frequency, and where the "strength parameter" Φ is the rms phase fluctuation due to internal waves. Φ is a measure of the strength of the sound-speed fluctuations induced by internal wave vertical displacements, according to

$$\Phi^2 = \left\langle \left(\frac{\omega}{C_0} \int \frac{\Delta c(\mathbf{x}, t)}{C_0} ds \right)^2 \right\rangle . \tag{4.4.2}$$

The integral is over the equilibrium ray path in the absence of internal waves. These fluctuations are of low frequency, and they peak at the inertial frequency; they are referred to as "wander." Because Φ is proportional to ω, the wander τ_{rms} is independent of acoustic frequency. Φ is proportional to $\sqrt{r}$. For $r = 1$ Mm, numerical calculations give τ_{rms} of order 10 ms for purely refracted rays, and of order 1 ms for surface-reflected rays (fig. 4.6).

At sufficiently long range and/or high frequency, internal-wave-induced sound-speed perturbations not only cause pulse travel times to fluctuate but also cause the equilibrium ray path to break up into *micropaths*. The received signal is then not simply a replica of the transmitted signal, but is the sum of unresolved signals propagating over many micropaths, all with slightly different travel times. This gives rise to a pulse "spread." To predict when pulse spreading is significant, it is necessary to define another parameter, the "diffraction parameter" Λ. It is a weighted average along the equilibrium ray of $(R_F/L_V)^2$, where R_F is the *Fresnel-zone* radius[5] and L_V is the vertical correlation length of the fluctuations. The reader should consult the previously cited references for a detailed definition and discussion of the diffraction parameter. For our purposes, the principal point is that Φ and Λ can be used to determine whether the propagation is *unsaturated, partially saturated*, or *saturated* (fig. 4.7).

[5]The Fresnel-zone radius is a measure of the width of the *ray tube* making up the equilibrium ray. Travel times of rays within this tube differ by less than $2\pi/\omega$.

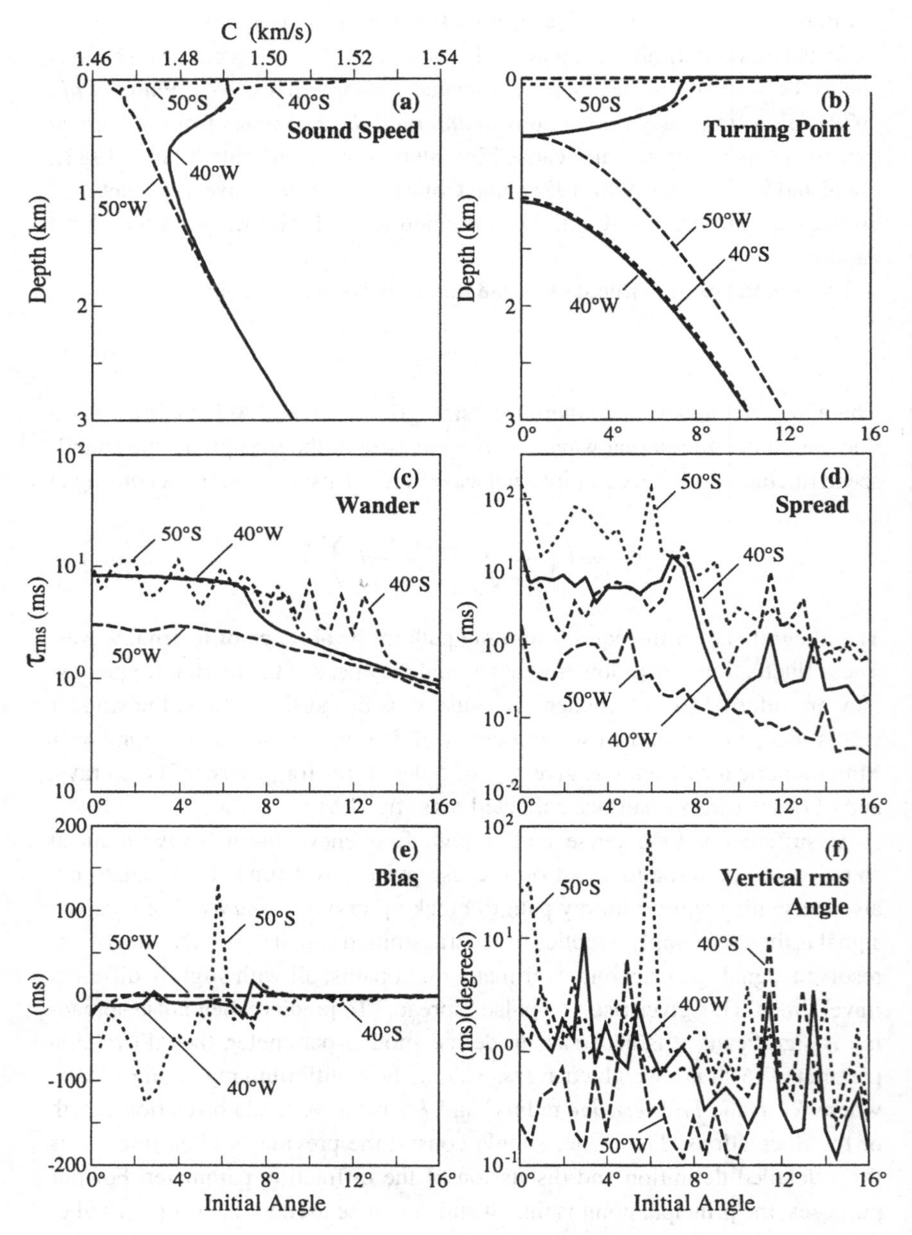

Fig. 4.6. Internal-wave-induced acoustic fluctuation parameters at 1 Mm range for a source at 500 m depth in the North Pacific: (a) sound-speed profiles from the North Pacific at 40°N and 50°N, in winter and summer; (b) upper and lower turning depths of the rays; (c) rms travel-time fluctuations; (d) characteristic time spread of an acoustic pulse; (e) average bias in the arrival time of an acoustic pulse; (f) fluctuations in vertical wave-front tilt. (Adapted from Flatté and Stoughton, 1988.)

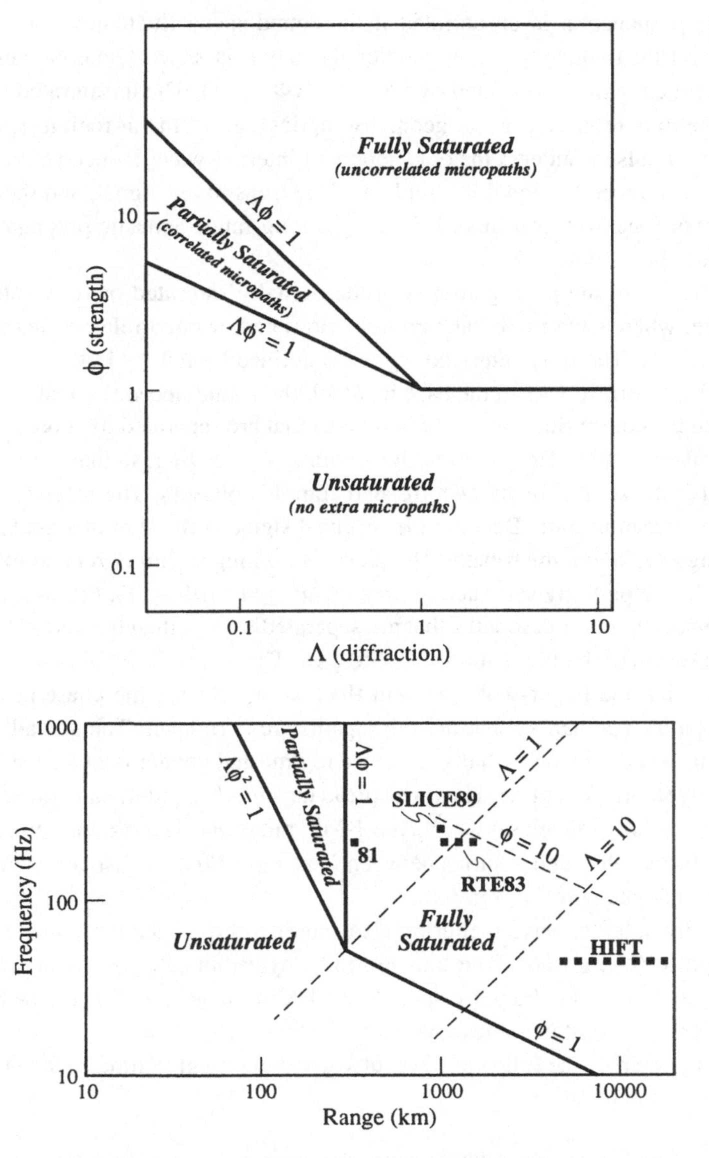

Fig. 4.7. Sound transmission regions in $\Lambda - \Phi$-space (top) and translated into range-frequency space (bottom). The translation assumes a long-range, steep ray in the temperate sound channel given in chapter 2 and scattering dominated by internal waves. The unsaturated regime is also referred to as the geometric-optics regime; as drawn, it includes the so-called Rytov extension (Flatté *et al.*, 1979.) Points refer to major tomography experiments (see table A.1 for references).

The propagation is unsaturated if the sound-speed fluctuations are weak, the acoustic frequency is low, and/or the range is short. Quantitatively, the unsaturated regime is defined by $(\Lambda \leq 1,\ \Lambda\Phi^2 \leq 1)$. The unsaturated regime is sometimes referred to as the geometric-optics regime. In this regime, spread is zero, and pulse wander is the only source of internal-wave-induced travel-time noise. The received signal is a replica of the transmitted signal, and the phase is proportional to travel time. It is rare for long-range acoustic propagation to fall into this regime.

More often, the propagation is in the partially saturated or fully saturated regime, where each ray breaks up into correlated or uncorrelated micropaths, respectively. The fully saturated regime is defined by $\Lambda\Phi \geq 1,\ \Phi \geq 1$. Qualitatively, it corresponds to the case in which the sound-speed fluctuations have broken the equilibrium ray into micropaths that are separated by more than the vertical correlation length of the fluctuations ($R_F \gg L_V$), so that the received signal is the sum of many uncorrelated complex phasors. The received signal is then spread in time. Because the received signal is the sum of signals propagating over many micropaths, the phase is no longer simply related to travel time. In the partially saturated regime $(\Lambda\Phi^2 \geq 1,\ \Lambda\Phi \leq 1)$, the equilibrium ray breaks up into micropaths that are separated by less than the vertical correlation length of the fluctuations ($R_F \ll L_V$). The received signal is still spread in time, but the larger-scale medium fluctuations change the phase of all the micropaths together, so that the micropaths are correlated. The spread is less than the wander in the partially saturated regime and greater than the wander in the fully saturated regime. The time spread depends logarithmically on acoustic frequency, and it increases as r^2. At a 1-Mm range and 100 Hz, the spread is of order 10 ms, although it varies between 1 ms and 100 ms for small changes in the sound-speed profile (fig. 4.6).

Finally, internal wave scattering also causes a shift in the mean arrival time of a pulse (*i.e.*, a bias). The bias depends logarithmically on frequency and increases as r^2 (like the pulse spread). At 1 Mm range and 100 Hz, the bias is of order 10 ms, but in some cases reaches 100 ms (fig. 4.6).

To summarize, the following formulae give very rough estimates for a typical saturated condition:

$$\begin{aligned} \text{wander } \tau_{\text{rms}} &= \tau_0 \, (r/r_0)^{\frac{1}{2}} , \\ [\text{spread, bias}] &= \tau_0 \, (r/r_0)^2 \, \ln{(f\tau_{\text{rms}})} , \end{aligned} \tag{4.4.3}$$

for $f\tau_{\text{rms}} > 1$, with $r_0 = 1$ Mm and $\tau_0 = 0.010$ s for moderately steep rays. Wander, spread, and bias all increase with decreasing ray steepness θ_0 (fig. 4.6). At the onset of saturation, $f\tau_{\text{rms}} = 1$, giving zero spread and zero bias. (A more

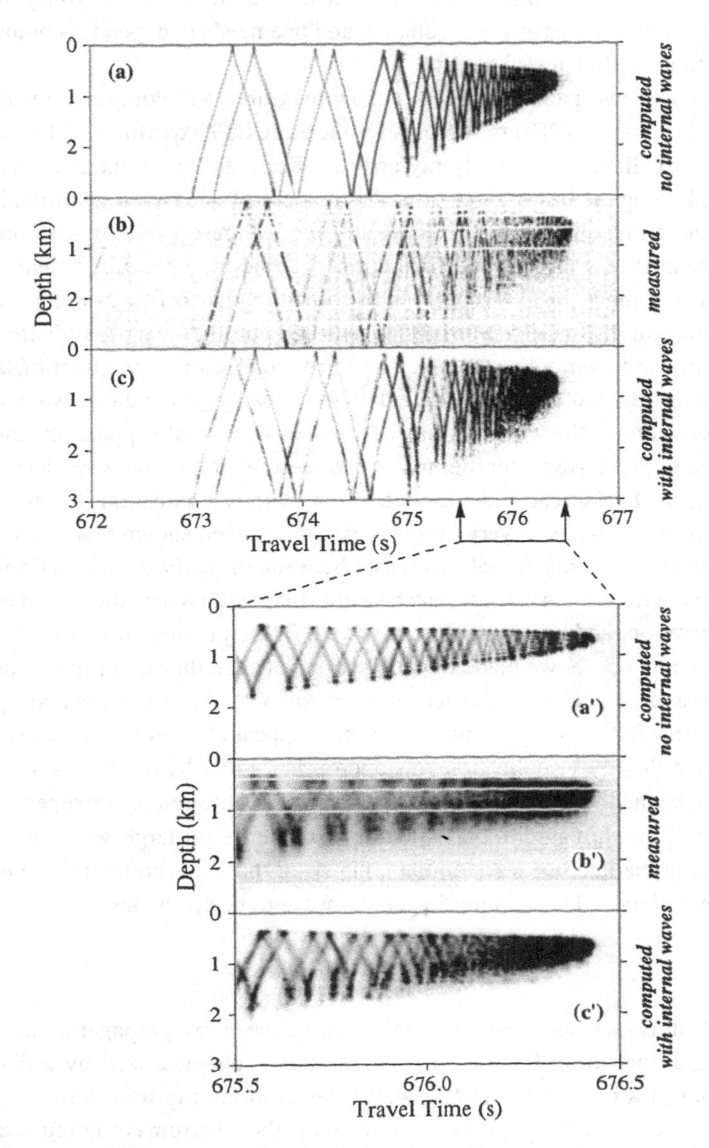

Fig. 4.8. A comparison between the simulation data without (a) and with (c) internal-wave sound-speed fluctuations, and a pulse from the SLICE89 experiment (b). The prominent horizontal data gaps in (b) and (b′) are the results of hydrophone failures. The lower panels show the last second on an enlarged scale, as indicated by the arrows. For the acoustic simulation, the center frequency is 250 Hz and the bandwidth is 102.2 Hz. Parameters for the internal wave field are $E_0 = 2.0$ and $j_m = 49$.

accurate theory gives finite bias.) In the near-axial limit, the foregoing formulation becomes increasingly unreliable, and one needs to depend on numerical simulations, as illustrated next.

The upper two panels of fig. 4.8 show measured and computed [using the parabolic equation (PE)] time fronts for the SLICE89 experiment.[6] The agreement is excellent for the early ray arrivals. There are two remaining discrepancies. During the last 0.3 s prior to the final cutoff, measured ray arrivals can no longer be resolved, whereas RI theory predicts resolvable ray-like arrivals until about 0.15 s prior to the final cutoff. Further, the measured final cutoff beneath and above the axis is delayed by something like 0.1–0.2 s relative to the computed cutoff. In other words, the measured cutoff "wedge" is *blunter* than that computed from either theory. The third panel shows the effect of taking internal waves into account in calculating a broadband parabolic wave equation, according to the work of Colosi *et al.* (1994). It would appear that the two discrepancies referred to earlier may be the results of internal wave scatter.

The growth of spread and bias with r^2 has serious implications for the application of tomography at very large ranges. It has been shown that the interval between adjoining ray arrivals decreases between the early steep rays (large θ_0) and late flat rays (2.6.4). There may be a time (angle θ_0^*) where the spread equals the separation, and rays cannot be resolved. In the previous ray/mode discussion of an RI ocean, we made the distinction between the early ray-like arrival patterns and the late mode-like termination. Ray visualization is not appropriate for the very flat rays (very small θ_0). It now appears that even for values of θ_0 for which the ray visualization was appropriate in an RI ocean, a distinction needs to be made on the basis of RD perturbations between rays steeper or flatter than θ_0^*. With regard to bias, there is no evidence for large-scale long-term changes in the internal wave climate, implying that one can treat these biases as time-invariant. This assumption could prove to be erroneous.

4.5. Ray Chaos

Fig. 4.9 illustrates an important distinction between ray propagation in an RI ocean and that in an RD ocean. The RI case is characterized by a discrete spectrum of a ray orbit $z(x)$ (fig. 4.10); neighboring ray trajectories diverge slowly, according to a power law. In the RD case the spectrum is continuous; rays are characterized by rapid, exponential-like divergence from their neighbors; the number of eigenrays grows exponentially with range, and their intensity decreases at the same exponential rate. This is called ray chaos.

[6] A similar representation has been shown in fig. 2.21.

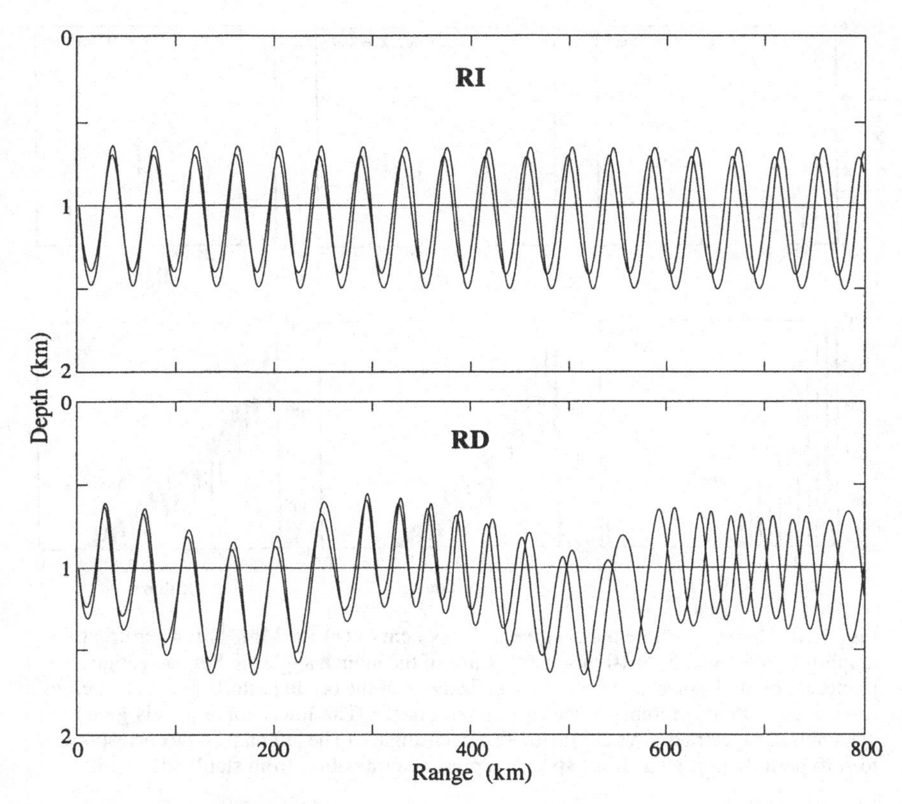

Fig. 4.9. Ray trajectories for axial launch angles of 3.00° and 3.01° in an RI temperate ocean (upper panel) and in the temperate ocean with RD Rossby-wave perturbation (lower panel). The perturbations consist of a random-phase superposition of the first four baroclinic modes, with a sound-speed perturbation of 5 m/s at 1 km depth, in qualitative agreement with observations at temperate latitudes. (From Smith *et al.*, 1992*b*.)

Figs. 4.9 and 4.10 are taken from Smith *et al.* (1992*a,b*). The RI state is represented by a temperate sound-speed profile. Range dependence is associated entirely with a long ($\lambda > 250$ km) random phase perturbation superimposed on the RI temperate profile. In accordance with the work of Smith and associates and the literature cited in their papers, *any* RD profile is chaotic; the only question is how far from the source one can go until chaos is manifest (the examples cited all refer to a range of 3277 km). Accordingly, these authors define a "predictability horizon": within this horizon the forward problem is well posed, and the concepts previously developed apply. It is not clear to us what the situation is beyond the predictability horizon. In the preceding section we have discussed

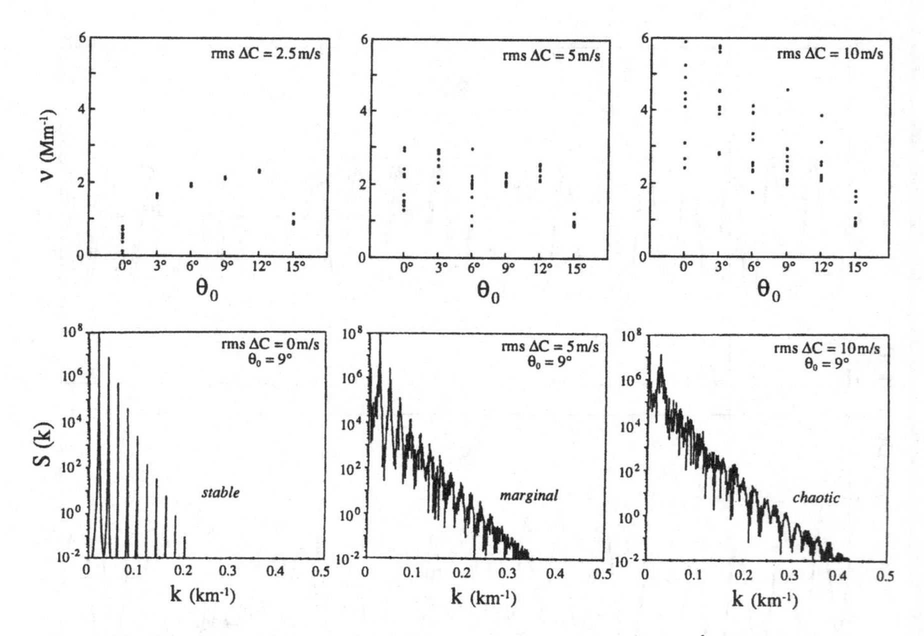

Fig. 4.10. The upper three panels give the exponents $\nu(r)$ (in Mm^{-1}) for perturbation amplitudes $\Delta C = 2.5, 5, 10$ m/s as functions of the launch angle θ_0. For each angle, 10 points are plotted, corresponding to 10 realizations of the ocean perturbation, each being associated with a different choice of random phases. The lower three panels give the wavenumber spectra of rays $z(x)$ with 9° launch angles. (The left panel now corresponds to zero perturbation.) The three spectra represent a transition from stable to chaotic.

the statistical effects of internal waves in terms of a "saturation" concept. This concept may have an interpretation in the framework of chaos theory.

We require a quantitative estimate for the predictability horizon. Smith and associates define an exponent $\nu(r)$ that is a measure of ray divergence, with vertical launch angle at range r. This "Lyapunov exponent" is then estimated asymptotically in the limit $\nu_\infty = \nu(r)$ as $r \to \infty$. Fig. 4.10 shows the values of $\nu(r)$ at $r = 3277$ km for 10 realizations of the mesoscale perturbation. A close grouping of points indicates that the asymptotic values have not been reached, and all that can be said is that ν_∞ is less (possibly much less) than $\nu(r)$. On the other hand, a *spread* of values indicates that the approximate asymptotic values have been reached. Each value is then an estimate of ν_∞, and it is reasonable to assign to ν_∞ the average value $\bar{\nu}$ of the estimates. To summarize:

$$\begin{aligned} \text{realizations clumped}: &\quad \nu_\infty < \bar{\nu}(r)\,, \\ \text{realizations spread}: &\quad \nu_\infty = \bar{\nu}(r)\,. \end{aligned} \qquad (4.5.1)$$

Table 4.3. *Predictability horizons for a temperate ocean with baroclinic Rossby-wave perturbations ΔC at 1 km depth, for rays with axial inclinations θ_0*

rms ΔC at 1 km	θ_0		
	0° (axial)	6°	12° (SLR)
2.5 m/s	10 Mm	> 2.5 Mm [13 Mm][a]	> 1.5 Mm [17Mm]
5 m/s	2 Mm	2.5 Mm [2.5 Mm]	> 1.5 Mm [3.3 Mm]
10 m/s	1.2 Mm	1.5 Mm	2 Mm

[a] Values in brackets are guesses based on extrapolation of the first column and third line of the table.

The "predictability horizon" is given by

$$r_{ph} = \text{constant} \times \nu_{\infty}^{-1} , \tag{4.5.2}$$

with the constant rather larger than 1 (Smith and associates set the constant equal to 5). Using the results shown in fig. 4.10, a greatly oversimplified summary is given in table 4.3. The middle lines correspond to $\Delta C = 5$ m/s (1°C) perturbation amplitude, a value representative of temperate latitudes. The ray $\theta_0 = 12°$ is the surface-limited ray (SLR). As expected, r_{ph} increases with decreasing amplitude of the perturbation (left column). Table 4.3 also shows that r_{ph} increases with increasing ray steepness (bottom line). The three lower limits in the remaining table are then so low that they do not give useful limits. Extrapolating the measured trend with perturbation amplitude (left column) and with ray steepness, the appropriate prediction horizons *might* have the values given in brackets. At megameter range in a variable climate of mesoscale activity, we expect that moderately steep rays (usually predictable) will be intermittently unpredictable and that moderately flat rays (usually unpredictable) will be intermittently predictable. Steep rays in a weak mesoscale field have prediction horizons so large that they are predictable even at antipodal ranges.

The implications of these results for ray tomography are not clear (Collins and Kuperman, 1994). With regard to modes, we note that ray chaos is, by definition, a manifestation of the asymptotic formalism used to define ray equations. The acoustic field equations underlying the modal representation do not admit chaotic solutions.

4.6. Modes in a Range-Dependent Profile

We require a generalization of the RI derivation in (2.10). The wave equation (2.31)

$$\left(\nabla^2 - S^2(z;r)\frac{\partial^2}{\partial t^2}\right) p = 0 \tag{4.6.1}$$

has a separable solution[7]

$$p(r,z,t) = \sum_m Q_m(r)\, P_m(z;r) e^{i\omega t}, \tag{4.6.2}$$

where $r = r(x, y)$. Writing

$$k(z;r) = \omega S(z;r), \quad \kappa_m(r) = \omega s_{p,m}(r), \tag{4.6.3}$$

for the scalar wavenumber and its horizontal projection (previously k_H), the wave equations can be separated as follows:

$$\left[\frac{\partial^2}{\partial z^2} + k^2(z;r)\right] P_m(z;r) = \kappa_m^2(r) P_m(z;r), \tag{4.6.4}$$

$$\left[\frac{\partial^2}{\partial r^2} + \kappa_n^2(r)\right] Q_n(r) = \sum_m \left[A_{mn} Q_n + B_{mn} \frac{\partial}{\partial r} Q_n\right], \tag{4.6.5}$$

where

$$A_{mn} = \int dz\, P_n \frac{\partial^2}{\partial r^2} P_m, \quad B_{mn} = 2\int dz\, P_n \frac{\partial}{\partial r} P_m. \tag{4.6.6}$$

The integration is over the entire water column.

For the RI case, the right-hand side of (4.6.5) vanishes, and we are back to the solutions discussed in chapter 2. For RD variations with scales large compared with the double-loop scale,[8] the right-hand side can again be set to zero, yielding the adiabatic mode equations. Under adiabatic conditions, a mode generated at the source propagates independently of other modes, maintaining its energy to any range r (Milder, 1969; Desaubies *et al.*, 1986; Brekhovskikh and Lysanov, 1991). The modal wave function is modified along the propagation path in accordance with the *local* sound-speed profile. The approximation is useful for megameter propagations in regions without sharp frontal zones, and in the absence of intense mesoscale activity (section 8.4).

[7] The notation $(z;r)$ implies a relatively weak dependence on r.

[8] The double-loop scale can be given a modal interpretation in terms of the interference of modes of neighboring orders (Brekhovskikh and Lysanov, 1991, p. 137).

In the cases where the right-hand side of (4.6.5) cannot be neglected, modes of like frequency (ω) interact. Each mode travels at its own group velocity. As modes propagate from the source, they become separated in space. This is the normal modal dispersion that takes place without consideration of mode coupling. When coupling does occur, any mode n can become the source of another mode m, and since the modes are already dispersed in space, this will lead to a spread of mode m in arrival times.[9] The recorded spread is then a measure of ocean perturbation and could be used as a tomographic observable.

The "exact adiabatic" (EA) travel-time perturbation is given by

$$\tau_m - \tau_m(-) = \int_0^r dr\ [s_{g,m} - s_{g,m}(-)]\,, \tag{4.6.7}$$

where $s_{g,m}$ and $s_{g,m}(-)$ are the perturbed and unperturbed RD group slownesses for mode m, computed *locally* in accordance with RI formulae (2.10.7) and (2.10.8). Similarly, a linearized adiabatic (LA) perturbation treatment leads to

$$\Delta\tau_m = \int_0^r dr\ \Delta s_{g,m}\,, \tag{4.6.8}$$

where $\Delta s_{g,m}$ is the expression in (2.14.8).

A very efficient method of solving the RD wave equation for waves propagating close to the horizontal is given by Brekhovskikh and Lysanov (1991, sect. 7.4) following Tappert (1977). This is called the parabolic-wave-equation (PE) method, and it has found widespread use in underwater acoustics. The PE solution is applicable in strongly range-dependent environments in which mode coupling is severe.

Mesoscale eddies. Table 4.4 gives the LA, EA, and PE results for perturbations by a single mesoscale eddy at midrange and middepth embedded in a temperate sound channel (Shang and Wang, 1993*a,b*; Taroudakis and Papadakis, 1993). As presented in the table, the numbers are in the order of increasing accuracy. We note the following:

(a) For a typical eddy strength (5 m/s), the PE and EA perturbations are within 1 ms, but for a very strong eddy (12.5 m/s) the difference can be several tens of milliseconds. (The strong warm eddy splits the sound channel into a double channel.)
(b) The LA perturbation is off by several tens of milliseconds for the typical eddy, but in error by hundreds of milliseconds for the strong eddy.

[9] This is over and above the frequency dispersion of any one mode generated by a broadband source.

Table 4.4. *Travel-time perturbations (in ms) for a weak and a strong warm eddy and a strong cold eddy, using the linearized adiabatic theory LA (4.6.8), the "exact adiabatic" theory EA (4.6.7), and the parabolic equation PE*

Eddy strength		+5 m/s	+12.5 m/s	−12.5 m/s
Mode 1	LA	−392	−980	980
	EA	−385.7	−717	990
	PE	−386.3	−680	
Mode 5	LA	−340	−849	849
	EA	−316.1	−714	925
	PE	−315.9	−699	
Mode 10	LA	−246	−616	616
	EA	−237.1	−582.4	722
	PE	−236.7	−582.1	

Note: One eddy of horizontal scale 100 km and vertical scale 0.5 km centered at 1 km depth and 300 km range. Total range $r = 600$ km. Frequency 50 Hz.

Table 4.5. *PE amplitudes at the receiver for single-mode generation at the source and 600-km range*

	$\Delta C = 5$ m/s					$\Delta C = 10$ m/s				
m	1	2	3	4	5	1	2	3	4	5
1	.95	.25	0	0	0	.58	.64	.43	.17	.03
2	.25	.95	.01	0	0	.64	.22	.64	.30	.05
3	0	.01	.98	0	0	.43	.64	.42	.44	.09
4	0	0	0	.98	0	.17	.30	.44	.72	.35
5	0	0	0	0	.98	.03	.05	.09	.35	.67

(c) The magnitude of the perturbation is larger for a cold eddy than for a warm eddy, suggesting that the nonlinearity is the result of the deformation of the wave functions.

Table 4.5 shows the result of mode coupling, with the same model parameters as in table 4.4. A single mode of unit amplitude is generated at the source; the table shows the distribution of modal amplitudes at the receiver. For the adiabatic case, this results in a unit diagonal matrix. The weaker eddy results

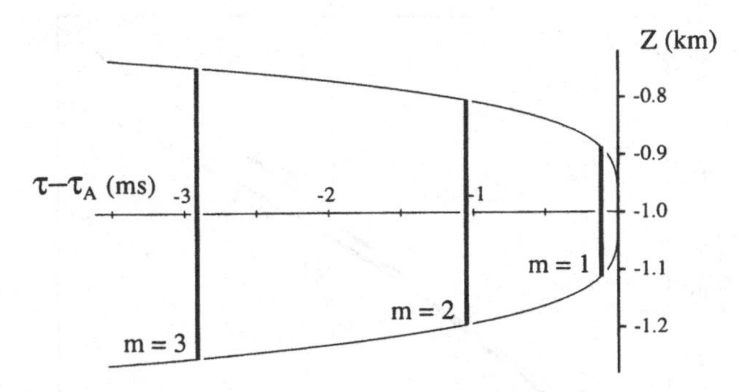

Fig. 4.11. Arrival times τ_m of 250 Hz modes in a temperate profile at 1 Mm range, with modes represented by vertical lines between $\tilde{z}_m^{\pm}$.

in nearest-neighbor coupling at low modes. For the strong eddy, the coupling is significant everywhere. In the case of an initial generation of mode 2 only, 80% of the energy at the receiver is associated with modes 1 and 3. Details are, of course, highly model-dependent.

The implication for megameter tomographic transmissions is significant (the tables are computed for a 600-km range). Fig. 4.11 sketches the modal arrival pattern at 1 Mm in the absence of mode coupling. For low modes (small $\tilde{\phi}$) in a temperate profile,

$$\tau_m = r s_{g,m} = r S_A[1 - \tfrac{1}{48}\,\gamma_a h\,\tilde{\phi}_m^4], \quad \tilde{\phi}_m^2 = 6\,(m - \tfrac{1}{2})/(f/F_T),$$
$$\tilde{z}_m^{\pm} = z_A \pm 2^{-\frac{1}{2}} h\,\tilde{\phi}_m, \tag{4.6.9}$$

from (2.11.8*m*) and (2.5.14), respectively. Eliminating $\tilde{\phi}$, the expression for the $\tau, \tilde{z}$-wedge is given by

$$\tilde{z}_m^{\pm} = z_A \pm K(\tau_A - \tau_m)^{\frac{1}{4}}, \quad K^4 = 12\,h^3/(\gamma_a\,\tau_A), \tag{4.6.10}$$

where $\tau_A = r S_A$. The small exponent $\frac{1}{4}$ is associated with the bluntness of the wedge. For a distributed eddy system with nearest-neighbor coupling only, the unscattered mode 1 arrives at $\tau_1 = s_{g,1} r$. Scattering of mode 2 into mode 1 takes place all along the path. For scattering close to the source, the arrival time is nearly τ_2; for scattering near the receiver, it is nearly τ_1. Thus the arrival is spread between τ_2 and τ_1, and the average arrival is advanced. For mode 2, the earliest arrival is at τ_3, and the latest at τ_1. Thus the scattering has the potential for smearing the late, closely spaced, modal arrivals (and associated ray arrivals) and for further blunting the final arrival wedge (fig. 4.8).

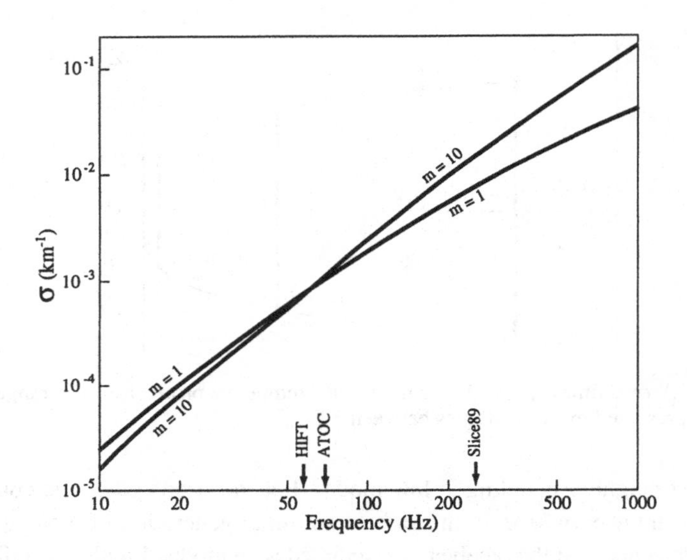

Fig. 4.12. Internal-wave scattering in a temperate ocean: transition probability σ_m of energy transfer between modes m and $m \pm 1$ per kilometer. (From Dozier and Tappert, 1978*b*.)

Internal waves. Dozier and Tappert (1978*a,b*) have computed the acoustic mode coupling associated with a GM internal wave perturbation in a temperate ocean. The coupling is predominantly a nearest-neighbor coupling, with only weak dependence on the mode number m (fig. 4.12).

The mode coupling is surprisingly strong. The transition probability σr over the range r is of order 1 in a 1-Mm transmission at 70 Hz, and in a 100-km transmission at 250 Hz. Suppose all the initial energy is in mode 1. The relaxation range for distributing this energy among modes 1 to M is $2000M$ km at 70 Hz, and $30M$ km at 250 Hz. (The analytical and numerical procedures of the Dozier and Tappert papers are very complex, and one would like to have the results confirmed by an independent analysis.)

Consider the case of scattering among neighboring modes generated with the same energy E. The range is now taken sufficiently short that the probability of multiple scatter is small. For some range interval between x and $x + \delta x$, a fraction of energy $\sigma E\, \delta x$ is scattered from $m + 1$ into m, giving a travel time

$$x\, s_{g,m+1} + (r - x)\, s_{g,m},$$

and similarly for $m - 1$ into m. Mode m loses $\sigma E\, \delta x$ energy to both $m + 1$ and $m - 1$, hence $E(1 - 2\sigma r)$ arrives at $x = r$ after travel time $r s_{g,m}$. Integrating over all such range intervals δx yields the energy-weighted mean group slowness

$$\begin{aligned}\overline{s_{g,m}} &= \frac{\sigma}{r}\int_0^r dx\left[x s_{g,m+1} + x s_{g,m-1} + 2(r-x)s_{g,m}\right]\\ &\quad + (1-2\sigma r)\, s_{g,m} \\ &= s_{g,m} - \sigma r\left[s_{g,m} - \tfrac{1}{2}\left(s_{g,m+1} + s_{g,m-1}\right)\right].\end{aligned} \tag{4.6.11}$$

The result is consistent with a probability distribution in travel time

$$\begin{aligned} p_m(\tau) &= (1-2\sigma r)\ \delta(\tau - \tau_m) + \frac{\Delta_m\, \sigma r}{\tau_m - \tau_{m+1}} + \frac{\Delta_{m-1}\, \sigma r}{\tau_{m-1} - \tau_m}, \qquad m \geq 2, \\ p_1(\tau) &= (1-\sigma r)\ \delta(\tau - \tau_m) + \frac{\Delta_1\, \sigma r}{\tau_1 - \tau_2}, \end{aligned} \tag{4.6.12}$$

where $\delta(x)$ is the delta-function, and $\Delta_m = 1$ for $\tau_{m+1} \leq \tau \leq \tau_m$, and zero otherwise. Thus the scattered energy-averaged travel time is given by

$$\begin{aligned}\overline{\tau_m} &= \int_{\tau_{m+1}}^{\tau_{m-1}} d\tau\, \tau\, p_m(\tau) = \tau_m - \sigma r\left[\tau_m - \tfrac{1}{2}\left(\tau_{m+1} + \tau_{m-1}\right)\right], \\ \overline{\tau_1} &= \tau_1 - \tfrac{1}{2}\sigma r(\tau_1 - \tau_2),\end{aligned} \tag{4.6.13}$$

in agreement with (4.6.11). The smearing of the modal arrival times by internal waves might be related to the observed loss of resolution of late ray arrivals (fig. 4.8).

We shall evaluate (4.6.13) for a temperate profile. From (4.6.9),

$$\overline{\tau_m} = \tau_m - L\,\sigma\, r^2\, S_A, \qquad L = \frac{3\,\gamma_a h}{4(f/F_T)^2},$$

for all m, including the special case $m = 1$. Thus, travel time is equally diminished for all modes alike, and the arrival pattern is uniformly advanced by $L(\sigma\, r)\,(r\, S_A)$. For reference, the interval between the last two modes is $2L\, r\, S_A$, which equals 1 ms at $f = 250$ Hz at a range $r = 1$ Mm (fig. 4.11). Setting $\sigma = 10^{-3}$ to 10^{-2} km^{-1} in accordance with fig. 4.12 yields a bias $\Delta\tau = -12$ ms to -10 ms. This estimate goes well beyond the assumption of single scattering. The foregoing result applies only to the very special assumptions of (i) no multiple scattering, (ii) nearest-neighbor scattering, (iii) equal excitations of all modes, and (iv) a temperate profile. Dropping any one of these assumptions might lead to a blunting of the final arrival wedge.

From our view, these effects will be taken as time-invariant and therefore as not contributing to time-dependent inverse problems.

4.7. Horizontal Refraction

So far it has been assumed that the transmission paths remain in the vertical plane between source and receiver. Yet it is well known that horizontal gradients in sound-speed cause horizontal deflections in ray paths. Here the mesoscale eddies and their powerful cousins, such as Gulf Stream rings, are probably the most important features. We shall show that, in general, the deflections are small, a few degrees at most. But at large ranges, eddies can produce horizontal multipaths, and these might have an important effect. Horizontal multipaths have not been taken into account in tomographic work so far.

In long-range tomographic experiments it is difficult to avoid seamounts and other bathymetric features. The refractive effects of the bathymetric features generally exceed those of the hydrographic features and play crucial roles in acoustic propagation studies. We shall give but a brief account of bathymetric refraction at the end of the chapter. The emphasis here is on the oceanic temporal perturbations, and bathymetric effects are generally beyond the scope of this monograph.

Mesoscale eddies. Here we consider a hybrid model with modes in the vertical and rays in the horizontal (Dysthe, 1991). The phase velocity c_p is determined by the *local* profile in accordance with the adiabatic approximation and equals the sound-speed $\widetilde{C}$ at the local turning depth $\widetilde{z}$ (2.10.4). For axial rays and modes, we have approximately $c_p = C_A$. In general, the modal phase velocity depends on mode number and frequency. Thus, for a given mode number, there will be a bundle of horizontal rays subtending a beam width that will depend on the frequency bandwidth Δf; further, there will be separate ray bundles for different modes m. The angular departures from a straight path are generally quite small, but measurable. The effects become important for global transmissions.

Snell's law in cylindrical coordinates can be written

$$r\, s(r) \cos \gamma(r) = a\, s_0 \cos \gamma_0, \quad a = r_0\,, \tag{4.7.1}$$

where $s(r) = 1/c_p$ is the phase slowness, and the subscript 0 refers to conditions at the entrance point A (fig. 4.13). The total angular deflection is given by (Munk, 1980)

$$\beta = 2 \int_0^{\gamma_0} d\gamma \, \frac{1}{\tan \gamma} \, \frac{1}{s} \, \frac{ds}{d\gamma}\,. \tag{4.7.2}$$

For a simple model, we take

$$\frac{s}{s_0} = 1 + \nu \, \frac{a^2 - r^2}{a^2}, \tag{4.7.3}$$

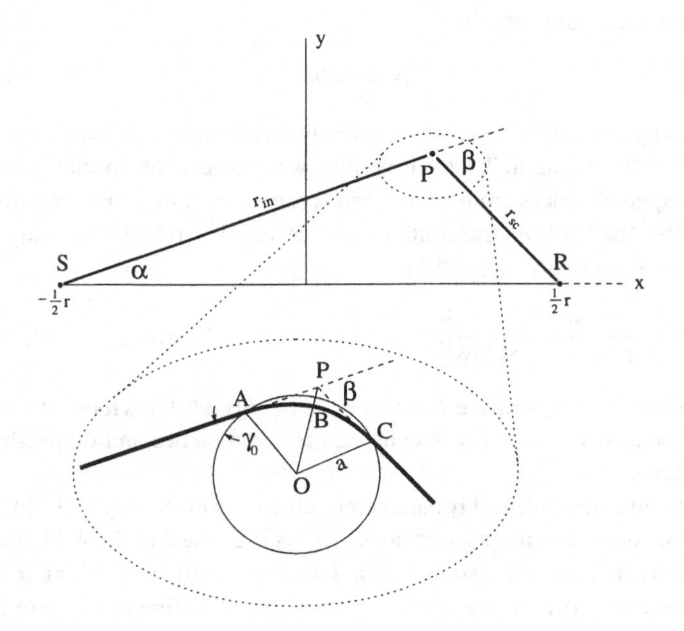

Fig. 4.13. Ray paths S-A-B-C-R through a cold eddy at P. The ray forms an angle γ with the local circular contours s = constant. γ varies from γ_0 at the entrance A to $\gamma = 0$ at the "zenith" B.

so that $s = s_0(1 + \nu)$ is the fractional sound-speed *defect* at the eddy center. Thus $\nu = +10^{-2}$ refers to –15 m/s (–3° C) at the eddy center relative to the surrounding waters. From (4.7.1),

$$\frac{1}{s}\frac{ds}{d\gamma} = -2\,\nu\,\cos^2\gamma_0\;\sec^2\gamma\;\tan\gamma + O\,(\nu^2)\,.$$

Substituting into (4.7.2), we have simply

$$|\beta| = 2\nu\,\sin 2\gamma_0, \tag{4.7.4}$$

regardless of eddy size. There is no deflection for a ray $\gamma_0 = 90°$ through the eddy center, nor for a tangent ray $\gamma_0 = 0$. A maximum deflection

$$|\beta_{\max}| = 2\nu \tag{4.7.5}$$

occurs for $\gamma_0 = 45°$, which corresponds to OB $= a/\sqrt{2}$. The result (4.7.5) holds approximately for a variety of eddy models, and we shall use the model (4.7.3) for further reference.

An important parameter is

$$\rho = \nu r/a\,. \tag{4.7.6}$$

The quantity a/ν can be regarded as the eddy focal length, and so ρ is a measure of range to focal length. To illustrate this, we consider the special case of an eddy at equal distances from source and receiver, with its center at a distance Y from the line between transmitter and receiver. The relative intensity at the receiver is given by (Munk, 1980)

$$I = \left|\frac{\sin\gamma_0}{\sin\gamma_0 + \rho\,\cos 2\gamma_0}\right|, \quad \frac{Y}{a} = (\rho\,\sin\gamma_0 - 1)\,\cos\gamma_0\,. \tag{4.7.7a, b}$$

For $\gamma_0 = 90°, Y = 0$, we have $I = |1-\rho|^{-1}$: a cold eddy has a focus at $r = a/\nu$. Quite generally, the intensity is enhanced for cold eddies and diminished for warm eddies.

For a fixed Y and given eddy parameters, eigenrays must obey (4.7.7b). There may be one or several roots γ_0, or none. This is illustrated in fig. 4.14. Take the case of a weak, cold eddy (solid lines), with $\nu = +0.01$, $r = 1$ Mm, $a = 100$ km, hence $\rho = +0.1$. As the northward traveling eddy first comes into line of sight, the ray is deflected southward (away from the eddy center), soon reaching a maximum deflection $\beta_{\max}$. It drifts back to the undisturbed direction as the eddy is centered, and so forth. There is only one ray path provided $\rho < 1$.

For a very intense small eddy we take $\nu = 0.1$ and $a = 50$ km, corresponding to a focal length $a/\nu = 500$ km. Then for $r = 1$ Mm, we have the case $\rho > 1$, and there can be one or three separate paths, depending on the eddy location. There are multipaths even when the eddy has just crossed the line of sight (but not beyond).[10]

The excess in travel time is given by (Munk, 1980)

$$\Delta\tau = \tau - r s_0 = 2(\nu a\, s_0)\ \sin^2\gamma_0\,(\tfrac{2}{3}\ \sin\gamma_0 + \rho\,\cos^2\gamma_0)\,. \tag{4.7.8}$$

Thus $\Delta\tau$ scales as $\nu a\, s_0 = O(1\text{ s})$, as expected. It properly vanishes for $\nu = 0$, for $a = 0$, and for glancing incidence $\gamma_0 = 0$. It also vanishes for $\rho = -\frac{2}{3}\ \sin\gamma_0\ \sec^2\gamma_0$, when the additional eddy path length is just balanced by the increased speed in the warm eddy. $\Delta\tau$ is always positive for a cold eddy. The extreme for $\Delta\tau$ is $\frac{4}{3}\ \nu a s_0$ for head-on incidence when ρ is small, and $\frac{1}{2}\ \nu a s_0$ for $\gamma_0 = 45°$ when ρ is large.

[10]See also Itzikowitz *et al.* (1982*a,b*). Analogous work has been done previously using fluid-filled spherical and cylindrical acoustic lenses (Boyles, 1965; Sternberg, 1987).

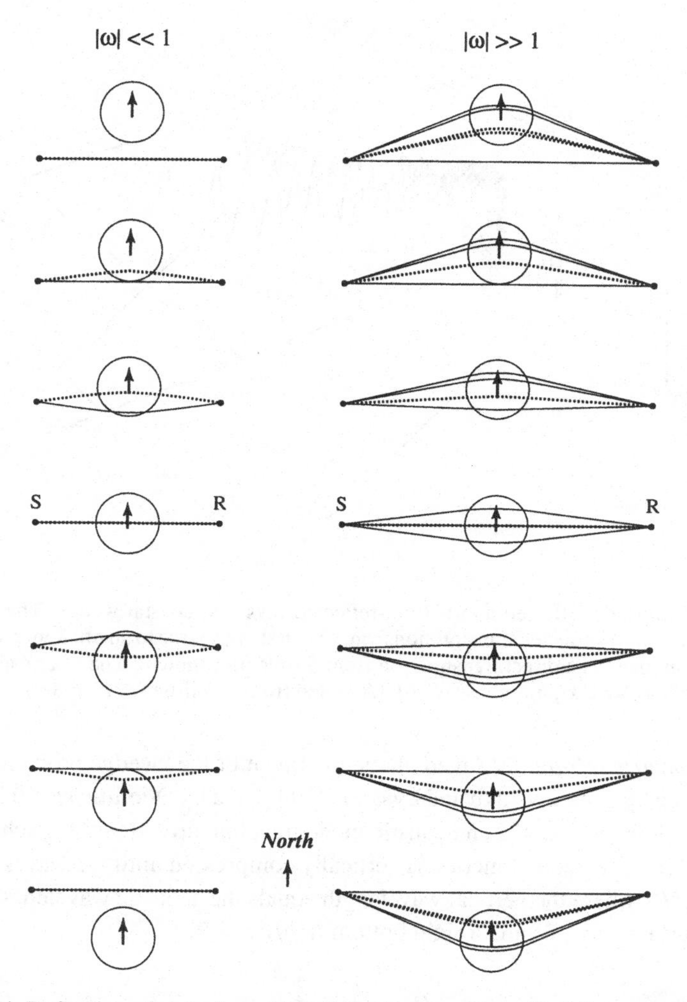

Fig. 4.14. Deflection of paths by cold (solid) and warm (dotted) eddies, or both (dash-dotted). The eddy moves northward from $Y = -\frac{3}{2}r_0$ to $+\frac{3}{2}r_0$. The left seven panels refer to short ranges $r \ll r_0/|\nu|$, and the right panels to long ranges $r \gg r_0/|\nu|$ (for a fixed eddy radius r_0 and the velocity defect $|\nu|$). For the short ranges (which is the usual situation) the apparent source direction at the receiver is deflected toward the eddy center for warm eddies, and away from the eddy center for cold eddies. There is only a single path, and the eddy has no effect once it is outside the line of sight. For long ranges (or small intense eddies) there are multiple paths, and these can persist for some distance outside the line of sight, as can be seen in the right panels.

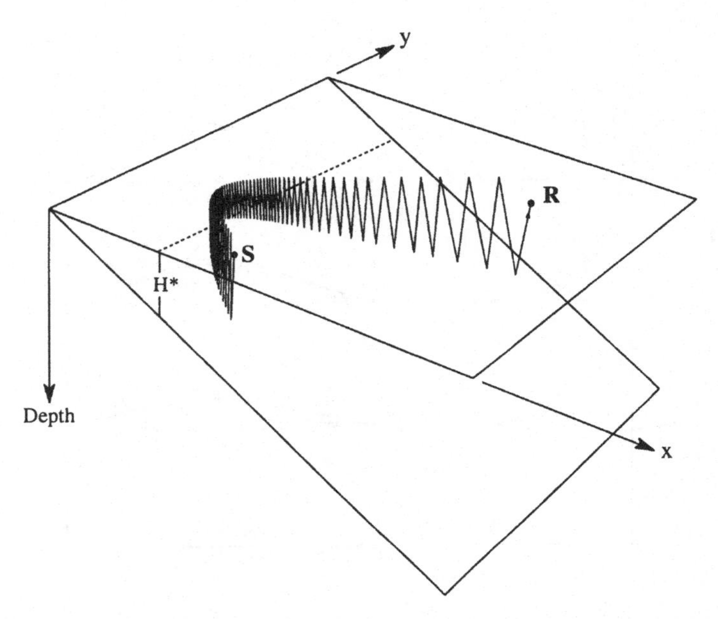

Fig. 4.15. Surface-reflected and bottom-reflected rays in a coastal wedge. The figure shows a coastally reflected transmission from a near-shore source to an offshore receiver. In addition, there is a direct transmission from S to R (not shown). The y-axis is along the shore line, with x pointing seaward. (Adapted from Doolittle *et al.*, 1988.)

Bathymetric refraction. An adiabatic treatment of the "wedge problem" has been given by Brekhovskikh and Lysanov (1991) and by Doolittle *et al.* (1988). Those authors considered an acoustic mode traveling up a sloping beach, with depth $H(x)$. The mode function is vertically compressed until the waves reach a depth H^*, where the vertical wavelength equals the acoustic wavelength. For a free surface ($p = 0$) and a rigid bottom ($dp/dz = 0$),

$$H^* = \frac{m - \frac{1}{2}}{2 f s_0}. \tag{4.7.9}$$

At this point, the phase slowness is zero, (2.10.3) and (2.10.4), and the waves cannot penetrate beyond this "barrier depth." In ray language, each "collision" with the sloping bottom steepens the rays, until the rays become vertical at H^* (fig. 4.15). In this limit the sound-speed profile $C(z)$ plays a negligible role, and the sound-speed might as well be taken as constant: the change in phase velocity is entirely the result of the converging surface and bottom boundaries. In the other limit of very deep water, the boundaries play a negligible role, and the waveguide is entirely determined by the sound channel $C(z)$. There is a

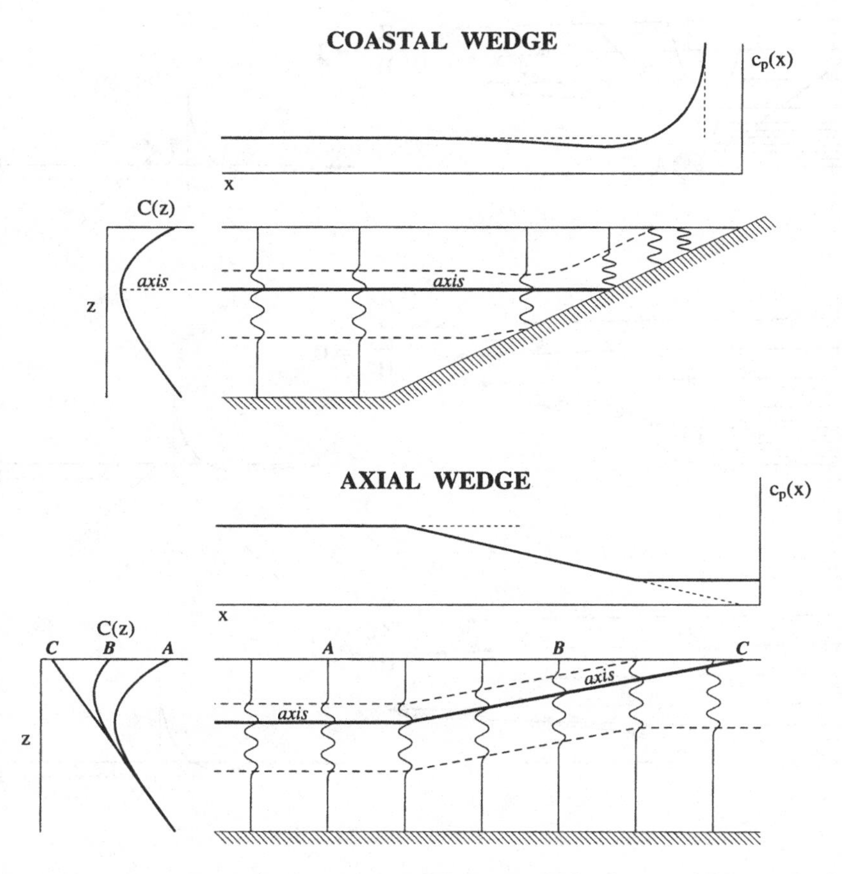

Fig. 4.16. Schematic representation of modal phase velocity for a variable depth of bottom (upper panel) and a variable depth of sound axis (lower panel). (From Munk, 1991.)

slight dip in the modal phase velocity as the shoreward-traveling mode first encounters the sloping bottom.

Thus a coastal shelf acts as a repulsive barrier. An important result is that an offshore source on a sloping bottom will transmit two rays toward a fixed offshore receiver: a slightly refracted direct ray and a coastally repelled indirect ray (fig. 4.15). Doolittle *et al.* (1988) have observed the dual arrivals.

Fig. 4.16 compares two models of wave refraction: (i) a coastal wedge with rising sea bottom, but a fixed sound axis, and (ii) a rising sound axis, but fixed sea bottom, as found at high (northern and southern) latitudes (see fig. 4.1). Transmission across any major current system is associated with a transition

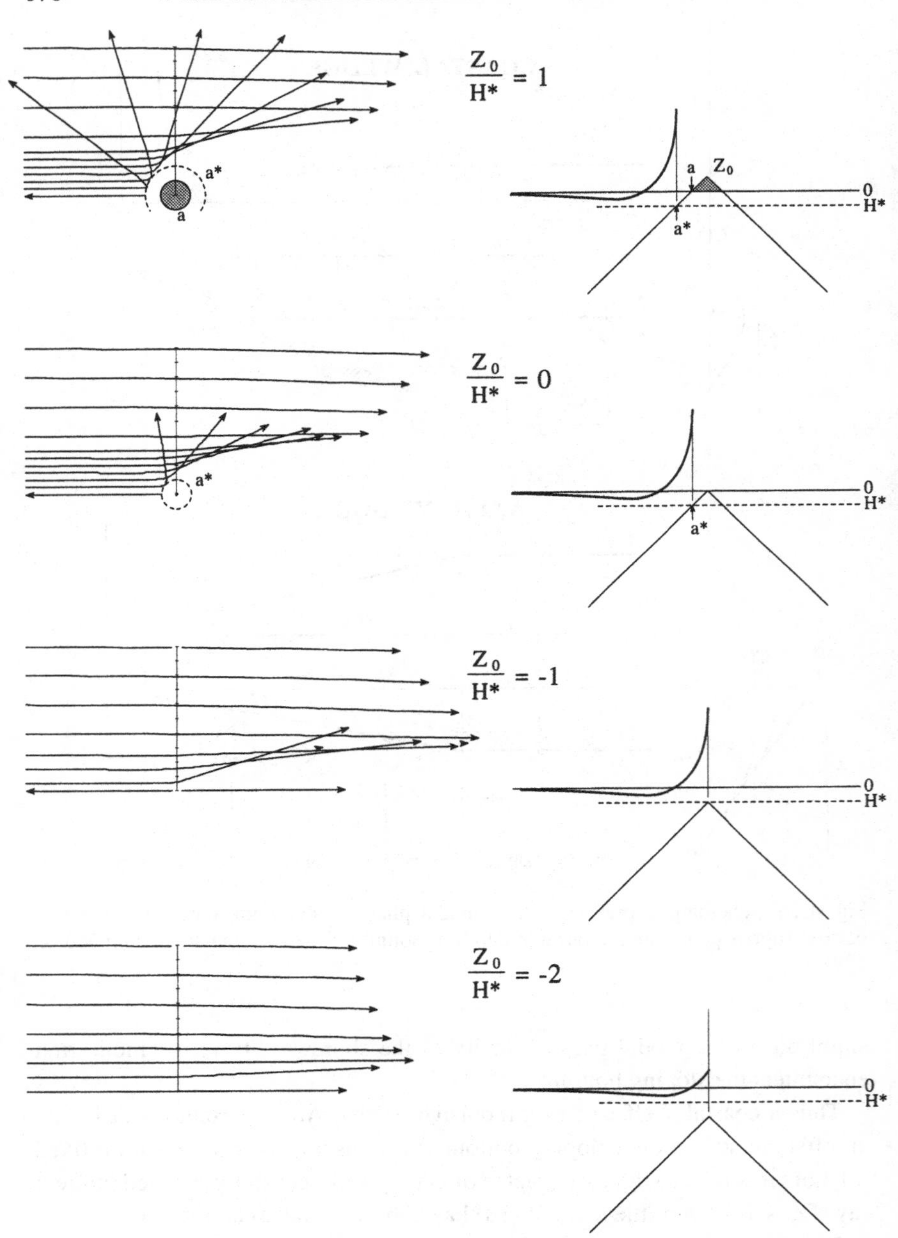

Fig. 4.17. Top view (left) and side view (right) for $H^* = 100$ m, bottom slope = 0.05, and summits at $z_0 = 100$ m, 0, -100 m, -200 m, respectively. Phase velocity goes from infinity at the barrier depth H^* through a slight minimum to an offshore asymptote, as shown to the right. (Adapted from Munk and Zachariasen, 1991.)

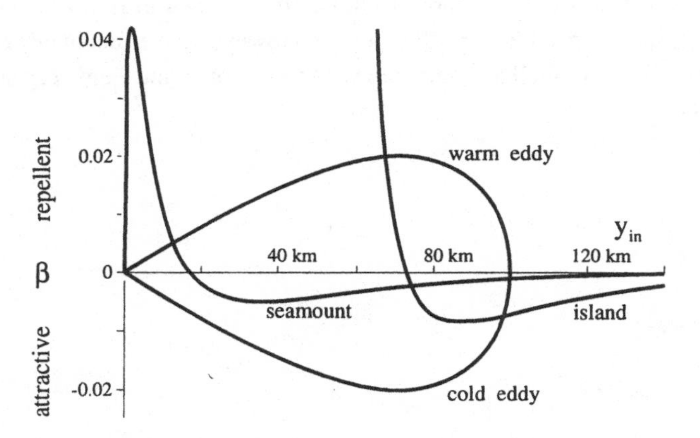

Fig. 4.18. Scattering angle β in radians (defined in fig. 4.13) as a function of the incidence distance y_{in} (the closest distance of the unrefracted ray from the eddy or island center). For the island (radius 50 km, elevation 2.5 km) and seamount (elevation –0.3 km) we have taken $H^* = 0.1$ km. For the eddies we have taken a 100-km radius, with $\Delta C = \pm 15$ m/s for the warm and cold eddies, respectively.

in axial depth (but not an outcropping of the axis). Upwelling in shallow water gives rise to a change in axial depth, thus combining the two models.

The assumption of a rigid bottom boundary is not realistic, expecially for low-order modes. A more realistic description is the so-called Pekeris model of a fluid bottom underlying the water column, with density and sound-speed being two or three times those of sea water. The situation is highly dependent on details and needs to be calculated separately for each experimental situation, taking into account the rigidity and steepness of the bottom. In general, one finds that for glancing incidence, the energy is totally reflected seaward at H^*, whereas for normal incidence the energy is transmitted into the bottom, and almost none of it is reflected, with a sharp transition between the two situations (Chapman *et al.*, 1989; Brekhovskikh and Lysanov, 1991, chap. 4).

Fig. 4.17 shows the refractions by circular islands and seamounts[11] (Munk and Zachariasen, 1991). The important distinction is not that between an island and a seamount, but between a seamount above and one beneath the barrier depth H^*. In the latter case, the refractive effect is usually negligible. The critical depth is very small for low modes: $H^* = 5$–101 m for $m = 1$–10 at $f = 70$ Hz, and even less at higher frequency. Thus, low modes will pass over ordinary seamounts without being appreciably refracted. Scattering angles from

[11]The adiabatic approximation does not apply to steep seamounts.

bathymetric features are compared to those from eddies in fig. 4.18. Islands, seamounts, and warm eddies are all repellent. However, the effect of even strong warm eddies is very small as compared with those for islands and very shallow seamounts.

CHAPTER 5

OBSERVATIONAL METHODS

The fundamental experimental requirement of ocean acoustic tomography is to make precise measurements of acoustic propagation in the ocean. Any of a variety of characteristics of acoustic propagation potentially can provide useful information about the ocean through which the sound has traveled, provided that the forward problem is thoroughly understood. The task of the measurement system is to provide estimates of whatever parameters are desired, with useful precision, together with estimates of the errors of the measurements.

In this chapter we focus on the use of broadband acoustic signals to measure the impulse response of the ocean with sufficient *resolution* (in time and/or vertical angle) to separate individual ray arrivals. The *precision* with which travel times and other parameters of the individual ray arrivals can be estimated is limited by the ambient acoustic noise in the ocean. In addition, and more important in most cases, small-scale ocean variability causes fluctuations in ray amplitudes, travel times, phases, and arrival angles. Travel time is the most robust observable and is the one we emphasize in this book. The inversion of travel-time data to obtain information on the ocean sound-speed and current fields was outlined in chapter 1. Because the expected magnitude of the travel-time perturbations is $O(100 \text{ ms})$, travel times need to be measured with a precision of a few milliseconds, corresponding to a few parts per million over 1 Mm range.

The conceptually simplest approach to measuring the impulse response of the ocean is to transmit a single loud, short-duration pulse and record the resulting arrival structure at the receiver. We showed in chapter 2 that to resolve the multipath structure requires roughly 10-ms resolution, independent of range, which, as we shall show in this chapter, means that the signal must have roughly a 100-Hz bandwidth. Unfortunately, the power levels that can be transmitted by existing sources are inadequate to generate such short-duration pulses that would exceed the ambient acoustic noise level at long ranges. (The source level is fundamentally limited by cavitation at high power levels at shallow depths, and by nonlinear propagation effects at greater depths, but other practical problems often restrict low-frequency sources to much lower power.)

An adequate signal-to-noise ratio (SNR) can be achieved without sacrificing travel-time resolution, however, by transmitting modulated signals that extend the signal duration without reducing the bandwidth. The received signal is correlated against a replica of the transmitted waveform (matched filtering). The modulation is chosen such that the autocorrelation of the transmitted signal will be substantially shorter than the transmitted signal itself (pulse compression). The SNR achieved with matched-filter processing is proportional to the total transmitted energy. We shall show later that to measure travel times to $O(1\text{ ms})$ requires a 20-dB postprocessing SNR. The source level needed to achieve this depends on the signal duration used. Internal-wave-induced phase fluctuations limit the duration over which the received signal can be coherently processed. At 1 Mm range and 250 Hz, for example, internal waves impose a decorrelation time of a few minutes. Taking account of the propagation loss due to geometric spreading and volume attenuation, the minimum source level needed to give about a 20-dB SNR can be shown to be 192 dB relative to 1 μPa at 1 m (132 W of acoustic power).

Internal waves also cause travel-time fluctuations, as discussed in chapter 4. Internal-wave-induced fluctuations typically are larger than the travel-time uncertainty imposed by ambient noise. At a 1-Mm range, travel times fluctuate by $O(10\text{ ms})$ rms due to internal waves. Because internal-wave-induced fluctuations decorrelate in time, internal-wave-induced errors can be reduced by averaging over a number of independent transmissions.

In addition to the fundamental limits imposed by ambient noise and high-frequency ocean variability, travel-time accuracy is limited by a number of practical problems. The precise nature of these problems will depend on the experimental geometry. For autonomous, moored systems, measured travel-time fluctuations are contaminated by errors in timekeeping and by travel-time signals associated with the motion of moored sources and receivers. Autonomous, low-power timekeeping procedures accurate to a few milliseconds over a year have been developed. The relative motion of moored instrumentation can be measured using long-baseline acoustic navigation systems with bottom-mounted acoustic transponders. The corrections required are commonly of order 100 ms, corresponding to relative mooring displacements of 150 m, which is comparable to the expected signal.

The number of transmissions that can be generated and averaged to reduce internal-wave-induced travel-time noise is limited by the amount of source energy that can be stored in an autonomous instrument. One can view this problem as arising owing to the low efficiency of the existing broadband, low-frequency acoustic sources capable of operating at the depth of the sound-channel axis. Equivalently, one can view the limited energy storage achievable in moored instruments as the problem.

Ship-suspended instruments are not constrained by energy storage, and time-keeping is easily accomplished. Relative positions must be known to the order of 10 m, requiring a combination of satellite positioning of the ship and high-frequency acoustic navigation to locate the suspended instruments relative to the ship. It is difficult to generate ray paths that adequately sample a large horizontal area using only instruments suspended from two ships, however, in part because the survey must be completed before the ocean changes significantly, so that achievement of mesoscale resolution using moving-ship tomography appears practical only when carried out in conjunction with some number of fixed sources or receivers.

Timekeeping and energy storage are not problems for transceivers that are cable-connected to shore, but the advantage gained has to be balanced against the loss of flexibility caused by having to be near land. For sub-basin-scale experiments, the decision has been in favor of the flexibility of autonomous moored systems, sometimes in conjunction with ship-suspended instruments. For basin and global experiments, with their requirements of high power and long duration, the decision largely goes the other way. Various mixtures of autonomous, ship-suspended, and cable-connected instruments are of course possible.

Section 5.1 presents the "sonar equation," used to make rough-and-ready estimates of the expected SNR when designing an experiment. The pulse-compression techniques needed to obtain an adequate SNR from peak-power-limited sources are outlined in section 5.2. The precision with which travel times and arrival angles can be estimated in the presence of ambient noise and internal-wave-induced fluctuations is discussed in sections 5.3 and 5.4. Source and receiver velocities, whether the instruments are moored or suspended from shipboard, affect the signal processing. Typically, although not always, the associated Doppler shift is a nuisance parameter that must be estimated even though it contains little useful information (section 5.5). In sections 5.6 and 5.7 we discuss the practical problems associated with accurate timekeeping and with positioning moored or ship-suspended systems. Errors in timekeeping and/or positioning largely cancel in some situations. Finally, the sequence of steps required to convert raw acoustic data into the time-series data on travel times used in the inversion procedure is summarized in section 5.8. Spindel (1985) has previously reviewed much of the material discussed in this chapter.

5.1. The Sonar Equation

The sonar equation is simply a systematic way of estimating the expected SNR at a distant receiver, taking into account the source characteristics, geometric spreading with range, attenuation, boundary effects, ambient noise, and the receiver characteristics. Simple formulae are commonly used to estimate geometric spreading, although one should ultimately use a suitable propagation model to do so more accurately. Received signal levels are quite variable at long ranges, requiring rather conservative experimental design, so simple estimates are often adequate. We shall find that for realistic values of the parameters, the received signal amplitude is well below the noise level. In the next section we shall describe the signal processing techniques used to overcome this problem.

The intensity I of an acoustic wave is defined as the power per unit area normal to the direction of propagation. For a plane wave,

$$I = p^2/\rho C, \tag{5.1.1}$$

where p is acoustic pressure, ρ is density, and ρC is the specific acoustic impedance of the medium. By convention, intensities in underwater acoustics are expressed in logarithmic units (decibels, dB) referred to the intensity $I_{\text{reference}}$ of a plane wave with rms pressure 10^{-6} newton/m^2, or 1 micropascal (μPa). The acoustic source level, for example, is defined to be

$$\text{SL} \equiv 10 \log [I_s/I_{\text{reference}}] = 20 \log [p_s/p_{\text{reference}}] \tag{5.1.2}$$

where I_s is the rms source intensity 1 m from the source and on its acoustic axis. (The source level must be measured in the far field, which exceeds 1 m for large, low-frequency sources; spherical spreading is used to compute the equivalent source level at 1 m.) Thinking in terms of pressure, the units for source level are commonly expressed as decibels relative to 1 μPa at 1 m (written dB re 1 μPa at 1 m).

The sonar (*so*und *na*vigation and *r*anging) equation models the expected SNR for transmission from a source to receiver:

$$\text{SNR} = \text{SL} - \text{TL} - (\text{NL} - \text{AG}) \quad \text{dB}, \tag{5.1.3}$$

where TL is transmission loss, NL is noise level at the receiver, and AG is (receiving) array gain, all expressed in decibels. Urick (1983) discusses each of these terms at length. A brief discussion follows.

Transmission loss. Transmission loss includes attenuation and geometric spreading, $\mathrm{TL} = \mathrm{TL}_a + \mathrm{TL}_g$, in decibels. Attenuation is linearly proportional to range, $\mathrm{TL}_a = \alpha r$. Francois and Garrison (1982*a,b*) and Garrison *et al.* (1983) provide comprehensive summaries of what is known about sound absorption in the ocean. An approximate expression for the attenuation coefficient α, valid for low frequencies (below about 8 kHz) and at the depth of the sound-channel axis, is

$$\alpha(f) = 0.79\, A \frac{f^2}{(0.8)^2 + f^2} + \frac{36 f^2}{5000 + f^2} \quad \mathrm{dB/km}, \tag{5.1.4}$$

where f is in kilohertz (Fisher and Simmons, 1977; Lovett, 1980). The first term is due to boric acid relaxation, which depends on ocean pH through the coefficient A, and the second term is due to magnesium sulfate relaxation, which is independent of pH. At frequencies below 1 kHz, the first term is dominant. Lovett (1980) provides charts of the coefficient A for the Atlantic, Indian, and Pacific oceans. It varies by a factor of 2 between the North Pacific ($A = 0.055$) and the North Atlantic ($A = 0.11$).

The geometric spreading loss is more problematic. In ocean acoustic tomography, one requires the intensity of *individual* ray arrivals. The correct approach is to use a propagation model to compute the expected arrival pattern for the geometry and sound-speed field of interest. More often, simple rules of thumb are used. A conservative approach is to assume that each ray spreads spherically, as would be the case in a homogeneous, unbounded, lossless medium. The total power crossing any spherical surface surrounding the source must then be constant,

$$P = 4\pi r_0^2 I(r_0) = 4\pi r^2 I(r), \tag{5.1.5}$$

and the geometric spreading loss is

$$\mathrm{TL}_g = 10 \log \left[I(r_0)/I(r) \right] = 20 \log (r/r_0) \quad \mathrm{dB}, \tag{5.1.6}$$

where $r_0 = 1$ m with the source level SL defined 1 m from the source. An alternate approach is to assume that (i) the total power summed over all ray paths spreads spherically out to a distance r_1 of the order of the water depth (10 km, say) and then spreads cylindrically (since the signal is confined between the top and bottom of the ocean), and (ii) the signal is apportioned among n ray arrivals (reducing the intensity per ray arrival). For ranges in excess of several convergence zones, the result is

$$\mathrm{TL}_g = 20 \log (r_1/r_0) + 10 \log (r/r_1) + 10 \log (n) \quad \mathrm{dB}. \tag{5.1.7}$$

Equations (5.1.6) and (5.1.7) yield the same answer if $n = r/r_1 = 0.1r$ (r in km). It was shown in section 2.6 that the number of ray arrivals increases linearly with range (rule 3). The rate of increase is not necessarily sufficiently rapid to give spherical spreading, however. For the temperate sound-speed profile, $n = 0.02r$ (r in km). At 1 Mm range, spherical spreading (5.1.6) gives $\mathrm{TL}_g =$ 120 dB, and (5.1.7) with $n = 0.02r$ gives $\mathrm{TL}_g = 113$ dB. The intensities of individual ray arrivals are, of course, not all the same. Assuming spherical spreading for experiment design assures that the weaker refracted paths will be usable.

Ambient noise. The ambient noise level NL is defined to be the ratio of the expected value for the noise intensity in a 1-Hz band to the previously defined reference intensity:

$$\mathrm{NL} \equiv 10 \log \left(\frac{\langle I_{\mathrm{noise}} \rangle}{I_{\mathrm{reference}}} \right) . \tag{5.1.8}$$

Thinking in terms of noise pressure, noise level defined in this way has units of decibels relative to 1 μPa in a 1-Hz band. By convention, this is written dB re 1 $\mu\mathrm{Pa}/\sqrt{\mathrm{Hz}}$. [The rather odd-looking $\sqrt{\mathrm{Hz}}$ is used to indicate that in spectral terms the appropriate units are $(\mu\mathrm{Pa})^2/\mathrm{Hz}$.]

At frequencies of a few hundred hertz, noise generated by distant ships and noise generated at the sea surface both make significant contributions (fig. 5.1). The noise generated at the sea surface increases with increasing wind speed and with increasing local wave size (which are strongly correlated). Breaking waves generate more noise than nonbreaking waves. Noise levels can vary by 20–30 dB for different sea states (*i.e.*, wave size) and different shipping densities; sea state 3, corresponding roughly to wind speeds of 5.5–7.9 m/s (11–16 knots) (Bowditch, 1984), is a common design criterion. Because the ambient noise level decreases while attenuation loss increases with increasing frequency, there is an optimum frequency that minimizes (TL + NL) and maximizes the SNR at each range. The optimum frequency at 1 Mm is near 100–150 Hz, but the minimum is not sharp (fig. 5.2). Frequencies somewhat above the nominal optimum frequency are often used, because sources become smaller and generally more efficient as frequency increases.

Assuming that the ambient noise is approximately independent of frequency over the receiver bandwidth (a locally white-noise spectrum) gives, for the total noise in the equivalent receiver bandwidth BW (in Hz),

$$\mathrm{NL}_{\mathrm{total}} = \mathrm{NL} + 10 \log \mathrm{BW} \quad \mathrm{dB\ re\ 1}\ \mu\mathrm{Pa} . \tag{5.1.9}$$

(Equivalent bandwidth is defined to be the bandwidth of a rectangular filter that passes the same noise power as the actual receiver filter.)

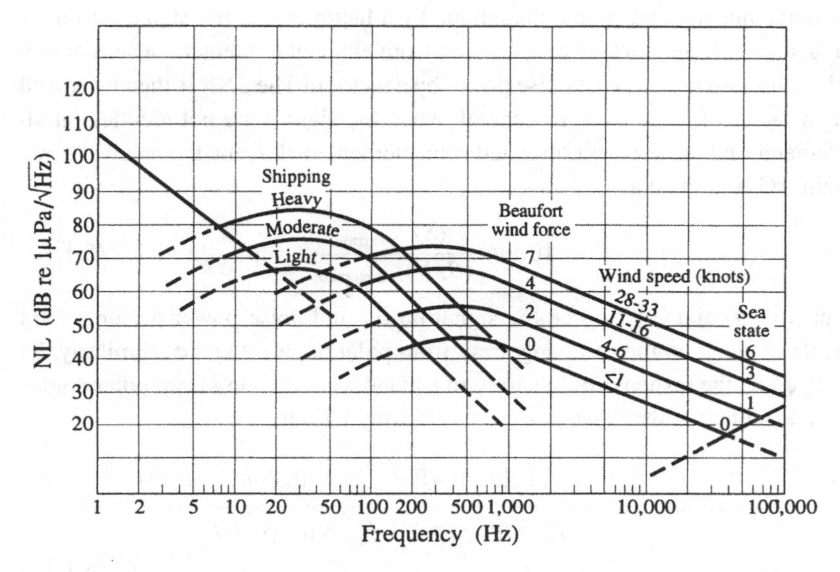

Fig. 5.1. Average deep-water ambient-noise spectra. (From Urick, 1983.)

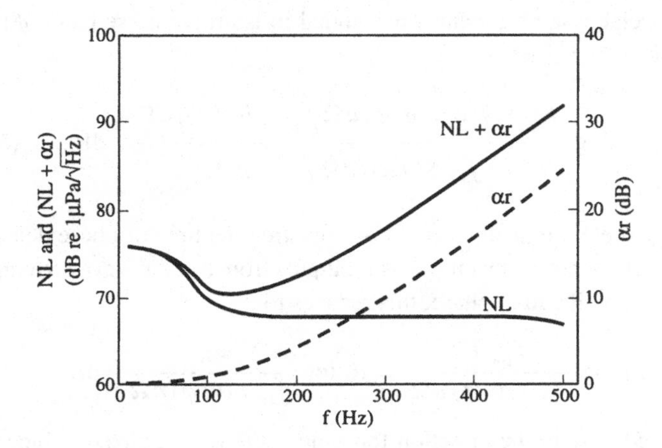

Fig. 5.2. Attenuation αr (dB) at $r = 1$ Mm for the North Atlantic ($A = 0.11$), noise level NL (dB re 1 μPa/$\sqrt{\text{Hz}}$), and the sum $\alpha r +$ NL. Frequencies of 100–150 Hz give maximum SNRs.

Array gain. Combining the signals from an array of receiving hydrophones, either horizontal or vertical, improves the SNR over that obtained from a single sensor, provided that the signal is coherent over the array while the noise is not. For a signal that is constant over an array of n elements, summing the element outputs increases the signal magnitude by a factor n, and the signal power by a factor n^2. If the noise is uncorrelated from element to element, a sum over n elements will increase the noise power by a factor n. The SNR is then improved by a factor of $n^2/n = n$. In general, when the signals are not identical at all elements and the noise is correlated from element to element, the receiver array gain AG is defined as

$$\mathrm{AG} \equiv 10 \log \frac{(S/N)_{\mathrm{array}}}{(S/N)_{\mathrm{one\ element}}} \quad \mathrm{dB}\,. \tag{5.1.10}$$

Let $S(\theta, \phi)$ and $N(\theta, \phi)$ be the signal power and noise power per unit solid angle incident on the receiving array from polar angles θ and ϕ. Similarly, let $b(\theta, \phi)$ be the response of the array to a plane wave arriving from polar angles θ and ϕ. (This is the *beam pattern* of the array.) Then

$$\mathrm{AG} = 10 \log \frac{\int_{4\pi} S(\theta,\phi) b(\theta,\phi)\, d\Omega \Big/ \int_{4\pi} N(\theta,\phi) b(\theta,\phi)\, d\Omega}{\int_{4\pi} S(\theta,\phi)\, d\Omega \Big/ \int_{4\pi} N(\theta,\phi)\, d\Omega} \quad \mathrm{dB}. \tag{5.1.11}$$

For a single nondirectional element, $b(\theta, \phi) = 1$, giving AG = 0 dB, as it must. For the special case of a plane-wave signal in isotropic noise $[N(\theta, \phi) = 1]$, the array gain becomes

$$\mathrm{AG} = 10 \log \frac{\int_{4\pi} S(\theta,\phi) b(\theta,\phi)\, d\Omega \Big/ \int_{4\pi} b(\theta,\phi)\, d\Omega}{\int_{4\pi} S(\theta,\phi)\, d\Omega \Big/ \int_{4\pi} d\Omega} \quad \mathrm{dB}. \tag{5.1.12}$$

If the array is electronically steered by adjusting the time or phase delays[1] for each of the elements, so that the sensor outputs from a plane wave arriving from direction θ, ϕ will add in phase, this reduces to

$$\mathrm{AG} = 10 \log \frac{\int_{4\pi} d\Omega}{\int_{4\pi} b(\theta,\phi)\, d\Omega} = 10 \log \frac{4\pi}{\int_{4\pi} b(\theta,\phi)\, d\Omega} \quad \mathrm{dB}, \tag{5.1.13}$$

since $b(\theta, \phi) = 1$ in the direction for which $S(\theta, \phi)$ is nonzero. Under these conditions the array gain is called the directivity index, DI. For narrowband sound of wavelength λ, a line of n elements of equal spacing d has (*e.g.*, Urick, 1983)

$$b(\theta,\phi) = \left[\frac{\sin[(n\pi d/\lambda)\sin\theta]}{n \sin[(\pi d/\lambda)\sin\theta]} \right]^2, \tag{5.1.14}$$

[1] Phase-delay beam-forming is adequate for CW signals, but true time-delay beam-forming is required for the broadband signals normally used in tomography.

$$\mathrm{DI} = 10 \log \frac{n}{1 + \dfrac{2}{n} \displaystyle\sum_{\rho=1}^{n-1} \frac{(n-\rho)\sin(2\rho\pi d/\lambda)}{2\rho\pi d/\lambda}} \text{ dB.} \qquad (5.1.15)$$

When d is an integer multiple of $\lambda/2$, this simplifies to $\mathrm{DI} = 10 \log n$. Somewhat surprisingly, this result is the same as that obtained for the special case in which the noise is uncorrelated from element to element.

Two caveats are in order. The calculations given earlier implicitly assume a continuous single-frequency (CW) signal. With broadband signals this assumption is not adequate, but the formulae still provide useful guidance provided that the fractional bandwidth is not too large. Second, the assumption of isotropic noise does not accurately model the vertical noise distribution at frequencies below a few hundred hertz. At these frequencies, much of the noise is generated by distant ships, rather than by local surface processes. It might be expected, therefore, that the noise energy would be confined to surface-limited rays, crossing the sound-channel axis at roughly $\pm 12°$ to $\pm 15°$ from the horizontal in mid-latitudes. Fig. 5.3 shows that the measured noise distribution is usually not bimodal, however, but is rather uniformly distributed within $\pm 15°$ of the horizontal (Wales and Diachok, 1981; Dashen and Munk, 1984). (Scattering from continental slopes converts the noise energy to low angles.) The signal is concentrated in exactly the same angular range, because these angles correspond to low-loss rays that do not interact with the surface and bottom. The array gain achieved by vertical arrays at low frequencies is therefore substantially less than would be expected for isotropic noise.

The horizontal noise distribution is often more nearly isotropic, and horizontal array gains close to that expected for isotropic noise can be realized. (When the major noise is due to a few ships, however, the noise spectrum is far from flat, and the angular pattern is far from isotropic. In that case, *adaptive* beam-forming, in which the element weighting is adjusted to place nulls in the directions of the major noise sources, may be able to virtually eliminate the noise from a few discrete ships, leaving just the low-level true ambient background.)

Sound-speed perturbations associated with internal wave vertical displacements cause rays arriving at vertically and/or horizontally separated receivers to become decorrelated, limiting the maximum aperture that can be used for beam-forming. Section 5.4 discusses this further.

Sample SNR calculation. One of the acoustic sources that has been used in ocean acoustic tomography experiments has source level SL = 192 dB re 1 μPa at

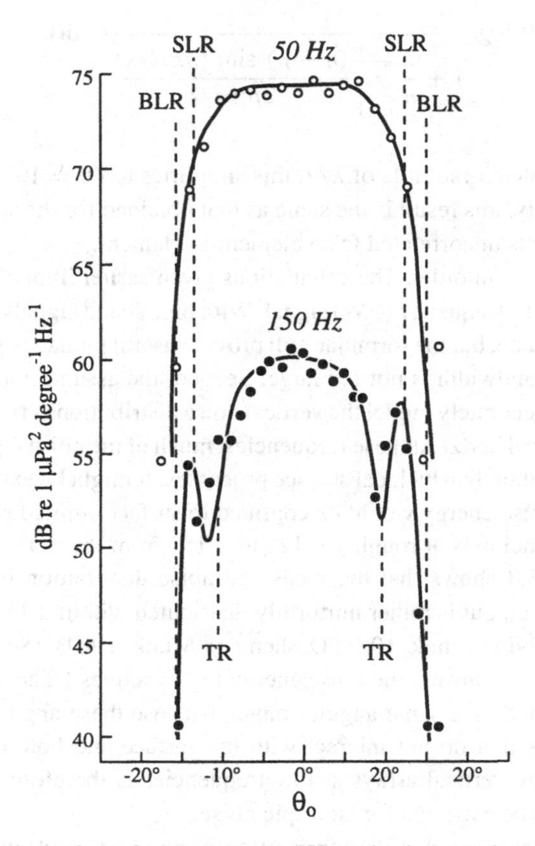

Fig. 5.3. Vertical angular noise spectrum at 650 m depth (near the sound axis) as a function of the inclination θ_0, taken by Fisher and Williams 350 miles west of San Diego. The spectrum at 150 Hz is at a lower level than at 50 Hz and shows gaps at about $\theta_0 = \pm 9°$ that are not present at 50 Hz. Vertical dashed lines correspond to surface-limited rays ($\theta_0 = \pm 13°$), to bottom-limited rays ($\theta_0 = \pm 15°$), and to "thermocline rays" ($\theta_0 = \pm 10°$), with an upper turning point at a depth of 80 m in the middle of a steep thermocline. (From Dashen and Munk, 1984.)

1 m, center frequency 250 Hz, and (full) bandwidth 83.3 Hz ($Q = f/\Delta f = 3$).[2] This source can transmit short pulses of duration $\Delta t \approx 1/\Delta f$, or 12 ms. The receiver bandwidth is chosen to match the signal bandwidth. The predicted SNR for a resolved arrival at 1 Mm range with a single hydrophone is then well below zero decibels (table 5.1). The solution is to transmit broadband coded

[2]A hydraulically driven HLF-5 acoustic source manufactured by Hydroacoustics, Inc., Rochester, NY (Bouyoucos, 1975; Spindel and Worcester, 1986).

Table 5.1. *SNR at 250 Hz and 1 Mm range in the North Atlantic* ($A = 0.11$) *assuming spherical spreading*

Source level	192 dB re 1 μPa at 1 m
Spreading loss (spherical)	−120 dB
Attenuation	−8 dB
Received signal intensity	64 dB re 1 μPa
Noise, SS3	68 dB re 1 μPa/$\sqrt{\text{Hz}}$
Bandwidth, $Q = 3$ (83.3 Hz)	19 dB
Receiver noise level	87 dB re 1 μPa
SNR (single hydrophone)	−23 dB
Periodic average (16 periods)	12 dB
Sequence removal (1023 digits)	30 dB
Total processing gain	42 dB
SNR (single hydrophone)	19 dB

signals and to correlate the received signal against a replica of the transmitted signal, as described in the next section. Over 40 dB of processing gain is needed to achieve +19 dB SNR.

5.2. Pulse Compression

Pulse-compression techniques are needed to obtain adequate SNRs from peak-power-limited sources without degrading the resolution of the signal. In this section we first define what is meant by the *complex envelope* of a signal, because we shall use this construct in what follows. We then show that the *time-bandwidth product* (TBW) of any signal (*i.e.*, the product of the duration of a signal and its bandwidth) must exceed a minimum value. This is important because it specifies the signal bandwidth required to obtain the resolution needed in the time domain. With this background in place, we describe the signal coding and processing used to improve the SNR over that which can be achieved with a simple short pulse.

Complex envelope. For signals with bandpass spectra, it is convenient both theoretically and experimentally to define the complex envelope of the signal.

A modulated signal with carrier frequency $\omega_0 = 2\pi f_0$ may be written

$$p(t) = a(t) \cos\left[\omega_0 t + \theta(t)\right], \tag{5.2.1}$$

where $a(t)$ is the *amplitude modulation*, $\theta(t)$ is the *phase modulation*, and $d\theta/dt$ is the *frequency modulation*. Expanding the cosine,

$$\begin{aligned} p(t) &= a(t)\left[\cos\theta(t)\, \cos\omega_0 t - \sin\theta(t)\, \sin\omega_0 t\right] \\ &\equiv 2p_x(t)\, \cos\omega_0 t - 2p_y(t)\, \sin\omega_0 t\,, \end{aligned} \tag{5.2.2}$$

where $p_x(t)$ and $p_y(t)$ are the in-phase and quadrature components of the signal. Combining them into a complex number gives the complex envelope

$$p_z(t) \equiv p_x(t) + ip_y(t) = \tfrac{1}{2}\, a(t)\, \exp i\theta(t)\,. \tag{5.2.3}$$

The signal $p(t)$ can then be written

$$p(t) = p_z(t)e^{i\omega_0 t} + p_z^*(t)e^{-i\omega_0 t}\,, \tag{5.2.4}$$

where $p_z^*(t)$ is the complex conjugate of $p_z(t)$. The complex envelope is obtained experimentally by multiplying the signal $p(t)$ by $\exp(-i\omega_0 t)$,

$$\begin{aligned} p(t)e^{-i\omega_0 t} &= \left[p_z(t)e^{i\omega_0 t} + p_z^*(t)e^{-i\omega_0 t}\right]e^{-i\omega_0 t} \\ &= p_z(t) + p_z^*(t)e^{-i2\omega_0 t} \\ &\to p_z(t)\,, \end{aligned} \tag{5.2.5}$$

where the last line indicates that the signal has been low-pass-filtered to remove the double-frequency components. There is some freedom in the choice of demodulation frequency; selection of the carrier frequency for carrier-borne signals simplifies the analysis.

From (5.2.4), the frequency spectrum of $p(t)$ is

$$P(\omega) = \int_{-\infty}^{\infty} p(t)e^{-i\omega t}dt = P_z(\omega - \omega_0) + P_z^*(-\omega - \omega_0), \tag{5.2.6}$$

where

$$P_z(\omega) = \int_{-\infty}^{\infty} p_z(t)e^{-i\omega t}\, dt\,, \tag{5.2.7}$$

and the Fourier-transform shifting theorem has been used. If $|P_z(\omega)|$ is significant only over a frequency band much smaller than the carrier frequency, the two components on the right-hand side of (5.2.6) do not overlap, and the signal is *narrowband*. For broadband signals, the spectral distortion introduced by the overlap of the two components needs to be properly allowed for (Metzger, 1983).

A number of different implementations of complex demodulation (5.2.5) are commonly used (Grace and Pitt, 1970; Pridham and Mucci, 1979; Horvat *et al.*, 1992; Menemenlis and Farmer, 1992). A conceptually and computationally simple approach is possible using digital techniques if the input sample rate is chosen to be $4f_0$ (*i.e.*, four samples per carrier cycle). Multiplication by

$$\exp(-i\omega_0 t_k) = \exp\left(-i\,2\pi f_0 \frac{k}{4f_0}\right) = \exp\left(-i\frac{\pi}{2}k\right) = (-i)^k \quad (5.2.8)$$

for integer k corresponds to multiplication by $\{1, -i, -1, i\}$. Low-pass filtering to remove the double-frequency components can be performed using a block average over N values. Nonoverlapping block averages lead to a sample rate reduction from $4f_0$ real numbers per second to $4f_0/N$ complex numbers per second. Sliding block averages preserve the sample rate while changing from real to complex, doubling the data storage, but may remove the need for subsequent interpolation. The choice $N = 4$ gives one demodulate per carrier cycle and has the virtue of placing zeros in the transfer function of the low-pass filter at $-\omega_0$ and $-2\omega_0$. This removes any zero-frequency (DC) offset in the analog-to-digital converter, because the frequency-shift operation moves this to $-\omega_0$. It also places a zero in the center of the band of the negative frequency components that have been shifted from $-\omega_0$ to $-2\omega_0$. Other parameter choices are possible, as discussed in detail by Metzger (1983).

Minimum time-bandwidth product. One of the critical experimental parameters is the source bandwidth required to transmit sufficiently short pulses to resolve individual ray arrivals. Resonant, high-Q sources are typically quite efficient, but by definition are narrowband, with bandwidth f_0/Q. Broader-band, low-Q sources typically are more difficult to design and build and usually are less efficient (Decarpigny *et al.*, 1991). The relation

$$\Delta f \approx 1/\Delta t \quad (5.2.9)$$

was used earlier to relate the bandwidth Δf and duration Δt of a simple pulse, although the definitions of Δf and Δt were deliberately left vague. Simple dimensional considerations indicate that some such relation must exist. If the signal $p(t)$ has the Fourier transform $P(\omega)$, then the scaling theorem for Fourier transforms,

$$ap(at) \leftrightarrow P\left(\frac{\omega}{a}\right), \quad (5.2.10)$$

states that if the time scale of a waveform is changed, the frequency scale of its spectrum will change reciprocally. A signal cannot simultaneously have an arbitrarily short duration and arbitrarily narrow bandwidth.

The most common rule of thumb is (5.2.9), written

$$\Delta t \Delta f \geq 1\,, \tag{5.2.11}$$

but the actual minimum value of the time-bandwidth product depends on the definitions of duration and bandwidth.

The classic derivation of the minimum time-bandwidth product uses the rms signal duration and bandwidth (Helstrom, 1968; Papoulis, 1977). The rms signal bandwidth $\Delta\omega_{\rm rms}$ is

$$(\Delta\omega_{\rm rms})^2 \equiv \overline{(\omega - \overline{\omega})^2} = \overline{\omega^2} - \overline{\omega}^2\,, \tag{5.2.12}$$

where the mean and mean square frequency deviations are defined in terms of the spectrum of the complex envelope,

$$\overline{\omega} \equiv \int \omega |P_z(\omega)|^2\, d\omega \Big/ \int |P_z(\omega)|^2\, d\omega\,, \tag{5.2.13}$$

$$\overline{\omega^2} \equiv \int \omega^2 |P_z(\omega)|^2\, d\omega \Big/ \int |P_z(\omega)|^2\, d\omega\,. \tag{5.2.14}$$

All integrals are from $-\infty$ to ∞. Similarly, the rms signal duration is defined to be

$$(\Delta t_{\rm rms})^2 \equiv \overline{(t - \overline{t})^2} = \overline{t^2} - \overline{t}^2, \tag{5.2.15}$$

where

$$\overline{t} \equiv \int t |p_z(t)|^2\, dt \Big/ \int |p_z(t)|^2\, dt\,, \tag{5.2.16}$$

$$\overline{t^2} \equiv \int t^2 |p_z(t)|^2\, dt \Big/ \int |p_z(t)|^2\, dt\,. \tag{5.2.17}$$

It can then be shown, using Schwarz's inequality, that

$$\overline{\omega^2} \cdot \overline{t^2} - \overline{\omega t}^2 \geq \tfrac{1}{4}, \quad \overline{\omega} = 0, \quad \overline{t} = 0, \tag{5.2.18}$$

where we have chosen the time origin to make $\overline{t} = 0$ and the demodulation frequency to make $\overline{\omega} = 0$. This is equivalent to

$$\Delta t_{\rm rms} \cdot \Delta\omega_{\rm rms} \geq \tfrac{1}{2} \quad \text{or} \quad \Delta t_{\rm rms} \cdot \Delta f_{\rm rms} \geq \frac{1}{4\pi}\,. \tag{5.2.19}$$

The equality holds for Gaussian signals with complex envelopes of the form

$$p_z(t) = a\ \exp(-bt^2/2), \qquad \mathrm{Re}\,(b) > 0\,. \tag{5.2.20}$$

The minimum time-bandwidth product obtained using rms duration and rms bandwidth is significantly less than unity, the common rule of thumb. The reason for this is that the rms duration and rms bandwidth are significantly smaller than more physically meaningful measures of duration and bandwidth, such as the time between the half-amplitude points of a pulse (resolution width) and the bandwidth between −3-dB points (half-power bandwidth). This is not surprising, because one standard deviation is a rather small width unit. One standard deviation encompasses only 39% of a normal distribution, for example. [The concept of the rms bandwidth (duration) will reappear later as the appropriate definition to determine the precision with which travel-time (Doppler-shift) estimates can be made.]

Papoulis (1977) derives minimum time-bandwidth products for other definitions of signal duration and bandwidth. Suppose, for example, that the duration Δt and bandwidth $\Delta\omega$ are defined by

$$\alpha = \int_{-\Delta t/2}^{\Delta t/2} |p_z(t)|^2\,dt \Big/ \int_{-\infty}^{\infty} |p_z(t)|^2\,dt\,, \tag{5.2.21}$$

$$\beta = \int_{-\Delta\omega/2}^{\Delta\omega/2} |P_z(\omega)|^2\,d\omega \Big/ \int_{-\infty}^{\infty} |P_z(\omega)|^2\,d\omega\,, \tag{5.2.22}$$

where α and β are two *given constants*. Δt and $\Delta\omega$ correspond to the time duration and bandwidth that encompass specified fractions of the total signal energy. Papoulis (1977) derives a general expression for the minimum time-bandwidth product as a function of α and β. When $\alpha = \beta = 0.9$, the result is

$$\Delta t\,\Delta\omega \geq 4.8 \quad \text{or} \quad \Delta t\,\Delta f \geq \frac{2.4}{\pi}\,, \tag{5.2.23}$$

which is much closer to the usual rule of thumb (5.2.11) than is (5.2.19).

It is difficult to derive a general relation for the minimum time-bandwidth product using the half-peak-amplitude resolution width and the half-power bandwidth, although these are probably the physically most meaningful definitions. Computing the product for a number of simple pulses reveals that the time-bandwidth product with these definitions is always of $O(1)$, however.

One of the important consequences of (5.2.11) is that it specifies a minimum source bandwidth required to achieve a given travel-time resolution. Because the time interval between ray arrivals is independent of center frequency in the ray-theory approximation (chapter 2), lower Q values are needed as one goes to lower frequencies.

Matched-filter processing. We are now ready to describe the signal coding and processing techniques required to achieve an adequate SNR with peak-power-limited sources. Assume that the ocean sound channel can be modeled as a linear, time-invariant system for which a transfer function and associated impulse response can be defined (for the duration of a transmission). The ray approximation leads to the additional assumption that the received signal $x(t)$ is a sum of delayed replicas of the transmitted signal $p(t)$,

$$x(t) = \sum_i a_i p(t - \tau_i), \tag{5.2.24}$$

where a_i is the amplitude and τ_i the travel time of ray i. Propagation along each ray path is then both linear and nondispersive. Under these conditions, and assuming that the channel adds Gaussian white noise, matched-filter processing will maximize the SNR and will be optimum for estimating signal amplitude and time delay.

Suppose, for simplicity, that the received signal consists of a single ray arrival $x(t) = a_0 p(t - \tau_0)$ embedded in additive Gaussian white noise, with zero mean and spectral density N_0. (The noise need not be truly white, but only flat over the band of interest.) The noise density versus frequency is plotted in fig. 5.1. This is the one-sided, or unilateral, spectral density confined to positive frequencies. The corresponding spectral density defined over both positive and negative frequencies is $N_0/2$. Applying an arbitrary linear filter with impulse response $h(\tau)$ to the received signal gives

$$y_s(T) = \int_0^T h(\tau) p(T - \tau)\, d\tau \tag{5.2.25}$$

for the output at time T due to signal only (the amplitude factor a_0 has been dropped). The output time T is chosen to include all of the (time-limited) signal, $p(t) = 0$ for $t < 0$ and $t > T$. The matched filter is derived by finding the impulse response $h(\tau)$ that will maximize the ratio of the peak output signal power $[y_s(T)]^2$ to the variance of the noise. The result is (*e.g.*, Turin, 1960; Helstrom, 1968)

$$h(\tau) = p(T - \tau), \quad 0 < \tau < T. \tag{5.2.26}$$

[More generally, for complex signals, $h(\tau) = p^*(T - \tau)$.] Because the impulse response is simply a time-reversed version of the transmitted signal, matched-filter processing corresponds to correlating the received signal with a replica of the transmitted signal. The output of the matched-filter is proportional to the autocorrelation of the transmitted signal. The output is a maximum when the replica is exactly aligned with the received signal. The resulting SNR can be shown to be

$$\text{SNR} \equiv \frac{[y_s(T)]^2}{\langle y_n^2 \rangle} = \frac{2}{N_0} \int_0^T [p(T-\tau)]^2 d\tau = \frac{2E}{N_0}, \tag{5.2.27}$$

where $\langle y_n^2 \rangle$ is the output variance due to noise only. Here we use the convention that E is the energy dissipated during the observation interval in a 1-Ω resistor, if the signal $p(t)$ is the voltage across the resistor. The SNR is proportional to the energy in the transmitted signal. With a peak-power-limited source, the only way to improve the output SNR after matched-filter processing (which was derived to give the maximum possible SNR) is to transmit more energy by increasing the signal duration, because $E = \textit{average power} \times \textit{signal duration}$. One implication is that the transmitted signal should have a constant envelope (*i.e.*, constant power level) to minimize the signal duration required to obtain a given SNR. A second implication is that long-duration signals are required to obtain an adequate SNR for realistic source levels, as we showed at the end of section 5.1 that the predicted SNR for a single short pulse is well below 0 dB. [These results are easily extended to the case of colored noise (Turin, 1960).]

Equation (5.2.27) gives an alternate perspective on the sample SNR calculation in table 5.1. Instead of computing the additional noise 10 log BW admitted by a receiver matched to the transmitted signal bandwidth (BW), we can compute the SNR for a signal lasting 1 s (*i.e.*, for a 1-Hz bandwidth) and then add $10 \log (\Delta t/(1\ \text{s}))$, where Δt is now the total signal duration in seconds, rather than the rms duration, to adjust for the amount the signal energy differs from the energy in a signal lasting 1 s (*i.e.*, to adjust for the amount the signal energy differs from the signal power). For a single short pulse of the type described at the end of section 5.1, the SNR decrease due to the amount the receiver bandwidth exceeds 1 Hz, $-10 \log \text{BW}$, where $\text{BW} = f/Q$ when the receiver bandwidth is matched to the transmitted signal bandwidth, is exactly equivalent to the SNR decrease due to the amount the pulse duration is shorter than 1 s, $10 \log \Delta t$, where $\Delta t = 1/\Delta f = Q/f$. We recover the SNR given previously. [For the modulated signals to be described in the next section, the SNR increase associated with matched-filter processing is greater than the SNR decrease due to the amount the receiver bandwidth exceeds 1 Hz. The matched-filter gain, $10 \log (\Delta t/(1\ \text{s}))$, then combines the entries for bandwidth and total processing gain in table 5.1.]

In the frequency domain, the matched-filter transfer function is

$$H(\omega) = \int_{-\infty}^{\infty} h(\tau) e^{-i\omega\tau} d\tau = \int_{0}^{T} p(T-\tau) e^{-i\omega\tau} d\tau$$
$$= \int_{0}^{T} p(u) e^{-i\omega(T-u)} du = e^{-i\omega T} P^*(\omega) . \tag{5.2.28}$$

The factor $\exp(-i\omega T)$ corresponds to a delay of T s in the filter output, called a realizability delay, because a physical filter cannot respond until after the signal arrives. (Digital filters operating on recorded samples of the incoming signal do not have this limitation, and one can design the filters to have $T = 0$.)

It can be proved that matched-filter processing is also optimum for estimating signal amplitude and time delay in added Gaussian noise of known spectrum (Turin, 1960; Helstrom, 1968).

Factor-inverse matched filtering. Fig. 5.4a shows an idealized view of a system for making ocean propagation measurements using matched-filter processing. A probe signal $P(\omega)$ is transmitted through an ocean with frequency response $O(\omega)$. The received signal is contaminated by additive noise $N(\omega)$. Matched-filter processing is implemented in the frequency domain by multiplying the Fourier transform of the received signal by $P^*(\omega)$. Because the ocean sound channel has been assumed linear and time-invariant, one can conceptually interchange the order of operations (fig. 5.4b). This makes it clear that what is being measured is the response of the ocean to a signal with spectrum $P(\omega)P^*(\omega)$, which in the time domain is the autocorrelation of the transmitted signal. When measuring a multipath channel using matched-filter processing, the goal is to choose the probe signal $p(t)$ such that its *autocorrelation* will be as narrow as possible. The simplest such signal is a rectangular pulse containing n cycles of a carrier frequency ω_0. The autocorrelation of a pulse with a rectangular envelope is a pulse with a triangular envelope whose base is twice the width of the original rectangular pulse.

Omitted from fig. 5.4 is complex demodulation of the received signal. A more detailed diagram of a typical receiver structure would show that the received signal is bandpass-filtered (to remove noise outside the band of interest and to prevent aliasing), digitized, complex-demodulated, and then processed. All processing occurs on the demodulated (*i.e.*, *baseband*) signal centered on 0 Hz. Fig. 5.4 can be thought of as representing the real signals and filters, centered on the carrier frequency, or as representing the baseband representations of the signals and filters, centered at 0 Hz. In the remainder of this chapter we shall typically work in terms of the baseband representation of the signals and filters, as this simplifies the notation and discussion.

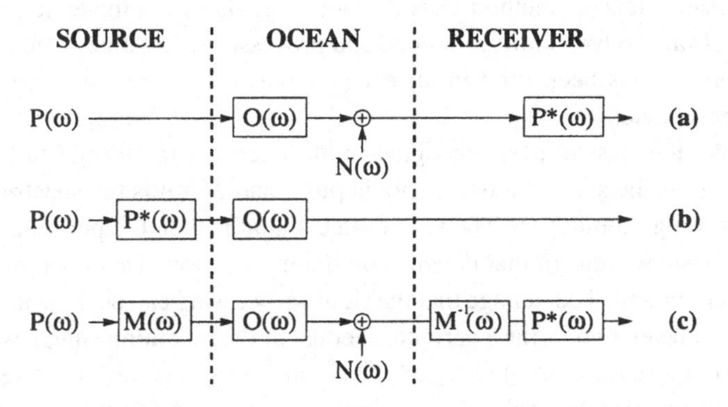

Fig. 5.4. Overview of channel measurement systems in the frequency domain. Signals propagating through the ocean are modified by a linear filter with transfer function $O(\omega)$ and corrupted by additive noise $N(\omega)$ (Birdsall and Metzger, 1986). (a) Matched-filter processing for a transmitted signal with spectrum $P(\omega)$ yields the autocorrelation. (b) Neglecting noise (and source limitations), one can conceptually think of the receiver processing as occurring prior to transmission for a linear system. One essentially probes the channel with a waveform that is the autocorrelation of the transmitted signal for matched-filter processing. (c) Factor-inverse matched filtering explicitly removes the modulating filter $M(\omega)$ by multiplying by its inverse in the receiver. $P(\omega)$ is then the spectrum of an unmodulated probe signal, typically a short pulse.

We have shown that for realistic source levels, a signal with enough energy E to achieve adequate SNR at long range and with enough bandwidth Δf to permit the resolution of individual ray arrivals must have a large time-bandwidth product, $\Delta f \, \Delta t \gg 1$. A simple rectangular pulse has $\Delta f \, \Delta t \approx 1$. This means that the transmitted signal must be modulated. The use of modulated signals that extend the signal duration may introduce undesired ripple in the time response after matched filtering, often called "self-clutter." The time-domain ripple associated with strong arrivals may mask weak arrivals.

Many different modulated signals have been used in sonar and radar applications, with great effort being devoted to finding signals with low self-clutter. These signals are designed primarily for use in monostatic geometries, in which the source and receiver are co-located. In a typical application, a probe pulse is transmitted, and the receiver then listens for returns for an extended time period. One implication of this is that the transmitted signal must be relatively short, because the instrument cannot transmit and receive simultaneously. Ocean acoustic tomography is bistatic, however, with separated sources and receivers. Longer transmissions are then possible. The remainder of this section is devoted to describing a particular approach to signal design and processing that was pioneered in underwater acoustics by T. Birdsall and K. Metzger.

The Birdsall–Metzger method uses periodic signals to eliminate self-clutter. The signals are conveniently generated and processed using digital techniques. This approach has been used in all ocean acoustic tomography experiments conducted to date.

Consider the class of periodic signals with spectra of the form $P(\omega)\,M(\omega)$, where $P(\omega)$ is the spectrum of a periodic pulse and $M(\omega)$ is the spectrum of a time-spreading modulation. The transmitted signal is then the periodic repetition of a basic waveform that depends on the modulation. This form of signal has the very practical advantage that the signal processing can be done in *layers*. For the first layer, one forms a periodic average of the incoming signal, with the same period as the transmitted signal. To do so does not require knowledge of the exact waveform, but only of its period. The output of this step is now only one period long, greatly reducing the amount of memory required for storage (compared with that which would have been required to store the entire incoming waveform) and greatly reducing the computational load for subsequent processing (compared with that required for a straightforward matched filter that would correlate the entire received signal against a replica of the entire transmitted signal). This form of signal has a second great advantage – one can remove the modulation with a filter that is the inverse of the modulating filter, that is, with a filter of the form $M^{-1}(\omega)$, provided that the modulation is chosen such that $|M(\omega)| \neq 0$ (fig. 5.4*c*). One then recovers exactly the signal that would have been received if the unmodulated prototype signal $P(\omega)$ had been transmitted. With $P(\omega)$ selected to be the spectrum of a cleanly time-limited periodic pulse, one obtains the signal that would have been received if a (much louder) periodic pulse train had been transmitted. Provided that the period is greater than the arrival spread, each period of the received signal is then equivalent to what would have been received if a single louder pulse had been transmitted. The processed signal has *no* side lobes in the time domain (*i.e.*, no self-clutter to mask weak arrivals).

This is called factor-inverse matched filtering (FIMF) if implemented as in fig. 5.4c, with a final multiplication by $P^*(\omega)$ to complete matched filtering of the unmodulated prototype signal. It is called factor-inverse filtering (FIF) if the final matched filtering by $P^*(\omega)$ is omitted (Birdsall, 1976; Metzger, 1983; Birdsall and Metzger, 1986). FIMF yields an output proportional to the autocorrelation of the prototype signal, whereas FIF yields an output proportional to the prototype signal itself. The penalty that one pays for using these, rather than simple matched-filter processing, is a lower SNR. The benefit that one gains is the ability to achieve the desired output waveform even when it is necessary to use energy-increasing modulations. FIMF and FIF can also be modified to correct for distortions introduced by the source and receiver transfer functions, although that is not explicitly shown in fig. 5.4 (Worcester, 1977*b*).

The reduction in SNR compared with that achieved using matched-filter processing (which gives the maximum SNR possible) can be minimized by appropriate choice of modulation. If $|M(\omega)|$ is independent of frequency, then $M^{-1}(\omega)$ is proportional to $M^*(\omega)$. FIMF is then exactly equivalent to a matched filter, with no loss of SNR. Construction of signals with this property is described later.

When digital techniques are employed to generate and process the signal, it is convenient to use digital modulations, in which the transmitted signal is composed of *digits*, each of which contains an integer number of cycles of the carrier frequency and is modulated appropriately. The minimum digit length possible with a transducer of given $Q = f/\Delta f$ is constrained by the relation between signal duration and bandwidth; the number of cycles per digit is normally selected to be Q. The bandwidth of a real transducer often restricts the ability to transmit pulses as short as might be desirable.

Although many types of modulation are possible, phase modulation, in which each digit is of the form $\cos(\omega_0 t \pm \theta_0)$, is the only type that will be examined here. Phase-modulated signals have constant power levels, satisfying the requirement enunciated earlier that the transmitted signal should have a constant power level to minimize the signal duration required to obtain a given SNR when the source is peak-power-limited.

The transmitted signal $g(t)$ is constructed by repeating a prototype signal, $p(t)$, L times, with time step (*i.e.*, digit length) T and modulated by a sequence $\{m_\ell\}$,

$$g(t) = \sum_{\ell=0}^{L-1} m_\ell \, p(t - \ell T) \,. \tag{5.2.29}$$

The prototype signal $p(t)$ is chosen to be the periodic repetition of a single digit (*i.e.*, a periodic pulse train, with period LT); $g(t)$ is then a continuing periodic signal with period LT. When this is done, the signal spectrum factors into the form $P(\omega)M(\omega)$ (section 5.9), and the FIMF and FIF processes are applicable. Phase modulation is achieved by selecting the modulating sequence $\{m_\ell\}$ to be $m_\ell = \exp(i s_\ell \theta_0)$, where s_ℓ takes on the values ± 1 (complementary phase modulation). The crucial step is to assign the s_ℓ using a binary finite-field algebraic code (Golomb, 1982; Lidl and Niederreiter, 1986). Then, $|M(\omega)|$ is independent of frequency if the phase angle θ_0 is chosen correctly, so that FIMF processing is fully matched to the transmission. It is shown in appendix section 5.9 that this choice of modulation introduces no self-clutter in the time domain, even for matched-filter processing.

Binary finite-field algebraic codes are often referred to as *binary m-sequences*, short for binary maximal-length sequences. They are also referred to as binary linear maximal shift-register sequences (LMSRS), because one hardware implementation to generate them uses shift registers. The $\{s_\ell\}$ are obtained from a (binary) m-sequence by replacing the ones in the sequence with the value -1 and the zeros with the value $+1$. The sequence $\{m_\ell\}$ is said to be a (binary) m-sequence of degree n if it satisfies a linear recurrence relation

$$m_\ell = \sum_{i=1}^{n} c_i m_{\ell-i} \qquad \text{(modulo 2)}, \tag{5.2.30}$$

and has period $L = 2^n - 1$ (Golomb, 1982). One immediate implication is that only the members of a discrete set of periods $(1, 3, 7, 15, 31, \dots)$ are possible when using binary m-sequences to control the signal modulation. The period of the signal, $T_{\text{period}} = LT$, must be chosen to be longer than the expected *arrival-time spread* of the transmission. If the period is shorter than the arrival-time spread (section 5.9), early and late arrivals will wrap around and overlap when the periodic average of the incoming signal is constructed, making identification of the arrivals difficult at best. With the digit length T chosen to be as short as possible, to maximize our ability to separate nearby arrivals, the m-sequence period L must be chosen large enough so that T_{period} will still exceed the expected arrival-time spread. In self-contained instruments it is desirable to minimize the sequence length L consistent with the constraint that LT be longer than the impulse response of the channel, in order to reduce the amount of memory and data storage required. Continuing the example begun at the end of section 5.1, a 250-Hz source with $Q = 3$ can transmit digits consisting of exactly three cycles at 250 Hz and lasting 12.0 ms. Although propagation models should be used to predict the expected arrival-time spread (appendix B), a rough rule of thumb is that the spread is 1% of the travel time. At 1 Mm range, this gives a spread of 6–7 s. Choosing $L = 1023$ (degree $n = 10$) gives $T_{\text{period}} = 12.276$ s, safely greater than the expected arrival-time spread. If L had been chosen to be 511, T_{period} would have been comparable to the expected arrival-time spread. Choosing $L = 2047$ makes T_{period} more than adequate, unnecessarily increasing memory and data storage requirements.

The discussion to this point has implicitly assumed that the transmission is a periodic sequence that continues forever. When processing finite transmissions, it is necessary to avoid end effects associated with the beginning and end of the transmission. The transmitted signal must therefore include more periods of the sequence than are processed. Some of the transmitted energy is discarded, and the SNR is lower than would be achieved with a receiver matched to the entire transmitted signal. The amount of energy that must be discarded will depend on the precision with which one can predict the time at which the signal from the slowest path will first arrive and the time at which the signal from the fastest path will finally end (including uncertainties caused by errors in the instrument positions). If the processing window is between those two times, then all of the paths will be insonified throughout the window. Typically, one or two more periods are transmitted than are processed, depending on the precision with which the processing window can be predicted, and processing begins one-half or one period after the signal is first expected to arrive. If one transmits a large number of periods, the SNR loss will be small. Transmitting 10 periods and processing 9 will reduce the SNR by $10 \log_{10}(9/10) = 0.5$ dB, for example. The maximum number of periods that can be transmitted in a given-length signal is limited by the fact that the sequence period LT must exceed the duration of the impulse response of the channel, as discussed earlier.

Although the computational techniques used to process the received signals will not be discussed in detail here, it is worth noting that extremely efficient algorithms exist for removing m-sequence modulations, based on the fast Hadamard transform (Cohn and Lempel, 1977; Borish and Angell, 1983). Using this approach, cross-correlation processing for m-sequences is the fastest method known for any signal of comparable time-bandwidth product, requiring only $2SL \log_2(L)$ add/subtract integer arithmetic operations, where S is the number of samples per digit. The real and imaginary parts of the complex demodulates are processed in parallel.

Minimum SNR. The SNR that the measurement system must be designed to achieve is set by the dual requirement that the SNR be adequate to detect individual ray arrivals with an acceptably low false-alarm rate *and* to measure the arrival times of the ray arrivals with a precision of a few milliseconds. Both requirements lead to a design goal of at least 20 dB SNR. The precision with which travel times can be measured for a given SNR is discussed in the next section. The false-alarm rate can be computed by realizing that arrivals are simply peaks in the envelope of the processed signal that stand out above peaks due to noise alone. The distribution function of the envelope squared (*i.e.*, intensity) of complex Gaussian noise is

$$F(x) = 1 - \exp(-x/x_0), \qquad x \geq 0, \qquad E(x) = x_0, \tag{5.2.31}$$

where x is the intensity. Given a series containing M independent and identically distributed such variables, the probability P that one of the values will exceed a threshold x is

$$P = 1 - F^M(x). \tag{5.2.32}$$

Conversely, given a desired false-alarm rate probability P, the threshold is

$$x/x_0 \approx \ln(M/P)\,. \tag{5.2.33}$$

For example, if the sequence period length $L = 1023$, and 6 demodulates/digit have been retained, then $M = 6138$. Selecting $P = 10^{-4}$ gives $10 \log_{10}(x/x_0) = 12.5$ dB. Peaks more than 12.5 dB greater than the mean intensity would be identified as probable signal peaks. Ambient acoustic noise is not precisely Gaussian, however, and experience has shown that thresholds of 14–17 dB are usually required to achieve a low false-alarm rate. SNR values of about 20 dB are required to be compatible with these threshold levels.

One cannot arbitrarily increase the signal duration to achieve 20 dB SNR, however. The overall signal length and the processing gain that can be achieved are limited by the requirement that the beginning and end of the signal be *phase-coherent*. Correlation processing of signals constructed from phase-modulated digits can be viewed as a simple sum of all of the digits in the transmitted signal, with the phases suitably adjusted to remove the phase modulation. When thought of in these terms, it is obvious that if some process shifts the phase of the last digit by more than about $\pi/2$ radians (rad) relative to the phase of the first digit, the digits will no longer add constructively, and the processing gain will be reduced. (With a phase shift of π rad, the digits will be exactly out of phase and will cancel.) In general, the peak signal integration gain is reduced by $\text{sinc}^2(\Delta\theta/2\pi)$, where $\Delta\theta$ is the end-to-end phase change, and $\text{sinc}(x) \equiv \sin(\pi x)/\pi x$. A phase change $\Delta\theta = \pi/2$ rad will cause a 0.9-dB loss; a phase change of $\Delta\theta = \pi$ rad will cause a 3.9-dB loss.

A phase change of $\pi/2$ rad will occur if the source and receiver are approaching or receding from one another with sufficient velocity to cause a change in range by $\lambda/4$ during the transmission, for example. (Processing techniques to handle the case of constant relative velocity are discussed in section 5.5.) Sound-speed perturbations caused by internal waves also cause signals arriving at different times to become decorrelated, even with fixed source and receiver, limiting the maximum signal duration that can be coherently processed. At 250 Hz and 1 Mm range, coherent integration times are limited to a few minutes, for example (Flatté and Stoughton, 1988). Section 5.5 discusses this further.

SNR improvement using pulse compression. At 1 Mm range, a 12-ms pulse (exactly three cycles at 250 Hz) is predicted to have SNR = –23 dB for realistic source parameters (table 5.1). To achieve SNR = +19 dB requires that the transmitted energy be increased by 42 dB. The transmitted signal therefore needs to contain approximately 16,000 digits of 12-ms duration each, lasting 200 s overall. One can achieve this using signals of the type described ealier by transmitting 17 periods of a 1023-digit (degree-10) m-sequence and processing 16 of them (196.416 s). Discarding 1 out of 17 periods reduces the SNR by 0.26 dB. Each period lasts 12.276 s, which is more than adequate to exceed the impulse response at 1 Mm. In a self-contained instrument, the processing should be set to start approximately one-half period (*i.e.*, 6 s) after the slowest ray is predicted to arrive. Travel times can easily be predicted to much better than 6 s. (As a practical matter, moorings usually can be deployed within 2–3 km of a preset location, so that the travel-time error caused by a range error will be 1–2 s.) It is shown in appendix section 5.9 that no self-clutter is introduced, and FIMF processing is fully matched to the transmission, if the modulation angle is chosen to be $\theta_0 = \arctan(L^{1/2}) = 88.209°$.

5.3. Travel Time

Travel time is the most commonly used datum in ocean acoustic tomography. The precision with which arrival times can be measured is fundamentally limited by (i) ambient acoustic noise, (ii) internal wave scattering, and (iii) interference between unresolved ray paths. Each of these limitations will be discussed in turn. Less fundamental, but nonetheless important, limitations are also set in some situations by clock error and the precision with which the source and/or receiver positions are measured. These limitations will be discussed in sections 5.6 and 5.7.

Ambient acoustic noise. In the ray approximation, the received signal is the sum of delayed replicas of the transmitted signal (5.2.24). Matched-filter processing can then be shown to be the optimum procedure for estimating the arrival time of a single, resolved arrival embedded in Gaussian noise (Helstrom, 1968). Using the time at which the envelope is a maximum yields rms error

$$\sigma_\tau = \left[(\Delta\omega)_{\text{rms}}\sqrt{2E/N_0}\right]^{-1}, \tag{5.3.1}$$

where $(\Delta\omega)_{\text{rms}}$ is the rms bandwidth (5.2.12), and $2E/N_0$ is the SNR obtained with a matched filter (5.2.27).

If the propagation is unsaturated, as defined in chapter 4, the signal carrier phase is simply $\omega_0\tau$, and the phase can be used to give an improved travel-time estimate with rms error

$$\sigma_\tau = \left[\omega_0\sqrt{2E/N_0}\right]^{-1}, \tag{5.3.2}$$

where ω_0 is the carrier frequency. [One simply replaces the rms bandwidth of the complex envelope in (5.3.1) with the rms bandwidth of the undemodulated signal about zero frequency, which is approximately the carrier frequency for a narrowband signal.] Using the phase decreases the error by $(\Delta\omega)_{\rm rms}/\omega_0$. But the estimate using the phase is ambiguous by an integer number of cycles. The ambiguity can be resolved if the estimate using the complex envelope has an rms error smaller than the time corresponding to 1 rad of carrier-phase change. The appropriate cycle can then be determined from the envelope, and the phase provides a vernier scale within the cycle. Unfortunately, however, the carrier phase is generally not simply $\omega_0\tau$, but includes phase changes associated with caustics and boundary interactions. Interpretation of the carrier phase is further complicated by the effects of internal waves on acoustic propagation in the ocean. As discussed in chapter 4, for realistic acoustic frequencies and geometries, internal-wave-induced sound-speed perturbations tend to break each ray into micromultipaths, so that the measured phase becomes the sum of the phases from a number of ray paths. The measured phase is then no longer interpretable in terms of travel time, and the travel-time precision computed from (5.3.2) is not meaningful.

Even the errors computed from (5.3.1) are rarely achievable in practice, because of the greater travel-time uncertainties due to internal-wave-induced travel-time fluctuations, as will be discussed later.

For illustration, consider a pulse $p_z(t)\exp(i\,\omega_0 t)$, with a rectangular envelope of duration Δt,

$$\begin{aligned} p_z(t) &= 1, \quad -\Delta t/2 \le t \le \Delta t/2\,, \\ &= 0, \quad \text{otherwise}\,. \end{aligned} \tag{5.3.3}$$

The Fourier transform is

$$P_z(\omega) = \Delta t\ \operatorname{sinc}(\omega\Delta t/2\pi)\,. \tag{5.3.4}$$

If the spectrum is sharply band-limited at the first zeros of the sinc function (by the source and receiver transfer functions, for example), (5.2.13) gives the mean frequency

$$\overline{\omega} = \int_{-2\pi/\Delta t}^{2\pi/\Delta t} \omega \left|P_z(\omega)\right|^2 d\omega \Bigg/ \int_{-2\pi/\Delta t}^{2\pi/\Delta t} \left|P_z(\omega)\right|^2 d\omega\,, \tag{5.3.5}$$

which is zero, because the numerator vanishes (an odd function is being integrated over an even domain). After some algebra, (5.2.14) gives the rms bandwidth

$$(\Delta\omega)_{\mathrm{rms}} = \left(\frac{2\pi}{\Delta t}\right)\frac{1}{\sqrt{2\pi\,\mathrm{Si}(2\pi)}} \approx \left(\frac{2\pi}{\Delta t}\right)\frac{1}{\pi}\,,$$
$$\mathrm{Si}(x) \equiv \int_0^x \frac{\sin(u)}{u}\,du\,, \tag{5.3.6}$$

where Si (2π) has been approximated by its asymptotic value $\pi/2$ as x approaches infinity. This result can be written $\Delta f_{\mathrm{rms}} \cong 1/(\pi\,\Delta t)$. [This is not directly comparable to the time-bandwidth relation (5.2.19), because Δt is not the rms signal duration.] The example considered at the end of the last section had SNR = 19 dB, digit duration $\Delta t = 0.012$ s, and carrier frequency $f_0 = 250$ Hz. The rms bandwidth $(\Delta f)_{\mathrm{rms}}$ is then approximately 27 Hz. (Note that the rms bandwidth of 27 Hz is significantly less than the bandwidth obtained from the commonly used rule of thumb $\Delta f = 1/\Delta t \approx 83$ Hz, because the rms bandwidth is a rather small unit of measure for the full bandwidth, as discussed previously.) The rms error using the magnitude of the complex envelope is $\sigma_\tau = 0.6$ ms, which is about 1 rad at 250 Hz. If the propagation were unsaturated, the phase could then be used to reduce the rms error to $\sigma_\tau = 64\ \mu$s, although typically this is possible only at short range.

Internal wave scattering. As discussed in chapter 4, travel-time fluctuations at frequencies above one cycle per day are predominantly caused by sound-speed perturbations associated with internal wave vertical displacements (although internal wave currents contribute significantly to high-frequency fluctuations in differential travel times from reciprocal transmissions). For purposes of mesoscale mapping or making long-range integrating measurements, the internal-wave-induced travel-time wander is measurement noise. The wander is of order $10\sqrt{r}$ ms (r in Mm), typically exceeding the variance due to ambient noise for refracted paths. Pulse spread also limits the precision with which pulse travel times can be measured, by distorting the pulse shape. There is no point to transmitting digits much shorter than the pulse spread.

In fixed geometries, time averaging is typically used to reduce the travel-time noise caused by internal waves. A common strategy is to transmit a number of pulses separated by more than the decorrelation time of the internal-wave-induced fluctuations during a 24-hour period. Averaging the travel times of n independent pulses reduces the rms travel-time noise by $\sqrt{n}$. The average should be taken over a period greater than or comparable to the inertial period,

$$T_{\mathrm{inertial}} = (12\,\mathrm{hours})/\sin(\mathrm{latitude})\,, \tag{5.3.7}$$

which is the period of the lowest-frequency, most-energetic internal waves. For example, averaging over pulses transmitted at hourly intervals for a day will reduce the travel-time wander at 1 Mm range from 10 ms to 2 ms. In situations for which time averaging is not possible, receptions at receivers separated vertically by more than the internal-wave-induced vertical decorrelation length can be used to obtain an equivalent averaging effect. Whereas the internal-wave-induced travel-time variance can be reduced by computing daily averages or using an equivalent low-pass filtering operation, the internal-wave-induced travel-time bias is not reduced by averaging. If the bias can be accurately computed, it should be subtracted from the measured travel times prior to inverting. If the bias is not known, it needs to be explicitly allowed for in the inversion procedure.

Travel-time fluctuations caused by internal-wave-induced sound-speed fluctuations have been found to largely cancel in the computation of differential travel times out to the longest ranges for which data are available, about 1 Mm (table 3.2), so that differential travel times have significantly lower high-frequency variances than do one-way travel times. The cancellation of internal-wave-induced noise is important because the differential signal associated with large-scale currents is an order of magnitude smaller than the one-way signal associated with sound-speed perturbations. The degree of cancellation is reduced by nonreciprocity of ray paths due to large-scale current shear (*i.e.*, the extent to which the oppositely traveling signals do not follow the same ray path) and by changes in the internal wave field during the travel-time interval (because the field changes between the times that oppositely traveling pulses traverse a given point). Situations in which the current shear is large where the sound-speed shear is small (*e.g.*, in the 18° C water in the western North Atlantic) can lead to large nonreciprocity (Sanford, 1974), although calculations for an actual 300-km experiment west of Bermuda gave vertical separations of order 10 m between oppositely traveling paths in a reciprocal pair, far less than the vertical decorrelation scale for internal waves, which is of order 100 m (Stoughton *et al.*, 1986). The degree of nonreciprocity due to finite propagation times increases with increasing range. At 300 km range, the travel time is approximately 200 s, which is short compared with typical internal wave periods. At 1 Mm range, the travel time is 670 s, which is comparable to the periods of high-frequency internal waves, but shorter than the periods of the most energetic waves.

Travel-time fluctuations caused by internal wave currents are present in the differential travel times and can be important when travel-time fluctuations due to internal-wave-induced sound-speed fluctuations largely cancel. Even though the one-way travel-time variance due to internal wave currents is only 1% of the variance due to internal wave displacements, because $u/\Delta C$ due to internal waves is about 0.1, internal wave currents were found to be the largest contributors to the differential travel-time variance in the 300-km experiment mentioned earlier (Stoughton *et al.*, 1986).

Ray interference. Interference between overlapping, but formally resolved, ray arrivals degrades the precision with which the travel time of either arrival can be measured (Cornuelle *et al.*, 1985). When pulse travel times are measured with an error significantly smaller than the width of the pulse, in accord with (5.3.1), the high SNR is exploited to determine precisely where the pulse peak is located. Adjacent pulses are defined to be formally resolved when they overlap at their −6-dB (half-amplitude) points. When that is the case, two arrival peaks can usually be seen. Nonetheless, the pulses still overlap, causing significant interference. The precise locations of the peaks of the two arrivals can be shifted by the interference, with the shifts dependent on the relative phases of the pulses. Because the relative phase is more or less random from transmission to transmission, the interference effectively increases the uncertainty in the arrival-time determination. The ideal solution is to transmit digits that are sufficiently short so as to cleanly separate adjacent arrivals. There will, however, always be arrivals too closely spaced in time to be resolved with realistic pulse lengths. Degeneracies in the ray arrival pattern give distinct rays with identical travel times. For axial source and receiver, rays with equal numbers of upper and lower loops and opposite source angles will have identical travel times, for example. As rays become more nearly axial, the arrival-time separation becomes steadily smaller, but the digit length used to resolve the arrivals cannot be arbitrarily reduced, because of internal-wave-induced pulse broadening. (A vertical receiving array will increase the number of resolved arrivals by adding spatial resolution.) An alternative procedure is to explicitly allow for overlapping arrivals in the signal processing (Ehrenberg *et al.*, 1978; Ewart *et al.*, 1978). One models the reception as a sum of (unresolved) arrivals as in (5.2.24) and minimizes the mean square difference between the data and the model by *simultaneously* adjusting all of the amplitudes and travel times. When using this approach, it is important to have a good model for the received pulse shape, including any distortions caused by the source and receiver, by differential attenuation across the signal bandwidth (Jin and Worcester, 1989), and by spreading due to internal waves.

5.4. Vertical Arrival Angle

Although a single hydrophone receiver is the simplest configuration, a small vertical receiving array will provide significant benefits, at the cost of somewhat increased receiver complexity (Worcester, 1981; Worcester *et al.*, 1985*b*). Vertical receiving arrays improve the SNR, as described in section 5.1. (In terms of overall system cost, often it is less expensive to improve the SNR using a receiving array than by using a louder acoustic source.) Vertical arrays enable the separation, in terms of vertical angle, of some arrivals that are not resolved in the time domain. The measured arrival angles assist in ray identification (fig. 5.5). Tracking algorithms that take the raw acoustic data and generate time series of travel times for resolved ray paths perform significantly better using both travel time and arrival angle, rather than travel time alone. Joint inversions combining travel time and arrival angle may be feasible, although no work has been done in this area.

The signal bandwidths in ocean acoustic tomography typically are too great to permit the application of simple phase-shift beam-forming. True time-delay beam-forming, in which the hydrophone outputs are shifted in time prior to summing, must be used. This can be implemented either before or after complex demodulation (Pridham and Mucci, 1979; Metzger, 1983; Horvat *et al.*, 1992).

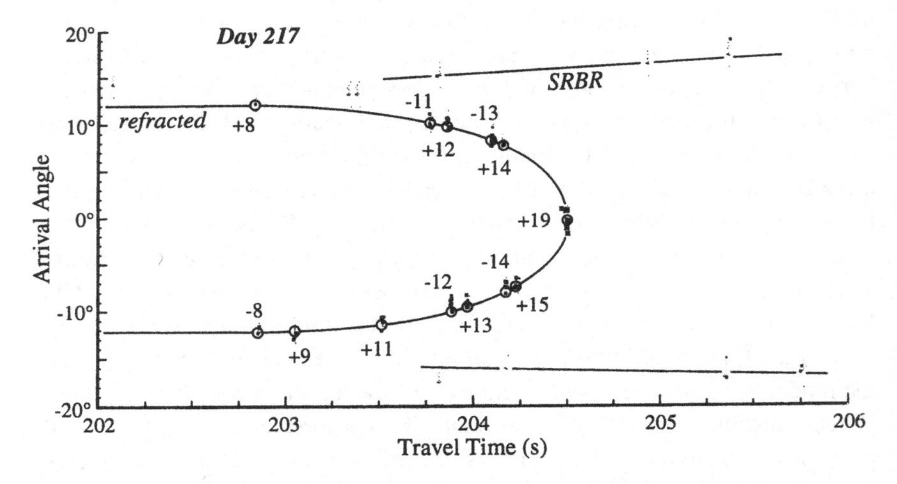

Fig. 5.5. Measured and predicted ray arrivals at 300 km range plotted against travel time and vertical arrival angle. The predicted arrivals are plotted as open circles connected by the smoothly drawn solid curve. The measured one-way arrivals for the 10 receptions on year day 217, 1983, are plotted as points, with size proportional to SNR. Geometric arrivals are labeled $\pm n$, where n is the total number of turning points. SRBR refers to surface-reflected, bottom-reflected rays. (From Howe *et al.*, 1987.)

Just as was the case for travel time, one must distinguish between vertical resolution and the precision with which vertical arrival angles can be measured for resolved arrivals. Vertical resolution, defined as the ability to separate, in terms of angle, two arrivals with nearly identical travel times, depends on the beam width of the receiving array. Whereas (5.1.14) gave the formula for the beam pattern of a uniform line array, a rough rule of thumb, analogous to $\Delta t \approx 1/\Delta f$ in the time domain, is

$$\Delta\theta \approx \frac{1}{(nd/\lambda)} \quad \text{rad}, \tag{5.4.1}$$

where θ is the vertical arrival angle, n is the number of elements, and d is the spacing between elements. For a four-element array, with $d/\lambda = 1.5$, $\Delta\theta \approx 0.17$ rad $= 9.5°$. This is adequate to separate upward- and downward-traveling rays, but little more. With a small array, the bulk of the resolution must occur in the time domain.

The precision with which the arrival angles of resolved paths can be measured is limited by the same factors that limit travel time: (i) ambient acoustic noise, (ii) internal wave scattering, and (iii) interference between unresolved ray paths (in the vertical).

Ambient acoustic noise. The rms error inherent in measuring the arrival angle of an isolated arrival is given by a formula analogous to (5.3.1):

$$\sigma_\theta = \left[\beta_x \sqrt{2E/N_0}\right]^{-1}, \tag{5.4.2}$$

where

$$\beta_x = 2\pi \left[\int_{-\infty}^{\infty} x^2 I(x)\, dx\right]^{-1/2} \tag{5.4.3}$$

is the rms length of the antenna illumination function $I(x)$, and x is in wavelengths (Rihaczek, 1969; Worcester *et al.*, 1985*b*). For example, four discrete hydrophones separated by d/λ give $\beta_x = \sqrt{5}\pi(d/\lambda)$. For $d/\lambda = 1.5$ and SNR = 20 dB, the rms arrival-angle fluctuation is $\sigma_\theta = 0.01$ rad $= 0.6°$. Just as was the case in the time domain, the high SNR is being exploited to precisely locate the peak of a relatively broad pulse.

Internal wave scattering. Sound-speed perturbations associated with internal wave vertical displacements cause arrival-angle fluctuations as well as travel-time fluctuations. Flatté *et al.* (1979) define a vertical coherence length z_0 such that the internal-wave-induced phase difference between rays arriving at receivers vertically separated by z_0 has a variance of (1 rad)^2. This can be interpreted as an rms arrival-angle fluctuation

$$\sigma_\theta \equiv \left(\frac{2\pi}{\lambda} z_0\right)^{-1} ; \tag{5.4.4}$$

σ_θ is nearly independent of frequency. Numerical calculations for 1 Mm range give σ_θ between roughly 1° and 10°, corresponding to vertical coherence scales of 55 m and 5.5 m, respectively, at 250 Hz (see fig. 4.6). Just as internal-wave-induced phase fluctuations over time limit the maximum interval for which signals can be coherently processed, internal-wave-induced phase fluctuations in the vertical limit the maximum vertical extent for which signals can be coherently processed. The fluctuations are roughly proportional to $\sqrt{r}$, although rapid variations with range are superposed because of focusing and defocusing of each ray tube. The largest values of σ_θ are associated with rays that end near a caustic. Even for a small four-hydrophone array at 250 Hz, with $d/\lambda = 1.5$ and 4.5 λ overall length, internal-wave-induced fluctuations at 1 Mm range are comparable to or exceed the fluctuations caused by ambient noise.

5.5. Doppler

The most general definition of the Doppler effect is as a rate of change in travel time. It is caused by relative motion of the source and receiver, and by variability in the ocean sound-speed and current fields sufficiently rapid to change the travel time during a transmission. For the case of constant relative velocity of the source and receiver, the effect is to uniformly compress or expand the time axis of the received signal. For narrowband signals, the principal result is the familiar frequency shift

$$\frac{\Delta f}{f} = \frac{v}{C}, \tag{5.5.1}$$

where v is the velocity at which the source and receiver are approaching. For broadband signals of the type used in ocean acoustic tomography, the envelope is also significantly compressed by a factor $(1 + (v/C))^{-1}$. Doppler is normally, although not always, a nuisance parameter that needs to be appropriately accounted for during signal design and processing, but in which we are not interested.

As was pointed out at the end of section 5.2, Doppler limits the time over which a signal can be coherently processed, unless the signal processing explicitly accounts for Doppler. To get a feeling for the magnitude of the effect, consider a source and receiver with constant relative velocity v. Using the criterion that for coherent processing the phase at the end of the received signal cannot differ from the phase at the beginning by more than $\pi/2$, the maximum signal duration, T_{signal}, can be found from

$$\left(\frac{2\pi}{\lambda}\right) v T_{\text{signal}} = \pi/2\,; \qquad (5.5.2)$$

that is, T_{signal} is the time required for the range to change by $\lambda/4$. Outside of major currents, the maximum velocity observed for tautly moored instruments is roughly 1 cm/s, giving $T_{\text{signal}} = 150$ s at 250 Hz. This is close to the integration time of 196 s computed for the example at the end of section 5.2, so that Doppler can be largely ignored (although a shorter integration time would be more conservative). For sources or receivers suspended from surface drifters or floats, however, typical velocities are 10 cm/s, reducing the integration time without Doppler correction to 15 s. If sources or receivers are suspended directly from shipboard, typical velocities might be 100 cm/s (2 knots), further reducing integration time to 1.5 s at 250 Hz. Although the use of lower frequencies would help, it is clear that the signal processing must be designed to compensate for Doppler in applications such as moving-ship tomography, where sources and/or receivers are suspended from shipboard or from surface floats.

For constant Doppler (*e.g.*, constant relative velocity of the source and receiver), the solution is to process for a range of possible Dopplers and to select the output with the maximum value. One proceeds by selecting a mesh of uniformly spaced Doppler compression ratios (*i.e.*, relative speeds). For each hypothesized speed, the data are interpolated and resampled to obtain samples at the times that would have been sampled in the absence of Doppler. This can be done directly on the complex demodulates. For periodic signals of the type described in section 5.2, the resampling must be done prior to forming the periodic average. Finally, the resampled signal for each Doppler compression ratio is processed, and the one with the largest peak is selected. [In the event that the relative velocity is not constant, but the relative acceleration is, one can expand the search space to include a mesh of uniformly spaced accelerations at each velocity (*e.g.*, Rihaczek, 1969).]

Just as for travel time and vertical arrival angle, one must distinguish between Doppler resolution and the precision with which Doppler can be measured for resolved arrivals. Resolution in Doppler is important not because we expect to receive signals with greatly different Dopplers, but because the resolution in Doppler sets the maximum grid spacing that can be used. A rough rule of thumb is that the Doppler resolution is

$$\Delta f_{\text{Doppler}} = 1/T_{\text{signal}}\,. \qquad (5.5.3)$$

$\Delta f_{\text{Doppler}}$ is approximately the bandwidth of a simple, unmodulated pulse of duration T_{signal}, even when the signal is modulated.

The precision with which the Doppler shift of resolved paths can be measured is limited by ambient acoustic noise and internal wave scattering.

Ambient acoustic noise. The rms error inherent in measuring the Doppler shift of an isolated arrival is given by a formula analogous to (5.3.1) and (5.4.2):

$$\sigma_f = \left[\Delta t_{\text{rms}} \sqrt{2E/N_0}\right]^{-1}, \tag{5.5.4}$$

where Δt_{rms} is the rms signal duration. This relation is applicable to simple pulses and to phase-modulated signals of the type described in section 5.2. It is not valid if the signal is frequency-modulated. Estimates of arrival time and Doppler shift are correlated for frequency-modulated signals, introducing a fundamental *ambiguity* (*e.g.*, Helstrom, 1968).

Internal wave scattering. As mentioned at the end of section 5.2, time-dependent internal wave vertical displacements cause signals arriving at different times to become decorrelated, limiting the maximum signal duration that can be coherently processed. Decorrelation in time corresponds to a broadening in frequency, commonly referred to as Doppler broadening. A continuous-wave (CW) transmitted signal consisting of a single frequency will be broadened upon reception to a finite bandwidth after propagating through a time-varying internal wave field. At 1 Mm range, Flatté and Stoughton (1988) give the order-of-magnitude estimate

$$\Phi\, t_0 \approx 1 \text{ hour}, \tag{5.5.5}$$

where t_0 is the decoherence time. At 250 Hz, $\Phi \approx 16$ (section 4.4), giving a decoherence time $t_0 \approx 225$ s. The corresponding Doppler broadening is then roughly

$$\nu = t_0^{-1} \approx 0.004\,\text{s}^{-1}. \tag{5.5.6}$$

At 1 Mm range and 250 Hz center frequency, the internal-wave-induced Doppler broadening is fortuitously close to the Doppler resolution of the signals typically used.

5.6. Timekeeping

To measure travel times with millisecond precision for time periods of a year or more requires highly accurate clocks. A frequency error of

$$\Delta f/f = 3 \times 10^{-11} \tag{5.6.1}$$

results in a time error of 1 ms after 1 year. To maintain time to better than 1 ms is not a problem for shore-connected instruments, surface drifters, or other configurations in which it is possible to have a satellite time-code receiver. The NAVSTAR Global Positioning System (GPS), for example, provides time to better than 1 μs worldwide. Other time-dissemination systems are also available. Maintaining time to better than 1 ms is also not a problem when adequate power is available to operate an atomic frequency standard continuously. Cesium frequency standards, for example, have a long-term fractional frequency stability of 10^{-13}–10^{-14}. They are primary frequency standards. Rubidium frequency standards are secondary standards requiring calibration, but once calibrated the fractional frequency stability is typically 10^{-11} per month.

Timekeeping becomes a problem in low-power, autonomous systems, such as moored tomographic transceivers that operate from batteries. Even in this case, however, clock errors cancel for sum travel times and for vorticity measurements (Munk and Wunsch, 1982*a*). With the clock errors explicitly shown, the travel time from mooring i to mooring j can be written

$$\tau_{ij} = (t_j + \delta t_j) - (t_i + \delta t_i), \tag{5.6.2}$$

where t_i is transmission clock time and δt_i is clock error for a transmitter at mooring i, and t_j, δt_j refer to a receiver at j. True time is $t + \delta t$. Similarly, for a transmitter at j and a receiver at i,

$$\tau_{ji} = (t'_i + \delta t_i) - (t'_j + \delta t_j)\,. \tag{5.6.3}$$

The clock error drops out of the sum travel time:

$$\tau_{ij} + \tau_{ji} = (t'_i - t_i) - (t'_j - t_j)\,. \tag{5.6.4}$$

One can think of j as a transponder with a known delay $t'_j - t_j$ between the time of reception t_j and the time of transmission t'_j. The delay needs to be subtracted from the measured interval at mooring i between transmission time t_i and response time t'_i.

The clock error also drops out of the sum of the travel times between n moorings around the periphery of an area:

$$\tau_{ijk\cdots ni} = \tau_{ij} + \tau_{jk} + \cdots + \tau_{ni} = t'_i - t_i, \tag{5.6.5}$$

where the transponder delays, $t'_j - t_j$, $t'_k - t_k$, ..., have been set to zero for simplicity. The difference in travel times around the periphery in opposite directions,

$$\tau_{ijk\cdots ni} - \tau_{in\cdots kji}, \tag{5.6.6}$$

therefore has no clock error. This difference is related to the circulation around the periphery and therefore to the area-average relative vorticity over the enclosed region.

The clock errors do not cancel for the differential travel times

$$\tau_{ij} - \tau_{ji} \, . \tag{5.6.7}$$

The differential travel-time signal due to ocean currents is roughly one order of magnitude smaller than the sum travel-time signal due to sound-speed perturbations, so that precise timekeeping is especially important.

Two basic approaches have been used to achieve adequate timekeeping precision in low-power, autonomous instruments. In one scheme, a two-oscillator system is employed (Spindel *et al.*, 1982; Worcester *et al.*, 1985*a,b*). A low-power oscillator runs continuously to drive the clock chain and other circuitry requiring a stable frequency input, such as the acoustic receivers. It is not sufficiently stable to give millisecond timing for a year, however. A second oscillator, which is much more stable, but which also consumes much more power, is turned on periodically so that the frequency difference between the low-power oscillator and the high-stability oscillator can be measured and recorded. At the conclusion of the experiment, the frequency offsets are integrated to yield a clock correction as a function of time. Compact rubidium frequency standards that will return to their previous frequency to within 2 parts in 10^{10} within 10 min after power is applied and that consume about 13 watts are available for use as reference standards. As usually implemented, this approach is not quite adequate to give 1-ms precision after 1 year. Pre- and post-cruise clock checks are therefore used to make a final correction.

A second scheme uses low-power crystal oscillators whose output frequency is temperature-compensated using digital techniques. (Historically, temperature compensation has been rather crudely done using analog techniques.) The approach is to calibrate the crystal by measuring its output frequency as a function of temperature. A microprocessor then combines this calibration information with the *in situ* temperature of the crystal, as measured with a thermistor or other temperature sensor, to determine the required frequency correction. In one approach, the frequency correction is then applied by adjusting the number of periods of the oscillator frequency counted to give a predetermined time interval, such as 1 s. Alternatively, the frequency of the oscillator circuit can be digitally adjusted. This digital approach gives much better temperature compensation than conventional analog techniques. The long-term stability achieved will depend on the inherent stability characteristics of the crystal, however. Although unusually stable crystal cuts have been used, that approach currently is not as accurate as the two-oscillator scheme using a rubidium frequency standard as a reference.

Clock performance of moored instruments can be checked *in situ* using high-frequency acoustic techniques (Worcester *et al.*, 1985*a*). To do so, the subsurface transceiver sends out interrogation pulses at predetermined times (once an hour, say). These pulses are received on board a ship in the vicinity, the precise reception time is recorded, and a reply pulse is transmitted. The subsurface instrument in turn detects the reply and records the reception time. After recovery of the instrument, the times recorded on board ship can be compared with the midpoint between the times of transmission and reception by the subsurface instrument. With suitable corrections for instrument delays, the two times should be identical; the difference gives the offset of the subsurface clock relative to the shipboard clock, which would normally be set using satellite time codes or other precise techniques. Clock offsets can be measured to better than 1 ms using this technique.

In the event that timekeeping of the desired accuracy has not been achieved, whether due to instrument malfunction or other causes, clock offsets can easily be included as unknowns in the inversion procedure, because they affect all travel times involving a given instrument identically (Cornuelle, 1983, 1985). If this is done, part of the information in the acoustic travel times is used to determine the clock offsets, rather than to learn about the ocean.

5.7. Positioning

To compute travel time with a precision of a few milliseconds requires that the range between the source and receiver be known to a few meters, because changing the range by 1.5 m will change the computed travel time by 1 ms. This seems to suggest that the positions of tomographic instruments must be measured quite precisely. Although that is true in some situations, the problem is more complex. One needs to distinguish between experiments in which absolute travel times are the basic data and experiments in which it is adequate to measure changes in travel time over time. One must also distinguish between errors due to changes in mooring position from measurement to measurement and errors due to mooring motion during a measurement.

Fixed systems. The simplest geometry is one in which both source and receiver are fixed. The only way to achieve this in the ocean is to mount the instruments on, or close to, the sea floor, accepting the possibility that bottom interaction may complicate the interpretation of the received signal. To measure

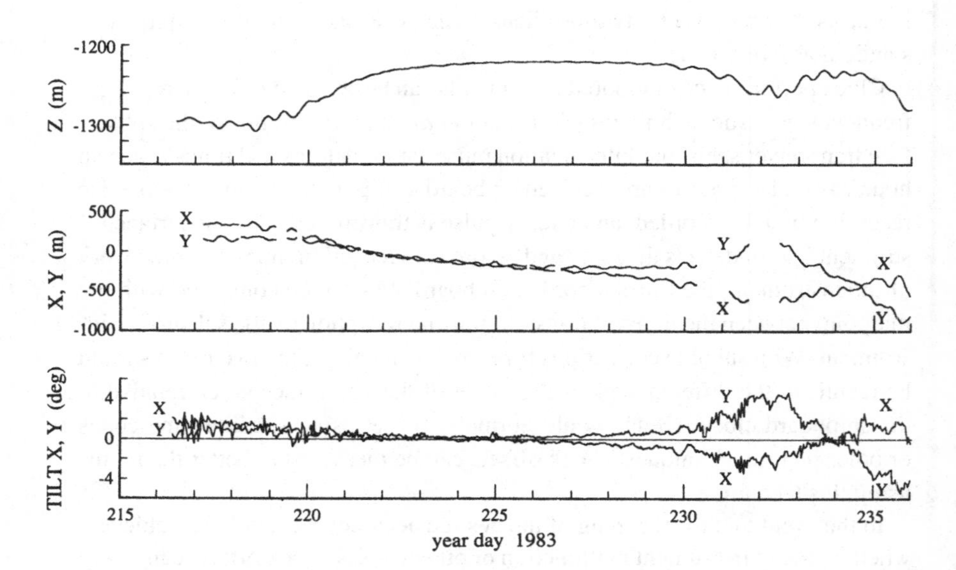

Fig. 5.6. Acoustic transceiver position and mooring tilt for the northern mooring in the 1983 Reciprocal Transmission Experiment, located at about 32°41′N, 68°57.8′W. The top panel shows the transceiver depth as measured by a pressure sensor. The second panel shows the east (x) and north (y) excursions of the transceiver relative to an arbitrary origin as measured by the acoustic navigation system. Gaps are due to missing data. The third panel shows the east (x) and north (y) tilt of the mooring at the depth of the transceiver. (Adapted from Worcester *et al.*, 1985*b*.)

time-dependent changes in the ocean, which is usually the case of interest, it is unnecessary to locate the source and receiver precisely.

Moored systems. The situation is more complicated for moored systems. Even with taut moorings, the instruments will move in response to barotropic tidal currents and currents associated with the ocean eddy field. Mooring velocities up to about 1 cm/s are observed for mooring tensions of about 2000 pounds (8.9 kN). Displacements of hundreds of meters in the horizontal and tens of meters in the vertical can occur, as shown in fig. 5.6 for a mooring near Bermuda (Spindel *et al.*, 1982; Worcester *et al.*, 1985*b*). To evaluate the impact of mooring motion, consider two transceivers at horizontal positions $x_1(t)$ and $x_2(t)$ at time t, where the x-axis is drawn through the transceivers from 1 to 2 (Worcester, 1977*b*; Munk and Wunsch, 1982*a*). Writing sound-speed as $C = C_0 + \Delta C(x)$, and the component of flow in the positive x direction as $u(x)$, the ray-averaged sound-speed perturbation and the ray-averaged current are

$$\overline{\Delta C_{12}} = \frac{1}{x_2 - x_1} \int_{x_1}^{x_2} \Delta C(x)\, dx \tag{5.7.1}$$

and

$$\overline{u_{12}} = \frac{1}{x_2 - x_1} \int_{x_1}^{x_2} u(x)\, dx, \tag{5.7.2}$$

where propagation along the x-axis has been assumed for simplicity. Computing the travel time from transceiver i to transceiver j to first order in the Mach number (u/C) gives

$$\tau_{12} = [R(t)/C_0]\left[1 - (\overline{\Delta C_{12}} - u_2 + \overline{u_{12}})/C_0\right] \tag{5.7.3}$$

and

$$\tau_{21} = [R(t)/C_0]\left[1 - (\overline{\Delta C_{12}} + u_1 - \overline{u_{12}})/C_0\right], \tag{5.7.4}$$

where $R(t) = |x_2(t) - x_1(t)|$, and u_1, u_2 are the instrument velocities. The sum and difference travel times are then

$$\tau_{12} + \tau_{21} = \frac{2R(t)}{C_0}\left[1 - (\overline{\Delta C_{12}} - \tfrac{1}{2}(u_2 - u_1))/C_0\right] \tag{5.7.5}$$

and

$$\tau_{12} - \tau_{21} = -\frac{2R(t)}{C_0}\left[(\overline{u_{12}} - \tfrac{1}{2}(u_1 + u_2))/C_0\right]. \tag{5.7.6}$$

The sum travel time depends on mooring motion through the absolute separation of the transceivers, $R(t)$, and through their relative velocity, $u_2 - u_1$. The difference travel time depends on mooring motion through the absolute separation and through the average velocity, $(u_1 + u_2)/2$. Rewriting gives

$$\overline{\Delta C_{12}} - \tfrac{1}{2}(u_2 - u_1) = \frac{C_0^2}{R(t)}\left[\frac{R(t)}{C_0} - \tfrac{1}{2}(\tau_{12} + \tau_{21})\right] \tag{5.7.7}$$

and

$$\overline{u_{12}} - \tfrac{1}{2}(u_1 + u_2) = -\frac{C_0^2}{R(t)}\left[\tfrac{1}{2}(\tau_{12} - \tau_{21})\right]. \tag{5.7.8}$$

The sound-speed perturbation $\overline{\Delta C_{12}}$ is relative to the difference in the velocities of the two instruments. This is usually a small effect and can be neglected, because sound-speed perturbations typically exceed mooring velocities by one to two orders of magnitude (m/s vs. cm/s). The current $\overline{u_{12}}$ is relative to the mean velocity of the two instruments. A tomographic measurement of current is no different from that by a moored current meter in this respect. This is often, but not always, a small effect, because typical current velocities are 10 cm/s, and maximum mooring velocities are about 1 cm/s. To achieve maximum accuracy when measuring barotropic tidal currents, which are of order 1 cm/s, one must correct for mooring velocity (just as one should when using current meters).

To see the effect of an error in the absolute separation, write $R(t) = R_0 + \delta R$, where δR can be considered to be an error in the assumed range or a change in range from one measurement to the next. Then

$$\overline{\Delta C_{12}} - \tfrac{1}{2}(u_2 - u_1) = \frac{C_0^2}{R_0}\left(1 - \frac{\delta R}{R_0} + \cdots\right)\left[\frac{R_0}{C_0} + \frac{\delta R}{C_0} - \tfrac{1}{2}(\tau_{12} + \tau_{21})\right] \tag{5.7.9}$$

and

$$\overline{u_{12}} - \tfrac{1}{2}(u_1 + u_2) = -\frac{C_0^2}{R_0}\left(1 - \frac{\delta R}{R_0} + \cdots\right)\left[\tfrac{1}{2}(\tau_{12} - \tau_{21})\right]. \tag{5.7.10}$$

The terms proportional to $\delta R / R_0$ represent small errors. Even a large error in range of 1 km out of 300 km will change $\overline{\Delta C_{12}}$ and $\overline{u_{12}}$ by only 0.3%, because of the $\delta R / R_0$ term. This is the only effect on $\overline{u_{12}}$, and the range error is therefore not important when using differential travel times to measure current. The term $\delta R / C_0$ in (5.7.9) is of the same order as the difference

$$R_0 / C_0 - \tfrac{1}{2}(\tau_{12} + \tau_{21}). \tag{5.7.11}$$

Changing the range by 150 m will change the travel time by about 100 ms, for example. Errors in range are therefore important when using sum travel times to determine sound-speed.

Summarizing, sum travel times are sensitive to range error, but not especially sensitive to mooring velocity during a measurement. Differential travel times are sensitive to mooring velocity (but no more so than currents measured with a standard moored current meter), but not to range error. As was the case for fixed source and receiver, however, if one is interested only in changes with time, one does not need to know the absolute separation of the instruments, but only the changes in range from measurement to measurement (*i.e.*, the relative mooring displacement). Acoustic long-baseline navigation systems can measure relative mooring displacements with precisions of about 1 m, which is adequate to correct travel times (Spindel *et al.*, 1982; Creager and Dorman, 1982; Milne, 1983; Worcester *et al.*, 1985*b*). In a typical system, a high-frequency acoustic interrogator is placed on the mooring near the acoustic transceiver. The interrogator periodically transmits a pulse to acoustic transponders placed on the sea floor about the base of the mooring. The distance from the mooring to the transponders is made to be about the same as the distance from the acoustic transceiver to the bottom. The transponders detect the interrogate pulse and transmit reply pulses at different frequencies, which in turn are detected by the interrogator. Frequencies near 10 kHz are commonly used, as they are near optimum at ranges of a few kilometers. The interrogator measures the round-trip travel times, which can be converted to slant ranges. Triangulation then gives the instrument position. The relative positions of the transponders must be accurately determined after deployment, but the absolute positions are not important to determine relative motion. A survey is required to locate the transponders. To do so, the ship interrogates the bottom transponders from a variety of locations, recording the round-trip travel times, which are subsequently used to determine the relative positions of the ship stops and the transponders. The ship would normally record its position at each ship stop, but this does not need to be done with a precision of a few meters unless the absolute positions of the transponders are also needed.

Ray travel-time corrections could be computed from the measured source and receiver mooring displacements, $\Delta x_{s,r}(t)$, $\Delta y_{s,r}(t)$, and $\Delta z_{s,r}(t)$, by retracing rays for the new source and receiver coordinates. For the mooring displacements typically encountered in practice, however, the assumption that the ray wave fronts are locally plane waves normal to the ray path is an excellent approximation (Cornuelle, 1983). Travel-time corrections are computed by defining fixed reference positions for both the source and receiver (given by the mean instrument positions, for example, although the precise locations are not critical as long as they are close to the source and receiver positions). The receiver correction is found by calculating the difference between the time that the wave front reaches the receiver reference location and the time that it reaches the receiver. This requires that the ray arrival be identified, so that its vertical arrival angle is known. A similar calculation gives the source correction. The measured travel times are then corrected using these time differences to give the travel times for source and receiver at the respective reference locations.

An alternative to directly measuring mooring displacement is to incorporate instrument positions as unknowns in the inversion procedure (Cornuelle, 1983, 1985; Gaillard, 1985; Cornuelle *et al.*, 1989). (This is analogous to the inclusion of unknown earthquake locations in the seismological problem.) Numerical simulations indicate that unrealistically high travel-time precision and/or large numbers of redundant ray paths are required to avoid significant degradation of the inverse solutions for sound-speed perturbations in the absence of mooring displacement information (Cornuelle, 1985). The fundamental uncertainty in the problem of estimating mooring parameters is the leakage of ocean energy into mooring displacement energy and vice versa.

Moving systems. The most difficult geometries are those encountered in moving-ship tomography or in the use of drifting tomographic instruments. In both of these cases the transmission geometry is continuously changing. It is therefore no longer possible to use changes in travel time over time to determine changes in the ocean. Absolute travel times must be inverted to give absolute $\overline{\Delta C_{12}}$. Absolute instrument positions (ranges) are then needed to an accuracy of a few meters to give travel times to within a few milliseconds. This is a difficult, but not impossible, task. The NAVSTAR GPS gives absolute antenna position to within a few meters when used in the differential mode. This requires that a reference GPS receiver be operated sufficiently near the ship or drifting instrument to record signals from the same satellites, for use in post-processing the signals recorded on board the ship or drifter. This is possible out to separations of roughly 1–2 Mm, although at the longer ranges ionospheric effects become important, degrading the positioning accuracy unless dual-frequency receivers are used. Precise satellite positioning of the antenna on the surface must be combined with precise positioning of the subsurface source or receiver relative to the antenna. In a surface drifter, with receiving hydrophones suspended below, a combination of tilt, heading, and pressure sensors distributed along the cable from the surface to the hydrophone may be adequate to give the receiver location relative to the antenna on the surface to within a few meters, although this has not been experimentally verified. For moving-ship tomography experiments in which a receiving array is lowered from shipboard every few hours to receive signals from a moored source (or, equivalently, in which a source is lowered to transmit to moored receivers), acoustic techniques can be used to locate the subsurface instrument relative to the ship. Ultra-short-baseline acoustic navigation systems measure the travel time and direction from a shipboard receiving array 12–15 cm across to a pinger on the subsurface instrument. To achieve an accuracy of 1 m at a range of 1000 m requires angular measurements accurate to 1 milliradian (mrad). This means that the orientation (pitch, roll, and heading) of the ultra-short-baseline array must be measured to about 1 mrad, as well. The greatest difficulty is with heading, as typical gyroscopes are accurate only to about 1°; systems using multiple GPS antennas allow the orientation to be measured to the required accuracy, however. An alternative approach that avoids having to determine ship orientation with high precision is to use a Floating Acoustic Satellite Tracking (FAST) system (Howe *et al.*, 1989*a,b*). This is essentially a long-baseline acoustic navigation system, analogous to the system used to track mooring motion, in which transponders (or precisely timed pingers) are attached to surface floats whose locations are determined using GPS. Two to three floats are deployed at each location where a tomographic measurement is to be made, and then are recovered prior to proceeding to the next location.

Cornuelle *et al.* (1989) used numerical simulations of moving-ship tomography experiments to determine the degradation incurred as a function of the precision with which the absolute positions were measured, when the source and receiver positions were included as unknowns in the inversion procedure. They found that the inverse solutions started to degrade when the range uncertainty was greater than the travel-time uncertainty due to internal-wave-induced travel-time fluctuations (*i.e.*, when the range uncertainty was roughly ± 10 m).

5.8. Data Treatment

The signal processing described in section 5.2 is only the first step in generating the time series of travel times to be used in the inversion procedures described in chapters 6 and 7. In this section we briefly summarize the additional steps required.

The usual first step is to *resolve* individual arrivals by constructing a *dot plot*. Every arrival peak exceeding a preset SNR is located, and its travel time is plotted as a function of reception time (fig. 5.7). The dot size plotted at each arrival time is normally made proportional to SNR to emphasize the loudest peaks. The SNR threshold is selected by considering the false-alarm rate, as discussed in section 5.2. For plotting purposes, the threshold is set somewhat low, giving additional small dots due to noise peaks, but ensuring that all of the ray paths are shown. Ray paths are evident in this plot from their continuity from reception to reception, while noise peaks are random. Correcting the dot plot for clock error (which affects all ray arrivals the same), and to first order for mooring motion, assuming that all of the rays are horizontal, sometimes makes the ray paths easier to recognize, because the travel times then change quite slowly, on the time scales with which the ocean changes. (Using data from a single hydrophone, only ray arrivals that are resolved in time will be evident, as in fig. 5.7. If a vertical receiving array is available to compute arrival angle, then encoding the arrival angles using the colors of the dots will sometimes reveal additional paths that are resolved in travel-time–arrival-angle space, but not using travel time alone.)

The next step is to *identify* the ray arrivals evident in the dot plot with specific ray paths. An estimate of the sound-speed field between the source and receiver, either from historical data or from concurrent measurements, is needed to predict the expected arrival pattern. Comparing the measured and predicted

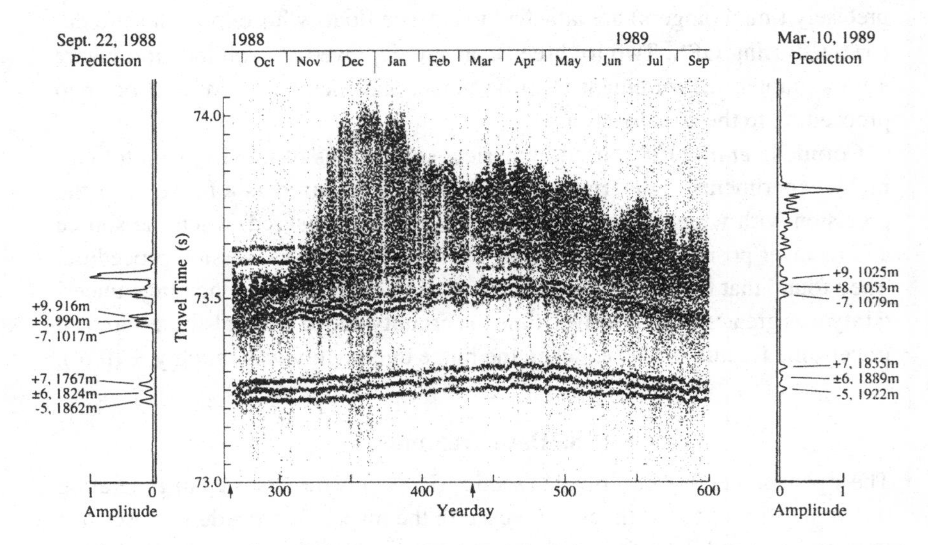

Fig. 5.7. Absolute travel times as functions of year day for transmissions from mooring 1 on the northern edge of the Greenland Sea tomography array to mooring 6 in the center, approximately 105 km distant. Each arrival peak is plotted as a dot, with size proportional to SNR. The travel times have been corrected for instrumental clock drift and to first order for mooring motion (assuming horizontal rays). Predicted arrival patterns constructed from environmental data collected on 22 September 1988 (left) and 10 March 1989 (right) using the WKBJ propagation algorithm are in good agreement with the measurements at the times indicated by arrows. The first and second sets of rays have lower turning points near 1800 m and 1000 m depths, respectively. The third set of rays, arriving between 73.55 and 73.65 s, are bottom-reflected paths. (Adapted from Worcester *et al.*, 1993.)

arrival patterns in travel-time space is usually adequate to identify the observed arrivals (fig. 5.7), although comparisons in travel-time–arrival-angle space are more robust (fig. 5.5). The geometric ray arrivals are labeled $\pm n$, where n is the total number of ray turning points, and $+(-)$ refers to rays that start upward (downward) at the source. One or more nongeometric arrivals are often observed, typically associated with caustics (*e.g.*, Worcester, 1981; Brown, 1981). Diffracted energy on the shadow-zone side of a caustic can be detectable up to several hundred meters from the caustic, at high SNR. Inadvertent errors in identifying ray arrivals become obvious when inversions are performed. Incorrect identifications lead to large travel-time residuals, because the assumed ray path samples the ocean differently than the actual ray path.

The most difficult task is to *track* the identified paths from one reception to the next, thus generating the required time series. The tracking programs that perform this function are essentially pattern-recognition algorithms. If only travel-time information is available, small windows in travel time centered on the arrival times in the first reception are defined. The algorithm then selects corresponding peaks in subsequent receptions, allowing the windows to translate as a whole (due to mooring motion or clock drift) and to move by small amounts relative to one another. Quality criteria, such as the peak width, can be used to reject peaks that do not correspond to clean arrivals. If information on both travel time and vertical arrival angle is available, the procedure is similar, except that the windows are then rectangles in travel-time–arrival-angle space. If large clock drifts or mooring motions are present, it often helps to correct the

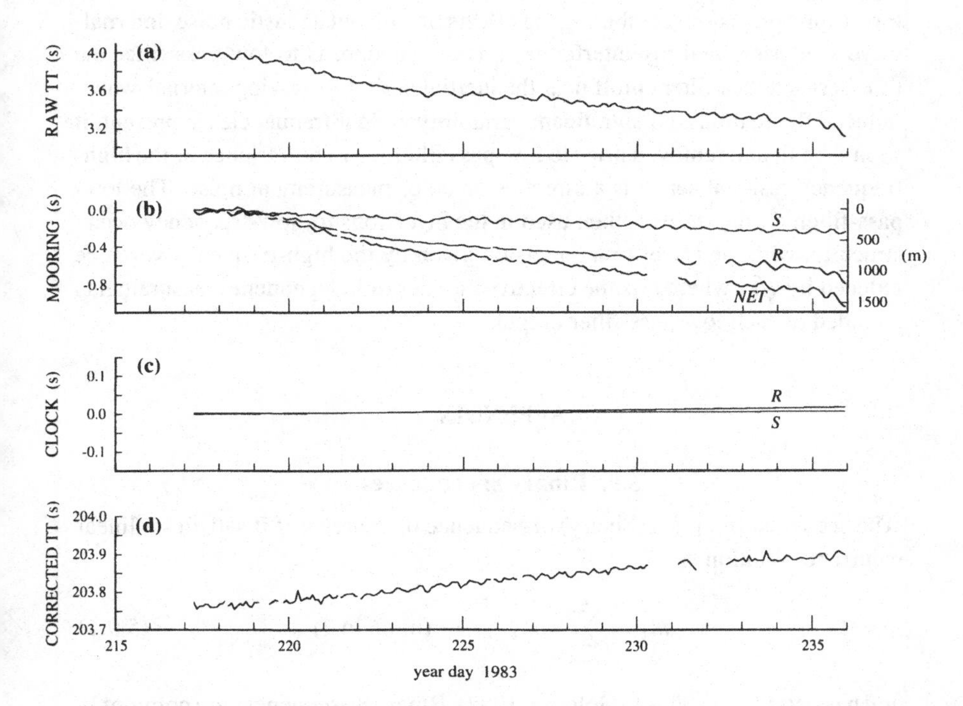

Fig. 5.8. Travel-time corrections applied to path −11 in the 1983 Reciprocal Transmission Experiment. The range is about 300 km; the ray upper-turning point depth is about 739 m. (a) The raw tracked peak series, with travel times relative to the beginning of the sequence period. (b) The corrections for source motion S, receiver motion R, and the net motion of the two. The right-hand vertical axis of (b) shows the equivalent displacement along the ray path. (c) Source and receiver clock corrections. (d) Final corrected travel-time series. This is equal to the raw travel time minus the corrections in (b) and (c) and a constant time offset giving the travel time at the beginning of the sequence period. The vertical scales of panels (c) and (d) are expanded by a factor of 4 relative to panels (a) and (b). (From Worcester *et al.*, 1985*b*.)

data for clock drift, and to first order for mooring motion (assuming all rays are horizontal), prior to tracking. The first-order mooring-motion correction is then removed from the tracked paths, and the full three-dimensional correction is applied, as described in section 5.7, using the appropriate vertical arrival angles (fig. 5.8). (The clock correction is also applied to the tracked paths, if that was not done prior to tracking.) Variants on this basic procedure have been developed (*e.g.*, Hippenstiel *et al.*, 1992; Send, in press).

At that point, the corrected time series of travel times are ready for use in the inversion procedure. The other data needed for the inversions are estimates of the precision of the measured travel times, as discussed in section 5.3. The high-frequency variance of the measured travel times gives a direct estimate of travel-time precision, combining the effects of ambient acoustic noise, internal-wave scattering, and ray interference. The procedure is to low-pass filter the time series with a filter cutoff near the inertial period, removing internal-wave-induced fluctuations. (If significant variability at tidal frequencies is present, it should be fit and removed prior to low-pass filtering.) The variance of the high-frequency residual series is a direct estimate of measurement noise. The low-pass-filtered time series is then used in the inversions for low-frequency ocean structure, with the travel-time precision given by the high-frequency variance reduced by $\sqrt{n}$, where n is the effective number of independent measurements included in each low-pass-filter output.

APPENDIX

5.9. Binary *m*-sequences

The sequence $\{m_\ell\}$ is a (binary) m-sequence of degree n if it satisfies a linear recurrence relation

$$m_\ell = \sum_{i=1}^{n} c_i m_{\ell-i} \qquad \text{(modulo 2)} \tag{5.9.1}$$

and has period $L = 2^n - 1$ (Golomb, 1982). Binary m-sequences are commonly named using the octal representation of the coefficients c_i written as a binary number in order of decreasing i. For example, the sequence of degree $n = 7$ whose coefficients are

$$\{c_7, c_6, \ldots, c_0 \equiv 1\} = \{1, 0, 0, 1, 1, 1, 0, 1\} \tag{5.9.2}$$

is named 235_8. Ordered tables of coefficients that result in m-sequences of various degrees are available (Metzger and Bowens, 1972). The n-values required for initialization are arbitrary, except they may not all be zero. An m-sequence of degree n includes all possible combinations of ones and zeros of length n, except for n zeros. The choice of initialization therefore determines only the location at which the periodic series begins; it does not alter the sequence.

There are two key properties of m-sequences that we shall exploit: (i) An m-sequence of degree n contains 2^{n-1} ones and $2^{n-1}-1$ zeros. (ii) If $\{s_i\}$ is the sequence of values obtained by taking a binary linear maximal sequence and replacing the binary ones with the value -1 and the binary zeros with the value $+1$, then the unnormalized periodic correlation function of $\{s_i\}$ is

$$\begin{aligned} r_k = \sum_{i=0}^{L-1} s_i s_{i-k} &= L \quad \text{if} \quad k = 0\,, \\ &= -1 \quad \text{if} \quad k \neq 0\,. \end{aligned} \tag{5.9.3}$$

(The same result is obtained if the binary ones are replaced by $+1$ and the binary zeros by -1.) The two-level autocorrelation function of $\{s_i\}$ is the key to its desirable properties as a modulating sequence.

We can now examine the implications of the particular signal choice (5.2.29), closely following Metzger (1983). Because the signal is periodic, the appropriate spectral representation is a Fourier series

$$G_k = \frac{1}{LT} \int_{-LT/2}^{LT/2} \left[\sum_{\ell=0}^{L-1} \exp(i s_\ell \theta_0) p(t - \ell T) \right] \exp\left(-i\, 2\pi \frac{kt}{LT} \right) dt, \tag{5.9.4}$$

where G_k is the complex amplitude of the spectral line at frequency k/LT. After some algebra, this gives

$$\begin{aligned} G_k = &\left\{ \sum_{\ell=0}^{L-1} \exp(i s_\ell \theta_0) \exp\left(-i\, 2\pi \frac{k\ell}{L} \right) \right\} \\ &\cdot \left\{ \frac{1}{LT} \int_{-LT/2}^{LT/2} p(t) \exp\left(-i\, 2\pi \frac{kt}{LT} \right) dt \right\}. \end{aligned} \tag{5.9.5}$$

The spectrum factors into L times the digital Fourier transform of the coefficients $\{m_\ell\}$ and the Fourier series of the waveform $p(t)$. FIMF and FIF are applicable because the spectrum has factored into the form $P(\omega)M(\omega)$.

We compute the power spectrum $|M(\omega)|^2$ by first computing the normalized periodic autocorrelation of m_ℓ:

$$\begin{aligned} R_k &= \frac{1}{L}\sum_{\ell=0}^{L-1} m_\ell m^*_{\ell-k} \\ &= \frac{1}{L}\sum_{\ell=0}^{L-1} (\cos\theta_0 + i\, s_\ell\, \sin\theta_0)(\cos\theta_0 - i\, s_{\ell-k}\sin\theta_0)\,. \end{aligned} \tag{5.9.6}$$

Using (5.9.3) and the fact that the sequence $\{s_i\}$has one more -1 than $+1$ gives

$$R_0 = 1 \quad \text{and} \quad R_{k\neq 0} = \frac{L\,\cos^2\theta_0 - \sin^2\theta_0}{L}\,. \tag{5.9.7}$$

If $\tan^2\theta_0 = L$, then $R_{k\neq 0} = 0$. The autocorrelation function of the modulating sequence is identically zero at all nonzero lags. Transforming the squared magnitude of (5.9.5), we see that the autocorrelation of (5.2.29) consists of the convolution of the autocorrelation of the sequence with the autocorrelation of the prototype pulse (because the Fourier transform of a product is the convolution of the Fourier transforms). The autocorrelation of (5.2.29) is then simply L times the autocorrelation of the prototype pulse when $\tan^2\theta_0 = L$. The modulation has introduced no self-clutter in the time domain even for matched filtering.

This result may also be examined in the frequency domain. The digital Fourier transform of (5.9.7) yields the power spectrum of $\{m_\ell\}$:

$$\begin{aligned} |M_k|^2 &= \frac{R_0 + (L-1)R_{k\neq 0}}{L}, \qquad k = 0\,, \\ &= \frac{R_0 - R_{k\neq 0}}{L}, \qquad k \neq 0\,. \end{aligned} \tag{5.9.8}$$

For $\tan^2\theta_0 = L$ this yields $|M_k|^2 = 1/L$ for all k. Because $|M_k|$ is independent of frequency, FIMF and ordinary matched filtering are identical. For this case, the inverse filter $M^{-1}(\omega)$ in fig. 5.3c is sometimes referred to as a "phase-only filter," since its magnitude is independent of frequency. The modulating filter, $M(\omega)$, modifies the phases of the frequency components of the prototype signal, $P(\omega)$, but does not change their amplitudes. The inverse filter then simply adjusts the phases back to those corresponding to the prototype signal. (FIF still results in lower SNR. In return, the output is proportional to the prototype signal, rather than to the autocorrelation of the prototype signal, as is the case for FIMF. This usually gives better time-domain resolution.)

For other choices of modulation angle, the power spectrum is still white, with the exception of the zero-frequency (carrier) line. FIMF and FIF are then no longer phase-only filters, since the gain at zero frequency is different from the gain at other frequencies, and the SNR is reduced relative to that achievable with matched-filter processing. The reduction is called the *nonflatness loss* (NFL) (Metzger, 1983; Birdsall and Metzger, 1986). But the output waveform is still proportional to the autocorrelation of the prototype pulse or to the prototype pulse itself. It is sometimes desirable to intentionally increase the amount of power in the carrier line relative to that obtained with the "ideal" modulation angle. For $\theta_0 = 45°$, for example, one-half the power is in the carrier, and one can effectively conduct CW and pulse experiments simultaneously. The SNR for the pulse experiment is reduced by 3 dB relative to that which would have been obtained using the ideal phase-modulation angle, however, because the inverse filter must remove essentially all of the excess power placed in the carrier line. It can also be shown that the SNR is reduced by 3 dB if $\theta_0 = 90°$. For ideal modulation angles near 90°, the asymmetry with angle in the NFL means that it is important that any approximations used when constructing the transmitted signal cause the phase angle to be slightly less than ideal, rather than greater.

CHAPTER 6

THE INVERSE PROBLEM: DATA-ORIENTED

6.1. Introduction

The preceding chapters have demonstrated that a variety of measurable acoustic features, including ray travel time, amplitude, and inclination, mode group velocity, and carrier phase, are integral functions of the oceanic sound-speed field. As discussed in previous chapters, sound-speed is intimately related to the oceanic density field, which is, in turn, a dynamic variable related to the oceanic flow field. Under many circumstances, knowledge of the density field alone is adequate to compute the oceanic flow field to a high degree of approximation. Reciprocal tomographic measurements are direct weighted averages of the flow field in the plane of the source and receiver. Thus, determinations of C and u carry immediate implications for the ocean circulation and must be consistent with known physics.

The forward problem has been presented in detail: Given C (or S) and u, and the characteristics of a sound source, compute the detailed structure of the signal as recorded at a receiver of known characteristics. This problem is labeled "forward" mainly as a reflection of its connection to the classic problem of finding solutions to the wave equation.

The "inverse" problem demands calculation of the ocean properties, C and/or u, given the measured properties of the arriving signal. At this stage, the problem becomes a matter of intense oceanographic interest.

Oceanographers are mostly familiar with point value data (*e.g.*, a current meter reading or a thermometer measurement). In contrast, tomographic data are weighted integrals through the oceanic field. It is the integrating property that makes the data uniquely valuable, and their analysis interesting. For some problems, the tomographic integrals could represent precisely what is required; more commonly, one requires different averages than those generated by the sound field (ray and mode trajectories produce a complex vertical/horizontal weighting). In other cases, one seeks estimates of the spatial variations along the ray paths.

In both cases, one has the problem of determining a three-dimensional field, $C(x, y, z)$ or $u(x, y, z)$, from knowledge of a set of integrals along specified paths $\Gamma(x, y, z)$. The problem may remind the reader of many situations involving integral transforms; this connection was made explicit in chapter 2 using the Abel-transform relations.

The purpose of this chapter is to provide a general discussion of the problem of using tomographic integrals to make inferences about the oceanic structure. Because the spatial and temporal variabilities in the ocean are very small fractions of the background field, so-called linear inverse methods become the central focus. As always when linearizing a problem, we must remain alert to potential failure.

The approach taken here is to describe inverse methods beginning with the simple and familiar method of least-squares, gradually modifying the basic formulation to accommodate the investigator's prior knowledge of the ocean. This method is powerful and efficient, but in the resulting solutions it is not always apparent which of the data are controlling which solution elements, why some elements are very uncertain, nor why the results can give the appearance of being arbitrary. The singular-value decomposition (SVD) is therefore introduced as a version of least-squares that is practical and that produces unmatched insight into the solution structure.

In contrast with least-squares, a statistical approach – we call it the Gauss-Markov method – is recommended as the default analysis approach – but we present it last because we have found in practice that the nature of the solutions is most readily understood after the method of least-squares is reviewed. Readers thoroughly familiar with least-squares methods may wish to skip sections 6.3 and 6.4, but we would urge everyone to at least skim them, as there are some novel features in the development, and the concepts and definitions are first laid out there.

Little or nothing in this chapter will be new to professional statisticians. Our excuse for the somewhat discursive approach is that we could not find a clear account suitable for those engaged in tomography. Consistent with the discussion of the observational technology in chapter 5, we take the view that the estimate of the noise present in tomographic measurements should always be regarded as part of the solution, and therefore that all inverse problems involving real data must be regarded as underdetermined.[1]

6.2. Representation

In chapter 2 it was shown that the most useful acoustic properties, ray or mode travel times, and so forth, could all be represented in the form

$$y_i = \int_{\text{source}}^{\text{receiver}} ds\, w_i(\mathbf{r})\, S(\mathbf{r}) \tag{6.2.1}$$

[1] True also of forward problems that are integrated using observed boundary or initial conditions.

where w_i is a weighting function, and $S(\mathbf{r})$ is a continuous function of the spatial coordinates. The measurements y_i, whatever their nature, are discrete and finite in number. The mathematical inverse problem, which is not of central interest to us, but which needs to be understood, requires the inference of a *continuous* function, $S(\mathbf{r})$ (containing thereby an uncountable infinity of degrees of freedom with potentially arbitrarily rapid spatial variability), from y_i, in which i is at best countably infinite. The oceanographic problem, the one that is our focus, differs from this idealized problem in two major ways: (i) the measurements, as described in chapter 5, are finite in number and *always* noisy[2] and (ii) we exploit our physical understanding of the ocean to remove from the inverse problem the need to infer structures that are already known.

Linearization. The most fundamental piece of independent prior information is the knowledge that the actual sound-speed profiles in the ocean are well represented as $S(\mathbf{r}) = S(\mathbf{r}, -) + \Delta S(\mathbf{r})$, where $|\Delta S| \ll S(-)$. $S(\mathbf{r}, -)$ is available from climatological data bases, or from shipboard observations at the time of an array deployment, or from a realistic ocean model, or from a previous set of tomographic measurements.

We saw in chapter 2 that the forward problem can be linearized [*e.g.*, (2.8.4)], so that the travel times can be written explicitly in terms of ΔS. If we then take the separate and distinct step of writing ΔS as a function of a discrete set of parameters ℓ, the forward problem can be stated in the form (2.15.1)

$$\mathbf{E}\boldsymbol{\ell} = \Delta\boldsymbol{\tau}, \tag{6.2.2}$$

or

$$\mathbf{E}\mathbf{x} = \mathbf{y}, \tag{6.2.3}$$

that is, the relationship of a discrete set of measurements ($\Delta\tau_i$ or y_i) to a finite discrete number of parameters ($\boldsymbol{\ell}$ or $\mathbf{x}$).[3] Earlier chapters have shown that perturbations to prior slowness profiles can be represented in any number of ways: as layers, dynamic modes, empirical orthogonal functions, or, for the *chic*, wavelets. Any representational basis will do. Ideally the choice should

[2]Much mathematical theory is directed at inference from infinite numbers of perfect measurements. These assumptions permit the use of very powerful methods not often applicable with real data.

[3]We shall focus primarily upon this discrete-data/discrete-solution problem. If one seeks to retain the discrete-data/continuous-solution representation of ΔS, the form of inverse theory based in functional analysis, as pioneered by Backus and Gilbert (1967, 1968, 1970) and Parker (1977, 1994), is available. The monograph by Bennett (1992) is useful in this context. Eisler *et al.* (1982) and Eisler and Stevenson (1986) have applied these methods to the acoustic tomography problem. Attention is also called to the general Bayesian approach of Tarantola (1987), which subsumes many of the methods we discuss, albeit in a somewhat formal way.

reflect insights into the most efficient representation of the perturbations, so as to maximize the use of prior knowledge and to minimize the number of quantities that need to be estimated. But the efficiency of representation is not critical, and the available methods are capable of compensating for all but the most misguided choices. Even the apparent distinction between representation in layers and representation in mode-like functions is more apparent than real: Hadamard transforms or Walsh functions, which are periodic functions with value ± 1 (Huang, 1979), or the Haar wavelet (Daubechies, 1992), can produce layer or block-like representations.

A reasonably general representation is a separable one, in which

$$\mathbf{x} = \sum_n F_n(z)\, G_n(x, y),$$

where F_n, G_n are chosen for convenience and efficiency. The pedagogically attractive, and practically useful, representation in horizontally uniform layers is given by

$$\begin{aligned} F_n(z) &= 1, \quad z_n \leq z \leq z_{n+1} \\ &= 0, \quad \text{otherwise}, \end{aligned}$$

with $G_n = \alpha_n =$ constant. In a range-dependent ocean, one can write $G_n(x, y) = \sum_{i,j} \alpha_{ij}\, g_i(x)\, h_j(y)$, with $g_i(x)$ and $h_j(y)$ given by tabulated values on a horizontal grid. A horizontal grid is conceptually just the limit of representation in a set of boxes as the size of the boxes becomes arbitrarily small, and their number arbitrarily large. Thus, vertical layers and a horizontal grid are the same as Munk and Wunsch's (1979) use of a set of three-dimensional rectangular solids; they are readily combined with a vertical representation in modes or other functions. Variations on uniform layers are readily used (linear or cubic splines, *etc.*), and they can have advantages in acoustic computations by controlling the order of the inevitable discontinuities at layer boundaries. A large number of representations is both possible and useful, as will be shown.

To proceed, we make one essential modification to (6.2.3), writing

$$\mathbf{E}\mathbf{x} + \mathbf{n} = \mathbf{y}, \tag{6.2.4}$$

where $\mathbf{n}$ is introduced to represent both the noise component of $\mathbf{y}$ and errors resulting from the linearization and any other inaccuracies in the representation (6.2.3). These latter components are commonly referred to as "model errors," as (6.2.4) is a model of the observations. The solutions to noisy systems of linear simultaneous equations have very broad application, and the methods that follow apply to a wide variety of scientific problems. $\mathbf{E}$ is sometimes called the "design" matrix; we call it the "observation" matrix.

The problem is to infer $\mathbf{x}$, representing the perturbation parameters of chapters 2–4 (*i.e.*, layer perturbations of sound-slowness or water velocity, or modal amplitudes, *etc.*). In the presence of noise elements, which necessarily have a partially stochastic character, the problem becomes one of *estimation*, which leads inevitably to a discussion of the statistical uncertainty of the results.

Expression (6.2.4) represents a potentially complicated model of the ocean structure and of its relationship to any measurable acoustic properties. In some situations there may be elaborate additional constraints relating the elements of $\mathbf{x}$, written symbolically as $\mathbf{L}(\mathbf{x}) = \mathbf{d}_m$, where $\mathbf{d}_m$ are known values. The operator $\mathbf{L}$ need not be linear, and the observations, $\mathbf{y}$, do not appear. Such relationships are often generated by dynamic models carrying expressions of oceanic physics that are not easily incorporated into $\mathbf{E}$. If $\mathbf{L}(\cdot)$ is linear (*e.g.*, a matrix $\mathbf{A}$), it is straightforward to append the relations to (6.2.4) without changing its structure; an example is the geostrophic relationship between the velocity and density fields (section 6.8). But the problems of oceanic physics are generally quite complex, and these extra constraints may greatly exceed in number those appearing in (6.2.4). It proves convenient, therefore, to divide discussion of inverse methods into two separate, but overlapping, types: "data-oriented" methods are based mainly on equations such as (6.2.4), where the data appear explicitly, and any ancillary constraints are few in number. In contrast, "model-oriented" methods involve ancillary constraints that greatly outnumber those in (6.2.4) and may be of great complexity. The latter methods are taken up in chapter 7. The reader is cautioned that the distinction is blurry: data are essential to both methods, and both involve explicit ocean models.

We can now be wholly generic and discuss the solution of linear simultaneous equations, exploring some of the tools available for understanding the solution. Several distinct issues arise. Because the measurements are noisy, the elements of $\mathbf{y}$ will contain errors. For a square matrix $\mathbf{E}$, the conventional inverse is

$$\widehat{\mathbf{x}} = \mathbf{E}^{-1}\mathbf{y}; \tag{6.2.5}$$

this solution cannot be correct,[4] as it implies $\mathbf{E}\widehat{\mathbf{x}} = \mathbf{y}$, with the unacceptable conclusion that $\widehat{\mathbf{n}} = \mathbf{0}$.

More generally, situations arise in which there are more equations than unknowns, or vice versa. A simple example of the first situation is the vertical slice problem in section 2.8, where the polar profile parameter γ is perturbed by $\Delta\gamma$. If the arrival times of more than one ray, or mode, are measured, one

[4] A subtle point is that (6.2.5) is an unbiased estimate for $\widehat{\mathbf{x}}$ if the noise has zero mean. It is here rejected because it produces an unacceptable noise variance.

has several observations available to infer $\Delta\gamma$. In contrast, if the ocean perturbations are represented by $1 \leq j \leq N$ layers of constant slowness perturbation ΔS_j, or N vertical modes of unknown amplitudes α_j, and we have available only $M < N$ travel times of acoustic rays or modes (or both), there are fewer equations available than unknowns.

We shall explore several different, but intimately related, procedures for inferring $\mathbf{x}$; in particular, we seek methods that have the property of being useful whatever the relative numbers of equations and formal unknowns and that will be capable of providing a determination of the accuracy of the result. (The methods differ from the usual procedure for solving the "just-determined" case and the classic least-squares procedure for the overdetermined case.) We label these techniques as "inverse methods" – a terminology that is conventional, but unfortunate because of its confusion with "inverse problems." For any problem reducible to a set of noisy simultaneous equations like (6.2.4), whether arising from a forward or an inverse problem, linear inverse methods can be used to find solutions.

The reader is reminded that sections 6.3 and 6.4 focus on least-squares. They are followed by a discussion of the Gauss-Markov method. Although the derivation of the Gauss-Markov estimate differs considerably from that using least-squares, the results are often identical, which is both extremely confusing (judging from the published literature) and very useful. Other methods directed at finding solution extremes (linear programming), and those suitable when nonlinearities are present, are then briefly described. The chapter ends with discussion of several published applications of inverse methods to real tomographic data.

Matrix notation and identities. In what follows, the reader is assumed to have a basic familiarity with matrices and vectors. Unless otherwise stated, all vectors are column vectors and are denoted by boldface lowercase letters, either Latin or Greek (*e.g.*, $\mathbf{f}$ or $\boldsymbol{\alpha}$). Matrices are generally boldface uppercase ($\mathbf{A}$, $\boldsymbol{\Gamma}$). $\mathbf{A}^T$ is the transpose of $\mathbf{A}$, and $\mathbf{A}^{-1}$ is the conventional inverse when the matrix is square. $\mathbf{I}$ is the identity matrix, sometimes written $\mathbf{I}_K$, to show its dimensions ($K \times K$). It is helpful to recall the relations $(\mathbf{AB})^T = \mathbf{B}^T\mathbf{A}^T$, $(\mathbf{AB})^{-1} = \mathbf{B}^{-1}\mathbf{A}^{-1}$, when the inverses exist, and

$$\frac{\partial}{\partial \mathbf{q}}(\mathbf{q}^T\mathbf{r}) = \frac{\partial}{\partial \mathbf{q}}(\mathbf{r}^T\mathbf{q}) = \mathbf{r}\,, \qquad \frac{\partial}{\partial \mathbf{q}}(\mathbf{q}^T\mathbf{A}\mathbf{q}) = 2\mathbf{A}\mathbf{q}\,. \qquad (6.2.6a, b)$$

6.3. Least-Squares

Consider the polar profile of fig. 2.13, with 50 rays passing through 12 perturbed layers, with the travel-time anomalies and **E** matrix described in section 2.15. This problem is to infer $N = 12$ unknown x_i from a set of $M = 50$ equations and is therefore readily solved by conventional least-squares. In that approach, we seek to minimize the sum of the squares of the residuals:

$$J = \sum_i^M \left(y_i - \sum_j^N E_{ij} x_j \right)^2 = (\mathbf{y} - \mathbf{E}\mathbf{x})^T (\mathbf{y} - \mathbf{E}\mathbf{x}) = \sum_i^M n_i^2 = \mathbf{n}^T \mathbf{n} \,. \quad (6.3.1)$$

Taking the derivatives of J with respect to the x_i and setting them to zero results in a set of "normal equations" with solution

$$\widehat{\mathbf{x}} = (\mathbf{E}^T \mathbf{E})^{-1} \mathbf{E}^T \mathbf{y} \,. \quad (6.3.2)$$

We have written $\widehat{\mathbf{x}}$ to distinguish (6.3.2) from the true solution, $\mathbf{x}$. An estimate of the noise is

$$\widehat{\mathbf{n}} = \mathbf{y} - \mathbf{E}\widehat{\mathbf{x}} = \mathbf{y} - \mathbf{E}(\mathbf{E}^T \mathbf{E})^{-1} \mathbf{E}^T \mathbf{y} = (\mathbf{I} - \mathbf{E}(\mathbf{E}^T \mathbf{E})^{-1} \mathbf{E}^T)\mathbf{y} \,. \quad (6.3.3)$$

Traditionally, this problem would be described as solving M equations in N unknowns. In what follows, however, it is important to notice that estimates are made not only for the N elements of $\mathbf{x}$ but also for the M elements of $\mathbf{n}$ – that is, we really solve M equations in $M + N$ unknowns: the problem is underdetermined.

The forward problem illustrated in fig. 2.13 produces travel times, $\Delta\tau_i \equiv y_i$, for the situation in which the perturbations vanish in all layers except 6–8, where the perturbation is the constant -6.6667×10^{-4} s/km. With 50 perfect observations, the noise-free case is uninteresting unless the inverse of $\mathbf{E}^T\mathbf{E}$ should fail to exist (it does exist here), and it is readily confirmed that the perturbations are essentially perfectly recovered from (6.3.2).

To consider a slightly more interesting example, the travel times $\Delta\tau_i$ were corrupted with pseudorandom numbers, n_i, to produce $y_i = \Delta\tau_i + n_i$, shown in fig. 6.1. The n_i were chosen to have an expected value of zero and an rms value of about 10 ms, consistent with the expected internal wave noise (chapters 4 and 5). The expressions (6.3.2) and (6.3.3) were then used to find $\widehat{\mathbf{x}}$ and $\widehat{\mathbf{n}}$, and the results are plotted in fig. 6.2: $\widehat{\mathbf{n}}$ is acceptable, but $\widehat{\mathbf{x}}$ is poor. Although the perturbations in layers 6–8 are reasonably estimated (though invisible in the figure), the large values near the surface are wrong. Even with a small noise contribution and the familiar nature of the method, something more needs to be understood about this solution.

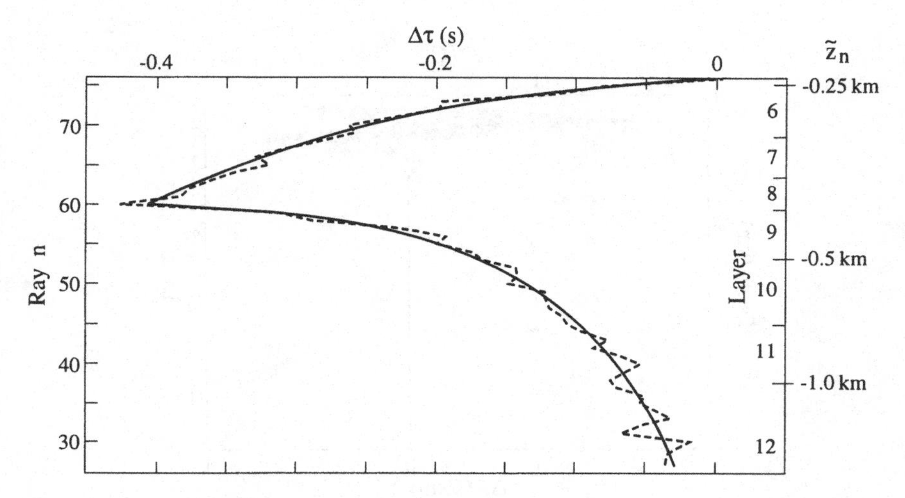

Fig. 6.1. Ray travel-time perturbations for the polar profile with 12 layers and 50 rays (see fig. 2.13, top). Layers 6, 7, and 8 are perturbed by $\Delta C/C_0 = +10^{-3}$. Perfect travel-time perturbations (solid) have been corrupted by random noise (dashed line). The noisy curves are used in the inversions.

We shall come to understand why this least-squares solution is so poor, but first, as with any such problem, we must determine the expected uncertainty of the result. In realistic problems, one does not know the right answer, and any sensible investigator will fear the possible trap of accepting the result shown in fig. 6.2 as "truth." Fortunately, least-squares methods provide protection against the need for blind acceptance of poor solutions. Define $\langle q\rangle$ to be the expected value (ensemble average) of any variable q. Then,

$$\langle \widehat{\mathbf{x}} - \mathbf{x}\rangle = \langle (\mathbf{E}^T\mathbf{E})^{-1}\mathbf{E}^T(\mathbf{y}_0 + \mathbf{n}) - (\mathbf{E}^T\mathbf{E})^{-1}\mathbf{E}^T\mathbf{y}_0\rangle = (\mathbf{E}^T\mathbf{E})^{-1}\mathbf{E}^T\langle\mathbf{n}\rangle = 0,$$

under the assumptions that the inverse exists, that $\langle\mathbf{n}\rangle = \mathbf{0}$, and hence that $\mathbf{y}_0$ are perfect travel times. Subject to this behavior of the noise, (6.3.2) produces a solution that *on average* is the correct one and that is thus termed "unbiased." The covariance about the true solution, referred to as the "uncertainty," is readily found as

$$\mathbf{P} = \langle(\widehat{\mathbf{x}} - \mathbf{x})(\widehat{\mathbf{x}} - \mathbf{x})^T\rangle = (\mathbf{E}^T\mathbf{E})^{-1}\mathbf{E}^T\langle\mathbf{n}\mathbf{n}^T\rangle\mathbf{E}(\mathbf{E}^T\mathbf{E})^{-1} = \sigma_n^2(\mathbf{E}^T\mathbf{E})^{-1}. \tag{6.3.4}$$

The last step depends on the so-called white-noise assumption $\langle n_i n_j\rangle = \sigma_n^2\delta_{ij}$ (also usefully written as $\langle\mathbf{n}\mathbf{n}^T\rangle = \sigma_n^2\mathbf{I}$); that is, there is no correlation between the noise in the equations, and the variances of the noise in all equations are

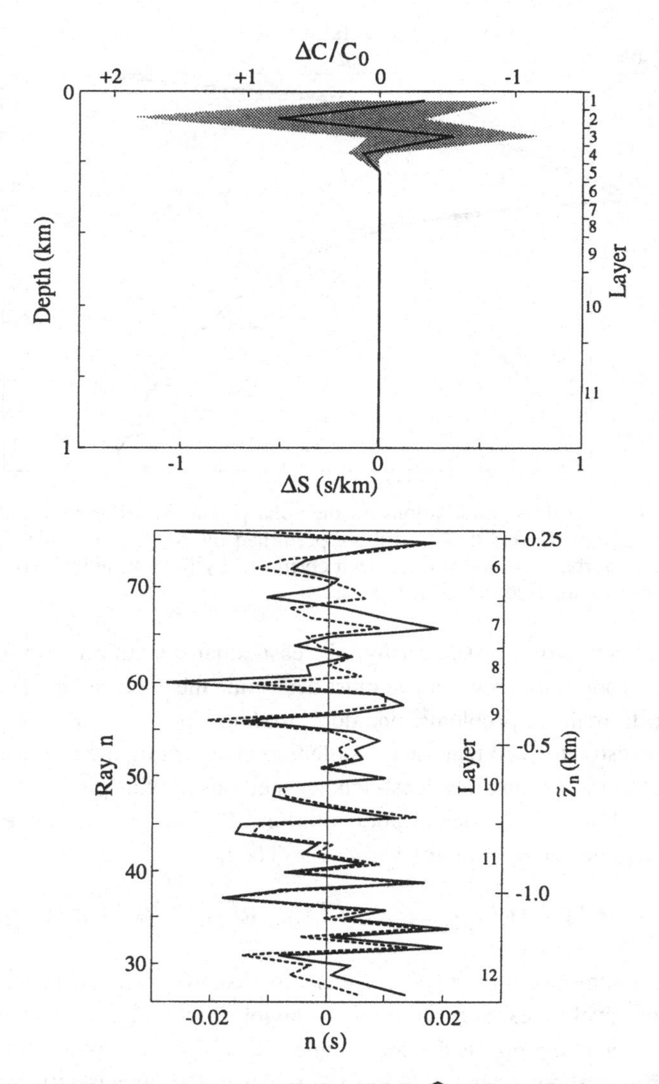

Fig. 6.2. Top: Ordinary least-squares solution (6.3.2), $\widehat{\mathbf{x}}$, with standard errors (gray) for the layered polar profile. The solution is of order 1 near the surface, 1000 times the "true" solution, which is $\Delta C/C_0 = 10^{-3}$ in layers 6, 7, and 8 (too small to be visible), and zero elsewhere. The standard errors are also very large, and the solution, though very poor, is consistent with the true solution within two standard deviations. The poor performance is associated with an extreme sensitivity to the rather slight noise in $\mathbf{y}$. Bottom: Estimated noise $\widehat{\mathbf{n}}$ in the ray travel-time perturbations (dashed line) from the least-squares solution (6.3.3) compares favorably to the true noise $\mathbf{n}$ (solid).

identical. If $(\mathbf{E}^T\mathbf{E})^{-1}$ is in some sense "large," the solution uncertainty can be large. The value of σ_n^2 can be taken either from the prior estimate of $\Delta\tau_i$ (rms about 10 ms) or from the estimated variance of $\widehat{\mathbf{n}}$ (which is here consistent with the prior estimate). In the present case, the square roots of the diagonal elements of $\mathbf{P}$ are shown by the gray band in fig. 6.2 (top). Apparently the presence of noise has rendered the solution unreliable in the upper ocean; nonetheless, the estimated solution is everywhere within two standard deviations of the correct one. Whether or not a solution with such large uncertainties in places is useful can be decided only in the context of a particular situation; it might well be regarded so uncertain as to be useless.

The noise level can be made sufficiently small that the correct solution will be recovered with arbitrary accuracy. But there is something very noise-sensitive about at least part of this solution. By way of contrast, consider the solution if the top three layers are simply omitted from the problem, reducing $\mathbf{E}$ to a 50×9 system, from the previous 50×12 system. In this case (not shown), the problem becomes quite insensitive to noise, and accurate solutions are produced even for much larger noise elements of $\mathbf{y}$. The reader may well wonder how the determination to drop these particular layers could be arrived at. There are two clues: these are the layers in which the solution uncertainty is largest, and, physically, none of the 50 observed rays have turning depths within these three layers.

The most powerful inverse methods are those that permit us to fully understand and anticipate this behavior, and to control it if necessary; with simple least-squares, it is not readily obvious how to do these things. A somewhat different method of solution will be introduced shortly – one that will provide nearly complete understanding of the structure of the solution. But for the time being, we remain with the simplest least-squares approach.

The function J in (6.3.1) is commonly referred to as an "objective" or "cost" function. The solution [(6.3.2) and (6.3.3)] along with its uncertainty (6.3.4), is evidently a rather special one – it renders J (*i.e.*, $\sum_i \widehat{n}_i^2$), as small as possible. Commonly, the investigator has the additional information that the noise in one or more equations is larger or smaller than that in others. Suppose the noise in one equation is expected to have a variance 1000 times smaller than the noise in any of the others. Rather than using (6.3.1), one prefers a solution $\widehat{\mathbf{x}}$, where the residual in that special equation is expected to be much smaller than the residual in any of the others. The objective function (6.3.1) would again be a reasonable choice, if each equation were first divided by an appropriate weight, removing the inhomogeneity of the variances. The result is equivalent to an objective function:

$$J = (\mathbf{y} - \mathbf{E}\mathbf{x})^T \mathbf{R}^{-1} (\mathbf{y} - \mathbf{E}\mathbf{x}) = \sum_i n_i R_{ii}^{-1} n_i, \tag{6.3.5}$$

where, for now, the weight matrix $\mathbf{R}^{-1}$ is diagonal, containing the appropriate squared weights. The minimization of J with respect to $\mathbf{x}$ produces the solution

$$\widehat{\mathbf{x}} = (\mathbf{E}^T\mathbf{R}^{-1}\mathbf{E})^{-1}\mathbf{E}^T\mathbf{R}^{-1}\mathbf{y}, \tag{6.3.6a}$$

$$\begin{aligned} \widehat{\mathbf{n}} &= \mathbf{y} - \mathbf{E}(\mathbf{E}^T\mathbf{R}^{-1}\mathbf{E})^{-1}\mathbf{E}^T\mathbf{R}^{-1}\mathbf{y}, \\ &= [\mathbf{I} - \mathbf{E}(\mathbf{E}^T\mathbf{R}^{-1}\mathbf{E})^{-1}\mathbf{E}^T\mathbf{R}^{-1}]\mathbf{y}, \end{aligned} \tag{6.3.6b}$$

$$\mathbf{P} = (\mathbf{E}^T\mathbf{R}^{-1}\mathbf{E})^{-1}\mathbf{E}^T\mathbf{R}^{-1}\langle\mathbf{n}\mathbf{n}^T\rangle\mathbf{R}^{-1}\mathbf{E}(\mathbf{E}^T\mathbf{R}^{-1}\mathbf{E})^{-1}, \tag{6.3.6c}$$

which differs from the solution of (6.3.2), (6.3.3), and (6.3.4) unless $\mathbf{R} = \mathbf{I}$.

The weights in the objective function can be anything one chooses; an appealing, but special, case is $\mathbf{R} = \langle\mathbf{n}\mathbf{n}^T\rangle$ and

$$\mathbf{P} = (\mathbf{E}^T\mathbf{R}^{-1}\mathbf{E})^{-1}. \tag{6.3.6d}$$

The use of weights based on the variances of the noise elements is best understood after we take up Gauss-Markov estimation.

The numerical value of the solution, $\widehat{\mathbf{x}}$, of the residuals, $\widehat{\mathbf{n}}$, and of the uncertainty, $\mathbf{P}$, is determined by the behavior of $(\mathbf{E}^T\mathbf{E})^{-1}$ or $(\mathbf{E}^T\mathbf{R}^{-1}\mathbf{E})^{-1}$ and is not otherwise subject to control. A tool permitting the investigator to use any available information about the relative sizes of $\widehat{\mathbf{x}}$ and $\widehat{\mathbf{n}}$ is to augment the objective function with an additional term

$$J = (\mathbf{y} - \mathbf{E}\mathbf{x})^T\mathbf{R}^{-1}(\mathbf{y} - \mathbf{E}\mathbf{x}) + \alpha^2\mathbf{x}^T\mathbf{x}, \tag{6.3.7}$$

where α^2 is a fixed positive constant. Minimizing (6.3.7) as before produces a new solution

$$\widehat{\mathbf{x}} = (\mathbf{E}^T\mathbf{R}^{-1}\mathbf{E} + \alpha^2\mathbf{I})^{-1}\mathbf{E}^T\mathbf{R}^{-1}\mathbf{y}, \tag{6.3.8a}$$

$$\widehat{\mathbf{n}} = [\mathbf{I} - \mathbf{E}(\mathbf{E}^T\mathbf{R}^{-1}\mathbf{E} + \alpha^2\mathbf{I})^{-1}\mathbf{E}^T\mathbf{R}^{-1}]\mathbf{y}, \tag{6.3.8b}$$

$$\mathbf{P}_n = (\mathbf{E}^T\mathbf{R}^{-1}\mathbf{E} + \alpha^2\mathbf{I})^{-1}\mathbf{E}^T\mathbf{R}^{-1}\langle\mathbf{n}\mathbf{n}^T\rangle\mathbf{R}^{-1}\mathbf{E}(\mathbf{E}^T\mathbf{R}^{-1}\mathbf{E} + \alpha^2\mathbf{I})^{-1}, \tag{6.3.8c}$$

where $\mathbf{P}_n = \langle(\widehat{\mathbf{x}} - \langle\widehat{\mathbf{x}}\rangle)(\mathbf{x} - \langle\widehat{\mathbf{x}}\rangle)^T\rangle$. It is apparent that if $\alpha^2 \to 0$, we recover (6.3.6), and if $\alpha^2 \to \infty$, $\widehat{\mathbf{x}} \to 0$ and $\widehat{\mathbf{n}} \to \mathbf{y}$. By varying the value of α^2, we can control the relative sizes of the sums of the squares of $\widehat{x}_i$ and $\widehat{n}_i$.

Equation (6.3.7) is the natural objective function for a set of equations

$$\mathbf{E}\mathbf{x} + \mathbf{n} = \mathbf{y}, \qquad \mathbf{x} + \mathbf{n}_1 = \mathbf{0},$$

or

$$\mathbf{E}_2\mathbf{x} + \mathbf{n}_2 = \mathbf{y}_2,$$

$$\mathbf{E}_2 = \{\mathbf{E} \quad \mathbf{I}_N\}, \quad \mathbf{n}_2 = \begin{bmatrix} \mathbf{n} \\ \mathbf{n}_1 \end{bmatrix}, \quad \mathbf{y}_2 = \begin{bmatrix} \mathbf{y} \\ \mathbf{0} \end{bmatrix},$$

which leads to an objective function

$$J = (\mathbf{y}_2 - \mathbf{E}_2\mathbf{x})^T \mathbf{R}_2^{-1} (\mathbf{y}_2 - \mathbf{E}_2\mathbf{x}), \qquad \mathbf{R}_2 = \begin{Bmatrix} \mathbf{R} & \mathbf{0} \\ \mathbf{0} & \alpha^2\mathbf{I} \end{Bmatrix}$$

identical with (6.3.7). Introduction of the term $\alpha^2\mathbf{x}^T\mathbf{x}$ into (6.3.7) is equivalent to the addition of N equations expressing a preference that $\mathbf{x} = \mathbf{0}$, with a confidence determined by the size of α^2 relative to the elements of $\mathbf{R}$. Any other preference (*e.g.*, $\mathbf{x} = \mathbf{x}_0$) can be imposed.

If information is available about the sizes of the individual elements of $\widehat{\mathbf{x}}$, it can be used in analogy to the method to control the sizes of $\widehat{\mathbf{n}}$ – through the introduction of a diagonal weight matrix, $\mathbf{S}$, such that the objective function becomes

$$J = (\mathbf{y} - \mathbf{E}\mathbf{x})^T \mathbf{R}^{-1} (\mathbf{y} - \mathbf{E}\mathbf{x}) + \alpha^2 \mathbf{x}^T \mathbf{S}^{-1} \mathbf{x}. \tag{6.3.9}$$

The minimizing solution now is

$$\widehat{\mathbf{x}} = (\mathbf{E}^T\mathbf{R}^{-1}\mathbf{E} + \mathbf{S}^{-1})^{-1}\mathbf{E}^T\mathbf{R}^{-1}\mathbf{y}, \tag{6.3.10a}$$

$$\widehat{\mathbf{n}} = [\mathbf{I} - \mathbf{E}(\mathbf{E}^T\mathbf{R}^{-1}\mathbf{E} + \mathbf{S}^{-1})^{-1}\mathbf{E}^T\mathbf{R}^{-1}]\mathbf{y}, \tag{6.3.10b}$$

$$\mathbf{P}_n = (\mathbf{E}^T\mathbf{R}^{-1}\mathbf{E} + \mathbf{S}^{-1})^{-1}\mathbf{E}^T\mathbf{R}^{-1}\langle\mathbf{n}\mathbf{n}^T\rangle\mathbf{R}^{-1}\mathbf{E}(\mathbf{E}^T\mathbf{R}^{-1}\mathbf{E} + \mathbf{S}^{-1})^{-1}. \tag{6.3.10c}$$

The ratios of the numerical values of S_{ii}, R_{ii} will partially determine the sizes of the elements of $\widehat{x}_i, \widehat{n}_i$. Determination is only partial because the values must be consistent with the equations.

Define

$$\mathbf{E}' = \mathbf{R}^{-T/2}\mathbf{E}\mathbf{S}^{T/2}, \quad \mathbf{y}' = \mathbf{R}^{-T/2}\mathbf{y}, \quad \mathbf{x}' = \mathbf{S}^{-T/2}\mathbf{x}, \quad \mathbf{n}' = \mathbf{R}^{-T/2}\mathbf{n}, \tag{6.3.11a}$$

where for diagonal matrices the matrix square root is found by taking the square root of each diagonal element. For a diagonal matrix, $\mathbf{R}^{T/2} = \mathbf{R}^{1/2}$. The transpose is introduced here for purposes that will emerge. With these definitions, (6.3.9) is now

$$J = (\mathbf{y}' - \mathbf{E}'\mathbf{x}')^T(\mathbf{y}' - \mathbf{E}'\mathbf{x}') + \alpha^2\mathbf{x}'^T\mathbf{x}'. \tag{6.3.11b}$$

In effect, the problem has been scaled so that no further weighting is required. The presence of α^2 is not strictly necessary, as it can be absorbed into either $\mathbf{R}$ or $\mathbf{S}$, but it may be useful to retain it for separate control. Division by $R_{ii}^{T/2}$ is sometimes called "row scaling," and multiplication by $S_{ii}^{T/2}$ is "column scaling." These scaled equations produce the solutions (dropping the primes)

$$\widehat{\mathbf{x}} = (\mathbf{E}^T\mathbf{E} + \alpha^2\mathbf{I})^{-1}\mathbf{E}^T\mathbf{y}\,, \qquad (6.3.12a)$$

$$\widehat{\mathbf{n}} = [\mathbf{I} - \mathbf{E}(\mathbf{E}^T\mathbf{E} + \alpha^2\mathbf{I})^{-1}\mathbf{E}^T]\mathbf{y}\,, \qquad (6.3.12b)$$

$$\mathbf{P}_n = (\mathbf{E}^T\mathbf{E} + \alpha^2\mathbf{I})^{-1}\mathbf{E}^T\langle\mathbf{n}\mathbf{n}^T\rangle\mathbf{E}(\mathbf{E}^T\mathbf{E} + \alpha^2\mathbf{I})^{-1}\,. \qquad (6.3.12c)$$

If a column scaling matrix $\mathbf{S}^{1/2}$ has been used, the variance of the original, unscaled solution is $\mathbf{S}^{1/2}\mathbf{P}_n\mathbf{S}^{T/2}$.

We return to the problem of the polar profile perturbation, which evidenced a considerable noise sensitivity. One might argue that the real difficulty with the solution depicted in fig. 6.2 (top) is not the large uncertainty near the surface, but rather the large numerical values there. The objective function with $\alpha^2 = 1000$ ($\mathbf{S} = \mathbf{I}$, $\mathbf{R} = \mathbf{I}$) leads to the solution depicted in fig. 6.3, showing acceptable variations in both $\widehat{\mathbf{x}}$ and $\widehat{\mathbf{n}}$. With this choice of α^2, the residuals have a variance consistent with the prior estimate. One can confirm that radically different values of α^2 will produce residual variances or solution variances, or both, which can be rejected as unacceptable on the basis of prior understanding. The solutions for $\alpha^2 = 100$ and 1000 are displayed in fig. 6.4. With $\alpha^2 = 100$, the "overshoot" in the upper ocean is apparent.

The question of how best to choose α^2 in any particular case falls under the general heading of "ridge regression," an important subject discussed, for example, by Hoerl and Kennard (1970*a,b*) and Lawson and Hanson (1974). As one might infer from the example just discussed, a range of values is tested to see if noise and solution variances are acceptable, and both must pass the test simultaneously. Further discussion is deferred until we can better understand the effects of the parameter α^2 on the solution in (6.3.8) or (6.3.12).

In the formally overdetermined situation just described, estimates were obtained not only for the $N = 12$ layer slowness perturbations but also for the $M = 50$ ray noise elements, for a total of $M + N = 62$ variables. The residuals of the problem had to be examined to make sure they were sensible in view of prior estimates. It becomes useful, especially in the tomographic problem, where measurement noise is the focus of complicated analyses, to regard the noise estimates as fully a part of the solution.

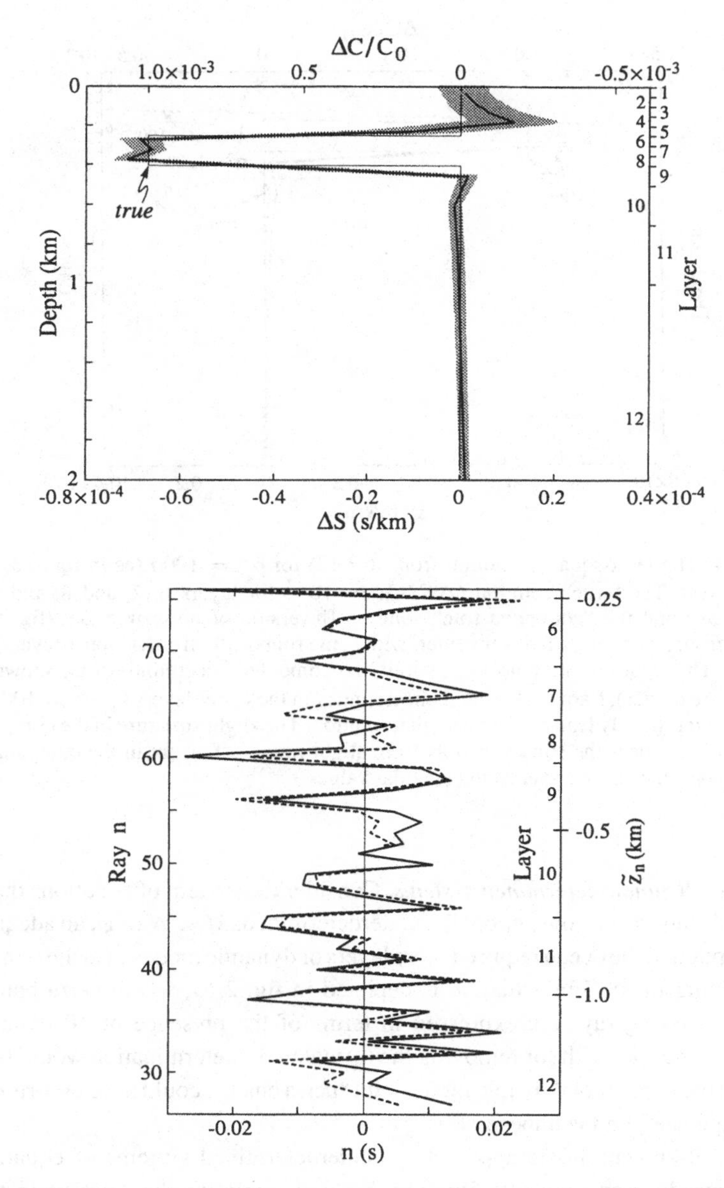

Fig. 6.3. As in fig. 6.2, but with the solution from (6.3.12), using $\alpha^2 = 1000$. The agreement with the true solution is satisfactory.

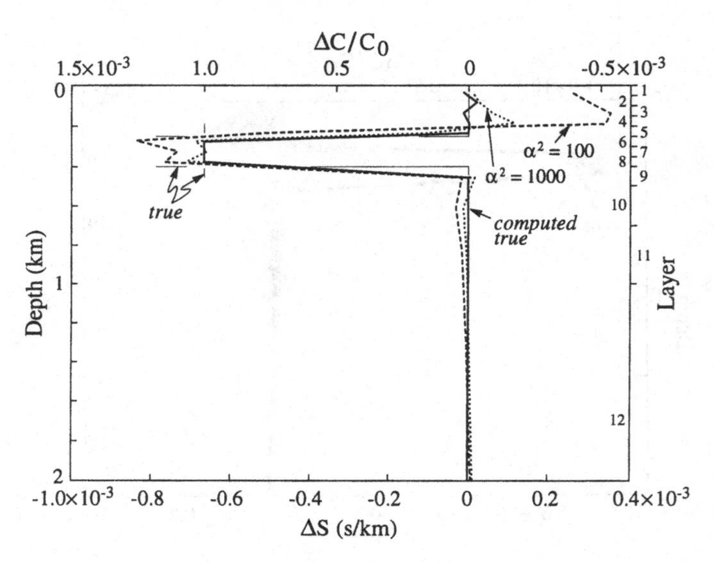

Fig. 6.4. The least-squares solution from (6.3.12) for $\alpha^2 = 1000$ (as in fig. 6.3) and $\alpha^2 = 100$. The "true" solution ($\Delta C/C_0 = 10^{-3}$ for layers 6, 7, and 8, and zero elsewhere) and the "computed true" solution [inversion of noise-free $\Delta\tau$ (fig. 6.1), drawn from layer center to layer center, with some round-off errors in upper layers] are shown. The solutions are consistent within the computed uncertainties (not shown, to reduce the clutter). Large values and uncertainties in the upper layers for $\alpha^2 = 100$ and less would probably lead one to reject that solution. The slight structure in the computed true solutions near the surface results from slight round-off errors in the computation and the presence of near-vanishing singular values.

Formally underdetermined systems. Consider the system of equations that is formally, and more conventionally, underdetermined, $M < N$ (*e.g.*, an adequate description of the ocean required more layers or dynamic modes than the number of arriving rays). This situation is depicted in fig. 2.15, where perturbations to five arriving rays are expressed in terms of the presence of 10 dynamic modes. One approach for removing the formal underdetermination would be to reduce the number of dynamic modes. But such a change could have unfortunate consequences, and is unnecessary.

A well-known classic approach to underdetermined systems of equations is to introduce an objective function that will minimize the solution instead of minimizing the residuals. For row- and column-scaled equations (dropping primes), consider

$$J = \mathbf{x}^T\mathbf{x} - 2\mu^T(\mathbf{Ex} - \mathbf{y}), \tag{6.3.13}$$

where $\boldsymbol{\mu}$ is a vector Lagrange multiplier (the 2 is purely for notational tidiness) that is treated as a new, unknown, independent parameter. Demanding stationarity of (6.3.13) with respect to $\mathbf{x}$ and $\boldsymbol{\mu}$ produces

$$\widehat{\mathbf{x}} = \mathbf{E}^T(\mathbf{E}\mathbf{E}^T)^{-1}\mathbf{y}, \tag{6.3.14a}$$

$$\widehat{\boldsymbol{\mu}} = (\mathbf{E}\mathbf{E}^T)^{-1}\mathbf{y}, \tag{6.3.14b}$$

$$\widehat{\mathbf{n}} = \mathbf{0}, \tag{6.3.14c}$$

$$\begin{aligned}\mathbf{P}_n &= \mathbf{E}^T(\mathbf{E}\mathbf{E}^T)^{-1}\langle\mathbf{n}\mathbf{n}^T\rangle(\mathbf{E}\mathbf{E}^T)^{-1}\mathbf{E}, \\ &= \mathbf{E}^T(\mathbf{E}\mathbf{E}^T)^{-2}\mathbf{E} \quad \text{if} \quad \langle\mathbf{n}\mathbf{n}^T\rangle = \mathbf{I},\end{aligned} \tag{6.3.14d}$$

when the matrix inverse exists. This would be a valid solution, except for one grave defect: zero residuals are unlikely ever to be acceptable in a real situation. Such a solution would have to be rejected; furthermore, the magnitude of $\widehat{\mathbf{x}}$ is out of our control. One remedy for the $\widehat{\mathbf{n}} = \mathbf{0}$ pathology is to use instead the objective function (6.3.11*b*); the resulting solution (6.3.12) exists whatever the relative magnitudes of M and N, as will emerge shortly. A "bridge" between solutions of the form (6.3.14), where the data were required to be fit perfectly by use of a Lagrange multiplier, and (6.3.12), where the data leave residuals, can be obtained by imposing the observations in the form (6.2.4), and adding a penalty on the residuals,

$$J = \alpha^2\mathbf{x}^T\mathbf{x} + \mathbf{n}^T\mathbf{n} - 2\boldsymbol{\mu}^T(\mathbf{E}\mathbf{x} + \mathbf{n} - \mathbf{y}). \tag{6.3.15}$$

One now sets the derivatives with respect to $\mathbf{x}$, $\mathbf{n}$, and $\boldsymbol{\mu}$ to zero and solves the resulting normal equations, producing

$$\widehat{\mathbf{x}} = \mathbf{E}^T(\mathbf{E}\mathbf{E}^T + \alpha^{-2}\mathbf{I})^{-1}\mathbf{y}, \tag{6.3.16a}$$

$$\widehat{\mathbf{n}} = \{\mathbf{I} - \mathbf{E}\mathbf{E}^T(\mathbf{E}\mathbf{E}^T + \alpha^{-2}\mathbf{I})^{-1}\}\mathbf{y}, \tag{6.3.16b}$$

$$\mathbf{P}_n = \mathbf{E}^T(\mathbf{E}\mathbf{E}^T + \alpha^{-2}\mathbf{I})^{-1}\langle\mathbf{n}\mathbf{n}^T\rangle(\mathbf{E}\mathbf{E}^T + \alpha^{-2}\mathbf{I})^{-1}\mathbf{E}, \tag{6.3.16c}$$

which is again controllable by choice of α^2.[5] The use of (6.3.15) makes it more evident that the separation of solution elements between $\mathbf{x}$ and $\mathbf{n}$ is arbitrary. In particular, we are free to rewrite the observations as

$$\mathbf{E}_1\boldsymbol{\xi} = \mathbf{y}, \quad \mathbf{E}_1 = \{\mathbf{E}\ \mathbf{I}_M\}, \quad \boldsymbol{\xi}^T = [\alpha\mathbf{x}^T\ \mathbf{n}^T], \tag{6.3.17}$$

[5] Meteorologists' jargon (Sasaki, 1970) decribes the exact imposition of relations as a "strong" constraint, and the imposition in a mean-square sense as a "weak" one. The distinction, while mathematically meaningful, is not physically so; although (6.2.4) is being imposed exactly in (6.3.15), it contains a noise term, and the "exact" nature of the constraint is illusory.

and the objective function (6.3.15) becomes

$$J = \boldsymbol{\xi}^T \boldsymbol{\xi} - 2\boldsymbol{\mu}^T (\mathbf{E}_1 \boldsymbol{\xi} - \mathbf{y}) . \tag{6.3.18}$$

It is possible to rewrite (6.3.16) in a form that will make clearer its connection to (6.3.12). The "matrix-inversion lemma" (*e.g.*, Liebelt, 1967) asserts

$$\begin{aligned} \{\mathbf{C} - \mathbf{B}^T \mathbf{A}^{-1} \mathbf{B}\}^{-1} &= \{\mathbf{I} - \mathbf{C}^{-1} \mathbf{B}^T \mathbf{A}^{-1} \mathbf{B}\}^{-1} \mathbf{C}^{-1} \\ &= \mathbf{C}^{-1} - \mathbf{C}^{-1} \mathbf{B}^T (\mathbf{B} \mathbf{C}^{-1} \mathbf{B}^T - \mathbf{A}^{-1}) \mathbf{B} \mathbf{C}^{-1} , \end{aligned} \tag{6.3.19}$$

provided the products and inverses exist. Application to (6.3.16) shows the two solutions to be identical. The dimensions of the matrices to be inverted are $N \times N$ in one case (6.3.12), and $M \times M$ in the other (6.3.16).

6.4. Singular-Value Solution and Decomposition

Least-squares has the advantage of being familiar and easy to use. One of its disadvantages is the difficulty of understanding exactly what is governing solutions such as (6.3.10). In particular, unless we are willing to select a value of α^2, or of a complete weight matrix $\mathbf{S}$ (and how should $\mathbf{S}$ be chosen?), there appears to be almost no control over the magnitudes of the elements of the solution or residuals; the behavior of the solution is buried in the somewhat opaque operator $(\mathbf{E}^T \mathbf{E})^{-1} \mathbf{E}^T$ or the related expression in (6.3.12). In any complex observational system, and tomography is somewhat complicated, it is worrisome not to fully understand the relationship between the individual measurements and the behavior of the best-estimate solution. Methods are available that permit one to "pick apart" solutions such as (6.3.10) to clarify what is happening.

Most of what follows is best understood by recalling the elementary problem of solving in eigenvectors a set of just-determined equations whose coefficient matrix, $\mathbf{E}$, is symmetric. The even more elementary problem of expanding an arbitrary vector in terms of an incomplete set of orthonormal vectors leads to the crucial concepts of "resolution" and solution variance, and some useful notation. Readers are urged to at least skim the next few pages leading up to the singular-value decomposition (SVD); the material, though familiar, lays a foundation for very powerful and general methods.

Consider the representation of an L-dimensional vector $\mathbf{f}$ as a sum of a complete set of L orthonormal vectors $\mathbf{g}_i$, $i = 1$ to L, $\mathbf{g}_i^T \mathbf{g}_j = \delta_{ij}$ (called a "spanning orthonormal set"). Without error,

$$\begin{aligned} \mathbf{f} &= \sum_{j=1}^{L} a_j \mathbf{g}_j , \quad a_j = \mathbf{g}_j^T \mathbf{f} , \\ \mathbf{f} &= \mathbf{G}(\mathbf{G}^T \mathbf{f}), \end{aligned} \tag{6.4.1}$$

where $\mathbf{G}$ is the matrix whose columns are the $\mathbf{g}_j$. If, for some reason, only the first K coefficients a_j are available, we can approximate $\mathbf{f}$ by its first K terms:

$$\widehat{\mathbf{f}} = \sum_{j=1}^{K} a_j \mathbf{g}_j = \mathbf{f} + \delta\mathbf{f}_1, \tag{6.4.2}$$

and there is an error, $\delta\mathbf{f}_1$. Define the "norm" of a vector $\mathbf{q}$ as the "ℓ_2 norm," or length:

$$\|\mathbf{q}\| \equiv (\mathbf{q}^T\mathbf{q})^{\frac{1}{2}} = \left(\sum_{j=1}^{L} q_i^2\right)^{\frac{1}{2}}. \tag{6.4.3}$$

From the orthogonality of the $\mathbf{g}_i$, $\delta\mathbf{f}_1$ will have minimum norm for fixed K if and only if it is orthogonal to the K vectors retained in the approximation, and if and only if a_j are given by (6.4.1). With vector sizes measured in this norm, the only way the error in the representation can be reduced is by increasing K. Let the vector of missing coefficients be denoted $\boldsymbol{\alpha}_Q$, and define the matrix $\mathbf{Q}_K$ to have columns made up of the "missing" vectors $\mathbf{g}_i$, $K+1 \leq i \leq L$. Then the uncertainty of $\widehat{\mathbf{f}}$ owing to the omitted terms is

$$\mathbf{P}_0 = \langle(\widehat{\mathbf{f}} - \mathbf{f})(\widehat{\mathbf{f}} - \mathbf{f})^T\rangle = \mathbf{Q}_K\langle\boldsymbol{\alpha}_Q\boldsymbol{\alpha}_Q^T\rangle\mathbf{Q}_K^T.$$

Define an $L \times K$ matrix $\mathbf{G}_K$ whose columns are the first K of the $\mathbf{g}_j$. Then $\mathbf{G}_K^T\mathbf{f}$ is the vector of coefficients $\mathbf{a} = [a_j] = [\mathbf{g}_j^T\mathbf{f}]$, $j = 1$ to K, and the finite representation (6.4.2) is

$$\widehat{\mathbf{f}} = \mathbf{G}_K\mathbf{a} = \mathbf{G}_K(\mathbf{G}_K^T\mathbf{f}) = (\mathbf{G}_K\mathbf{G}_K^T)\mathbf{f}, \tag{6.4.4}$$

where the last equality follows from the associative properties of matrix multiplication. This expression shows that the representation of a vector in an incomplete orthonormal set produces a resulting approximation that is a simple linear combination of the elements of the correct values (*i.e.*, a weighted average, or filtered version). $\mathbf{G}_K\mathbf{G}_K^T$ is often termed a "resolution matrix," as it represents the relationship between the correct $\mathbf{f}$ and the value computed from the truncated set of vectors. Because the columns of $\mathbf{G}_K$ are orthonormal, $\mathbf{G}_K^T\mathbf{G}_K = \mathbf{I}_K$; but $\mathbf{G}_K\mathbf{G}_K^T \neq \mathbf{I}_L$ unless $K = L$ (see any book on linear algebra). If $K < L$, $\mathbf{G}_K$ is "semiorthogonal." If $K = L$, it is "orthogonal."

Now suppose that the coefficients a_j are not perfectly known, but contain noise contributions δa_j such that $\mathbf{a} = \mathbf{a}_0 + \delta\mathbf{a}$, $\mathbf{a}_0$ being the correct value. Then the representation of $\mathbf{f}$ given by the finite representation (6.4.2) or (6.4.4) has an additional error, and

$$\widehat{\mathbf{f}} = \mathbf{f} + \delta\mathbf{f}_1 + \delta\mathbf{f}_2. \tag{6.4.5}$$

The two types of errors in **f** are distinct – the first owing to the failure to have a complete set of expansion coefficients, and the second owing to errors in the coefficients that are known. If the noise has zero mean, $\langle\delta\mathbf{a}\rangle = 0$, and the covariance matrix

$$\langle\delta\mathbf{a}\delta\mathbf{a}^T\rangle = \mathbf{R}_{aa}\,, \tag{6.4.6}$$

the second noise term in the representation of **f** has a mean and covariance given by

$$\begin{aligned}\langle\delta\mathbf{f}_2\rangle &= \left\langle\widehat{\mathbf{f}} - \mathbf{f} - \sum_{K+1}^{L} a_j\mathbf{g}_j\right\rangle = \langle\mathbf{G}_K\delta\mathbf{a}\rangle = \mathbf{G}_K\langle\delta\mathbf{a}\rangle = 0\,,\\ \mathbf{P}_n = \langle\delta\mathbf{f}_2\delta\mathbf{f}_2^T\rangle &\equiv \left\langle\left(\widehat{\mathbf{f}} - \mathbf{f} - \sum_{K+1}^{L} a_j\mathbf{g}_j\right)\left(\widehat{\mathbf{f}} - \mathbf{f} - \sum_{K+1}^{L} a_j\mathbf{g}_j\right)^T\right\rangle\\ &= \langle\mathbf{G}_K\delta\mathbf{a}(\mathbf{G}_K\delta\mathbf{a})^T\rangle = \mathbf{G}_K\mathbf{R}_{aa}\mathbf{G}_K^T\,.\end{aligned} \tag{6.4.7}$$

The total uncertainty of the representation of **f** is then

$$\mathbf{P} \equiv \langle(\widehat{\mathbf{f}} - \mathbf{f})(\widehat{\mathbf{f}} - \mathbf{f})^T\rangle = \mathbf{P}_0 + \mathbf{P}_n\,. \tag{6.4.8}$$

P is made up of two distinct parts – the first owing to missing coefficients, the second owing to noise in the existing coefficients.

Consider now a very special set of linear simultaneous equations of the form

$$\mathbf{E}\mathbf{x} + \mathbf{n} = \mathbf{y}, \tag{6.4.9}$$

where $M = N$ and **E** is *symmetric*. For symmetric **E**, the eigenvector/eigenvalue problem,

$$\mathbf{E}\mathbf{g}_i = \lambda_i\mathbf{g}_i\,, \tag{6.4.10}$$

is known to have the following special properties: (i) the eigenvectors $\mathbf{g}_i$, $1 \le i \le N$, are a spanning orthonormal set, and (ii) the eigenvalues, λ_i, are real. It is illuminating to recall the solution of systems of equations like (6.4.9) under these special circumstances. Because **x**, **n**, and **y** have the same dimension, they can all be written as

$$\mathbf{x} = \sum_i^N \alpha_i\mathbf{g}_i\,, \quad \mathbf{n} = \sum_i^N \gamma_i\mathbf{g}_i\,, \quad \mathbf{y} = \sum_i^N \beta_i\mathbf{g}_i\,, \quad \beta_i = \mathbf{g}_i^T\mathbf{y}, \tag{6.4.11a}$$

which is the same as

$$\mathbf{x} = \mathbf{G}\boldsymbol{\alpha}, \quad \mathbf{n} = \mathbf{G}\boldsymbol{\gamma}, \quad \mathbf{y} = \mathbf{G}\boldsymbol{\beta}, \quad \boldsymbol{\beta} = \mathbf{G}^T\mathbf{y}\,. \tag{6.4.11b}$$

The β_i are known. Substituting (6.4.11*a*,*b*) into (6.4.9) and invoking the orthonormality of the eigenvectors produces the requirement

$$\lambda_i \alpha_i + \gamma_i = \beta_i, \quad 1 \le i \le N. \tag{6.4.12}$$

If $\lambda_i \neq 0$,

$$\alpha_i = \beta_i/\lambda_i - \gamma_i/\lambda_i. \tag{6.4.13}$$

The β_i are all known, but the γ_i are not; we are free to choose, arbitrarily, $\gamma_i = 0$, and then the α_i are determined. But suppose one or more $\lambda_i = 0$ for $K + 1 \le i \le N$. Then the only possible way to satisfy (6.4.12) is to take the corresponding $\gamma_i = \beta_i$, leaving α_i unknown and having any value without restriction. We shall refer to the number, K, of nonzero λ_i as the "rank" of **E**.

If $K = N$, the system is said to be "full-rank," and the solution is

$$\mathbf{x} = \sum_{i=1}^{N} \frac{\beta_i}{\lambda_i} \mathbf{g}_i = \sum_{i=1}^{N} ((\mathbf{g}_i^T \mathbf{y})/\lambda_i)\mathbf{g}_i = \mathbf{G}^T \mathbf{y} \mathbf{\Lambda}^{-1} \mathbf{G}, \tag{6.4.14a}$$

$$\mathbf{n} = 0. \tag{6.4.14b}$$

$\mathbf{\Lambda}$ is the diagonal matrix whose elements are the λ_i, conventionally in descending order, which also fixes the order of the $\mathbf{g}_i$.

If $K < N$, the solution can be written

$$\hat{\mathbf{x}} = \sum_{i}^{K} \frac{\mathbf{g}_i^T \mathbf{y}}{\lambda_i} \mathbf{g}_i + \sum_{i=K+1}^{N} \alpha_i \mathbf{g}_i = \mathbf{G}_K^T \mathbf{y} \mathbf{\Lambda}_K^{-1} \mathbf{G}_K + \mathbf{Q}_K \boldsymbol{\alpha}_Q, \tag{6.4.14c}$$

$$\hat{\mathbf{n}} = \sum_{i=K+1}^{N} (\mathbf{g}_i^T \mathbf{y})\mathbf{g}_i = \mathbf{Q}_K \mathbf{Q}_K^T \mathbf{y}, \tag{6.4.14d}$$

where $\boldsymbol{\alpha}_K$ are defined as the coefficients of the $\mathbf{g}_i$ for which $\mathbf{E}\mathbf{g}_i = 0$. There is an infinite number of "solutions," because the α_i, $K + 1 \le i \le N$, can take on any value. We put "solutions" in quotation marks, and a caret over **x** in (6.4.14*c*), because there must be a residual, $\hat{\mathbf{n}}$, except in the rather special case when $\mathbf{g}_i^T \mathbf{y} = 0$, $K + 1 \le i \le N$ – relations known as "solvability conditions." The $\mathbf{g}_i$ corresponding to the zero eigenvalues are called the "nullspace" of **E**, and the others are the "range" of **E**. If there is a nullspace, there is not a unique solution whether or not the solvability conditions are met.

Before reverting to the more complicated situation that confronts us in tomographic inverse problems, we make a final point about (6.4.13). The arbitrary decision to set $\gamma_i = 0$, $1 \le i \le K$, which was used in (6.4.14), might be questioned if the y_i contain noise elements. Equations (6.4.14) are reasonable solutions only if it is believed that the noise components of $\mathbf{y}$ should satisfy the rigid conditions,

$$\mathbf{g}_i^T\mathbf{n} = 0, \; 1 \le i \le K, \tag{6.4.15}$$

of having no structure in the range of $\mathbf{E}$. Should there be any doubt about this condition, one might question the employment of solutions such as (6.4.14). We shall return to this issue later; in the meantime, note that use of (6.4.15) corresponds to finding a solution to (6.4.9) such that $\sum_i n_i^2 = \mathbf{n}^T\mathbf{n}$ is as small as possible (from the orthonormality of the $\mathbf{g}_i$) and therefore *is the least-squares solution when* $K = N$.

The "smallest" or "simplest" solution to (6.4.9) would set the nullspace coefficients in (6.4.14*c*) to zero, as there is no requirement for their presence, and any nonzero values would increase the norm of $\widehat{\mathbf{x}}$. If such a solution is used, then the expression for $\widehat{\mathbf{x}}$ in (6.4.14*c*) without the last term will have an uncertainty made up of the two terms in (6.4.8) – one part owing to the missing nullspace components, and the other part owing to noise in the coefficients of the range vectors. The uncertainty from the nullspace is again

$$\mathbf{P}_0 = \sum_{i=K+1}^{N}\sum_{j=K+1}^{N} \mathbf{g}_i\langle\alpha_Q\alpha_Q^T\rangle\mathbf{g}_j^T = \mathbf{Q}_K\langle\alpha_Q\alpha_Q^T\rangle\mathbf{Q}_K^T. \tag{6.4.16a}$$

The uncertainty owing to the coefficient noise is readily computed. Let $\mathbf{y} = \mathbf{y}_0 + \mathbf{n}$, where $\mathbf{y}_0$ is again the correct value:

$$\mathbf{P}_n = \left\langle \left(\frac{\sum_i \mathbf{g}_i^T\mathbf{n}}{\lambda_i}\mathbf{g}_i\right)\left(\frac{\sum_j \mathbf{g}_j^T\mathbf{n}}{\lambda_j}\mathbf{g}_j\right)^T \right\rangle = \sum_i^K\sum_j^K \mathbf{g}_i\mathbf{g}_i^T\frac{\langle\mathbf{n}\mathbf{n}^T\rangle}{\lambda_i\lambda_j}\mathbf{g}_j\mathbf{g}_j^T$$

$$= \sigma_n^2\sum_i^K\frac{\mathbf{g}_i\mathbf{g}_i^T}{\lambda_i^2} = \sigma_n^2\mathbf{G}_K\mathbf{\Lambda}_K^{-2}\mathbf{G}_K^T, \tag{6.4.16b}$$

where the last step requires $\langle\mathbf{n}_i\mathbf{n}_j\rangle = \sigma_n^2\delta_{ij}$.

The simple development leading to the discussion of the solution through eigenvectors was based on the theorem about the eigenvectors of a symmetric $\mathbf{E}$, so that they were guaranteed to be a complete orthonormal basis. The tomographic observation matrix is *not* of that form. But let us construct such a matrix from an arbitrary $\mathbf{E}$ of dimension $M \times N$. Define

$$\mathbf{D} = \begin{Bmatrix} \mathbf{0} & \mathbf{E}^T \\ \mathbf{E} & \mathbf{0} \end{Bmatrix},$$

which by definition is not only square (dimension $M+N$ by $M+N$) but also symmetric. Thus, **D** satisfies the theorem just alluded to, and the eigenvalue problem

$$\mathbf{D}\mathbf{q}_i = \lambda_i \mathbf{q}_i \tag{6.4.17}$$

gives rise to a complete $M+N$ orthonormal basis $\mathbf{q}_i$ whether or not the λ_i are distinct or nonzero. Write out the eigenvector relation (6.4.17),

$$\begin{Bmatrix} \mathbf{0} & \mathbf{E}^T \\ \mathbf{E} & \mathbf{0} \end{Bmatrix} \begin{bmatrix} q_{1i} \\ \cdot \\ q_{Ni} \\ q_{N+1,i} \\ \cdot \\ q_{N+M,i} \end{bmatrix} = \lambda_i \begin{bmatrix} q_{1i} \\ \cdot \\ q_{Ni} \\ q_{N+1,i} \\ \cdot \\ q_{N+M,i} \end{bmatrix}, \tag{6.4.18}$$

where q_{pi} is the pth element of $\mathbf{q}_i$. Taking note of the zero matrices, (6.4.18) can be rewritten as

$$\mathbf{E}^T \begin{bmatrix} q_{N+1,i} \\ \cdot \\ q_{N+M,i} \end{bmatrix} = \lambda_i \begin{bmatrix} q_{1i} \\ \cdot \\ q_{Ni} \end{bmatrix},$$

$$\mathbf{E} \begin{bmatrix} q_{1i} \\ \cdot \\ q_{Ni} \end{bmatrix} = \lambda_i \begin{bmatrix} q_{N+1,i} \\ \cdot \\ q_{N+M,i} \end{bmatrix}. \tag{6.4.19a, b}$$

Define

$$\mathbf{u}_i = \begin{bmatrix} q_{N+1,i} \\ \cdot \\ q_{N+M,i} \end{bmatrix}, \quad \mathbf{v}_i = \begin{bmatrix} q_{1i} \\ \cdot \\ q_{Ni} \end{bmatrix}, \quad \text{or} \quad \mathbf{q}_i = \begin{bmatrix} \mathbf{v}_i \\ \mathbf{u}_i \end{bmatrix}, \tag{6.4.20}$$

that is, defining the first N elements of $\mathbf{q}_i$ to be $\mathbf{v}_i$, and the last M to be $\mathbf{u}_i$. Then (6.4.19) are

$$\mathbf{E}\mathbf{v}_i = \lambda_i \mathbf{u}_i, \qquad \mathbf{E}^T \mathbf{u}_i = \lambda_i \mathbf{v}_i. \tag{6.4.21a, b}$$

If (6.4.21a) is left-multiplied by $\mathbf{E}^T$, and we use (6.4.21b), we have

$$\mathbf{E}^T \mathbf{E}\mathbf{v}_i = \lambda_i^2 \mathbf{v}_i. \tag{6.4.22a}$$

Similarly, left-multiplying (6.4.21b) by **E** and using (6.4.21a) produces

$$\mathbf{E}\mathbf{E}^T \mathbf{u}_i = \lambda_i^2 \mathbf{u}_i. \tag{6.4.22b}$$

These last two equations show that the $\mathbf{u}_i$, $\mathbf{v}_i$ each separately satisfy two independent eigenvector/eigenvalue problems of the square symmetric matrices

$\mathbf{E}\mathbf{E}^T$ and $\mathbf{E}^T\mathbf{E}$. If one of M, N is much smaller than the other, one need only solve the smaller of the two eigenvalue problems of (6.4.22*a,b*) for either of $\mathbf{u}_i$, $\mathbf{v}_i$, with the other set calculated from (6.4.21*a*) or (6.4.21*b*).

The $\mathbf{u}_i$, $\mathbf{v}_i$ are called "singular vectors," and the λ_i are the "singular values." By convention, the λ_i are ordered in decreasing numerical value. Also by convention, they are all nonnegative (taking the negative values of λ_i produces singular vectors differing only by a sign from those corresponding to the positive roots, and they are not independent vectors). Equations (6.4.21) and (6.4.22) provide a relationship between each $\mathbf{u}_i$ and each $\mathbf{v}_i$. But because, in general, $M \neq N$, there will be more of one set than of the other. The only way the equations can be consistent is if $\lambda_i = 0,\ i > \min(M, N)$, where $\min(M, N)$ is read as "the minimum of M and N."

Let there be K nonzero λ_i. Then

$$\mathbf{E}\mathbf{v}_i \neq 0, \qquad 1 \leq i \leq K\,, \tag{6.4.23a}$$

and these $\mathbf{v}_i$ are known as the "range of $\mathbf{E}$" or the "solution range vectors." For the remaining vectors,

$$\mathbf{E}\mathbf{v}_i = 0, \qquad K+1 \leq i \leq N\,. \tag{6.4.23b}$$

These $\mathbf{v}_i$ are known as the "nullspace vectors of $\mathbf{E}$" (or the "nullspace of the solution"). If $K < M$, there will be K of the $\mathbf{u}_i$, such that

$$\mathbf{E}^T\mathbf{u}_i = \mathbf{u}_i^T\mathbf{E} \neq 0, \qquad 1 \leq i \leq K\,, \tag{6.4.24a}$$

which are the "range vectors of $\mathbf{E}^T$," and $M - K$ of the $\mathbf{u}_i$, such that

$$\mathbf{E}^T\mathbf{u}_i = \mathbf{u}_i^T\mathbf{E} = 0, \qquad K+1 \leq i \leq M\,, \tag{6.4.24b}$$

which are the "nullspace vectors of $\mathbf{E}^T$" or the "data, or observation, nullspace."

Because the $\mathbf{u}_i$, $\mathbf{v}_i$ are complete orthonormal bases in their corresponding spaces, we can expand $\mathbf{x}$, $\mathbf{y}$, and $\mathbf{n}$ exactly:

$$\mathbf{x} = \sum_{i=1}^{N} \alpha_i \mathbf{v}_i, \quad \mathbf{y} = \sum_{j=1}^{M} \beta_i \mathbf{u}_i, \quad \mathbf{n} = \sum_{i=1}^{M} \gamma_i \mathbf{u}_i \tag{6.4.25}$$

where $\mathbf{y}$ has been measured, so that we know $\beta_i = \mathbf{u}_i^T\mathbf{y}$. To find the solution, we need α_i, and to find the noise, we need γ_i. Substituting (6.4.25) into the equations (6.4.9), and using (6.4.21*a*),

$$\sum_{i=1}^{N} \alpha_i \mathbf{E}\mathbf{v}_i + \sum_{i=1}^{M} \gamma_i \mathbf{u}_i = \sum_{i=1}^{K} \alpha_i \lambda_i \mathbf{u}_i + \sum_{i=1}^{M} \gamma_i \mathbf{u}_i = \sum_{i=1}^{M} \beta_i \mathbf{u}_i\,. \tag{6.4.26}$$

Notice the differing upper limits on the summations. By the orthonormality of the singular vectors, (6.4.26) can be solved as

$$\alpha_i \lambda_i + \gamma_i = \beta_i, \qquad 1 \le i \le M, \tag{6.4.27a}$$

or

$$\alpha_i = \mathbf{u}_i^T \mathbf{y}/\lambda_i - \gamma_i/\lambda_i, \quad \lambda_i \neq 0, \quad 1 \le i \le K. \tag{6.4.27b}$$

In these equations, if $\lambda_i \neq 0$, nothing prevents setting $\gamma_i = 0$, that is,

$$\gamma_i = \mathbf{u}_i^T \mathbf{n} = 0, \qquad 1 \le i \le K, \tag{6.4.27c}$$

which has the effect of making the noise norm as small as possible. Then (6.4.27*b*) produces

$$\alpha_i = \mathbf{u}_i^T \mathbf{y} \,/\, \lambda_i, \qquad 1 \le i \le K. \tag{6.4.28}$$

But because $\lambda_i = 0$, $i \ge K + 1$, the only solution (6.4.27*a*) for this range of indices is $\gamma_i = \mathbf{u}_i^T \mathbf{y}$, and α_i is indeterminate. These γ_i are nonzero, meaning that there is always a noise contribution, except in the event (unlikely with real data) that

$$\mathbf{u}_i^T \mathbf{y} = 0, \qquad K \le i \le M. \tag{6.4.29}$$

This last equation is the new solvability condition.

The solution obtained in this manner now has the following form:

$$\hat{\mathbf{x}} = \sum_{i=1}^{K} \frac{\mathbf{u}_i^T \mathbf{y}}{\lambda_i} \mathbf{v}_i + \sum_{i=K+1}^{N} \alpha_i \mathbf{v}_i, \tag{6.4.30a}$$

$$\hat{\mathbf{y}} = \mathbf{E}\hat{\mathbf{x}} = \sum_{i=1}^{K} (\mathbf{u}_i^T \mathbf{y}) \mathbf{u}_i, \tag{6.4.30b}$$

$$\hat{\mathbf{n}} = \sum_{i=K+1}^{M} (\mathbf{u}_i^T \mathbf{y}) \mathbf{u}_i. \tag{6.4.30c}$$

The coefficients of the last $N - K$ of the $\mathbf{v}_i$ in (6.4.30*a*), the solution nullspace vectors, are arbitrary. The nullspace vectors represent structures in the solution about which the equations provide no information, and so $\hat{\mathbf{x}}$ is nonunique. The simplest solution is the one with the nullspace coefficients all set to zero,

$$\hat{\mathbf{x}} = \sum_{i=1}^{K} \frac{\mathbf{u}_i^T \mathbf{y}}{\lambda_i} \mathbf{v}_i, \tag{6.4.31}$$

along with (6.4.30*c*). This "particular SVD solution," as we shall call it later, is the one that simultaneously minimizes the residuals and the solution norm.

The decision (6.4.27*c*) needs to be examined. For some other choice, the solution norm will decrease, but the residual norm will increase. Determining the desirability of such a trade-off requires an understanding of the noise structure – in particular, (6.4.27*c*) imposes rigid structures on the residuals.

Singular-value decomposition. The singular vectors and values have been used to provide a convenient pair of orthonormal spanning sets to solve an arbitrary set of simultaneous equations. The vectors and values have another use, however, in providing a powerful decomposition of **E**.

Define **U** as the $M \times M$ matrix whose columns are the $\mathbf{u}_i$, and **V** as the $N \times N$ matrix whose columns are the $\mathbf{v}_i$, and $\mathbf{\Lambda}$ as the $M \times N$ matrix whose diagonal elements are the λ_i, in order of descending values, and whose other elements are zero. As an example, suppose $M = 3$, $N = 4$; then

$$\mathbf{\Lambda} = \begin{Bmatrix} \lambda_1 & 0 & 0 & 0 \\ 0 & \lambda_2 & 0 & 0 \\ 0 & 0 & \lambda_3 & 0 \end{Bmatrix}.$$

Alternatively, if $M = 4$, $N = 3$,

$$\mathbf{\Lambda} = \begin{Bmatrix} \lambda_1 & 0 & 0 \\ 0 & \lambda_2 & 0 \\ 0 & 0 & \lambda_3 \\ 0 & 0 & 0 \end{Bmatrix},$$

extending the definition of a diagonal matrix to nonsquare ones.

Precisely as with the matrix **G** considered earlier, column orthonormality of **U**, **V** implies that these matrices are orthogonal,

$$\mathbf{U}\mathbf{U}^T = \mathbf{I}_M, \quad \mathbf{U}^T\mathbf{U} = \mathbf{I}_M, \quad \mathbf{V}\mathbf{V}^T = \mathbf{I}_N, \quad \mathbf{V}^T\mathbf{V} = \mathbf{I}_N. \quad (6.4.32a, b, c, d)$$

(It follows that $\mathbf{U}^{-1} = \mathbf{U}^T$, *etc.*) As with **G**, should one or more columns of **U**, **V** be deleted, the matrices would become semiorthogonal.

The relations (6.4.21) and (6.4.22) can be written in compact form as

$$\mathbf{E}\mathbf{V} = \mathbf{U}\mathbf{\Lambda}, \qquad \mathbf{E}^T\mathbf{U} = \mathbf{V}\mathbf{\Lambda}^T, \qquad (6.4.33a, b)$$

$$\mathbf{E}^T\mathbf{E}\mathbf{V} = \mathbf{V}\mathbf{\Lambda}^T\mathbf{\Lambda}, \qquad \mathbf{E}\mathbf{E}^T\mathbf{U} = \mathbf{U}\mathbf{\Lambda}\mathbf{\Lambda}^T. \qquad (6.4.33c, d)$$

Left multiply **E** by $\mathbf{U}^T$, and right-multiply it by **V**, invoking (6.4.33*a*)

$$\mathbf{U}^T\mathbf{E}\mathbf{V} = \mathbf{U}^T\mathbf{U}\mathbf{\Lambda} = \mathbf{\Lambda}. \qquad (6.4.34)$$

So **U**, **V** diagonalize **E** (with "diagonal" having the extended meaning, for a rectangular matrix, as defined previously).

Right-multiply (6.4.33*a*) by $\mathbf{V}^T$, and, using (6.4.32*c*),

$$\mathbf{E} = \mathbf{U}\mathbf{\Lambda}\mathbf{V}^T. \qquad (6.4.35)$$

This relationship is called the singular-value decomposition (SVD) of $\mathbf{E}$ into two orthogonal matrices and a diagonal one.

There is one further step. Notice that for a rectangular $\boldsymbol{\Lambda}$, as in the preceding examples, one or more rows or columns must be all zero, depending on the shape of the matrix. In addition, if any of the $\lambda_i = 0$, $i < \min(M, N)$, the corresponding rows and columns also will be all zeros. Let K be the number of nonvanishing singular values (the rank of $\mathbf{E}$). By inspection (multiplying it out), one finds that the last $N - K$ columns of $\mathbf{V}$ and the last $M - K$ columns of $\mathbf{U}$ are multiplied by zeros only. If these columns are dropped entirely from $\mathbf{U}$, $\mathbf{V}$, so that $\mathbf{U}$ becomes $M \times K$ and $\mathbf{V}$ becomes $N \times K$, reducing $\boldsymbol{\Lambda}$ to a $K \times K$ square matrix, then the representation (6.4.35) remains exact in the form

$$\mathbf{E} = \mathbf{U}_K \boldsymbol{\Lambda}_K \mathbf{V}_K^T, \tag{6.4.36}$$

with the subscript indicating the number of columns; $\mathbf{U}_K$, $\mathbf{V}_K$ are not square and only semiorthogonal. All of (6.4.33) remains valid for these reduced matrices, and $\boldsymbol{\Lambda}_K^T \boldsymbol{\Lambda}_K = \boldsymbol{\Lambda}_K \boldsymbol{\Lambda}_K^T = \boldsymbol{\Lambda}_K^2$. Subscripts are omitted when the context makes clear what is intended.

Either of (6.4.35) or (6.4.36) is the SVD, which for nonsquare matrices is due to Carl Eckart (Eckart and Young, 1939). Good accounts may be found in the work of Lanczos (1961), Noble and Daniel (1977), and Strang (1986) and in many recent books on applied linear algebra.

The solutions (6.4.30) can be written in compact form as

$$\tilde{\mathbf{x}} = \mathbf{V}_K \mathbf{U}_K^T \mathbf{y} \boldsymbol{\Lambda}_K^{-1} + \mathbf{Q}_v \boldsymbol{\alpha}_Q, \tag{6.4.37a}$$

$$\tilde{\mathbf{y}} = \mathbf{U}_K (\mathbf{U}_K^T \mathbf{y}), \tag{6.4.37b}$$

$$\tilde{\mathbf{n}} = \mathbf{Q}_u (\mathbf{Q}_u^T \mathbf{y}), \tag{6.4.37c}$$

where we define the two nullspace matrices

$$\mathbf{Q}_u = \{\mathbf{u}_i\}, \qquad K+1 \le i \le M, \qquad \mathbf{Q}_v = \{\mathbf{v}_i\}, \qquad K+1 \le i \le N.$$

The particular SVD solution sets $\boldsymbol{\alpha}_Q = 0$.

Solution of simultaneous equations by SVD has several important advantages. The same algebraic formulation applies to systems of equations, whether they are underdetermined, overdetermined, or just-determined. Unlike the eigenvalue/eigenvector solution for an arbitrary (nonsymmetric, or Hermitian) square system, the singular values (eigenvalues) are always nonnegative and real, and the singular vectors (eigenvectors) can always be made a complete orthonormal set. Neither of these statements is true for the conventional eigenvector problem. Furthermore, the relations (6.4.21) or (6.4.33) provide a specific, quantitative statement of the connection between a set of orthonormal structures in the data and the corresponding presence of orthonormal structures in the solution. These relations provide a powerful diagnostic method for understanding precisely why the solution takes on the form it does.

Resolution. Because it is a summation in an incomplete set of orthonormal vectors, the relationship between the basic SVD solution $\widehat{\mathbf{x}}$ and the true solution $\mathbf{x}$ follows from (6.4.4):

$$\widehat{\mathbf{x}} = \mathbf{V}_K \mathbf{V}_K^T \mathbf{x}, \qquad (6.4.38)$$

that is, $\widehat{\mathbf{x}}$ is a weighted average of the true solution. $\mathbf{T}_V = \mathbf{V}_K \mathbf{V}_K^T$ is called the "solution resolution matrix." It is most readily interpreted by considering its behavior if the true solution were unity in element j_0 and zero everywhere else. Then, from (6.4.38), the solution would be just column j_0 of $\mathbf{T}_V$. Of course, if $K = N$, then $\mathbf{V}\mathbf{V}^T = \mathbf{I}_N$, and the solution is said to be "fully resolved." More generally, if column q of the resolution matrix has unity on diagonal element $(\mathbf{V}_K \mathbf{V}_K^T)_{qq}$ and zero everywhere else, then x_q is fully resolved. Individual solution elements can be fully resolved while others are not resolved at all.

When the SVD solution is substituted back into the original equations, it does not reproduce $\mathbf{y}$: a residual, $\widehat{\mathbf{n}}$, is left. The solution produces an estimate of $\mathbf{y}$ that is

$$\widehat{\mathbf{y}} = \mathbf{U}_K \mathbf{U}_K^T \mathbf{y}, \qquad (6.4.39)$$

where $\mathbf{T}_U = \mathbf{U}_K \mathbf{U}_K^T$ is the "data resolution matrix." To interpret it, notice that we can write

$$\mathbf{E}\widehat{\mathbf{x}} = \mathbf{U}_K \mathbf{U}_K^T \mathbf{y}. \qquad (6.4.40)$$

Suppose an element of $\mathbf{y}$ is fully resolved, that is, some column j_0 of $\mathbf{U}_K \mathbf{U}_K^T$ is all zeros except for diagonal element j_0, which is unity. Then a unit change in y_{j_0} will produce a change in $\widehat{\mathbf{x}}$ that will leave unchanged all other elements of $\widehat{\mathbf{y}}$. If element j_0 is *not* fully resolved, then a change of unity in observation y_{j_0} will produce a new solution that will lead to changes in other elements of $\widehat{\mathbf{y}}$. Stated slightly differently, if y_i is not fully resolved, the system lacks adequate information to distinguish equation i from a linear dependence on one or more other equations. One can use these ideas to construct quantitative statements of which observations are the most important ("data ranking").

Along with the explicit relations between observation structure and solution structure contained in equation (6.4.33a), the resolution matrices are very powerful tools for understanding the behavior of the system. Although resolution matrices can be constructed for any of the solution methods, only the SVD makes it possible to understand their form in terms of orthonormal structures of both data and solution. More details about resolution matrices and the structures in the singular vectors can be found in the work of Wiggins (1972), Wunsch (1978, in press), or Menke (1989).

Solution variance. The SVD solution requires computation of the expansion coefficients (6.4.28); but because $\mathbf{y}$ contains noise, these numerical values must be regarded as partially uncertain. In particular, (6.4.28) is

$$\alpha_i = \frac{\mathbf{u}_i^T(\mathbf{n} + \mathbf{y}_0)}{\lambda_i}, \quad i = 1 \text{ to } K, \tag{6.4.41}$$

where $\mathbf{y}_0$ is again the value of $\mathbf{y}$ if there is no noise. If the noise mean vanishes, and the noise elements are uncorrelated,

$$\langle \mathbf{n} \rangle = 0, \quad \langle \mathbf{n}\mathbf{n}^T \rangle = \sigma_n^2 \mathbf{I}_M, \tag{6.4.42}$$

then

$$\begin{aligned}
\mathbf{P}_n &\equiv \left\langle (\widehat{\mathbf{x}} - \langle \widehat{\mathbf{x}} \rangle)(\widehat{\mathbf{x}} - \langle \widehat{\mathbf{x}} \rangle)^T \right\rangle, \\
&= \left\langle \left(\sum_i (\mathbf{u}_i^T \mathbf{n} / \lambda_i) \mathbf{v}_i \right) \left(\sum_j (\mathbf{u}_j^T \mathbf{n} / \lambda_j) \mathbf{v}_j \right)^T \right\rangle, \\
&= \sum_{i=1}^{K} \sum_{j=1}^{K} \mathbf{v}_i \langle \mathbf{u}_i^T \mathbf{n}\mathbf{n}^T \mathbf{u}_j \rangle / (\lambda_i \lambda_j) \mathbf{v}_j^T, \\
&= \sum_{i=1}^{K} \sigma_n^2 \mathbf{v}_i \lambda_i^{-2} \mathbf{v}_i^T, \\
&= \sigma_n^2 \mathbf{V}_K \mathbf{\Lambda}_K^{-2} \mathbf{V}_K^T
\end{aligned} \tag{6.4.43}$$

[recall (6.4.16*b*)]. Notice that the variances are taken about the expected value of the solution – not about the true value – because the nullspace vector portion is missing. $\sigma_n^2 \mathbf{V}_K \mathbf{\Lambda}_K^{-2} \mathbf{V}_K^T$ is the "solution covariance matrix." The value of σ_n^2 is commonly computed from the mean-square value of the residuals, the "sample variance" s^2. It is essential that s^2 prove statistically consistent with any prior estimate of its value. The square root of the diagonal elements of $\mathbf{P}_n$ is the standard error of the solution.

One can regard the unknown, suppressed, nullspace components as constituting an additional uncertainty owing to a "failure to resolve." Occasionally, one has an estimate of the expected variance of the solution, $\langle \mathbf{x}^T \mathbf{x} \rangle$. The quantity $1 - (1/N)\widehat{\mathbf{x}}^T \widehat{\mathbf{x}} / \langle \mathbf{x}^T \mathbf{x} \rangle$ is sometimes used as a measure of how much solution "energy" lies in the nullspace, and hence can be regarded, narrowly, as an error. It can be used element by element, as $1 - \widehat{x}_i^2 / \langle x_i^2 \rangle$, if $\langle x_i^2 \rangle$ is available.

Rank determination. The solution variance matrix shows that if some of the λ_i are very small, σ_n^2/λ_i^2 can become arbitrarily large and completely swamp the solution. *Mathematically*, any nonzero singular value is retained: mathematical problems normally suppose that $\mathbf{y}$ is known with perfect accuracy. In observational practice, such perfection is usually impossible.[6] The result (6.4.43) suggests defining an "effective rank," $K' < K$, which may be considerably less than the mathematical rank, beyond which the singular values are too small to be regarded as useful. As K' is reduced, fewer columns in the $\mathbf{U}$, $\mathbf{V}$ matrices are retained, and the resolution matrices differ even more from the desirable full resolution. Reducing the solution variance results in a reduction of resolution – a familiar trade-off in estimation problems – the nullspace grows, and the noise norm rises. Determination of the effective rank is an important problem that is commonly solved by studying the behavior of the solution norm, which grows with K', and the residual norm, which decreases with K', as well as considering the investigator's needs for resolution versus solution stability. From here on, it will be understood that K usually means the effective rank, rather than the mathematical one.

This solution method for the noisy simultaneous equations can now be briefly summarized. The solution is independent of whether the system is formally overdetermined, underdetermined, or just-determined (square and of full rank). In many practical problems, K is less than both M, N; then there are both solution and data nullspaces. If $K = M < N$, the "full-rank underdetermined case," there is no data nullspace (no residuals – usually an unphysical result), but there is a solution nullspace. If $K = N < M$, the "full-rank overdetermined case," there is no solution nullspace, but a data nullspace (*i.e.*, residuals). Because the solution, noise, observations, and governing equations are all described by two sets of orthonormal vectors, a complete and powerful piece of machinery is available for understanding the system.

Relation to least-squares. The SVD permits a complete description of the least-squares solutions. We suppose that all relevant row and column scaling has been performed first. Consider first (6.3.2), and substitute the SVD. Using the semiorthogonal behavior of $\mathbf{U}_k$,

$$\hat{\mathbf{x}} = (\mathbf{V}_K \mathbf{\Lambda}_K^2 \mathbf{V}_K^T)^{-1} \mathbf{V}_K \mathbf{\Lambda}_K \mathbf{U}_K^T \mathbf{y}.$$

If $K = N$, full rank, the inverse matrix can be obtained by inspection because of the orthogonality of $\mathbf{V}_N$,

$$(\mathbf{V}_N \mathbf{\Lambda}_N^2 \mathbf{V}_N^T)^{-1} = \mathbf{V}_N \mathbf{\Lambda}_N^{-2} \mathbf{V}_N^T,$$

[6]Exceptions occur when rigid requirements, perhaps arising from dynamics, are imposed on the solution, but even then such relationships often contain model errors.

and the least-squares solution is

$$\hat{\mathbf{x}} = \mathbf{V}_N \mathbf{\Lambda}_N^{-1} \mathbf{U}_N^T \mathbf{y},$$

which is identical with the SVD solution (6.4.37a) with $K = N$. It is readily confirmed by direct substitution that the uncertainty (6.3.4) is identical with (6.4.43).

Consider next the simple underdetermined solution (6.3.14a), with $\mathbf{S} = \mathbf{I}$. Substituting the SVD produces

$$\hat{\mathbf{x}} = \mathbf{V}_K \mathbf{\Lambda}_K \mathbf{U}_K^T (\mathbf{U}_K \mathbf{\Lambda}_K^2 \mathbf{U}_K^T)^{-1} \mathbf{y}.$$

If now $K = M$, the matrix inverse is, by inspection,

$$(\mathbf{U}_M \mathbf{\Lambda}_M^2 \mathbf{U}_M^T)^{-1} = \mathbf{U}_M \mathbf{\Lambda}_M^{-2} \mathbf{U}_M^T,$$

and the solution is

$$\hat{\mathbf{x}} = \mathbf{V}_M \mathbf{\Lambda}_M^{-1} \mathbf{U}_M^T \mathbf{y},$$

identical with the particular SVD solution, and an arbitrary summation of nullspace vectors could be added to $\hat{\mathbf{x}}$. Thus the SVD solution corresponds to both the formally overdetermined and the formally underdetermined least-squares solutions, where $K = N$ or $K = M$, respectively.

A number of other useful results are now readily available. In particular, we know when the crucial matrix $(\mathbf{E}^T\mathbf{E})^{-1}$ will exist – evidently we must have $K = N$. If any singular value vanishes, the ordinary least-squares solution does not exist. But the SVD solution always does.

We can also interpret the solutions (6.3.12) or (6.3.16) derived from the modified objective functions (6.3.11) or (6.3.15). Substituting the full SVD, and invoking the matrix-inversion lemma, produces

$$\begin{aligned}\hat{\mathbf{x}} &= \sum_{i=1}^{N} \frac{\lambda_i \mathbf{u}_i^T \mathbf{y}}{\lambda_i^2 + \alpha^2} \mathbf{v}_i ,\\ &= \mathbf{V}_N \mathbf{\Lambda}^T (\mathbf{\Lambda}\mathbf{\Lambda}^T + \alpha^2 \mathbf{I}_M)^{-1} \mathbf{U}_M^T \mathbf{y},\end{aligned} \tag{6.4.44a}$$

$$\begin{aligned}\hat{\mathbf{n}} &= \sum_{i=1}^{M} \mathbf{u}_i^T \mathbf{y} \left(\frac{\alpha^2}{\lambda_i^2 + \alpha^2} \right) \mathbf{u}_i ,\\ &= \alpha^2 \mathbf{U}_M [\mathbf{\Lambda}\mathbf{\Lambda}^T + \alpha^2 \mathbf{I}_M]^{-1} \mathbf{U}_M^T \mathbf{y},\end{aligned} \tag{6.4.44b}$$

$$\begin{aligned}\mathbf{P}_n =&\sigma_n^2 \sum_{i=1}^{N} \frac{\lambda_i^2}{(\lambda_i^2+\alpha^2)^2} \mathbf{v}_i \mathbf{v}_i^T , \\ =&\sigma_n^2 \mathbf{\Lambda}^T (\mathbf{\Lambda}^T\mathbf{\Lambda}+\alpha^2\mathbf{I}_N)^{-1}\mathbf{V}_N\mathbf{V}_N^T(\mathbf{\Lambda}^T\mathbf{\Lambda}+\alpha^2\mathbf{I}_N)^{-1}\mathbf{\Lambda},\end{aligned} \qquad (6.4.44c)$$

where it is assumed that $\langle \mathbf{n}\mathbf{n}^T \rangle = \sigma_n^2 \mathbf{I}_N$, and $\mathbf{\Lambda}$ is the full rectangular form. The presence of α^2 prevents any singularity owing to one or more vanishing singular values. The introduction of α^2 "tapers" the effect of small and vanishing λ_i [and hence (6.3.12) and (6.3.16) are called "tapered least-squares" solutions]. The form (6.4.44) is sometimes used rather than the truncation in the particular SVD solution; it is called the "tapered SVD" solution and is the same as the tapered least-squares solution. Hence the inverses in (6.3.8), (6.3.12), and (6.3.16) always exist. It is easy to construct tapered SVD analogues of the two resolution matrices.

The identity of the least-squares and the SVD permits us a full description of the ordinary least-squares solution (6.3.2). In particular, the minimization of the residuals in (6.3.1) is being obtained through the demand (6.4.27c) that there be no projection of the residuals onto the range vectors $\mathbf{u}_i$. Stating it another way, the residuals are forced to satisfy $\mathbf{u}_i^T \mathbf{y} = 0,\ 1 \le i \le K$, which, depending on the physics of the residuals, might be regarded as an overly stringent requirement. In contrast, the tapered forms (6.4.44b) do project some of the residuals onto the $\mathbf{u}_i$ range vectors.

Example: The polar profile. Let us apply these ideas to the inverse problem of the polar profile already solved in the least-squares context. The singular values of $\mathbf{E}$ are

$$\lambda_i = [2186.3,\ 1751.0,\ 1353.8, 979.3,\ 794.4, 630.3,\ 466.4,\ 420.0, 22.1,\ 2.1,\ 0.2,\ 0.01]$$

and are plotted in fig. 6.5; some of the singular vectors are in fig. 6.6. The spread in the λ_i is very large, but none of them actually vanish, and in a formal sense this is a "full-rank" system, $K = N$, with $\mathbf{T}_V = \mathbf{I}$, so the existence of a full-rank least-squares solution (fig. 6.2) is reasonable.

Using the y_i corrupted with the 1% variance noise, and the SVD at rank $K = 12$, produces the solution identical with that for ordinary least-squares, fig. 6.2, which is regarded as unsatisfactory, being 1000 times too large in the surface layers.

The SVD procedure provides an immediate explanation of what has gone wrong. The rapid drop in the singular values after the eighth value (fig. 6.5) is a strong suggestion that a rank-8 particular SVD solution is appropriate, because

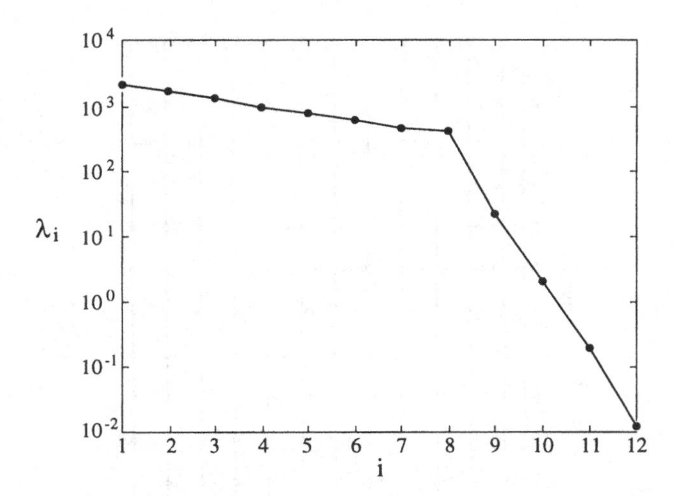

Fig. 6.5. Singular values λ_i of the observation matrix $\mathbf{E}$ for the layered polar profile. The rapid drop in the numerical values beyond $i = 8$ suggests a rank $K = 8$ system.

then the uncertainty owing to the last four terms of (6.4.4) will be removed. The resulting rank-8 solution and its standard error are depicted in fig. 6.7. The four $\mathbf{v}_i$ that are then relegated to the nullspace (fig. 6.6) will all have large values in the upper ocean, and the eight range vectors that have been retained will all be small there. $\mathbf{T}_V$ is no longer the identity: its diagonal is plotted in fig. 6.8 (top), and above 300 m there is almost no resolution.

The value of K chosen is clearly dependent in large part on the value of the noise variance – for larger noise we might drop even more terms. Dropping terms means that $\mathbf{T}_V$ will be even further from the identity. Alternatively, one can describe the reduction in rank as producing a reduction in the uncertainty of the particular SVD solution, but increasing the uncertainty of the nullspace components.

The diagonals of $\mathbf{T}_U$ are displayed in fig. 6.8 (bottom) at rank 8. A pattern is evident in which certain rays (measurements) dominate the solution. The dominant values tend to correspond to those rays for which the row norm of $\mathbf{E}$ is largest. The oscillatory character of the distribution is related to the penetration of rays (deeper for smaller i) across the layer boundaries. But the story is not quite that simple, because many of the rays are strongly correlated with each other. The data nullspace vectors $\mathbf{u}_i$ satisfy the relationship $\mathbf{u}_i^T\mathbf{E} = 0$, $K' + 1 \leq i \leq M$, which implies $\mathbf{u}_i^T\mathbf{y} = 0$, $K' + 1 \leq i \leq M$, the consistency relations that must have to hold if the measurements were perfect. These two

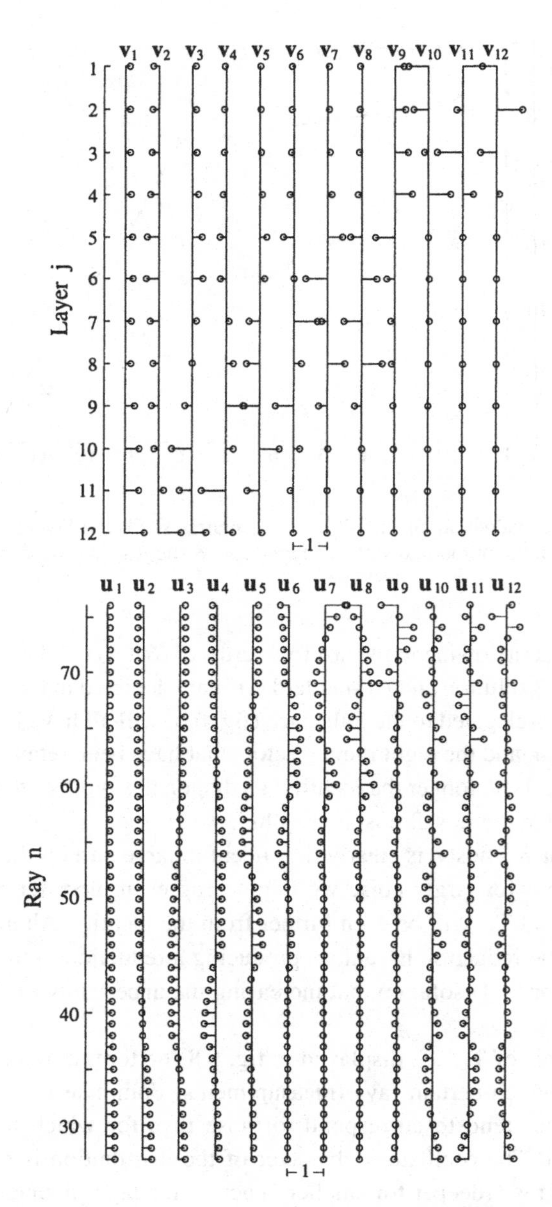

Fig. 6.6. Top: The 12 $\mathbf{v}_i$ for the layered polar profile. At rank $K = 12$ there is no nullspace. At rank $K = 8$, there are four nullspace vectors; these vectors tend to be large in the upper ocean and small at depth, with the eight range vectors having complementary structures. Bottom: The first 12 $\mathbf{u}_i$. The remaining $50 - 12 = 38$ $\mathbf{u}_i$ are in the mathematical nullspace of $\mathbf{E}^T$.

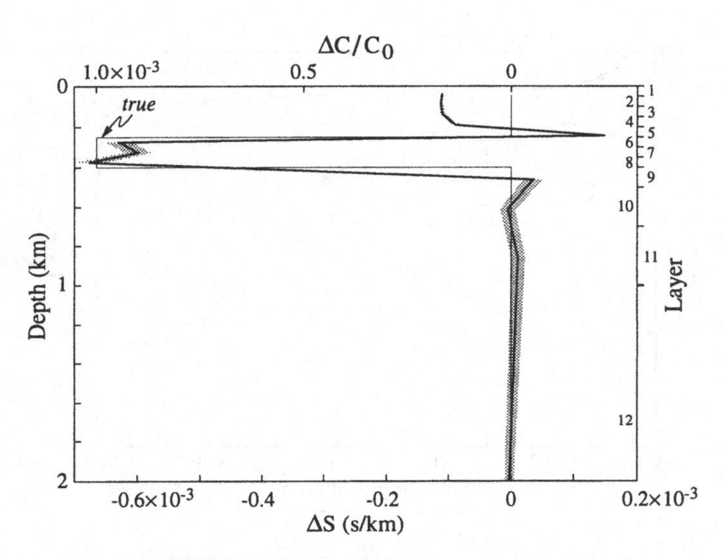

Fig. 6.7. Particular SVD solution (6.4.37) at rank $K = 8$, with standard errors (gray) for the layered polar profile, not including nullspace variance. Apart from an overshoot in the poorly resolved fifth layer, the solution is everywhere quite accurate. $\widehat{\mathbf{n}}$ (not shown) is also acceptable.

results become a complete description of the linear dependences of the rays (*i.e.*, the data redundancies) in the polar profile ocean.

The SVD provides complete understanding of the relationship between observations and the least-squares solution. For example, consider the $\mathbf{u}_1$, $\mathbf{v}_1$ vectors displayed in fig. 6.6: $\mathbf{u}_1$ is fairly uniform and is of the same sign for all ray numbers, and hence $\mathbf{u}_1^T \mathbf{y}$ represents a nearly uniform average of the travel times. $\mathbf{v}_1$ is similarly approximately constant with depth, and the physical interpretation is immediate: if there is a mean travel-time shift, one expects a nearly uniformly distributed perturbation in the sound-slowness to account for it. $\mathbf{v}_2$ reverses sign in the deepest two layers; the structure of $\mathbf{u}_2$ represents a differencing of the arrival times of the steep and shallow rays – which is, again, physically sensible, as it determines the difference between layer perturbations sampled by the steepest rays and by the shallower ones. [Munk and Wunsch (1982*b*) discuss further examples of this type.]

This simple example contains a number of features representative of a useful inverse method. Although parts of the solution appear indeterminate, other parts are extremely well determined. The structure of the indeterminate parts is explicitly known (the four nullspace vectors). The contribution to the solution of each measurement is also known through $\mathbf{T}_U$ and the redundant observations

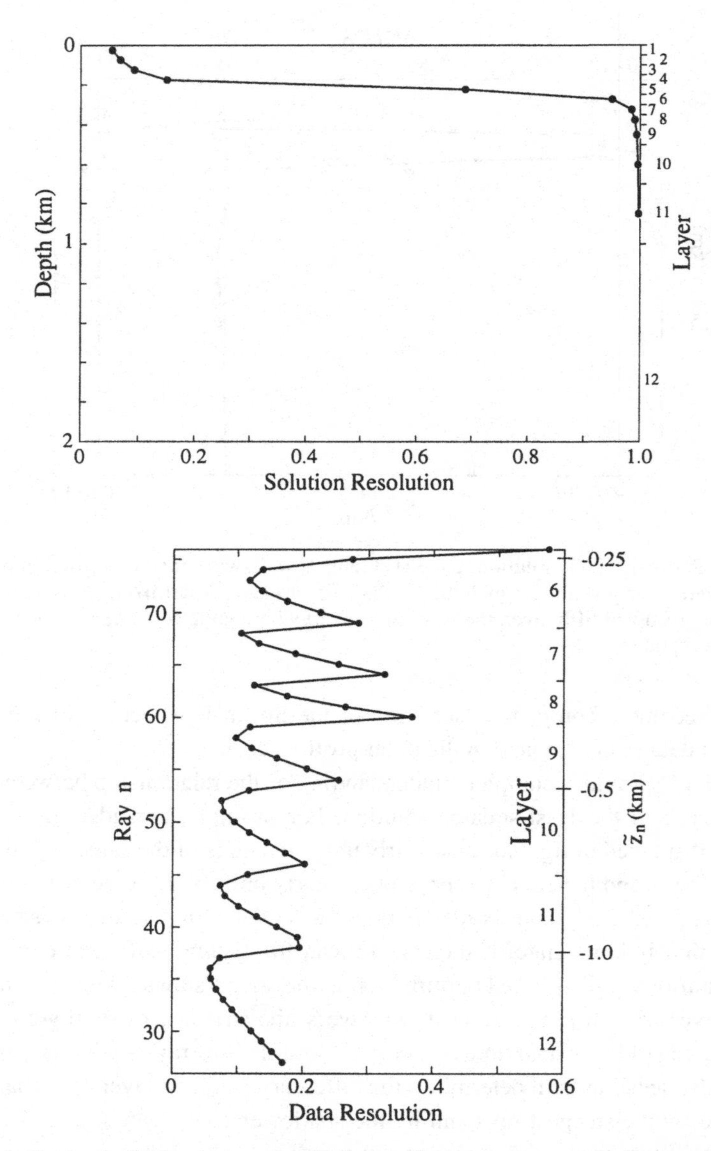

Fig. 6.8. Particular SVD solution of rank $K = 8$. There are no rays with turning points in the upper 300 m, and accordingly the diagonal elements of $\mathbf{T}_V$ give little or no resolution (top). Beneath 300 m, the resolution is nearly perfect, and the solution in fig. 6.7 is very well determined, despite the formal nondeterminism at rank 8. The diagonal elements of $\mathbf{T}_U$ (bottom) show the relative contribution of rays j to the solution in fig. 6.7. The oscillatory character of the distribution is related to the penetration of rays (deeper for smaller n) across the layer boundaries.

through the $\mathbf{u}_i$ in the nullspace. Truncating the SVD at $K' < K$ has the effect of simultaneously reducing the solution norm and its uncertainty. The tapered least-squares solution (6.3.8) and (6.3.12) has a similar effect, as was seen in figs. 6.3 and 6.4.

Example: The temperate profile. The perturbation for the temperate profile depicted in the lower part of fig. 2.13 provides an interesting contrast to the polar profile case. Apart from the change in the profile character, only five rays are available, yet 12 layers are believed necessary to properly describe the expected perturbations. None of the available rays even penetrates the top layer. This profile was analyzed at length by Munk and Wunsch (1982*b*), but it is worth a brief revisit. The true perturbation, as depicted in fig. 2.13, is identical with that for the polar profile, vanishing in all layers except 6 to 8, where $\Delta S = -6.667 \times 10^{-4}$ s/km. The singular values of the observation matrix are

$$\lambda_i = 10^3[1.52 \quad 0.19 \quad 0.10 \quad 0.06 \quad 0.05]\,.$$

The spread of singular values is somewhat smaller than that found for the polar profile case, leading to the expectation of a reduced noise sensitivity. In the absence of any observational noise, the rank-5 particular SVD and the least-squares solution (6.3.14) both produce the solution in fig. 6.9, which is, not unexpectedly, rather poor. The diagonal of $\mathbf{T}_V$ at rank 5 is also depicted in fig. 6.9. The resolution is zero in the top layer, as there are no rays entering it. In general, the resolution is greatest in the lower layers. As discussed by Munk and Wunsch (1982*b*), this deep resolution is a consequence of the tendency of the rays to spend most of their time in the deep water, leading to the up–down ambiguity described in chapter 2.

The SVD solution (fig. 6.9) was hardly affected when the perfect travel times were corrupted with white noise of rms 10 ms. This insensitivity to noise contrasts with the polar profile and is a consequence of the small range of singular values. If a very large noise is added to the travel times, the rank must be reduced, and even less solution structure can be inferred. The reader may find this result rather discouraging, but a great deal more can be said about making useful deductions from this situation using the SVD and least-squares. Before pursuing the problem, we shall introduce a little more inverse machinery that will simplify the discussion.

6.5. Gauss-Markov Estimation

Least-squares, including the basic SVD solution, is usually regarded by statisticians as a form of *approximation* or curve fitting (*e.g.*, Magnus and Neudecker,

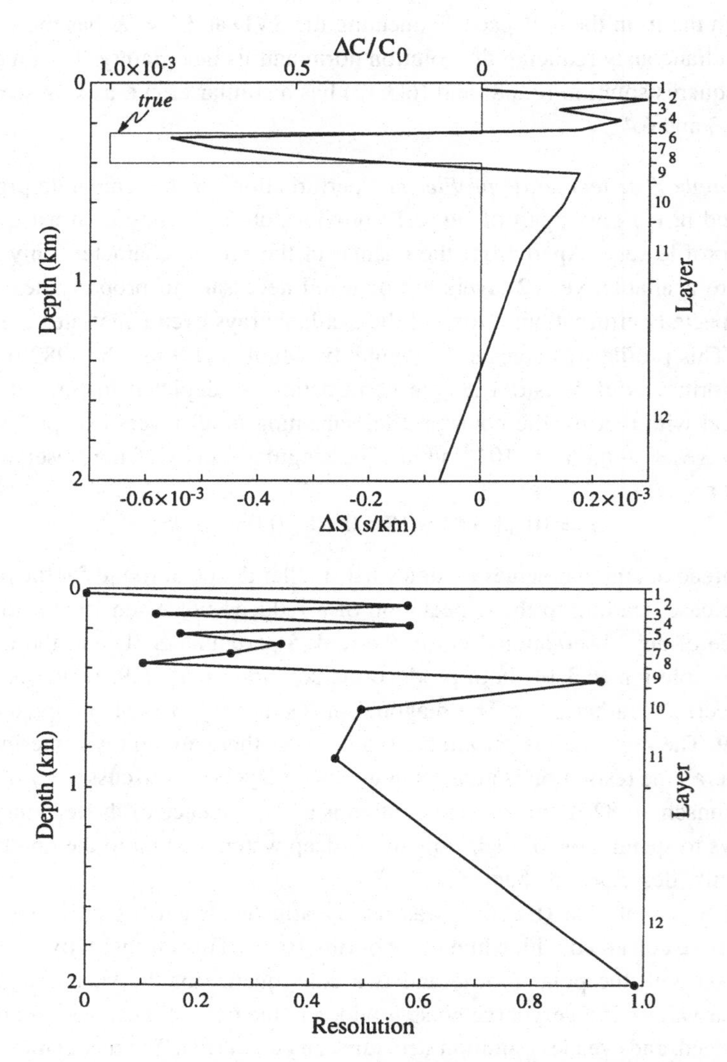

Fig. 6.9. Particular SVD solution (top) and diagonal resolution matrix (bottom) with perfect data for layered temperature profile of full rank. The dominant resolution is around 500 m and 2000 m as a result of the ray sampling structure. The result is not close to the "true" perturbation. Corruption of the travel times by 10 ms noise hardly changes this result.

1988). The discussion thus far has employed statistics only in the study of the uncertainty of $\hat{\mathbf{x}}$ as a function of the covariance of $\mathbf{n}$. The diagonal weight matrices $\mathbf{S}$, $\mathbf{R}$, employed either in the objective functions or for row- and column-scaling the observation matrix **E**, can be anything we please; choosing **R** as the noise variance is reasonable, but arbitrary.

We can instead approach the problem of finding solutions to the tomographic inverse problem by methods that are based in estimation theory. Somewhat confusingly, the solutions often turn out to be identical with those found from least-squares and the SVD. But the route to the answer is different, and the interpretation generally is more satisfactory.

We begin by attempting to minimize the expected uncertainty, the difference between the true value of x_j and the estimate $\widehat{x}_j$:

$$\text{minimize diagonal elements of}: \; \mathbf{P} \equiv \langle (\widehat{\mathbf{x}} - \mathbf{x})(\widehat{\mathbf{x}} - \mathbf{x})^T \rangle \,. \tag{6.5.1}$$

It is important that we seek to minimize the *individual* diagonal elements of **P**, not their sum of squares. The requirement of minimizing the expected square error between the estimated solution and the true one should be contrasted with the objective function (6.3.1) for the simplest form of least-squares, where the requirement was to render the square residuals $\sum_i n_i^2$ as small as possible. The slightly more complicated least-squares objective functions such as (6.3.9) that attempt to minimize weighted sums of squares of x_i, n_i also bear no resemblance to (6.5.1). There is no obvious reason why there should be any simple relationship between the least-squares solutions and whatever emerges from (6.5.1).

$\widehat{\mathbf{x}}$ will be taken to be a weighted average of the data:

$$\widehat{\mathbf{x}} = \mathbf{B}\mathbf{y}\,, \tag{6.5.2}$$

that is, each $\widehat{x}_i$ is a different linear combination of the data, given by the dot product of row i of **B** with **y**. The choice of an estimate that is a linear combination of the data can be understood in a number of ways. It can be demonstrated (Deutsch, 1965) that if the statistics are Gaussian, no nonlinear combination can do any better. Furthermore, apply **B** to **y** in (6.2.4):

$$\mathbf{B}\mathbf{y} = \mathbf{B}\mathbf{E}\mathbf{x} + \mathbf{B}\mathbf{n}\,. \tag{6.5.3}$$

Then if **B** is a true inverse of **E**, one will have **x** if $\mathbf{n} = 0$.

Substituting (6.5.2) into (6.5.1) yields the expected square error

$$\begin{aligned} \mathbf{P} &\equiv \langle (\widehat{\mathbf{x}} - \mathbf{x})(\widehat{\mathbf{x}} - \mathbf{x})^T \rangle\,, \\ &= \langle (\mathbf{B}\mathbf{y} - \mathbf{x})(\mathbf{B}\mathbf{y} - \mathbf{x})^T \rangle\,, \\ &= \mathbf{B}\,\mathbf{\Phi}_{yy}\mathbf{B}^T - \mathbf{\Phi}_{xy}\mathbf{B}^T - \mathbf{B}\mathbf{\Phi}_{yx} + \mathbf{\Phi}_{xx}\,, \end{aligned} \tag{6.5.4}$$

where

$$\boldsymbol{\Phi}_{xx} = \langle \mathbf{x}\mathbf{x}^T \rangle, \quad \boldsymbol{\Phi}_{yx} \equiv \langle \mathbf{y}\mathbf{x}^T \rangle, \quad \boldsymbol{\Phi}_{yy} = \langle \mathbf{y}\mathbf{y}^T \rangle .$$

Note

$$\boldsymbol{\Phi}_{yx} = \boldsymbol{\Phi}_{xy}^T, \quad \boldsymbol{\Phi}_{xx} = \boldsymbol{\Phi}_{xx}^T, \quad etc.$$

It is straightforward (*e.g.*, Liebelt, 1967) to prove that the value of **B** minimizing the diagonals of (6.5.4) is

$$\mathbf{B} = \boldsymbol{\Phi}_{xy}\boldsymbol{\Phi}_{yy}^{-1} . \tag{6.5.5}$$

This result is the Gauss-Markov theorem, and an estimate made with **B** is a "Gauss-Markov estimate," sometimes known as the "stochastic inverse" (*e.g.*, Aki and Richards, 1980) or the "minimum error variance" estimate. In our special case, $\mathbf{y} = \mathbf{E}\mathbf{x} + \mathbf{n}$, and so

$$\boldsymbol{\Phi}_{xy} = \langle \mathbf{x}\mathbf{x}^T\mathbf{E}^T \rangle = \boldsymbol{\Phi}_{xx}\mathbf{E}^T , \tag{6.5.6}$$

where it has been assumed that there is no covariance between **x** and **n** (this assumption is not necessary, but it simplifies the results). Similarly,

$$\boldsymbol{\Phi}_{yy} \equiv \langle \mathbf{y}\mathbf{y}^T \rangle = \langle (\mathbf{E}\mathbf{x} + \mathbf{n})(\mathbf{E}\mathbf{x} + \mathbf{n})^T \rangle = \mathbf{E}\boldsymbol{\Phi}_{xx}\mathbf{E}^T + \boldsymbol{\Phi}_{nn} . \tag{6.5.7}$$

Introducing the definitions $\mathbf{S} = \boldsymbol{\Phi}_{xx}, \quad \mathbf{R} = \boldsymbol{\Phi}_{nn}$, equations (6.5.5) and (6.5.2) reduce to

$$\begin{aligned} \mathbf{B} &= \boldsymbol{\Phi}_{xx}\mathbf{E}^T(\mathbf{E}\boldsymbol{\Phi}_{xx}\mathbf{E}^T + \boldsymbol{\Phi}_{nn})^{-1} \\ &= \mathbf{S}\mathbf{E}^T(\mathbf{E}\mathbf{S}\mathbf{E}^T + \mathbf{R})^{-1} \end{aligned} \tag{6.5.8a}$$

and

$$\hat{\mathbf{x}} = \boldsymbol{\Phi}_{xx}\mathbf{E}^T(\mathbf{E}\boldsymbol{\Phi}_{xx}\mathbf{E}^T + \boldsymbol{\Phi}_{nn})^{-1}\mathbf{y} = \mathbf{S}\mathbf{E}^T(\mathbf{E}\mathbf{S}\mathbf{E}^T + \mathbf{R})^{-1}\mathbf{y} , \tag{6.5.8b}$$

respectively. The solution uncertainty is

$$\begin{aligned} \mathbf{P} &= \boldsymbol{\Phi}_{xx} - (\boldsymbol{\Phi}_{xx}\mathbf{E}^T)(\mathbf{E}\boldsymbol{\Phi}_{xx}\mathbf{E}^T + \boldsymbol{\Phi}_{nn})^{-1}\mathbf{E}\boldsymbol{\Phi}_{xx} \\ &= \mathbf{S} - \mathbf{S}\mathbf{E}^T(\mathbf{E}\mathbf{S}\mathbf{E}^T + \mathbf{R})^{-1}\mathbf{E}\mathbf{S} . \end{aligned} \tag{6.5.9}$$

Using the matrix-inversion lemma, expressions (6.5.8) and (6.5.9) can usefully be rewritten in algebraically equivalent forms as

$$\hat{\mathbf{x}} = (\mathbf{S}^{-1} + \mathbf{E}^T\mathbf{R}^{-1}\mathbf{E})^{-1}\mathbf{E}^T\mathbf{R}^{-1}\mathbf{y} \tag{6.5.10}$$

and

$$\mathbf{P} = (\mathbf{S}^{-1} + \mathbf{E}^T\mathbf{R}^{-1}\mathbf{E})^{-1} . \tag{6.5.11}$$

A choice between the pairs (6.5.8)–(6.5.9) and (6.5.10)–(6.5.11) is usually made on the basis of the relative computational loads.

The choice (6.5.8*a*) for $\mathbf{B}$ is optimum; for any other value of $\mathbf{B}$, the error will be greater than this minimum. One is free to form any other linear combination of the data for an estimate of $\mathbf{x}$; the equivalent value of $\mathbf{B}$ can then be substituted into (6.5.4), and the expected square error, which can only exceed (6.5.9), can be evaluated.

The reader should compare (6.5.8) and (6.5.9) with (6.3.16), and (6.5.10) and (6.5.11) with (6.3.10). What has been shown is that the Gauss-Markov estimate is identical with the tapered, weighted least-squares estimate if the weight matrices in the least-squares objective functions are chosen to be the second-moment matrices of $\mathbf{x}$, $\mathbf{n}$. In general, and particularly in the tomographic context, these matrices are not usually diagonal. The discussion in section 6.3 of the equivalence of row and column scalings with the choice of weight matrices in the objective functions can be preserved in the nondiagonal case if we can generalize the square root to nondiagonal matrices. For that purpose, we introduce the so-called Cholesky decomposition (*e.g.*, Golub and Van Loan, 1989). Let $\mathbf{M}$ be any symmetric, positive definite matrix (all positive eigenvalues). Then there exists a new upper triangular matrix $\mathbf{M}^{1/2}$ such that $\mathbf{M}^{T/2}\mathbf{M}^{1/2} = \mathbf{M}$, and $\mathbf{M}^{-1/2}$ exists. All second-moment matrices have the property that they are at least positive semidefinite; if there are zero eigenvalues, remedies for the lack of an inverse exist. Thus the row and column scalings introduced in (6.3.11*a*) have a simple physical interpretation – they are rotating $\mathbf{x}$, $\mathbf{n}$ into new coordinate systems in which the scaled variables are uncorrelated with each other and have unit second moments (*i.e.*, $\langle \mathbf{n}'\mathbf{n}'^T \rangle = \mathbf{I}$, *etc.*). Recognizing that use of the covariance matrices in the tapered least-squares objective functions or for row/column scaling produces results identical witih those from the estimation-theory approach, one can readily switch between the two points of view.

The ability to specify prior knowledge of the covariances in $\mathbf{n}$ is very important. Among other issues, the clock and source/receiver position errors described in chapter 5 lead to strongly correlated errors in the acoustic measurements. For example, all ray or mode arrival times measured from a source whose clock is drifting or whose position is changing will show predictable covarying errors. Specification of these covariances can greatly reduce the uncertainty of the solution (Gaillard, 1985; Cornuelle *et al.*, 1989). Some investigators prefer to break up $\mathbf{n}$ into separate components, such as

$$n_i = n_{sc} + n_{sp} + \cdots + n_{ri}, \tag{6.5.12}$$

for travel-time measurement i, the total error consisting of the clock error of sources, the position error of sources s, and so forth, plus any residual noise elements, n_{ri}. Terms like n_{sc} will appear in all travel-time equations involving the same source, and so forth. Separate covariances can be specified for the separate noise terms. Whether the n_{sc} are to be regarded as elements of $\mathbf{x}$ or of $\mathbf{n}$ is arbitrary.

Solution acceptance. On first encounter with the Gauss-Markov method, it may appear that one is "pulling solutions arbitrarily out of the air," in part because the covariance matrices are rarely known accurately. But as with any estimation procedure, after a solution is generated, one must undertake a careful analysis of its acceptability. The fundamental tests are based on an analysis of the solution statistics, including those of the noise estimates. In particular, the Gauss-Markov solution is based on assertions of the validity of the simultaneous equations (6.2.4), plus the statistics embodied in $\mathbf{\Phi}_{xx}$, $\mathbf{\Phi}_{nn}$. If the solution fails the test of being consistent with these prior estimates (such tests are discussed in books on statistics and regression analysis), the reasons must be clearly understood. A judgment must be made as to whether the solution should be rejected because it shows inconsistencies with the known behavior of the ocean ($\mathbf{\Phi}_{xx}$) or because it shows inconsistencies with the combined behaviors of the ocean plus the observation system ($\mathbf{\Phi}_{nn}$). In a conflict between the imposed statistics and the measurement constraints (6.2.4), the constraints often dominate, but then one should ask why the statistics were apparently erroneous.

What oceanographers commonly refer to as "objective mapping," or as "objective interpolation" (OI), is a special case of the present problem (Bretherton *et al.*, 1976). In a typical situation, one observes x_i irregularly scattered in a one- or two-dimensional space and seeks estimates $\widehat{x_i}$ on a regular grid. Define $\mathbf{x}$ to consist of its values at all positions, both observed and gridded. The rows of $\mathbf{E}$ are either all zeros, meaning no observation of element i, or all zeros except for a 1 in position i, meaning a direct observation of x_i. $\mathbf{S}$ and $\mathbf{R}$ are usually specified as *functions* dependent only on the physical separation of the elements of $\mathbf{x}$ (thus implying spatial stationarity) and possibly independent of their orientation (implying isotropy). The combination of mostly zeros in $\mathbf{E}$ plus the analytic specification of $\mathbf{S}$, $\mathbf{R}$ simplifies the computer coding of (6.5.8) and (6.5.9) and avoids the construction and storage of large matrices.

Like the tomographic problem, objective mapping requires some decision concerning the basic state about which perturbations are to be described. Such basic states range in practice from (i) hydrographic climatologies (*e.g.*, the mean temperature at a given depth, from historical observations) or (ii) perturbations about a mean that is itself estimated from the data, to technically more complicated backgrounds, such as (iii) the forecast of a numerical model. Procedure (ii) is equivalent to the Gauss-Markov theorem used twice – first to estimate the mean (which can be spatially varying), and second to estimate the deviations about that mean. Such methods are usually called "kriging" (Ripley, 1981; Armstrong, 1989).

An important practical variation of objective mapping specifies the covariance matrix, $\mathbf{\Phi}_{xx}$ or $\mathbf{S}$, through its eigenvectors, rather than directly, and uses the eigenvectors to represent the field $\mathbf{x}$. Consider the problem of mapping (estimating) the vertical structure of the time-varying ocean. Form a matrix $\mathbf{M}$, which is $M \times N$, and whose columns are the temperatures as functions of depth at different times in one location. The Eckart-Young-Mirsky theorem (*e.g.*, Van Huffel and Vandewalle, 1991) asserts that the best representation

$$\mathbf{M} \approx \mathbf{a}_1 \eta_1 \mathbf{b}_1^T + \mathbf{a}_2 \eta_2 \mathbf{b}_2^T + \cdots + \mathbf{a}_L \eta_L \mathbf{b}_L^T \tag{6.5.13}$$

for fixed $L < \min(M, N)$ is obtained if the $\mathbf{a}_i$, $\mathbf{b}_i$, η_i are the singular vectors and singular values of $\mathbf{M}$. That is to say, we should choose the $\mathbf{a}_i$ to be the eigenvectors of $\mathbf{M}\mathbf{M}^T$ (the covariance matrix of the rows), and the $\mathbf{b}_i$ to be the eigenvectors of $\mathbf{M}^T\mathbf{M}$ (the covariance of the columns). In particular, $\mathbf{M}\mathbf{M}^T$ is a sample covariance that might be used to estimate $\mathbf{S}$, defined as the covariance of the temperature as a function of depth. The vertical field to be mapped is then written as a sum of unknown coefficients times the $\mathbf{a}_i$, which are called "empirical orthogonal functions" (EOFs).[7] The EOFs provide a particularly efficient representation of the vertical structure, but are here best regarded as simply another way of imposing a known prior covariance – defining the eigenfunctions and eigenvalues of a matrix being equivalent to defining the matrix itself. (The $\mathbf{b}_i$ are not used.)

Temperate-layer perturbation revisited. We return to the temperate-profile perturbation problem in the situation where only five ray travel times are available for a 12-layer representation. With the Gauss-Markov approach, we are free, indeed required, to specify the expected second moments of both the noise and sound-slowness perturbations. For the time being, the noise will continue to have zero mean, with the second moments given by $\sigma_n^2 \mathbf{I}_5$. The bias corrections owing to quadratic nonlinearities, discussed in section 4.1, are treated here as

[7] A full discussion of this method in a statistical context is available (Jolliffe, 1986). In other contexts, it is known as the Karhunen-Loève theorem, or the method of principal components.

being absorbed into the noise **n**. This approach is not completely consistent[8] with the assumption $\langle \mathbf{n} \rangle = \mathbf{0}$.

What about the slowness perturbations **x**? A great deal is known about the vertical profiles of density and hence of sound-slowness perturbations in the ocean (*e.g.*, Richman *et al.*, 1977; Fu *et al.*, 1982; Mercier and Colin de Verdière, 1985). A slightly oversimplified summary would be that for mesoscale perturbations, which tend to dominate tomographic measurements on time scales of weeks to months, $\Delta S(z_i) \propto N(z_i)$ (Munk and Wunsch, 1982*b*). The constant of proportionality is chosen to reflect the actual expected variance of **x**. Let us estimate **x**, **n** using the "prior statistics," embodied in **S**, **R**.

The minimum-norm solution (fig. 6.9) is not satisfactory. Use of the correct statistics (*i.e.*, **S** is diagonal with very small variances in all layers except 6–9, and **R** is diagonal with values appropriate to the assumed noise) yields an excellent solution,[9] as expected (fig. 6.10).

Suppose a more realistic statistic is used, one in which **S** represents an exponential decay in amplitude from the surface downward over a vertical scale of 500 m, and the surface value is $(0.667 \times 10^{-3}\ \text{s/km})^2$. The result is shown in fig. 6.11. Despite the incorrect **S**, the inversion is quite good, because important information (that the deep perturbations are very small) is provided. The combination of that information with the travel times is enough to produce an accurate solution. (The residuals are too small, but the elements estimated were too few for proper statistical tests.)

Dynamic constraints and models. When $y_i = s_i$ (the sum travel times for rays or modes), the inverse procedures provide estimates of the sound-speed perturbation parameters. Alternatively, when $y_i = d_i$ (the travel-time differences), we obtain estimates of the water velocities. Nothing precludes performing separate inversions for ΔS_i, u_i. Let $\mathbf{E}_S$, $\mathbf{E}_u$ be the appropriate observation matrices for sound-speed and water velocity, respectively. Then the combined problem

[8] A consistent result can be obtained by using the estimated, range-dependent fields from an initial inversion to compute the bias contribution to the travel time. If the bias appears significant compared with the overall noise level, a simple iterative method would use the computed bias to correct the travel times, and a new inversion would be done. A consistency test requires that the mean of $\widehat{\mathbf{n}}$ be indistinguishable from zero. Further iterations could be carried out, but in practice such corrections have not yet proved necessary.

[9] The example is somewhat artificial, because a perturbation confined to layers 6–9 at depths between 200 m and 400 m is unlike any in the real ocean, and the statistics from prior field measurements would represent an incorrect statement of the prior variances for the particular example.

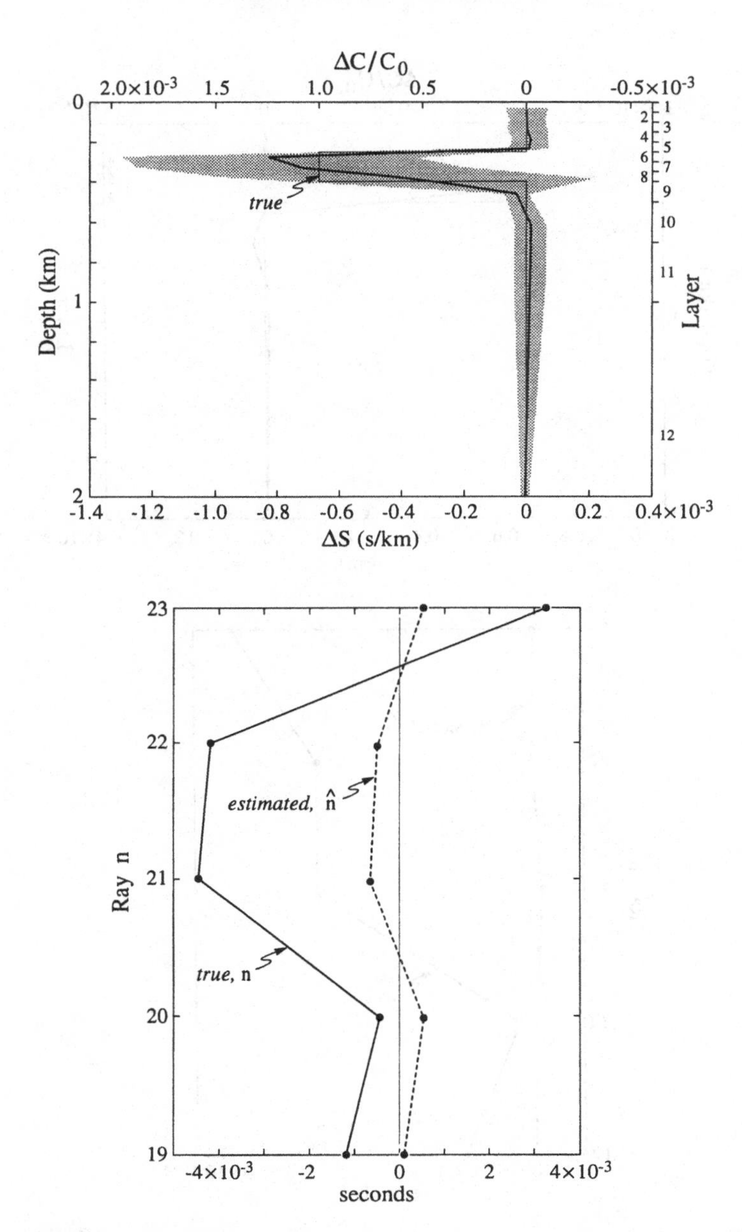

Fig. 6.10. Top: Gauss-Markov solution with standard errors (gray) for the layered temperate profile, using **S**, **R** appropriate to the known forward solution. The sound-slowness perturbation is within one standard error of the "true" value. Bottom: Estimated noise $\hat{\mathbf{n}}$ (dashed) is smaller than "true" noise **n** (solid), but there are too few values for a statistical test.

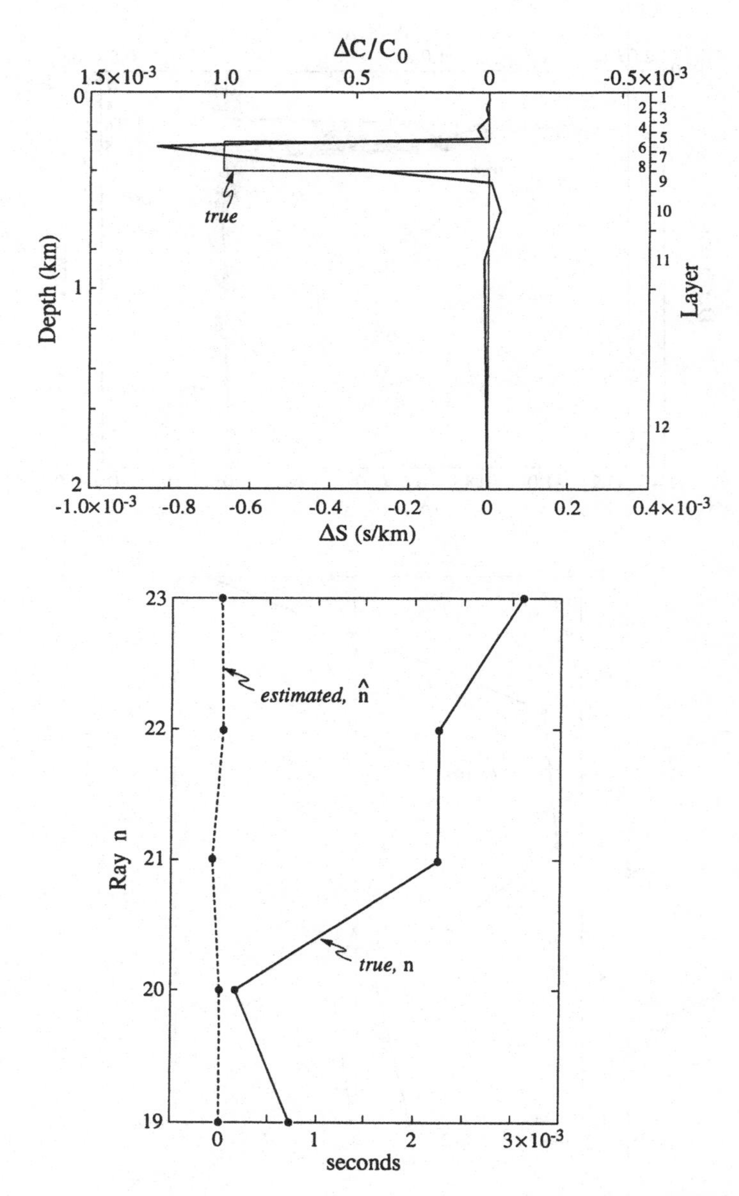

Fig. 6.11. As in fig. 6.10, but **S** chosen for the real ocean rather than the previous **S** appropriate to the known forward problem. But even with the "incorrect" **S** and the noisy travel times, the Gauss-Markov estimator does a fair job of obtaining the correct answer.

representing both simultaneously can again be written

$$\mathbf{y} = \mathbf{E}\mathbf{x} + \mathbf{n}\,, \tag{6.5.14}$$

$$\mathbf{y}^T = [\mathbf{y}_S^T \quad \mathbf{y}_u^T], \quad \mathbf{E} = \begin{Bmatrix} \mathbf{E}_S & \mathbf{0} \\ \mathbf{0} & \mathbf{E}_u \end{Bmatrix}, \quad \mathbf{x}^T = [\mathbf{x}_S^T \quad \mathbf{x}_u^T], \quad \mathbf{n}^T = [\mathbf{n}_S^T \quad \mathbf{n}_u^T].$$

$\mathbf{0}$ denotes zero matrices of appropriate dimension. Unless some information is provided linking the elements $\mathbf{x}_S$, $\mathbf{x}_u$, an inversion of (6.5.14) corresponds to inverting for $\mathbf{x}_S$, $\mathbf{x}_u$ separately and is computationally less efficient than separate inversions, because of the block-diagonal nature of $\mathbf{E}$. To the extent that one anticipates relationships between the sound-speed and velocity fields, such information can be exploited to provide improved (better resolved and/or less uncertain) solutions. The perturbation fields are expected to be in geostrophic balance, and there exist dynamic relationships between sound-speed (as a surrogate for density) and the velocity field. This problem was discussed by Munk and Wunsch (1982*b*), who wrote

$$\mathbf{x}_u = \mathbf{A}_1 \mathbf{x}_S, \tag{6.5.15a}$$

where $\mathbf{A}_1$ is a matrix of constants, asserting that the numerical vertical derivative of horizontal velocity is proportional to the horizontal derivative of density in the plane normal to the velocity component. An alternative form for (6.5.15*a*) is

$$\mathbf{A}\mathbf{x} = \mathbf{0}, \quad \mathbf{A} = \{-\mathbf{A}_1 \ \ \mathbf{I}\}, \tag{6.5.15b}$$

or, more generally,

$$\mathbf{A}\mathbf{x} = \mathbf{d}_m\,. \tag{6.5.15c}$$

The first issue is to decide whether or not these relationships are intended to be exact. Under the not entirely obvious assumption that they are, one method is to append (6.5.15*c*) to (6.2.4) and row-weight the result with very large values relative to $\mathbf{E}$, in effect stipulating that there is an arbitrarily small error. This procedure is called a "barrier method." The only limits on the weights come from numerical inaccuracies, which will be encountered when the system singular values differ by many orders of magnitude. But in practice, one can normally force (6.5.15) to apply with sufficient accuracy.

Another, perhaps more illuminating, approach is to impose (6.5.15*c*) on the objective function using a Lagrange multiplier,

$$J = (\mathbf{y} - \mathbf{E}\mathbf{x})^T \mathbf{R}^{-1} (\mathbf{y} - \mathbf{E}\mathbf{x}) - 2\boldsymbol{\mu}^T (\mathbf{A}\mathbf{x} - \mathbf{d}_m), \tag{6.5.16}$$

recognizing that the result is equivalent to the Gauss-Markov estimate constrained by (6.5.15), assuming that $\mathbf{R}$ describes the second moments of $\mathbf{n}$ and letting the norm of $\mathbf{S}$ become infinite.[10] The solution to the normal equations follows in straightforward fashion and can be written in several equivalent forms, one of which is (Seber, 1977)

$$\hat{\mathbf{x}} = \hat{\mathbf{x}}(-) + (\mathbf{E}^T\mathbf{E})^{-1}\mathbf{A}^T[\mathbf{A}(\mathbf{E}^T\mathbf{E})^{-1}\mathbf{A}^T]^{-1}(\mathbf{d}_m - \mathbf{A}\hat{\mathbf{x}}(-)),$$
$$\mathbf{P} = \mathbf{P}(-) - \sigma_n^2(\mathbf{E}^T\mathbf{E})^{-1}\mathbf{A}^T[\mathbf{A}(\mathbf{E}^T\mathbf{E})^{-1}\mathbf{A}^T]^{-1}\mathbf{A}(\mathbf{E}^T\mathbf{E})^{-1}, \qquad (6.5.17a, b)$$

where $\hat{\mathbf{x}}(-)$ is the least-squares/SVD solution without the use of (6.5.15*c*), and $\mathbf{P}(-)$ is its uncertainty. The expressions (6.5.17) show that the effect of the perfect constraints is to reduce the uncertainty of $\hat{\mathbf{x}}(-)$. If (6.5.17) is written in terms of the SVDs of $\mathbf{E}$, $\mathbf{A}$, it will be apparent that some structures in $\mathbf{x}$ corresponding to the range vectors of $\mathbf{E}$ are being replaced by the new (perfect) information made available by (6.5.15*c*).

The presence of products involving $\mathbf{A}^{-1}$ and real observations renders (6.5.17*a*) sensitive to small singular values in $\mathbf{A}$, and one must consider seriously whether or not relations such as (6.5.15) are meant to be exactly satisfied. The geostrophic relationship, for example, although a very good one, is not perfect – small deviations being essential to ocean physics (*e.g.*, Pedlosky, 1987). If noise element $\mathbf{n}_A$ is introduced, (6.5.15) is

$$\mathbf{A}\mathbf{x} + \mathbf{n}_A = \mathbf{d}_m, \quad \langle \mathbf{n}_A \rangle = 0, \quad \langle \mathbf{n}_A\mathbf{n}_A^T \rangle = \mathbf{Q}. \qquad (6.5.18)$$

Any weights used to introduce (6.5.18) into objective functions, or Gauss-Markov estimates, can reflect this "model noise" variance. The Lagrange-multiplier method is also readily adapted to the presence of $\mathbf{n}_A$ by introducing it into the objective function as

$$J = (\mathbf{y} - \mathbf{E}\mathbf{x})^T\mathbf{R}^{-1}(\mathbf{y} - \mathbf{E}\mathbf{x}) + \mathbf{n}_A^T\mathbf{Q}^{-1}\mathbf{n}_A - 2\boldsymbol{\mu}^T(\mathbf{A}\mathbf{x} + \mathbf{n}_A - \mathbf{d}). \qquad (6.5.19)$$

There is no limit to the use of available information concerning the model constraints that can be imposed on $\mathbf{x}$, $\mathbf{n}$. Schröter and Wunsch (1986) require that (simulated) inversions be consistent with a steady, nonlinear, baroclinic general circulation model (GCM). They use a form of "mathematical programming," with constraints written as inequalities. Malanotte-Rizzoli and Holland (1986) impose tomographic-like constraints, but with no observational error,

[10] A term in $\mathbf{x}^T\mathbf{S}^{-1}\mathbf{x}$ is readily included, but the solution is somewhat more complicated in appearance.

onto a steady GCM. With such sophisticated models, the number of constraints embodied in the dynamics through **A**, or equivalent, may greatly exceed those available directly through the data in the form of **E**.

The central point of these last remarks is that complex GCMs represent a large reservoir of prior knowledge. To the extent that one seeks a map of the density or velocity field, the combination of model constraints with direct observations represents a synthesis of existing knowledge along with the new information available from tomography. The density map produced from the combination of model and data will be more accurate, perhaps far more so, than that produced by either alone. In this context, what we have been calling "inversion" becomes what is widely known in many fields as the "state-estimation problem."[11] The properly constrained model can then be used to estimate almost any field of interest – potential vorticity or heat fluxes, wavenumber spectra, and so forth. This model-oriented tomography is the focus of chapter 7, where we shall generalize our discussion to include time-evolving models.

A dynamical example. Model physics can be imposed as constraints on the estimates explicitly through equations such as (6.5.15), or implicitly through judicious choice of the representation basis. As an example of the latter, consider the loop resonance problem described in section 4.2, where there is only a single, meridionally oriented source-receiver pair, and five rays are identified. Assume that the data have been accumulated over a sufficient duration that each identified ray travel time has been Fourier-analyzed, or filtered, so that we can treat all waves present as having a frequency of $\frac{1}{3}$ cycle/year. The dispersion relation requires all the wavenumbers to lie on the circle of fig. 4.3. An adequate representation basis is provided by the sines and cosines of (4.2.15), with $k = 0, \pm 1, \pm 2, \ldots, \pm 47$, where $\kappa_y = (2\pi/1000)k$ is the northward component of wavenumber in radians per kilometer. The dispersion relation associates each positive and negative κ_y to two values of κ_x (*i.e.*, there are four values of κ_x corresponding to any value of $|\kappa_y|$, with the exception of $\kappa_y = 0$, for which only two values of κ_x are possible). There are thus 190 waves present. The elements of **x** are the coefficients of each of the corresponding wavenumber components, implying 190 unknown amplitudes. Each wave has an in-phase and an out-of-phase component, for 380 total unknown coefficients (or 190 amplitudes and 190 phases). If measurements are available at only one time, there is no direct information about the out-of-phase components, and they can sensibly be omitted.

[11] A meteorologist would call it the problem of "assimilation." We prefer "state estimation" because of its wider context in engineering and mathematics.

The analysis of chapter 4 shows that the five rays are sensitive only to a restricted subset of the waves present: the mean in the y direction (two waves with $\kappa_y = 0$); four waves at each $|\kappa_y|$ corresponding to the loop periodicity of each of the five rays; and four waves corresponding to $|\kappa_y|$ for each of the first harmonics of the ray-loop periodicity. Higher harmonics of the rays correspond to $|\kappa_y|$ that are too large to satisfy the dispersion relation at $\frac{1}{3}$ cycle/year and can thus be excluded at the outset.

An investigator faced with this problem might despair: there are five pieces of noisy information, but 380 unknown amplitudes and five noise unknowns. There is, nevertheless, useful information in these measurements; the question is how best to exploit it.

Consider the fully agnostic case, using the SVD to analyze the structure of $\mathbf{E}$, with no externally prescribed prior covariances. There are five singular values, all of approximately the same size, and thus $K = 5$ is a plausible choice. The diagonal values of $\mathbf{T}_V$ are displayed in fig. 6.12a. Most of these values are zero – corresponding, as anticipated from the physics, to wavenumbers orthogonal to the ray structures. The first row (or column) of $\mathbf{T}_V$ is shown in fig. 6.12b. The two large peaks correspond to the waves with $\kappa_y = 0$.

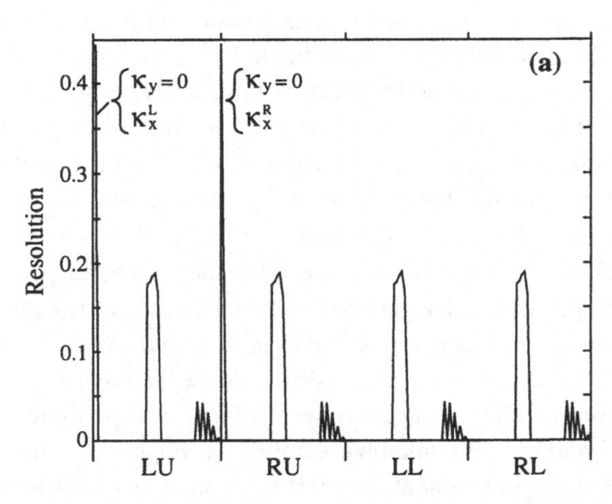

Fig. 6.12a. Diagonal element of the resolution matrix for the field of Rossby waves, given only a single meridional set of five ray travel times. No prior covariances have been imposed. The resolution is identically zero for all wavenumbers not contributing to the travel-time perturbations and is largest for the two waves with $\kappa_y = 0$, because all five rays are sensitive to perturbations in the mean. κ_x^L and κ_x^R denote the short- and long-wavelength waves in the left- and right-hand sides of the dispersion circle with $\kappa_y = 0$. Quadrants refer to the dispersion circle.

There is no information from the single meridional slice to permit separating the contributions from the two waves propagating due west, whose amplitudes are determinable only in linear combination. As an example of the remaining structure of the resolution matrix, fig. 6.12c shows the 20th row (or column) of $\mathbf{T}_V$, which corresponds to the 19th harmonic of the fundamental wave in the y direction. The ray-loop length corresponds to the scale of this wave. Here there are four peaks, each corresponding to one of the four waves on the circle in fig. 4.3. There is no information from a single meridionally oriented integral that would permit distinction between the contributions of these four waves to the 19th harmonic in the y direction. Without further information, the solution can determine only the sum of these four amplitudes. Using the travel times for $t = 0$ from table 4.2, the particular SVD solution is the one shown in fig. 6.12d, in which the solution is partitioned equally among the four waves available for each value of κ_y (two waves for $\kappa_y = 0$). The SVD analysis confirms the physical insight of chapter 4.

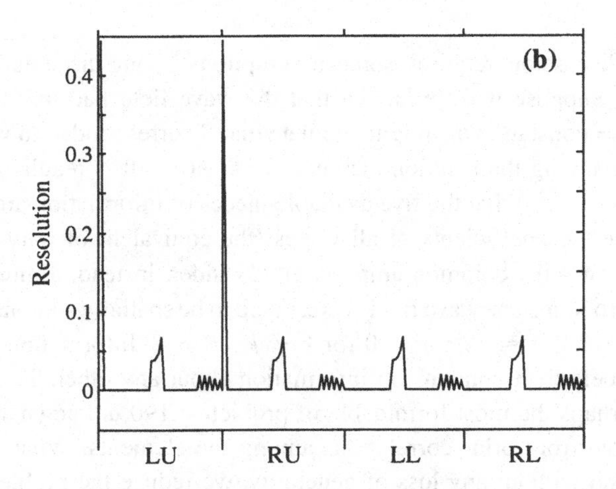

Fig. 6.12b. The first column of $\mathbf{T}_V$ at rank 5. The first element of $\mathbf{x}$ corresponds to the wavenumbers $\kappa_y = 0$, with $\kappa_x = \kappa_L$ having its maximum possible value at the left edge of the dispersion circle in fig. 4.3. The equally large peak at element 49 corresponds to the wave also with $\kappa_y = 0$, but $\kappa_x = \kappa_R$, the smallest possible x wavenumber at the right edge of the dispersion circle. The resolution matrix shows that the amplitudes of these two waves can be determined only in linear combination with each other and will be assigned equal weights in the solution, unless some prior variance information is assigned. Smaller peaks correspond to the loop fundamentals and first harmonics, to which the SVD or least-squares solution would assign smaller values, unless, again, prior variance information were provided.

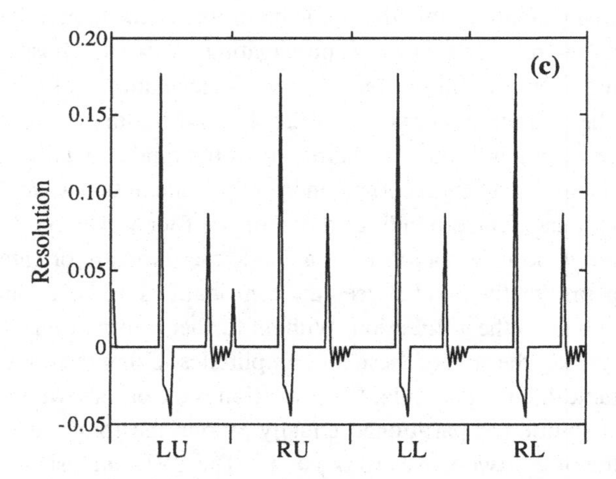

Fig. 6.12c. Twentieth column of $\mathbf{T}_V$ corresponding to upper left wavenumber of fig. 4.3, equal to the 19th harmonic of the integration distance. The four equal peaks correspond to the four waves all having the same value of $|\kappa_y|$. Without assigned prior variances, the simplest solution would uniformly partition the solution amplitude among them, assigning smaller values to the first harmonics of these waves.

Further discussion requires some assumptions about the extent of prior knowledge. Suppose it were known that the wave field had uniform amplitudes, $\ell(k) =$ constant. One might stipulate that $\mathbf{S}$ corresponded to very strong correlations among the solution elements $x_i \equiv \ell(\kappa_i)$. It is readily confirmed (but not shown here) that the five available pieces of information are adequate to determine the coefficients of all waves (the equivalent of only one piece of information – the common amplitude). Consider, instead, a more realistic stochastic problem. The wave field is presumed to be spatially stationary, equivalent to $\langle \ell(k)\ell(k')\rangle \equiv \langle x_i x_j\rangle = 0$ for $k \neq k'$, $i \neq j$. Information about any particular coefficient contains no information about any other. This situation presents perhaps the most formidable of problems: 190 unknown x_i, with no help available from prior correlations among the elements. What might one do? First, and without any loss of generality, we reduce the problem size by exploiting our insight into its structure: we might as well work with

$$x_i = \ell^L(k) + \ell^L(-k) + \ell^R(k) + \ell^R(-k) \quad \text{for} \quad k > 0,$$

$$x_i = \ell^L(k) + \ell^R(k) \quad \text{for} \quad k = 0 .$$

The prior variances S_{ii} are then simply the sums of the variances of the individual components.

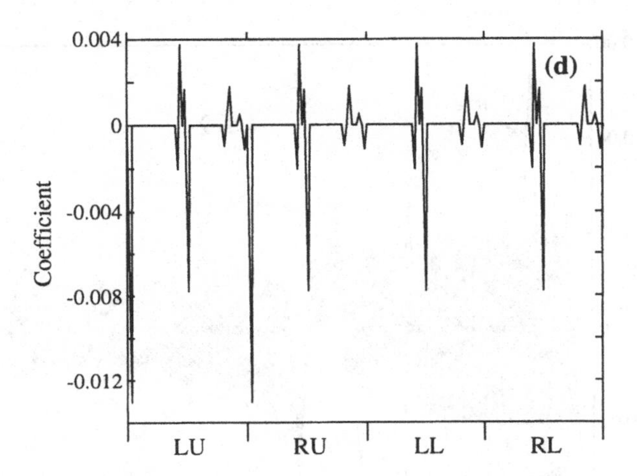

Fig. 6.12d. SVD solution at $K = 5$, without prior variances. The numerical amplitudes precisely reflect the structures visible in $\mathbf{T}_V$.

The resulting Gauss-Markov solution $\widehat{\mathbf{x}}$ and its uncertainty are shown in fig. 6.13. The uncertainty of the solution remains $\pm S_{ii}$ for those wavenumbers not affecting the travel time, and it is reduced from these prior values in the wavenumber bands contributing to the ray travel-time perturbations. The true solution is everywhere within two standard errors of the estimated solution. One could map the two-dimensional field in the restricted wavenumber bands $k = 0$; $k = 19, 20, \ldots, 23$, where loop resonance occurs and where the solution uncertainty is small. But it is somewhat unreasonable to demand a map of the horizontal field from only five pieces of information. For many purposes, one is not interested in such a detailed picture, preferring a statistical statement concerning the ocean variability. Suppose the diagonal values of $\mathbf{S}$ were believed to provide a useful prior estimate of the wavenumber spectrum, and the tomographic integrals were to be used to test the consistency of that prior estimate with actual observations. Observations are available at times $t = 0, 0.5, 1.0, 1.5$ years (table 4.2). As the field evolves according to the physics of (4.2.15), different combinations of the in-phase and out-of-phase components contribute to the travel times. Each measurement is thus nearly an independent realization. The Gauss-Markov estimate, $\widehat{x}_i(t)$, $t = 0, 0.5, 1.0, 1.5$ years, was squared and is displayed in fig. 6.13. The means of $\widehat{x}_i^2(t)$ and the estimated uncertainties of the means are shown.[12] The result passes the test of consistency

[12] For a Gaussian probability density of $\ell(k)$, the probability density for $\ell^2(k)$ is chi-square. To obtain the uncertainty of the mean of $\widehat{x}_i^2(t)$, it is treated as a chi-square variate with four degrees of freedom based on an underlying variance for each of P_{ii} determined from (6.5.9).

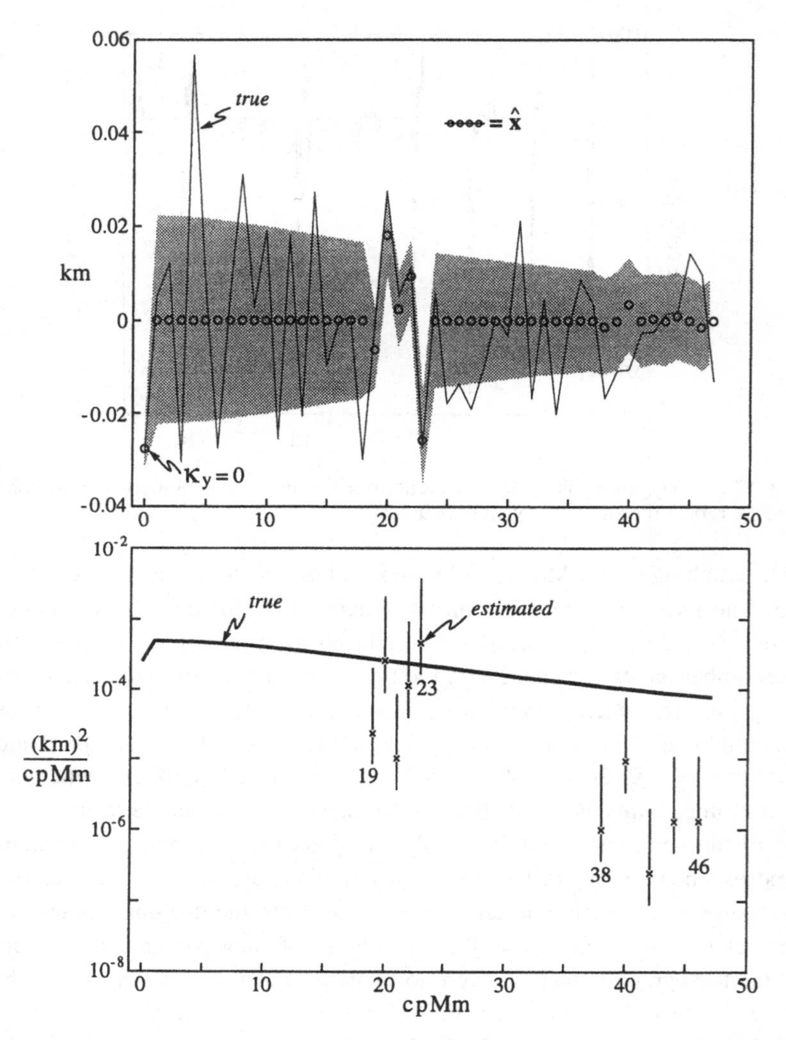

Fig. 6.13. Top: Gauss-Markov estimated solution for the field of Rossby waves as a function of wavenumber (cycles per Mm). The solution is for five noisy ray travel-time perturbations, a purely diagonal version of **S** chosen to reflect the spatial homogeneity of the random field, and the prior variances based on (4.2.13). Here, for convenience, the x_i represent the sums of the amplitudes of the four (two for $\kappa_y = 0$) waves that are otherwise indistinguishable. Gray area is the formal one standard deviation of the uncertainty, and the actual values of the random ℓ [defined in (4.2.16)] used to generate the travel times are shown. These are the "truth." Bottom: The *true* spectrum of x_i, *i.e.*, the diagonal of **S**, and the estimate of it, $\widehat{S}_{ii}$, were constructed using the assumption that each average of $\widehat{x}_i^2$ was distributed in χ^2 with four degrees of freedom and a variance of $\widehat{S}_{ii}$.

with the prior spectrum in the wavenumber bands in which observations are sensitive to the ocean, and one is entitled to conclude that the prior spectral estimate is consistent with the measurements and need not be modified. A more interesting example could have been constructed by choosing a prior **S** not consistent with the observations. This example will be encountered again in chapter 7, where we shall more fully exploit the information available in the model statement (4.2.15) about the expected time evolution in the observations.

6.6. Variant Linear Methods

There are many possible variations on the methods we have described. Each has advantages or conveniences, but many are simply ways of transforming one representation of the perturbations into another. For example, suppose we have a perturbation representation in layers, but choose to expand

$$\hat{\mathbf{x}} = \sum_{q=1}^{Q} \gamma_q \mathbf{f}_q, \tag{6.6.1}$$

where the vectors $\mathbf{f}_q$ are analytically prescribed functions, perhaps even chosen to be the dynamic mode representation. These methods are usually known as "universal kriging" (*e.g.*, Ripley, 1981; Davis, 1985). Cornuelle and Malanotte-Rizzoli (1986) represented the Gulf Stream front in terms of functions $\mathbf{f}_q$ carrying spatial discontinuities.[13] Such representations permit simple transformations between the representations; they also permit variations in the relative sizes of M, N, depending on how many functions are used to represent **x**.

In the least-squares context, one can choose the weight matrices **S** to achieve various goals. For example, the objective function

$$J = \mathbf{x}^T \mathbf{Z}^T \mathbf{Z} \mathbf{x} + (\mathbf{y} - \mathbf{E}\mathbf{x})^T \mathbf{R}^{-1} (\mathbf{y} - \mathbf{E}\mathbf{x}), \tag{6.6.2}$$

with

$$\mathbf{Z} = \begin{Bmatrix} 1 & -1 & 0 & 0 & 0 & \cdots & 0 \\ 0 & 1 & -1 & 0 & 0 & \cdots & 0 \\ 0 & 0 & 1 & -1 & \cdots & & 0 \\ \vdots & \vdots & & & & & \vdots \\ 0 & 0 & & & & & -1 \end{Bmatrix},$$

will produce estimates that are as "smooth" as possible given the observations. **Zx** is a numerical representation of first derivative of **x**; it is easy to write down objective functions that tend to minimize second and higher derivatives. Such "semi-norm" methods are discussed by Wahba (1990) and Bennett

[13] The actual fit was done, however, with a nonlinear procedure.

(1992) and are intimately linked to representations of $\mathbf{x}$ in spline functions. One anticipates, however, that the resulting estimate would be very close to a Gauss-Markov estimate, where the solution variance would be confined largely to small wavenumbers, and the noise variance to large wavenumbers.

The use of quadratic measures of misfit is somewhat arbitrary, but is widespread because of the resulting simple solutions and the close connection with Gaussian statistics (if the fields are Gaussian, then the Gauss-Markov solution is the maximum-likelihood solution). But the Gaussian assumption may fail, particularly for observational noise with frequent large outliers. A more "robust" measure of misfit (*e.g.*, the so-called ℓ_1-norm),

$$J = \sum_{i=1}^{M} \left| y_i - \sum_{j}^{N} E_{ij} x_i \right| + \alpha^2 \sum_{j} | x_j | , \tag{6.6.3}$$

is less sensitive to outliers (*e.g.*, Arthnari and Dodge, 1981).

Nonquadratic objective functions have other uses. Consider again the temperate-profile perturbation problem with 12 layers and five rays; as we have seen, the lack of resolution produces either a solution of great uncertainty or one strongly dependent on prior information. In such situations, it may again be prudent to make less demanding requirements than a detailed map (*e.g.*, upper and lower bounds on ocean properties of interest) (Parker, 1972). Such issues were discussed in an oceanographic, but nontomographic, context by Wunsch and Minster (1982) and Wunsch (1984). Suppose we are interested in the total heat stored in the water column, rather than the detailed vertical structure. We write an objective function

$$J = \sum_{j} a_j x_j , \tag{6.6.4}$$

where a_j are the numerical values required to calculate the heat content from ΔS_j (the products of layer thickness times the heat capacity of sea water times the conversion from sound-slowness to temperature). Subject to the data and to bounding statements on the maximum permitted perturbation, $|x_j| \leq b_j$, one can seek the maximum and minimum values of J. The ranges are useful bounds on the changing ocean.

Objective functions of the type (6.6.3) and (6.6.4) are most readily handled as problems in "linear programming." In this context, statements about the noise magnitudes in the observations are usually stated as "hard" inequalities:

$$\mathbf{b}^{-} \leq \mathbf{Ex} - \mathbf{y} \leq \mathbf{b}^{+} , \tag{6.6.5}$$

along with general hard bounding inequalities on $\mathbf{x}$:

$$\mathbf{x}^{-} \leq \mathbf{x} \leq \mathbf{x}^{+} . \tag{6.6.6}$$

The objective function is a general linear function of the **x**,

$$J = \mathbf{c}^T \mathbf{x}, \tag{6.6.7}$$

which can be either maximized or minimized. Transformations to put various objective functions and assertions about unknowns into such "canonical form" were discussed, for example, by Wagner (1969). Problems of this type are usually solved by the so-called simplex method (*e.g.*, Luenberger, 1984). For very large systems, the more recent Karmarkar algorithm (*e.g.*, Strang, 1986) may be employed.

By way of example, we use the data of the 12-layer, five-ray temperate example, corrupted by a 10-ms noise and subject to the general requirement $|\Delta S_j| \leq 10^{-3}$ s/km, and take the simple objective function

$$J = \Delta S_6 + \Delta S_7 + \Delta S_8$$

for total perturbation in layers 6–8. The simplex method yields $J_{\min} = -3 \times 10^{-3}$ s/km, $J_{\max} = +1.2 \times 10^{-3}$ s/km. The corresponding solutions are shown in fig. 6.14. To the extent that undesirable features emerge, they can

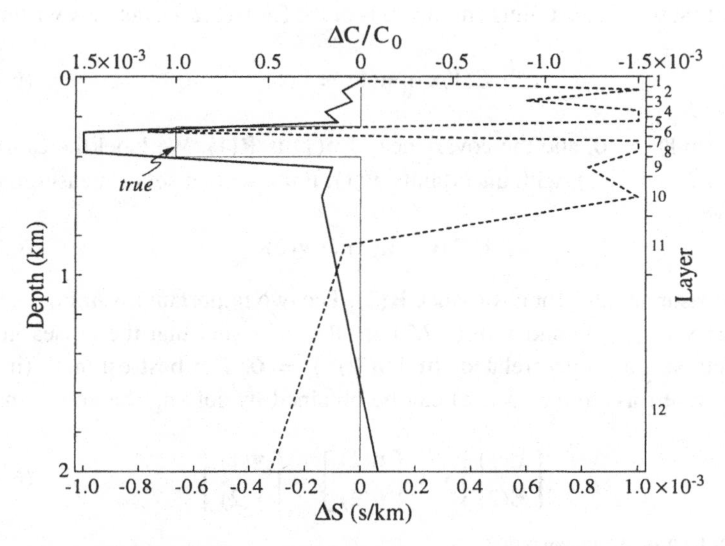

Fig. 6.14. Perturbations to the layered temperate profile that minimize (solid) and maximize (dashed) the sum of the perturbations in layers 6–8. Uniform bounds were placed on the perturbations in each layer, and the noise-corrupted ray travel times were given upper and lower bounds of 10 ms around the observations.

be controlled or removed by adding further constraints, (6.6.5) or (6.6.6), until the system becomes contradictory. It is characteristic of linear programming that the solution drives some of the bounding constraints hard up against their limits. A full analysis of the solution usually involves a study of the sensitivity of the objective function to perturbations in the hard-limited constraints. This sensitivity is directly provided by the "dual" solution (*e.g.*, Luenberger, 1984), which is intimately related to Lagrange multipliers. We shall not pursue this subject further, both because there are numerous good textbooks and papers describing linear programming methods and because there have not been any specific practical applications in the tomographic literature. The flexibility and power of these methods make them strong candidates for future use.

6.7. Recursive Solutions

If the data to make an estimate $\mathbf{x}$ are not acquired simultaneously, it becomes necessary to incorporate newly arrived information into a prior estimate of ΔS_j or u. Alternatively, the data load may grow to the point that all of the information available cannot be handled computationally in one pass, leading to a need for some form of recursive method. But most important, as will be seen in chapter 7, the employment of tomography with time-evolving dynamic models can lead to recursive solutions.

The first set of constraints employed is of the form (6.2.4), but now written as

$$\mathbf{E}(1)\mathbf{x} + \mathbf{n}(1) = \mathbf{y}(1), \tag{6.7.1}$$

where $\langle \mathbf{n}(1) \rangle = 0$, and the covariance of $\mathbf{n}(1)$ is $\mathbf{R}(1)$. We label the estimate from (6.7.1) as $\widehat{\mathbf{x}}(1)$, with uncertainty $\mathbf{P}(1)$. For a second set of measurements we have

$$\mathbf{E}(2)\mathbf{x} + \mathbf{n}(2) = \mathbf{y}(2), \tag{6.7.2}$$

with measurement-error covariance $\mathbf{R}(2)$. The two important assumptions here are that $\mathbf{x}$ in (6.7.1) and $\mathbf{x}$ in (6.7.2) are the same and that the noises in the two data sets are uncorrelated, $\langle \mathbf{n}(1)\mathbf{n}(2)^T \rangle = \mathbf{0}$. The best estimate (in the minimum-error-variance sense) can be obtained by solving the concatenated system

$$\left\{ \begin{matrix} \mathbf{E}(1) \\ \mathbf{E}(2) \end{matrix} \right\} \mathbf{x} + \begin{bmatrix} \mathbf{n}(1) \\ \mathbf{n}(2) \end{bmatrix} = \begin{bmatrix} \mathbf{y}(1) \\ \mathbf{y}(2) \end{bmatrix} \tag{6.7.3}$$

subject to a noise covariance,

$$\mathbf{R} = \left\{ \begin{matrix} \mathbf{R}(1) & \mathbf{0} \\ \mathbf{0} & \mathbf{R}(2) \end{matrix} \right\}. \tag{6.7.4}$$

But having labored to produce the previous estimate $\widehat{\mathbf{x}}(1)$ and its uncertainty, one can ask if it is possible to produce a best estimate from the combined system without having to "start from the beginning" by solving (6.7.3). The original solution, as well as the combined solution, can be regarded as one of ordinary weighted least-squares. By partitioning (6.7.3), and doing some elementary matrix manipulation (*e.g.*, Brogan, 1985), one can write the solution to the combined system as

$$\widehat{\mathbf{x}}(2) = \widehat{\mathbf{x}}(1) + \mathbf{K}(2)(\mathbf{y}(2) - \mathbf{E}(2)\widehat{\mathbf{x}}(1)), \tag{6.7.5}$$

$$\mathbf{K}(2) = \mathbf{P}(1)\mathbf{E}^T(2)\{\mathbf{E}(2)\mathbf{P}(1)\mathbf{E}^T(2) + \mathbf{R}(2)\}^{-1}, \tag{6.7.6}$$

$$\mathbf{P}(2) = \mathbf{P}(1) - \mathbf{K}(2)\mathbf{E}(2)\mathbf{P}(1). \tag{6.7.7}$$

The new estimate, $\widehat{\mathbf{x}}(2)$, is simply a weighted average of the old estimate and the *difference* between the new measurements and the value predicted for the new measurements by the old estimate; the weighting is inversely proportional to the errors in the old estimate and the errors of the new measurements. Equation (6.7.5) is also the minimum-error-variance recursive estimate. Notice that in (6.7.5)–(6.7.7), the origin of the $\widehat{\mathbf{x}}(1)$, $\mathbf{P}(1)$ is irrelevant – the original equations have wholly disappeared from the system. A mild generalization of (6.7.5)–(6.7.7) leads to the Kalman filter and is a central element of chapter 7.

6.8. Nonlinear Problems and Methods

The relationship (2.4.3) or (3.1.20) between travel time and the sound-slowness or water-velocity profile is nonlinear. Chapters 2–4 were devoted to obtaining useful linearizations. The inverse *problem* was then also linear, and the discussion of its solution has been in terms of linear inverse *methods*. As always, one must be alert to the need for testing the approximations, and to the possibility of failure. There are some circumstances, however, in which a nonlinear method is useful even when the inverse problem is linear. We must clearly distinguish such nonlinear inverse methods from the nonlinear inverse problem.

Nonlinear methods applied to linear problems. As an example, let

$$\Delta S(z, \mathbf{r}, t) = \sum_{n=1}^{N} \alpha_n F_n(z) \cos(\boldsymbol{\kappa}_n \cdot \mathbf{r} - \omega_n t - \phi_n), \tag{6.8.1}$$

and calculate the resulting travel-time perturbations from the linear expression (2.8.4),

$$\Delta\tau_i = L_i(\Delta S), \tag{6.8.2}$$

where L_i is a linear operator. The objective function is

$$J = \sum_i (y_i - L_i(\mathbf{x}))^2 . \tag{6.8.3a}$$

For fixed κ_n, ω_n, only the α_n, ϕ_n are variable, and the requirement of a minimum J results in the familiar set of linear normal equations. But for variable κ_n, ω_n, the conditions $\partial J/\partial \kappa_n = 0$, $\partial J/\partial \omega_n = 0$ result in a set of nonlinear normal equations.

Thus the linear inverse problem is being converted to one of nonlinear optimization. The change is analogous to a Fourier analysis of a time series where the best-fitting frequencies are to be determined, rather than being preassigned. The potential advantage of such a fit is that it can produce the most efficient representation of the observations. One can learn a great deal about the ocean structure if only small numbers of frequencies and wavenumbers, not necessarily harmonics, describe the data.

The determination of the minimum of (6.8.3*a*) falls into the general category of nonlinear regression (*e.g.*, Seber and Wild, 1989) and unconstrained optimization (*e.g.*, Gill *et al.*, 1981; Luenberger, 1984; Scales, 1985). In a geophysical context, it was called the problem of "total inversion" by Tarantola and Valette (1982). Much is known about solving such problems, usually by iterative search in which one seeks to go "downhill" from a starting guess, using procedures such as quasi-Newton, steepest-descents, and conjugate-gradient methods, as well as many others. In nonlinear problems, one usually must specify a starting position, $\widehat{\mathbf{x}}_0$. To the extent that $\widehat{\mathbf{x}}_0$ is regarded as a plausible solution, one may wish to introduce a penalty term into (6.8.3*a*):

$$J = \sum_i (y_i - L_i(\mathbf{x}))^2 + \alpha^2 \sum_i (x_i - \widehat{x}_{0i})^2 . \tag{6.8.3b}$$

As always, covariance or other weight matrices can be introduced into the objective function. For very large problems with complex objective functions, attention has recently turned to determining the minimum through combinatorial (Monte Carlo) methods; two such methods are "simulated annealing" (Kirkpatrick *et al.*, 1983) and "genetic algorithms" (Koza, 1992).

The potential advantages of the nonlinear approach are efficiency of representation and the ability to extend the algorithms to situations in which the data are nonlinear functions of $\mathbf{x}$. The potential disadvantages are twofold – first, the powerful analysis tools of solution uncertainty and resolution are all based on the linear relations between solution and data, such as (6.8.2), and second, the algorithms tend to be computationally intensive, and one may have difficulty in determining if a global, rather than a local, minimum has been found. As computer power relentlessly increases, many problems that have been computationally intractable are being reduced to modest proportions (but it will always prove possible to formulate optimization problems that can defeat the largest available computer). The problem of the detection of local rather than global minima is one of the reasons the combinatorial methods were invented; in general, they seem to work well.

The solution analysis for nonlinear optima is normally undertaken by linearizing the objective function about the apparent optimal solution, $\widehat{\mathbf{x}}_*$ (*e.g.*, Tarantola and Valette, 1982; Seber and Wild, 1989). In many such problems the objective function is locally an N-dimensional paraboloid and can be expanded as

$$J = \text{constant} + (\mathbf{x} - \widehat{\mathbf{x}}_*)^T \mathbf{H} (\mathbf{x} - \widehat{\mathbf{x}}_*) + \cdots . \tag{6.8.4}$$

The uncertainty and resolution analyses are then based on $\mathbf{H}^{-1}$ (the inverse "Hessian"), which we recognize from (6.3.1) as a local definition of $(\mathbf{E}^T\mathbf{E})^{-1}$.

In the ray travel-time approach, the largest potential linearization error is expected to arise from the frozen-ray approximation. In this and related contexts, such errors lie in the coefficient matrix $\mathbf{E}$. The problem is of the form

$$(\mathbf{E} + \Delta\mathbf{E})\mathbf{x} + \mathbf{n} = \mathbf{y} . \tag{6.8.5}$$

An estimate of $\Delta\mathbf{E}$ is needed, along with $\mathbf{x}$, $\mathbf{n}$ (the methods used thus far set $\Delta\mathbf{E} = \mathbf{0}$ except insofar as it is partially absorbed into $\mathbf{n}$). Such problems are often lumped under the title "total least-squares" (TLS). The presence of errors in $\mathbf{E}$ produces biases in the linear solution $\widehat{\mathbf{x}}$ that can be significant. Van Huffel and Vandewalle (1991) provide a general theory and discuss estimation of $\Delta\mathbf{E}$ through use of statistical methods. Their approach is interesting and enlightening, but the investigator's prior knowledge of statistical structures in the noise elements of $\mathbf{E}$ is difficult to impose.

In the tomographic case, we have more specific information available (the exact ray trace or modal structure, *etc.*) that can be made the basis of a simple iterative scheme. Taking the ray-trace problem again as an example, the linear solution $[\Delta\widehat{S}(z_i)] = \widehat{\mathbf{x}}$ can be used to construct a new profile $\widehat{S}(z) = \widehat{S}(z, -) + \Delta\widehat{S}$; a new ray trace or modal amplitude is then computed, leading to new estimated travel times, which are then compared with the observations. If the difference between predicted and observed times is larger than the error estimates, a new linearized inversion is done, and so forth. If the system converges, one at least has a consistent solution. Such numerical iterations have been carried out on both simulated data (*e.g.*, Spofford and Stokes, 1984) and real data (*e.g.*, Cornuelle *et al.*, 1993). In general terms, the corrections to the linearized tomographic inversions have been quite small. But in specific extreme instances (*e.g.*, short-range tomography across Gulf Stream rings) the correction becomes large (*e.g.*, Mercer and Booker, 1983).

Iteration is the most straightforward approach to the nonreciprocity problem in reciprocal tomography: after a first inversion, one simply retraces the rays in both directions using the newly estimated slowness and water velocities. The times will differ because of both water velocities and the resulting nonreciprocal trajectories. If the result is in agreement with the measurements, one stops. If not, the discrepancies are the basis for another inversion. In practice, the ability to calculate an accurate second-order quantity – the path perturbation – is dependent on the accuracy of the first-order model, including vertical model structure sufficient for an accurate path to be computed (*e.g.*, Cornuelle *et al.*, 1993). As with any method of successive linearization, some attention must also be paid to assuring that the linearization assumptions (including any statistical statement about the perturbations) remain valid if solutions move far from the starting position.

Nonlinear methods and nonlinear problems. A straightforward generalization of the objective-function approach permits one to address fully nonlinear inverse problems. A general, nonlinear relationship between the travel time and the sound-speed and water-velocity fields can be written as

$$\boldsymbol{\tau} = \mathbf{L}(S(\mathbf{x}),\ u(\mathbf{x})) + \mathbf{q} \qquad (6.8.6a),$$

where it is understood, in the spirit of our continuing discrete approach, that S, u are described and controlled by the vector of parameters, $\mathbf{x}$. We have written $\mathbf{q}$ to represent any lingering errors in the representation of $\boldsymbol{\tau}$. (It is desirable that $\langle \mathbf{q} \rangle = \mathbf{0}$, although that may not always be possible.) Such a relationship translates into a measurement set of the form

$$\mathbf{y} = \mathbf{E}(\mathbf{x}) + \mathbf{n} \qquad (6.8.6b),$$

where $\mathbf{E}$ is a *function* of $\mathbf{x}$, and $\mathbf{n}$ includes both the observation noise and the representation noise (or model noise). Consider an objective function

$$J = (\mathbf{y} - \mathbf{E}(\mathbf{x}))^T \mathbf{R}^{-1} (\mathbf{y} - \mathbf{E}(\mathbf{x})) + (\mathbf{x} - \widehat{\mathbf{x}}_0)^T \mathbf{P}_0^{-1} (\mathbf{x} - \widehat{\mathbf{x}}_0) . \qquad (6.8.7)$$

$\widehat{\mathbf{x}}_0$ represents a prior estimate (it could be zero). Possible choices for the weights $\mathbf{R}$, $\mathbf{P}_0$ are the noise variance and the uncertainty of $\widehat{\mathbf{x}}_0$, respectively. Such objective functions were considered by Tarantola and Valette (1982), who suggested a scheme for finding the minimum of J. But as with the objective function

(6.8.3), we prefer to regard the problem of minimizing J as part of the general class of nonlinear optimization and regression problems, so as to exploit the large body of expertise (and software) available. The usual uncertainty estimates are based on (6.8.4).

The general subject of mathematical programming permits the combination of inequality constraints, linear or otherwise, Lagrange-multiplier constraints, and so forth, with nonlinear objective functions. For example, the model, instead of being imposed in the mean square as in (6.8.7), could be imposed through hard inequalities, as in (6.6.5), or with a vector of Lagrange multipliers. The available mathematical machinery permits the treatment, in principle, of virtually any type of prior information, or of relationships in the solution. The main issues are computational (references previously cited provide reasonable entrée into this very large and interesting subject).

Analytic methods (Abel-Radon transforms). Two exact integral transforms are intimately connected to tomography. One is the Abel-transform relationship (2.5.9) between the refracted-ray travel times and the sound-speed profile, $C(z)$, in a range-independent sound channel. The other is the Radon-transform relationship between travel times along straight lines through a horizontal x, y-slice and the field $C(x, y)$. The Abel transform was discussed by Munk and Wunsch (1983) and extended by Jones *et al.* (1986*a*, 1993) and Jones and Georges (1994). In the seismic literature, a related problem is named for Herglotz and Weichert (Aki and Richards, 1980). Similar transform pairs exist for the action (Garmany, 1979). The Abel transform, in principle, permits a full finite-amplitude inversion, requiring no linearization in the range-independent case. Although it is a powerful and interesting theoretical tool that one might have thought would play a central role in tomography, it is unlikely to be of much use with real data. Linear theory permits the use of important prior knowledge (it is illogical to use hard-won tomographic information to reconstruct a profile dominated by known climatology), has recursive extensions, and contributes to an understanding of the solution through the complete resolution and variance estimates. Methods like the SVD also produce detailed understandings of which data determine which solution elements. None of these analyses is readily available from the transform pair (2.5.9). At the present time, the linearization, perhaps supplemented with iteration, seems to be a small price to pay for fuller understanding of the solution.

Moving ships and CAT scans. In horizontal slice tomography, one might confine an inversion discussion to the ray system depicted in fig. 6.15 (*i.e.*, the

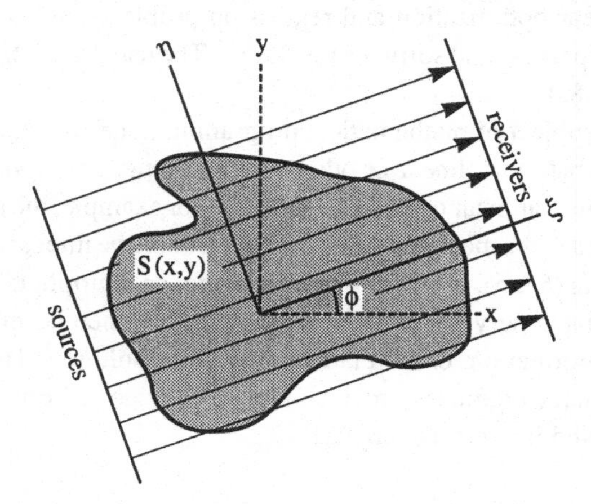

Fig. 6.15. Theoretical situation in which travel time is computed by integrating through the slice along parallel lines η = constant at angles ϕ, yielding $\tau(\eta, \phi)$. The Radon transform yields $S(x, y)$ from a perfectly known $\tau(\eta, \phi)$ for $0 \leq \phi \leq \pi$. In medical tomography, $S(x, y)$ is an absorption parameter, and $\tau(\eta, \phi)$ is an intensity. A collimated beam integrates through a slice of the skull at dense intervals in η. Source and receiver rotate about the patient, densely sampling ϕ for $0 \leq \phi \leq \pi$.

rays are straight lines between source and receiver). The source is a collimated beam, transmitting straight parallel rays through a slice $S(x, y)$ onto a receiving screen. Source and receiver are rotated through angle ϕ for many visualizations. It was the analogy between this situation and medical computer tomography, in which X-rays are passed through a patient by a surrounding "computer-aided tomography" or "computed axial tomography" (CAT) scanner, that led Munk and Wunsch (1979) to label the oceanic problem "ocean acoustic tomography." But the oceanic problem differs from the medical one in almost all aspects: space and time scales, technology, ray trajectories, and so forth, not to speak of the market demand. Perhaps the most basic difference is in the data density: the medical procedure (fig. 6.15) builds up an extremely dense set of integrals through the patient's cross-section. The high costs of work at sea mean that the oceanographer's array configuration is sparse. For this reason, medical and ocean acoustic tomography differ in that the former makes no use of prior information, whereas the latter is greatly dependent on it. An advantage of the oceanic case, however, is that we have dynamics, in the form of the Navier-Stokes equations, relating one part of the "patient" to another. Such equations are not yet available for the human interior.

In one situation, moving-ship tomography, as shown in fig. 1.6, the data density begins to approach that of the medical case. Consider fig. 6.16, from Cornuelle *et al.* (1989), who describe the combination of a shipborne acoustic receiver working with varying numbers of moored sources. The ship steams around the periphery of the volume shown in fig. 6.16, producing the acoustic paths depicted. Sound-speed perturbations are represented as

$$\Delta C(x, y) = \sum_n \sum_m \alpha_{mn} e^{i(mx+ny)/L} . \tag{6.8.8}$$

A physical realization is shown in the upper right panel of fig. 6.16. To the left is the tomographic mapping for the case of four moorings. The expected errors in both physical and wavenumber spaces are shown. The error is largest at the center of the physical array, amounting to 2.8 m/s for an rms perturbation of 7 m/s. The error variance is 3.8% of the perturbation variance. This error is sharply reduced to 1.1% in going to a five-point array.

Munk and Wunsch (1982*a*) had previously discussed a much cruder version of this system, using both ship-to-ship transmissions and the paths from ship to one or two moored receivers. In either case, but most clearly in the situation considered by Cornuelle *et al.* (1989), the path density is far higher than appears practical with purely moored systems. If the strong assumption is made that the ocean remains fixed during the measurement period, one achieves path-coverage geometries that begin to approach those available in medical X-ray and other types of tomography. A brief review of inversions under those circumstances is worthwhile.

Confining attention to the horizontal slice problem, and using the geometry of fig. 6.15, travel time can be written as in (6.2.1):

$$\tau(\eta, \phi) = \int_{\text{source}}^{\text{receiver}} d\xi \; S(\xi \cos\phi - \eta \sin\phi, \; \xi \sin\phi + \eta \cos\phi) . \tag{6.8.9}$$

That is, the travel time is computed by integrating through the volume along parallel lines at all angles $0 \leq \phi \leq \pi$. (The parallel assumption is unnecessary, and "fan beams" are sometimes used.) Radon (1917) produced an expression for the inverse transform to (6.8.9) yielding $S(x, y)$ (Rowland, 1979). Application of the Radon transform to the medical problem appears to have come initially through its use in radioastronomy by Bracewell (1956). A large literature has grown up subsequently, directed at numerical evaluation of the inverse transform, with much of it involved with the computational load arising in the medical context. Many of these methods are best described in wavenumber space and lead to "back-projection" methods and various approximations to

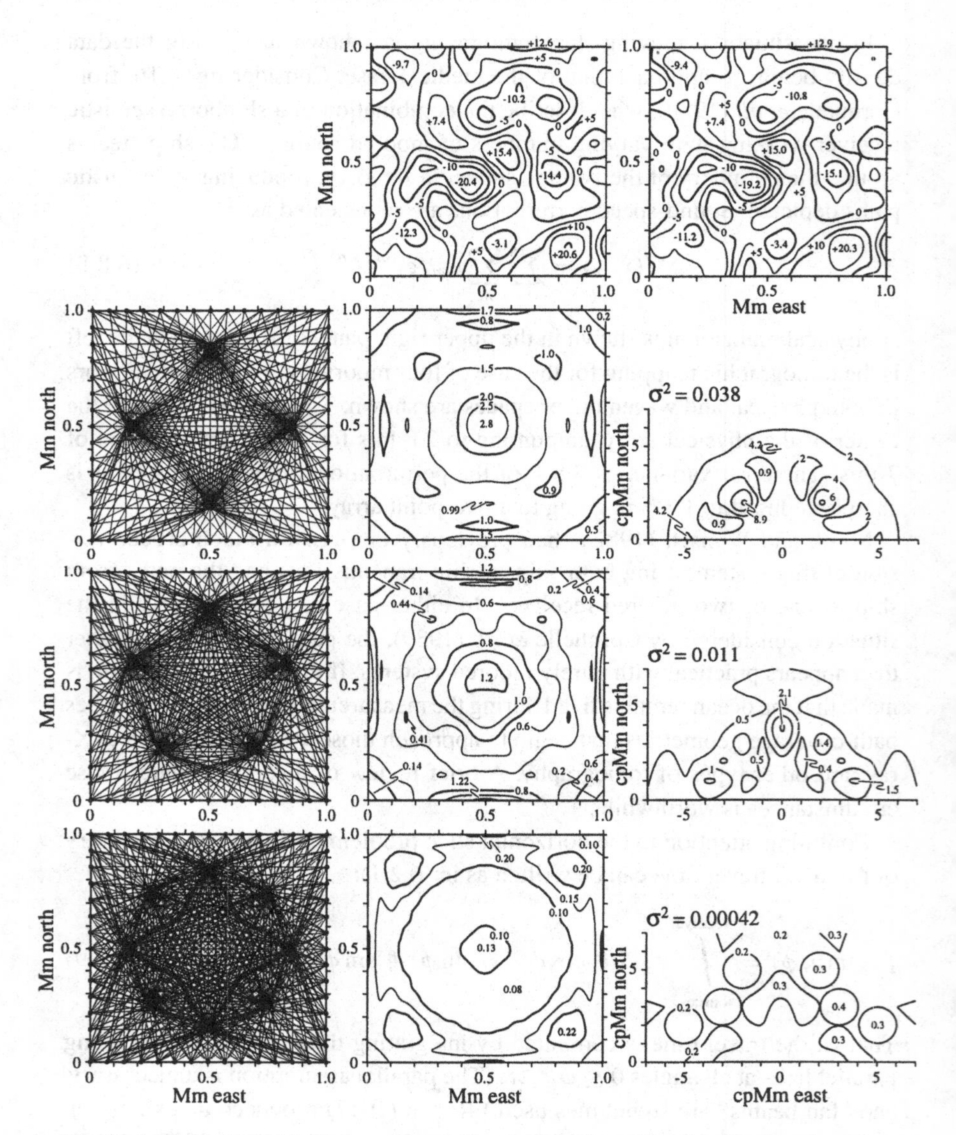

Fig. 6.16. Horizontal paths between moored sources and moving-ship receivers produce denser coverage than can be achieved by either moorings alone or ships alone. The right top panel shows a (frozen) realization of the "true" ΔC (m/s) contour map generated from (6.8.8), and to the left the tomographic reconstruction for the case of four source moorings, as shown below, together with the associated Gauss-Markov estimate of rms error in physical space and error spectrum in wavenumber space. The spectrum is in units of 100 $(\mathrm{m/s})^2$ per $(\mathrm{cycles/Mm})^2$. The error variance diminishes from 3.8% of the true variance for four source moorings to 1.1% for five moorings. (Adapted from Cornuelle *et al.*, 1989.)

rapid solutions of very large sets of simultaneous equations. Owing to the somewhat restricted applications in the oceanic case, we shall not pursue the details here. The interested reader can consult Herman (1979, 1980) for a wide-ranging discussion.

The idealized moving-ship geometry provides a vivid example of sampling and nullspace issues. Fig. 1.5 is constructed for the same perturbation field shown in fig. 6.16, but with transmissions from ship to ship (no source moorings) steaming along the boundaries. There is a vast nullspace (assuming no prior information), serving as an excellent example of a faulty observation strategy. The uppermost panels correspond to two ships steaming meridionally along the eastern and western boundaries. With no information available about covariances between meridional and zonal structures, inversions are based only on zonal integrals; zonal structures are left wholly in the nullspace of **E**. (With reference to fig. 6.15, this result corresponds to a single "snapshot" for $\phi = 0$, as compared with many hundreds of snapshots in medical practice.) The three panels beneath show the analogous situation when ships produce only meridional integrals, followed by a slowly increasing reality as more integrals become available. More complex sampling strategies, involving more ships, or adhering to the frozen-ocean assumption while the same ships maneuver, or employing moored instruments in the interior (as in fig. 6.16), can reduce the nullspace. The question of how best to deploy tomographic instrumentation to map the ocean is an example of the highly nonlinear "experiment design" problem, discussed for tomography by Barth and Wunsch (1989).

6.9. Inversions in Practice

In this section, we explore the issues raised by experiments. All real data raise practical issues that are rarely considered in discussions of hypothetical problems. In particular, we must consider the three-dimensional inverse problem, which has been regarded implicitly as a simple generalization of the two-dimensional case. In some cases the inversion must allow for such nuisance elements as clock drift and source or receiver location errors. The examples do not compose an exhaustive list of tomographic experiments. They were chosen to illustrate different features of real applications of inverse methods; more discussion of the scientific results will be found in the Epilogue following chapter 8.

(1) The 1981 tomography experiment. This experiment was the first attempt to demonstrate tomography in three dimensions at sea. The geometry of the

experiment is depicted in fig. 6.18. The source technology available at that time proved marginal; signal bandwidth was not adequate to provide satisfactory resolution. Nonetheless, as documented by the Ocean Tomography Group (1982) and by Cornuelle *et al.* (1985), the ability to produce a three-dimensional map using acoustics alone was demonstrated.

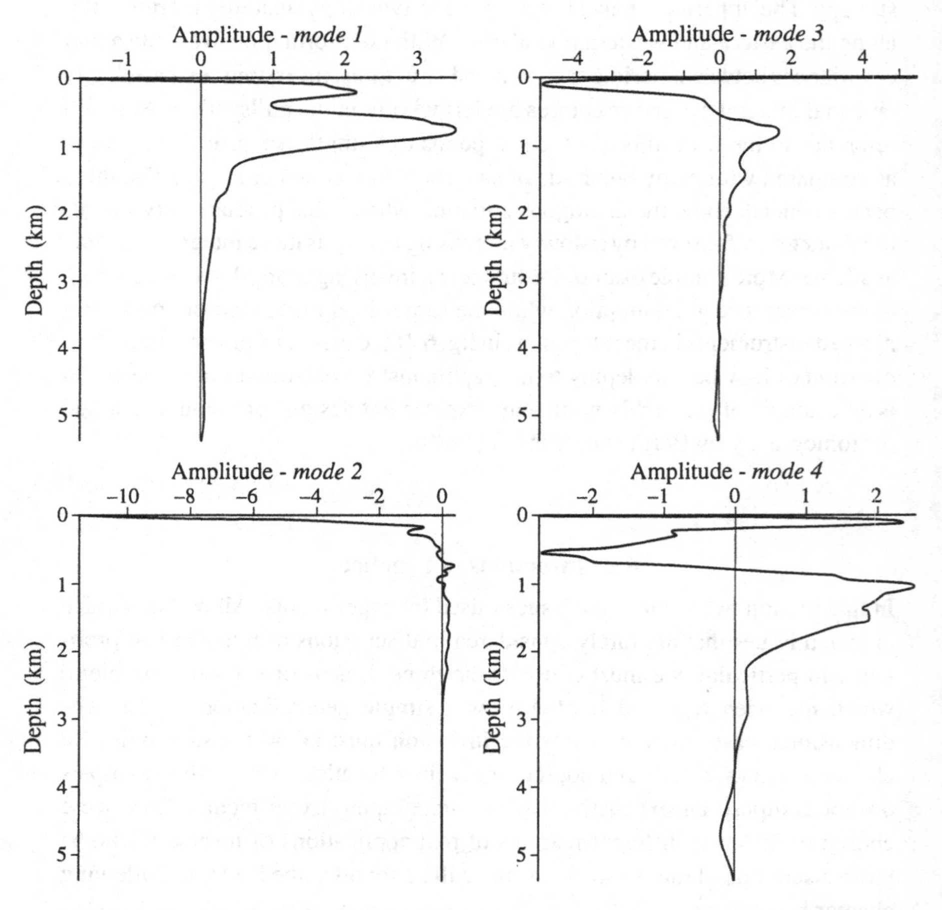

Fig. 6.17. Empirical orthogonal functions (EOFs) for sound-speed computed by Cornuelle *et al.* (1985) from the hydrographic modes of Richman *et al.* (1977). The EOFs 1, 3, and 4 are used by Cornuelle *et al.* to represent the perturbations in sound-speed in the 1981 experiment.

The prior sound-speed profile was based on a climatology from hydrographic measurements. Perturbations relative to the climatology are written as

$$\Delta C(x, y, z) = \sum_{n=1}^{N} F_n(z) G_n(x, y) \tag{6.9.1}$$

(the horizontal coordinate x should not be confused with the statevector $\mathbf{x}$). The vertical structures, $F_n(z)$, are the EOFs of the historical hydrography of the area and were denoted $\mathbf{a}_i$ previously, in (6.5.13). The first four sound-speed EOFs are depicted in fig. 6.17. These are related to the density or vertical-displacement EOFs through the scale factors described in section 2.15. Richman *et al.* (1977) estimated the relative energies in these hydrographic modes to be in the ratios 1.0 : 0.2 : 0.1. These ratios provide part of the prior statistics to determine $\mathbf{S}$ (by definition, the EOFs are uncorrelated, rendering $\mathbf{S}$ diagonal). Cornuelle *et al.* (1985) chose to use only RR rays, not employing any surface-reflected rays. Accordingly, the second mode, which is surface-intensified, is poorly resolved, and Cornuelle *et al.* (1985) simply dropped it from the representation.

The perturbations $\Delta C(x, y, z)$ are constructed using the Gauss-Markov estimator (6.5.8*b*) in three dimensions, with $\mathbf{x}$ defined as the coefficients $G_n(x_i, y_i)$ on a two-dimensional horizontal grid at 3-day intervals. $\mathbf{S}$ is constructed using

$$\langle G_n(x_i, y_i) G_m(x_j, y_j) \rangle$$

$$= \alpha_n^2 \delta_{nm} \left(1 + b_n^2 \exp\left\{ -\frac{1}{2} \frac{\sqrt{(x_i - x_j)^2 + (y_i - y_j)^2}}{(100\,\mathrm{km})^2} \right\} \right);$$

that is, the horizontal covariance for each EOF coefficient is isotropic and spatially stationary.

The error fields in the measured travel times include clock drift as well as lateral and vertical excursions of the instrument moorings. These and other sources of error are represented by the noise terms n_i in each of the travel-time measurements. The sub-elements of the error (clock, position, *etc.*), are denoted individually along with their covariances, as described in section 6.5.

Cornuelle *et al.* (1985) constructed independent maps at 3-day intervals and then combined them in a running 7-day average (fig. 6.18). Two CTD surveys, one at the beginning and one near the end of the tomographic experiment, are also displayed – but it took nearly 2 weeks to conduct each CTD survey, during which time the ocean changed.

Conventional oceanographic data, consisting of CTD and XBT surveys and moored temperature measurements, were withheld from the inversions so as to

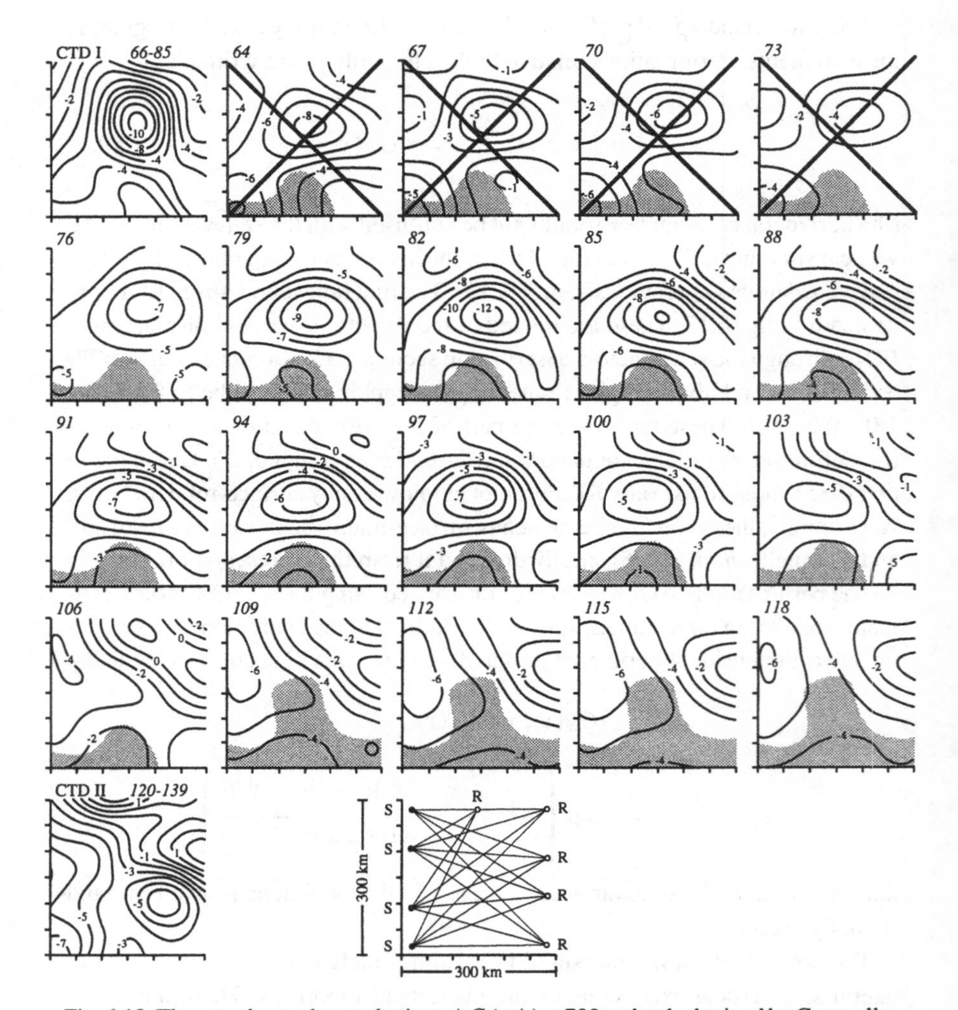

Fig. 6.18. The sound-speed perturbations ΔC (m/s) at 700 m depth obtained by Cornuelle *et al.* (1985) using the Gauss-Markov estimate. Separate estimates were made at 3-day intervals, and then a running 7-day average was formed and depicted every 3 days. (Numbers designate year days 1981 at center time.) Regions of high uncertainty are shaded, and times of large mooring displacements are "x-ed." The first and last panels are temperature estimates obtained from two shipborne CTD surveys. The inset shows the positions of the source (S) and receiver (R) moorings.

provide independent comparisons with the acoustic results. It will be apparent that employment of constraints on $\mathbf{x}$ demanding that the inversions also be consistent with the conventional measurements would produce a better solution in the sense of having a smaller uncertainty. In a combined inversion, the tests of consistency are made by showing that the resulting values of $\hat{\mathbf{x}}$, $\hat{\mathbf{n}}$ are consistent with the known errors in all types of observations and with any prior statistics for $\mathbf{x}$. These latter tests are often regarded as too subtle to be convincing to skeptical observers, and so data are withheld to provide concrete proof of new technologies.

Later, Gaillard and Cornuelle (1987) recomputed the inversions for this experiment, including the RSR rays, thus improving the vertical resolution. It was found, contrary to the original understanding, that the surface-reflected rays were no noisier than the RR rays.

(2) The 1981 experiment by nonlinear inversion. The same experimental data were reexamined by Chiu and Desaubies (1987), using dynamic modes (rather than EOFs) as the perturbation basis, but retaining only the first baroclinic mode. They permitted nonlinear interaction of waves with different horizontal wavenumbers and with a best-fitting mean-flow field. The frequencies and wavenumbers, as well as the amplitudes and phases of the modes, were regarded as parameters to be optimized. The combination of nonlinear interactions of the waves, and the use of wavenumbers and frequencies as parameters, rendered the optimization problem nonlinear in both of the forms considered earlier – a nonlinear model and a nonlinear parameter fit. Chiu and Desaubies (1987) included the position corrections as explicit parameters in the equations. In another difference from the inversions of Cornuelle *et al.* (1985), the fit included the conventional *in situ* observations among the constraints.

Their objective function required a minimum-square fit to the travel-time and temperature observations over the entire time duration, rather than fitting anew every 3 days as did Cornuelle *et al.* (1985). Thus the parameters governing the time-evolving field were themselves steady, with amplitudes, frequencies, and wavenumbers unchanging over the entire time history. The minima were found by the Fletcher and Powell (1963) search algorithm. Chiu and Desaubies found a convincing solution whose uncertainty was estimated by explicitly evaluating the Hessian matrix at the minimum. The best estimated sound-speeds at 700 m at 9-day intervals, along with the estimated error, are shown in fig. 6.19. The results are generally consistent with those of Cornuelle *et al.* (1985), although the estimated error is smaller, a result that Chiu and Desaubies ascribed to the use of nontomographic data, as well as to a fit over the entire time history of the data. The restriction to only one vertical mode greatly reduces the uncertainty compared with what it would be if it were believed, *a priori*, that many modes of unknown amplitudes and phases were present. Residuals suggested that the model was adequate.

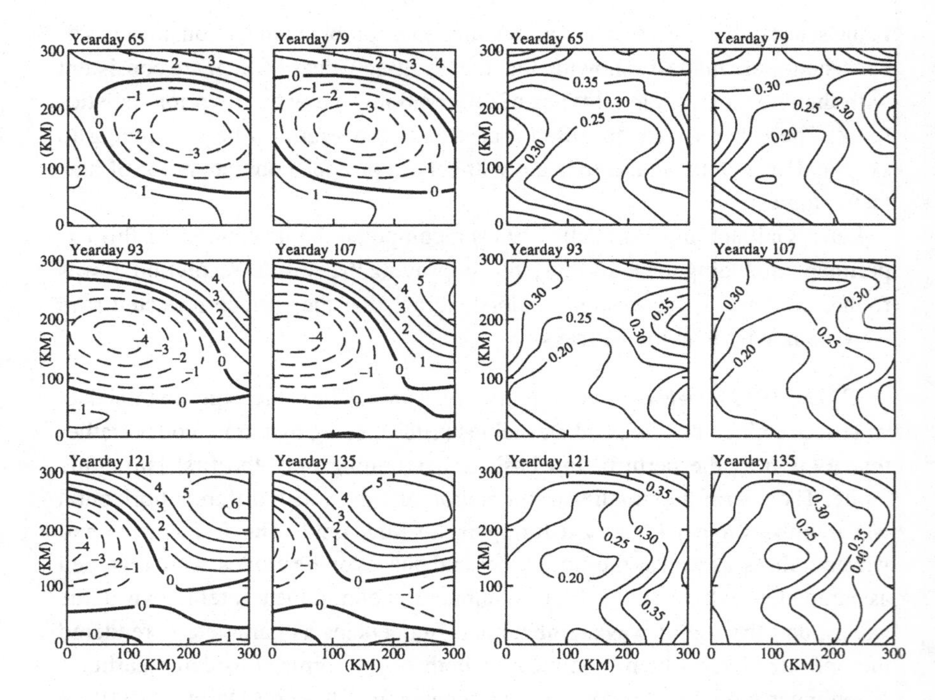

Fig. 6.19. Sound-speed perturbations ΔC (m/s) at 700 m depth (left) and the associated uncertainties (right) obtained by Chiu and Desaubies (1987) for the same ocean field data used by Cornuelle *et al.* (1985) in the previous figure. Chiu and Desaubies's maps are based on the best-fitting plane waves of the first baroclinic mode and their nonlinear interaction (unlike the Cornuelle *et al.* linear EOFs), in addition to the *in situ* CTD surveys.

(3) Range-dependent inversion. As discussed in section 4.2, single vertical slice tomographic receptions contain information on small-scale range-dependent features in the section. Cornuelle *et al.* (1993) described a Gauss-Markov inversion based on ray travel times between a source and vertical receiver array separated by about 1 Mm in the region north of Hawaii (fig. 6.20). The perturbation sound-speed is written

$$\Delta C(x, z) = \sum_i \sum_j \alpha_{ij} X_i(x) Z_j(z),$$

where x is the range coordinate. Five basis functions, $Z_j(z)$, were computed as the eigenvectors of a plausible (but partially guessed) prior vertical correlation matrix. In the horizontal, a truncated Fourier representation in sines and cosines containing all wavenumbers between 0 and $2\pi/17$ km was used. The

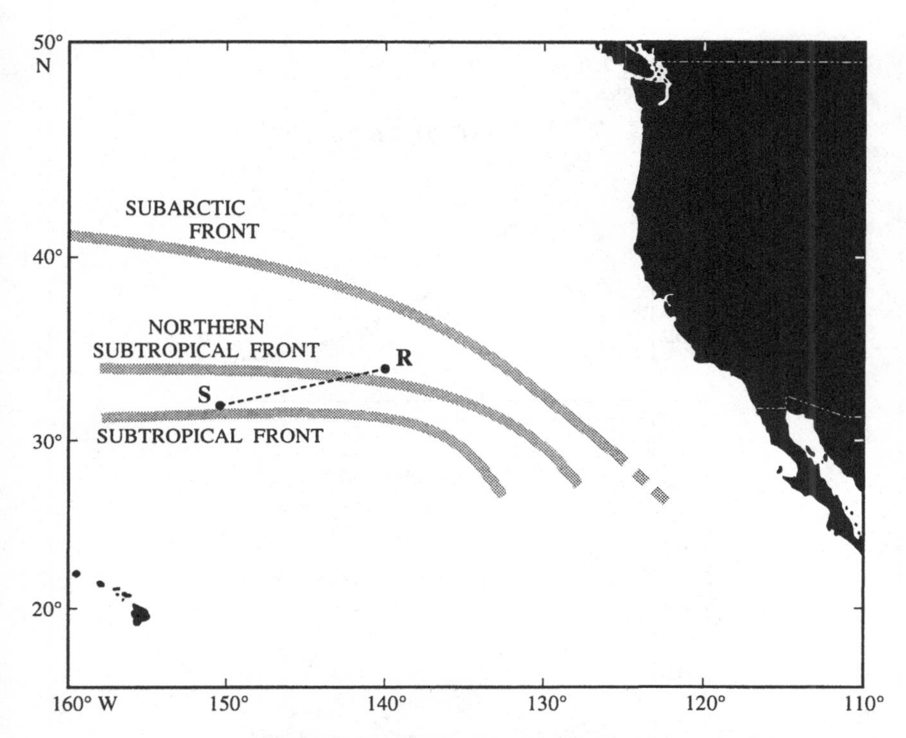

Fig. 6.20. Geometry of the vertical slice experiment discussed by Cornuelle *et al.* (1993). Some of the major fronts of this region are shown in their (very approximate) climatological positions.

prior covariance for the amplitudes was based on the eigenvalues of the prior vertical covariance matrix. The horizontal sinusoids were assumed to have a flat spectrum in wavenumber k, $0 \leq k \leq 2\pi/500$ km, decaying as k^{-2} for higher wavenumbers and cutting off at 17 km for all vertical modes. Data consisted of identified ray arrivals plus a suite of *in situ* measurements from XBTs and CTDs. The reference profile was taken to be range-independent.[14] The only error structure given special consideration was an offset in mean travel times, probably owing to a constant error in the source/receiver distance, but indistinguishable from an error in the mean sound-speed.

Because a focus of the experiment concerned the ability to determine along-track structures, Cornuelle *et al.* (1993) calculated the uncertainty of the solution as a function of the horizontal wavenumbers for the first vertical mode, with

[14]The paper also discusses inversions relative to a range-dependent reference profile constructed from an initial survey.

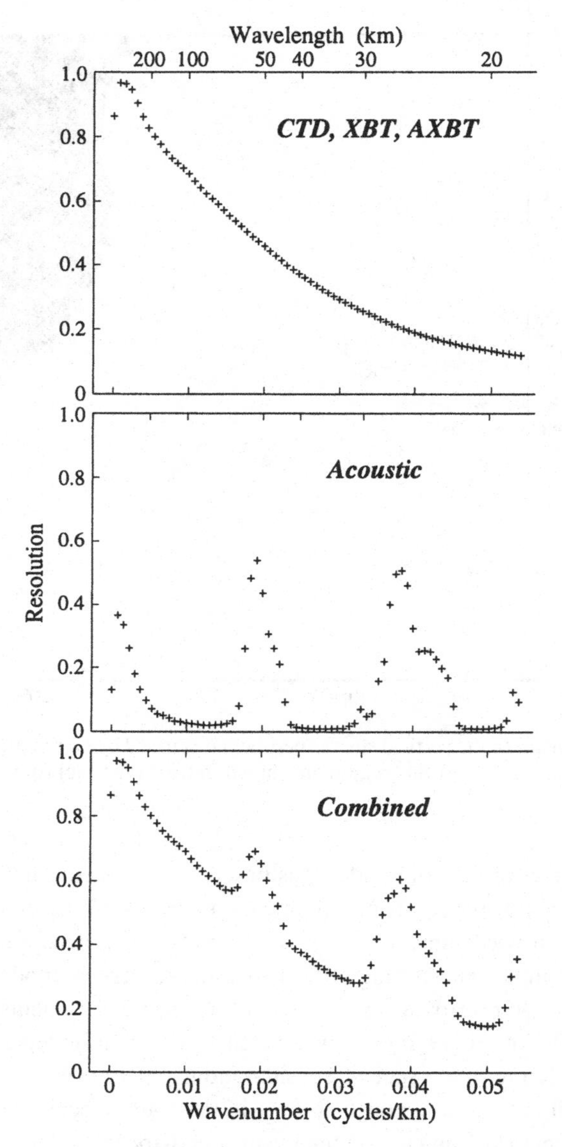

Fig. 6.21. Fraction of the *a priori* sound-speed variance in mode 1 resolved by the inversions of Cornuelle *et al.* (1993) as a function of horizontal ocean wavenumber using CTD, XBT, and AXBT data only (upper panel), using acoustic data only (middle panel), and combining all available data (bottom panel).

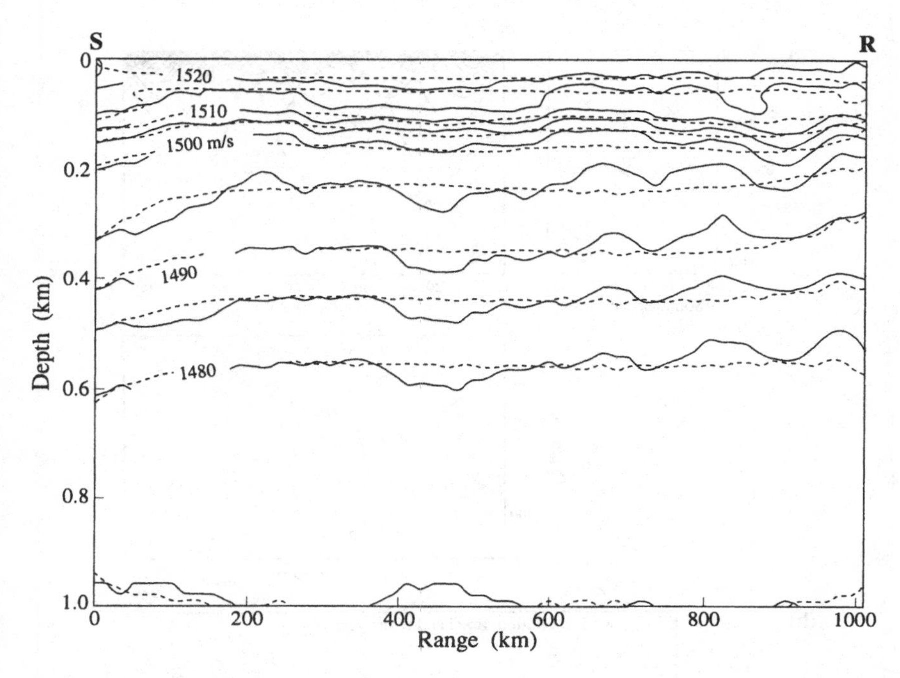

Fig. 6.22. Objectively mapped sound-speed from tomography alone (dashed) and from the combined tomography and *in situ* observations (solid). The pure tomography captures the large-scale structure, but not all of the range-dependent detail.

the result plotted in fig. 6.21 as a fraction of the total variance estimated in their solutions. As a measure of success, they computed the quantity $1 - P_{jj}/S_{jj}$ from $\mathbf{P}$, $\mathbf{S}$, where j refers to the jth wavenumber component. That is, if $P_{jj} = S_{jj}$, then the uncertainty after inversion is the same as the prior variance, and there is no success. At the other extreme, if $P_{jj} = 0$, there is no remaining uncertainty. Three cases were considered: *in situ* data alone, tomographic data alone, and the two data sets combined. For the acoustics alone, there was useful resolution for the mean (*i.e.*, the range independent component) and in two wavenumber bands corresponding to 25- and 50-km wavelengths, values related to the acoustic loop lengths discussed in chapter 4. Not surprisingly, the combined data produced the best estimates. Fig. 6.22 displays the vertical slice reconstructed from the tomography alone, using all the data.

(4) Inversion with acoustic modes. General inverse methods work with almost any physically sensible representation of the solution and any type of data

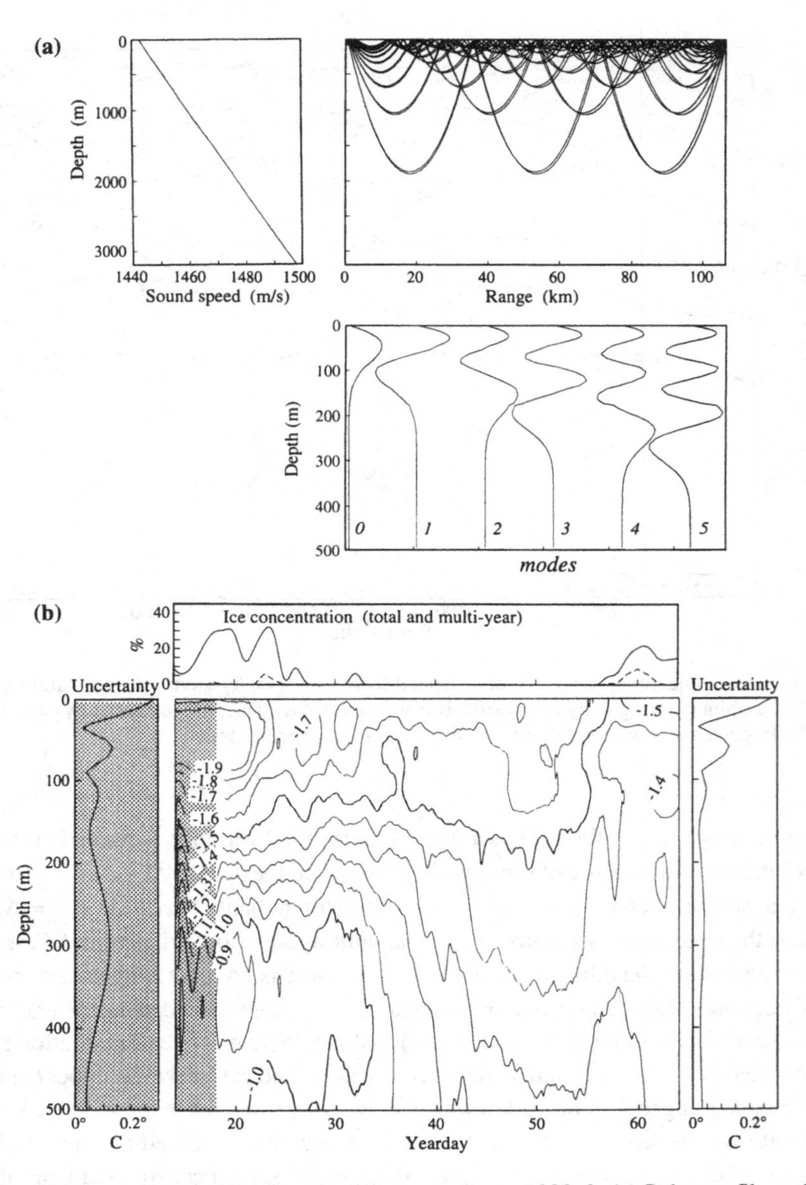

Fig. 6.23. Inversions for the Greenland Sea experiment 1988–9. (a) Polar profile and associated rays and acoustic modes. (b) Time history of range-averaged temperature section using both rays and modes. The shaded (left) and unshaded (right) uncertainties apply to the shaded and unshaded range-averaged temperature, respectively. (Adapted from Sutton, 1993.)

that are sensitive to the numerical values of $\mathbf{x}$. Sutton *et al.* (1994) inverted tomographic data from the Greenland Sea using travel times from both rays and acoustic modes (fig. 6.23). The perturbations were represented in a set of range-independent EOFs. Inversion was by least-squares, using the estimated diagonal noise and solution variances as weights. It was found that the surface temperature field lay in the nullspace of their observations, and so, for aesthetic reasons, an extra constraint was written, forcing its value to lie near that expected on the basis of climatological data. The resulting time-dependent vertical temperature structure is everywhere consistent with the withheld conventional observations.

(5) Inversions for heat content. The natural integrating property of tomography makes it a powerful technique for determining large-scale mean properties such as upper-ocean heat content. For some purposes, such as determining the heat stored as a function of season in the upper ocean, one seeks not the temperature change in individual layers but the integrated change over many layers (recall the simple example in chapter 1). If an inversion has been done layer by layer, producing $\mathbf{x}$, the Gauss-Markov theorem shows that the best estimate of any weighted sum of the solution is $\mathbf{a}^T\mathbf{x}$, with uncertainty $\mathbf{a}^T\mathbf{P}\mathbf{a}$, which often will be much less than the uncertainty of the individual elements of $\mathbf{x}$. Here $\mathbf{a}$ are the weights, which in the present case are the layer thicknesses times the heat capacity. Just such a calculation was done by Dushaw *et al.* (1993*c*) for the region north of Hawaii (fig. 6.24).

6.10. Summary Comments

We have tried to emphasize the essential unity and equivalence of a number of different inverse methods. These methods are applicable to a wide variety of problems, not just tomography, although tomography has some special features. Before moving on to the discussion of time dependence, some summary comments may be helpful.

Representation and sampling. Several different representations of the vertical perturbation structure have been applied, among them uniform and linear spline layers, vertical dynamic modes, and empirical orthogonal functions. There are other possibilities, such as cubic splines and wavelets. The choice is a matter of convenience rather than a fundamental issue. For example, the representation of the ocean requires a sufficient number of layers to capture the essential features of real disturbances. (A layer specification is identical with the use of a vertical mapping grid.) If only a few acoustic rays or modes are measured,

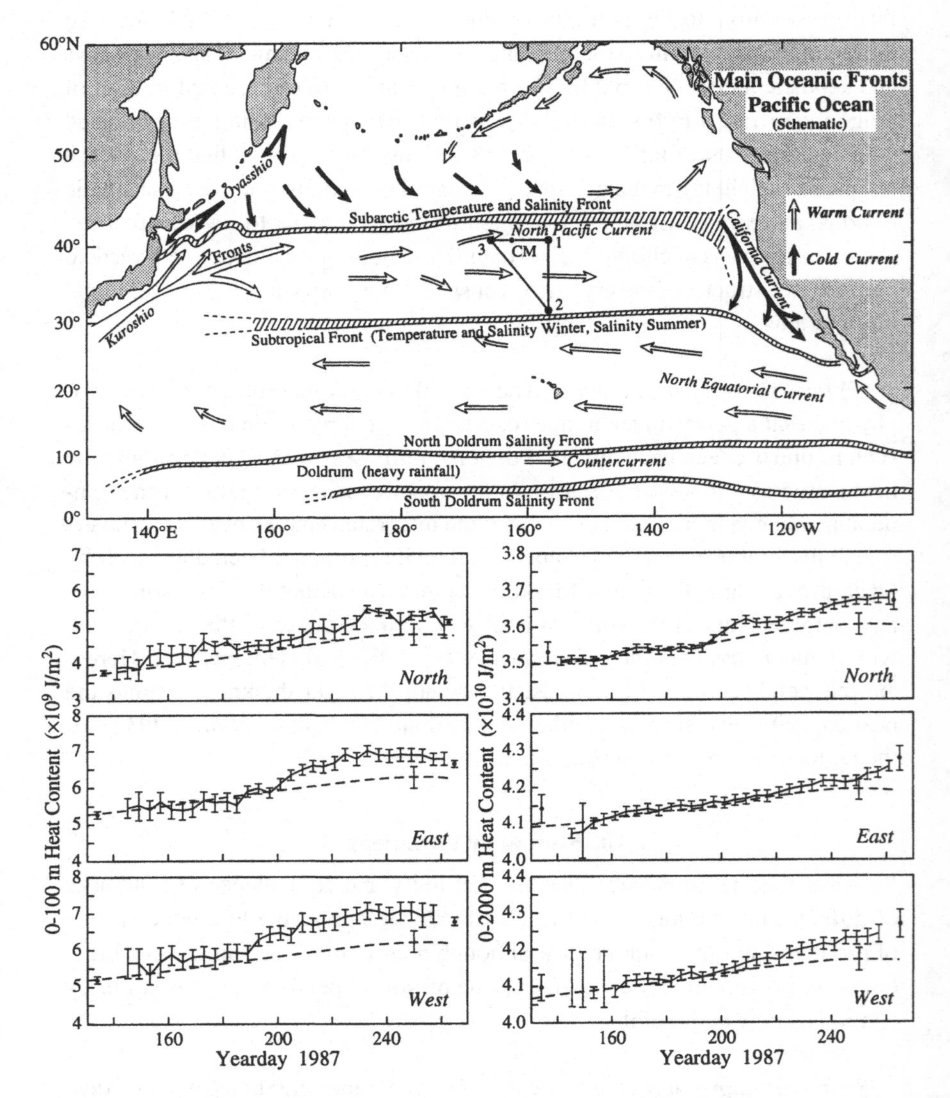

Fig. 6.24. Average heat content along the north, east, and west legs of the tomographic triangle shown in the upper panel. Left column shows the tomographic results from 0 to 100 m, right column from 0 to 2000 m. Dashed curve gives the heat content as inferred from air–sea exchange, aligned to the start of the time series. The points prior to and following the tomographic time series are based on XBT/CTD measurements (Dushaw *et al.*, 1993*c*).

the resulting formal underdetermination may appear very severe. But because the disturbances in nearby layers are strongly correlated, the employment of the covariance, **S**, effectively removes much, if not all, of the formal underdetermination.

In this situation, some investigators may opt to use the eigenvectors of **S** (the EOFs) as the expansion functions, with their eigenvalues giving the expected variance of each EOF coefficient. Specifying the eigenvectors and eigenvalues of a matrix is equivalent to specifying the matrix itself, and so the answers should be equivalent. The EOF approach is attractive because it provides a more efficient representation than a grid of values, but with modern computer power it is difficult to conclude that this consideration is paramount.

There is a variety of methods for imposing prior insights. For example, consider the ray travel times due to layered perturbations in the temperate profile (section 2.15). Munk and Wunsch (1982*b*) treated this problem as an illustration of "up–down ambiguity" (fig. 6.25). The layers were chosen so that almost all of the structure was in the upper ocean – there being only one layer beneath the sound-channel axis. That choice was not arbitrary. Each ray weights a layer by an amount proportional to the time or distance the ray spends in the layer.

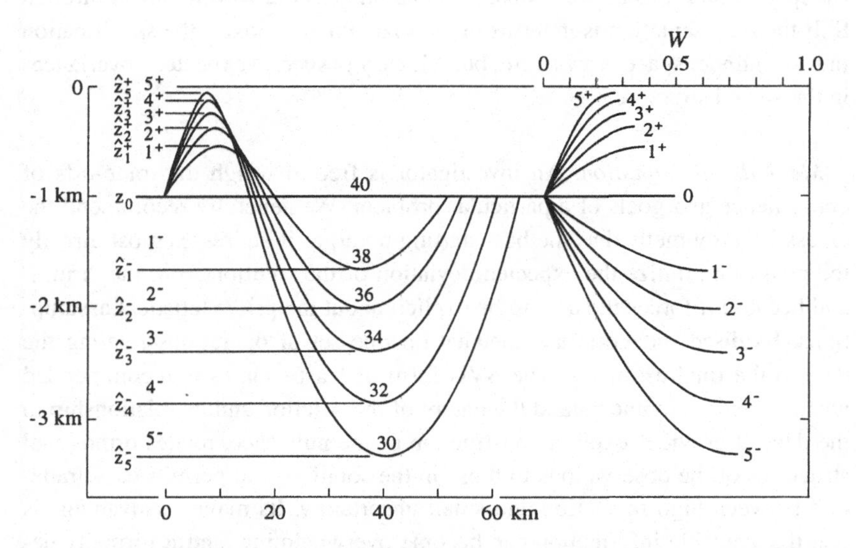

Fig. 6.25. Origins of the up–down ambiguity in vertical slice tomography (Munk and Wunsch, 1982*b*). The sound-channel axis is at z_0, and a series of layers above and below the axis are shown, along with the representative rays in a range-independent situation. The ray weighting, w_i, is proportional to the time or distance spent in each layer; the resulting integral of w, denoted by W for each ray, is shown on the right.

Rays spend more time in layers below the axis, and accordingly these are given greater weight than shallow layers. In the absence of prior information (*i.e.*, setting $\mathbf{S} \propto \mathbf{I}$), inverse solutions with many layers beneath the channel axis tend to place the perturbations preferentially in the deep ocean. Such a solution is acceptable in the absence of prior contrary knowledge. With knowledge to the contrary, one can employ structure in **S**, or one can remove the corresponding nullspace by combining layers below the axis. An additional possibility in a range-independent environment is to identify and use additional ray arrivals having one extra upper or lower loop. In summary, there are at least three ways to provide the information that disturbances are mainly in the upper ocean: introduce the correct **S**; introduce a representation in the eigenvectors of **S** (EOFs), or suppress the nullspace by reducing the number of deep layers.

The same considerations apply to horizontal representations: the sound field tends to integrate out certain structures and spatial scales, which consequently lie in the nullspace unless extra information is provided by introducing the correct **S** or a proper representation. In the loop-harmonic analysis of section 4.2, most of the horizontal wavenumbers did not contribute to the ray travel-time perturbations. If the horizontal representation is in terms of sines and cosines, this information permits one to (i) simply omit the unobserved wavenumbers or (ii) specify relations between observed and unobserved wavenumbers through **S**. If the representation is in terms of a horizontal grid/boxes, the specification in wavenumber space is awkward, but it is easy to specify expected covariances in the spatial structure.

Methods of estimation. An investigator is free to weigh the trade-offs of convenience and goals of a particular problem. We generally recommend the Gauss-Markov method as the best starting point, as it addresses most directly the goal to minimize the expected deviation of the solution from the "truth," and because it forces the user to be explicit about the prior statistical assumptions. Its disadvantage is a somewhat opaque set of operations relating the data to the final estimates. The SVD form of least-squares is recommended when one seeks to understand the nature of the solution and its relationship to the data. It provides explicit construction of the nullspace, relates orthogonal structures of the observations to those in the solution, and permits easy trade-offs between high resolution and small uncertainty. Its main disadvantage is that the available information can become overwhelming, and it formally demands going through the Cholesky decomposition steps for **S**, **R**. It is important to recognize, however, that the computational load of the Gauss-Markov method can itself become extremely burdensome. If, as is desirable, one is in a position to use full (*i.e.*, nondiagonal) covariance matrices for the solution, the very large number of estimation positions will generate spatial covariance matrices whose dimension may be overwhelming. Conventional least-squares methods are not generally recommended, because the results can be confused with those obtained from the Gauss-Markov solutions, but experienced users find the coincidence of the solutions to be a considerable convenience. There are situations where least-squares is the method of choice, as when seeking the imposition of nonstatistical structures, such as the "smoothest" solution or solutions in which some element is as large as possible. Least-squares is readily extended (Lawson and Hanson, 1974; Wunsch, in press) to impose inequality constraints, such as information that some solution elements are greater or less than a certain value.

Other inverse methods, such as linear programming, may prove preferable. The reader will have recognized that all the methods described involve an optimization step – the determination of stationary points of some objective function. Because optimization is computationally expensive, particularly if nonlinear models and constraints are used, one is led to explore still-novel approaches, such as the combinatorial methods known as simulated annealing and genetic algorithms. These and other methods appear highly promising, but we must leave their discussion to the references (Ripley, 1981; Press *et al.*, 1992*a,b*; Koza, 1992).

Uncertainty. The subject of uncertainties, whether prior or posterior to the inversion, is often regarded as rather boring. But the uncertainties are critical, and they often represent more important information than does the solution. The prior uncertainties, written in terms of covariances **S**, **R**, provide a convenient and powerful summation of knowledge about the solution and the ocean. Variances (the diagonal elements) that are taken too small or too large represent erroneous statements about the ocean or about the data and will distort the solutions. On the other hand, it is a mistake to be overly concerned about providing very accurate estimates of **S** and **R**; as the data volume grows, the direct impact of the uncertainty estimates diminishes. Furthermore, the residuals are quite sensitive to the prior statistics and readily lend themselves to "adaptive" methods for modifying **S**, **R**, if the data stream is a continuing one.

Similarly, the uncertainty, **P**, of $\widehat{\mathbf{x}}$ is a quantitative statement about how well different elements of the solution are determined and how they are correlated. **P** carries the vital information concerning which structures in the solution one can safely use for various purposes (*e.g.*, taking their derivatives or making a forecast) and which are in need of further attention. The poorly determined elements become the focus of future experiments.

Other acoustic data. The examples described here have emphasized acoustic travel times as the measured data, in either ray or mode representations. The inversion formalism is not in any way limited to these particular data, although they have tended to be used thus far in both theoretical and practical applications. But any measurable property of an arriving acoustic signal can be used to make inferences about the ocean, as long as there is a known functional relationship between what is measured and some property of the ocean through which the acoustic signal has passed. Obvious examples are ray angles or intensities.

The most general such approach is usually labeled "full-waveform inversion" (Brown, 1984) or "matched-field processing" (*e.g.*, Tolstoy *et al.*, 1991; Baggeroer and Kuperman, 1993). The complete arriving signal is matched in all its details of intensities and phases to the theoretically constructed form. The latter methods derive most directly from array-processing algorithms, where the problem traditionally has been to localize sources or reflectors. These locations are the parameters to be adjusted so as to best reproduce the arriving signals in as complete a manner as possible. In the tomographic context, one extends the adjustable parameters to include those describing the oceanic state and minimizes an objective function involving a fit to the entire received waveform. Success with such computations becomes the ultimate test of "solving" the forward problem, implying a full knowledge of source and receiver characteristics and an understanding of the intervening ocean. Difficulties with forward computations and data inadequacies have thus far prevented full exploitation of these possibilities with real tomographic data.

Brown *et al.* (1980) compared an approximate full-waveform solution to arriving signals, but did no actual inversion. A much simplified version of matched-field processing was used by Goncharov and Voronovich (1993) for CW signals in the Norwegian Sea.

Final steps. As the final step, one should add the estimated perturbation to the original basic state, recompute the estimates of the travel times or other acoustic data being used, and compare them with the observations, allowing for the uncertainty estimates. This step has the nature of what engineers often denote as a "sanity check": it tests the linearization, as well as all the other approximations that have been made. If all is well, the new state estimate will be consistent with the observations.

CHAPTER 7

THE INVERSE PROBLEM: MODEL-ORIENTED

7.1. Introduction: The Use of Models

The discussion in chapter 6 focused on inverse methods in which the observations were primary. Although such methods can readily accommodate information concerning the structure of **x**, through explicit constraints (6.5.15) or prior covariances **S** or through the representation basis, these considerations are secondary to finding solutions consistent with the observations. With the focus on (6.2.4), the methods were labeled "data-oriented."

There is another possibility, based on the rich reservoir of knowledge about the ocean contained in the equations of motion and in estimates of the forcing by the atmosphere. Suppose the Navier-Stokes equations are solved analytically or numerically, subject to atmospheric wind and buoyancy forcing, to produce an estimate of the oceanic state $\widehat{\mathbf{x}}(-)$, the minus in the argument again meaning that no tomographic data have been used. For an oceanic general circulation model (GCM), either global or spanning an ocean basin, the dimension of $\widehat{\mathbf{x}}(-)$ is very large (10^8 or more, as in fig. 7.1). Depending on the model sophistication and the accuracy and precision of the boundary conditions, $\widehat{\mathbf{x}}(-)$ may be a very good estimate indeed, implicitly containing a huge amount of information about the ocean. The model carries all the information that previously would have been provided by the solution covariance matrix **S**. In this situation, the tomographic observations (6.2.4) may directly involve only a minute fraction of the elements of **x**, and one seeks explicitly to use the model-based estimates of **x** in the best combination possible with the tomographic observations or other types of observations. We shall refer to this latter situation as "model-oriented" tomography, recognizing, however, that the methods represent a continuum between the two problem types.

The explicit use of steady GCMs with simulated tomographic measurements (including errors) was alluded to in chapter 6. Schröter and Wunsch (1986) considered how to force a steady GCM of arbitrary dimension in **x** to consistency with a comparatively sparse set of noisy tomographic measurements – essentially by employing the data-oriented methods already discussed.

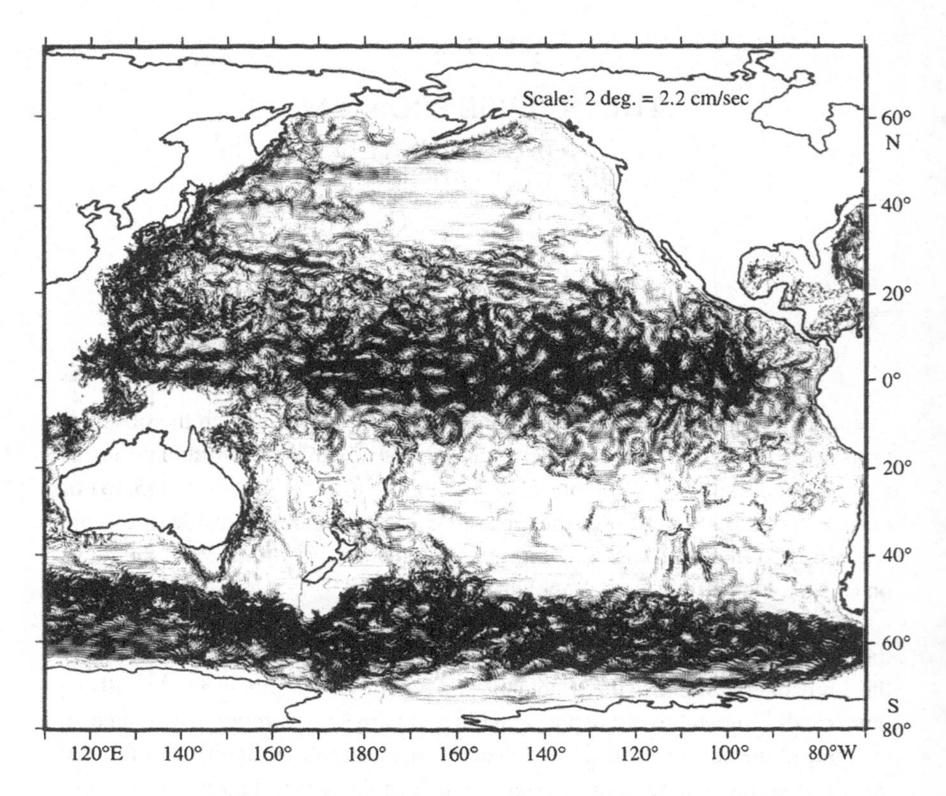

Fig. 7.1. Instantaneous deep-Pacific-Ocean flow field according to the global GCM of Semtner and Chervin (1992) with seasonal forcing. The velocity is a depth average from 1000 to 3300 m. Such models contain a very large variety of phenomena at all space and time scales. Inferences from tomographic and other data must be consistent with the physics of fluids; but this, and all other existing models, are simpler than the real ocean.

Model-oriented methods become essential when time-dependent GCMs are used. If the ocean is presumed to evolve through time t, the definition of $\mathbf{x}$ must be expanded. Let $\mathbf{x}(t)$ define the ocean state at t, and set the time interval $\Delta t = 1$. Then $\mathbf{x}$ encompasses all the defining elements over time:

$$\mathbf{x} = \begin{Bmatrix} \mathbf{x}(0) \\ \mathbf{x}(1) \\ \vdots \\ \mathbf{x}(t = N) \end{Bmatrix} . \tag{7.1.1}$$

The tomographic measurements at t are

$$y_i(t) = \sum_j E_{ij}(t)x_j(t) + n_i(t) . \tag{7.1.2}$$

Even if the number of elements of $\mathbf{x}(t)$ only modestly exceeds the number of observations, as time passes the cumulative number of unknown elements of $\mathbf{x}$ will ultimately dwarf the number of observations. One is driven to using knowledge of the dynamic evolution of $\mathbf{x}(t)$ to aid in estimating $\mathbf{x}$. Stated from another point of view, if a model of the temporal evolution is available, knowledge of the state of the ocean at one time t carries information about both its future and previous states. For example, knowledge of the initial conditions appropriate to a model might be inferred from observations of the state in which it exists at some considerable time later.

The complexity of oceanic motions can be appreciated from displays of the deep ocean flow in the Semtner and Chervin (1992) model (fig. 7.1). Such models are called eddy-resolving general circulation models (EGCMs) and represent the state of the art in oceanic modeling. These models are very much developmental, with many difficulties, and are so demanding of computer time that they can be run only for limited durations.

The seemingly inexorable growth in computer power and rapid advances in numerical techniques make it reasonable to believe, however, that ever more realistic versions of EGCMs will be available for oceanographic use. With this expected increase in modeling skill comes a parallel demand for increased observations. Some of the features seen in fig. 7.1 might well exist in the real ocean, but they are nearly unmeasurable by current seagoing techniques. Tomography is one of the few technologies capable of providing data on space scales and over time durations consistent with these features. The premise is that we shall progress most rapidly by combining the data with the models – thus simultaneously testing and constraining the models with the observations.

Until recently, oceanographers paid comparatively little attention to the issues of combining data with four-dimensional models, in part because so few data have been available. This situation contrasts sharply with that for meteorology, where the existence of the World Weather Watch and the demand for accurate weather forecasts have led to a data richness that oceanographers can only envy.[1] The situation in oceanography has been changing rapidly as new observing systems have appeared on the scene: satellites, drifters, floats, tracers, and acoustic tomography. But the actual practice of data/model combinations in oceanography remains comparatively primitive.

[1] This does not mean that meteorologists are content with their data base, merely that they are better off than oceanographers.

The economic stakes involved in weather forecasting have led meteorologists to sophisticated methods for combining data with models under the generic title of "assimilation" (Bengtsson *et al.*, 1981; Lorenc, 1986; Ghil and Malanotte-Rizzoli, 1991; Daley, 1991). This is an apt term, and the meteorological experience is an important guidepost as oceanographers begin to grapple seriously with the problems of combining data with EGCMs. But the heavy emphasis in meteorology on short-term forecasting and the specialized nature of atmospheric data have tended to narrowly focus the assimilation efforts. Forecasting is of some interest in oceanography (for short-term cruise planning, for military operations, and notably for El Niño problems), but the central issue now is to better describe the ocean for the purpose of understanding it. Data from times formally future to the time of estimation contain much useful information and distinguish oceanography from meteorology, at least as practiced in numerical weather prediction. For purposes of climate forecasting, one of the most pressing problems is "initialization" – determining the present state of the ocean with sufficient accuracy that the ultimate long-term computations will not become swamped by errors in the initial conditions. We therefore avoid the use of the term "assimilation," and instead employ the more general term "state estimation," borrowed from mathematical statistics and engineering; assimilation for forecasting is a special case of state estimation.

Much of the material in this chapter represents future directions rather than present methodologies. We present it as an indication of how powerful the combination of tomography with models can be. The procedures become essential for the basin and global problems described in chapter 8.

7.2. State Estimation and Model Identification

The problem of using tomographic data with time-evolving models does not differ in any fundamental way from the problem of using any other data. What follows is an abbreviated generic discussion, intended to show where we think future efforts must be directed.

We suppose there is available an oceanographic model in three space dimensions plus time, perhaps fully nonlinear, described by some approximate form of the Navier-Stokes equations, and written in a very condensed notation as

$$\mathbf{x}(t+1) = \mathbf{L}(\mathbf{x}(t),\ \mathbf{r},\ t,\ \mathbf{q}(\mathbf{r},t),\ \mathbf{w}(\mathbf{r},t))\,. \tag{7.2.1}$$

Without loss of generality, $\mathbf{q}$ describes all known boundary and initial conditions and forces acting on the fluid; $\mathbf{x}(t)$ is called the "statevector" and represents the state of the model ocean at time t (time units are chosen so that t is always an integer). Depending on the model, $\mathbf{x}(t)$ will comprise the three-dimensional velocity field at every grid point at all times, or the density field described by finite elements, or the pressure field in a modal expansion, or any or all of these things in combination. By definition, the statevector provides just the right amount of information to enable the model to compute its values one time step into the future; that is, $\mathbf{x}(t+1)$ is computable from $\mathbf{x}(t)$ along with the requisite boundary conditions and externally prescribed sources and sinks of energy and momentum. Equation (7.1.1) defines $\mathbf{x}(t)$ as a subvector of $\mathbf{x}$, as used in chapter 6.

The $\mathbf{w}(t)$ represent any model parameters that at some stage might be regarded as unknown; eddy mixing coefficients for momentum or heat are examples. For the time being, it is useful to separate these from the state variables, but the distinction is largely artificial. It is common also to use $\mathbf{w}(t)$ to represent unknown components in $\mathbf{q}(t)$. Of particular importance are the uncertainties of boundary conditions, whether at the sea surface or on the sides, especially if the model has "open-ocean" inflows and outflows. We append to (7.2.1) any information about the uncertainty in the model. For example, if the forcing contains the wind-stress curl, $\mathbf{f} \equiv \widehat{\mathbf{k}} \cdot \nabla \times \boldsymbol{\tau}$, there is a collection of statements

$$\mathbf{f} = \mathbf{f}_0 \pm \delta\mathbf{f}, \tag{7.2.2}$$

where $\mathbf{q} = \mathbf{f}_0$, $\mathbf{w} = \delta\mathbf{f}$, and we specify

$$\langle \mathbf{w}(t) \rangle = 0, \quad \langle \mathbf{w}(t)\, \mathbf{w}(t)^T \rangle = \mathbf{Q}(t)\,. \tag{7.2.3}$$

Similar uncertainties normally apply to all model elements, including initial and boundary conditions as well as the internal parameters. The general philosophy is that very little in a model can be claimed to be known with absolute certainty, and a general procedure should permit any aspect of the model to be regarded as subject to change in the light of experience. In the most general sense, $\mathbf{w}(t)$ also represent all model deficiencies, including missing or misspecified physics. The elements of $\mathbf{w}(t)$ are often called the "controls."

Finally, focusing specifically on tomography, we have a collection of integrals through the ocean, represented by (6.2.1) or (6.2.4). We must now distinguish measurements made at different times t over a finite time interval $t_1 \le t \le t_f$, although at any fixed time the tomographic measurements remain weighted averages over the statevector, as in (6.2.4). We must also accommodate the possibility that the observation matrix is time-evolving along with the ocean.[2] Writing (6.2.4) or (7.1.2) in the matrix notation

$$\mathbf{E}(t)\, \mathbf{x}(t) + \mathbf{n}(t) = \mathbf{y}(t)\,, \tag{7.2.4}$$

[2]For example, if a source is added, or a receiver fails, or moving ships are used, or the basic ocean state greatly changes.

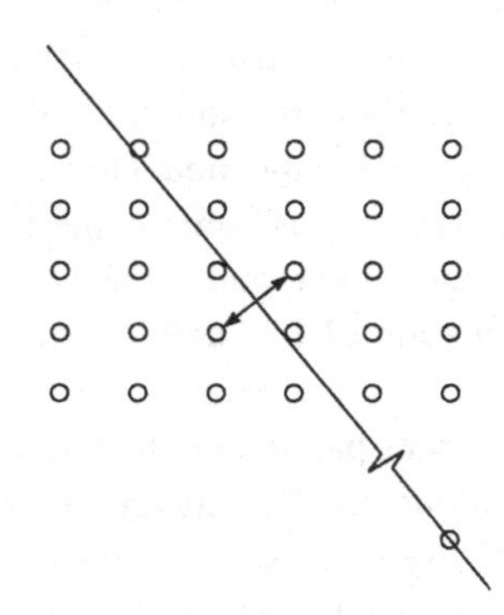

Fig. 7.2. Model grid in a plane in the vicinity of an acoustic ray. In one formulation, $x_j(t)$ is the sound-speed or velocity component at the jth grid point. If the ray trajectory (which can curve) passes directly through a grid point, then the corresponding row of $\mathbf{E}$ will have a nonzero element in column j. For grid points close to, but not intersected by, the ray, the row of $\mathbf{E}$ provides an interpolation rule from the $x_j(t)$ to the field value encountered by the ray.

many, if not most, of the elements of $\mathbf{E}(t)$ will vanish. Suppose the model is defined on grid points, as in fig. 7.2, and that the state variables consist of the pressure field and the three components of velocity. If the tomographic measurement at time t is the velocity field along the line shown, then all the elements of $\mathbf{E}(t)$ will vanish except those corresponding to the grid points intersected by the path, or close to it.[3]

The inverse computations of chapter 6 were directed at solving (7.2.4), for fixed t, without employment of (7.2.1), which relates estimates at different times. We now pose the state-estimation problem: What is the best estimate of $\mathbf{x}(t)$ given (7.2.1) and (7.2.4), and what is its uncertainty? For $t > t_f$ this is a forecasting problem; we focus on the problem of greater oceanographic interest, namely, $0 \leq t \leq t_f$.

In the spirit of least-squares and its relative, minimum-error-variance estimation, we seek estimates of $\mathbf{x}(t)$ and $\mathbf{w}(t)$ that will minimize an objective

[3]Ray and mode travel-time computations described in chapters 2–4 involve detailed specification of the structure of $S(z, x)$, while many general circulation models have few degrees of freedom, *e.g.*, 10 layers or levels in the vertical or coarse lateral resolution. It appears that sensible interpolation from the defining points and levels of the models to the denser grids of the acoustic calculations will suffice to compute $\mathbf{E}(t)$, *e.g.*, by spline or other interpolation.

function, over the whole time span of the data,

$$J' = \sum_{t=1}^{t_f} (\mathbf{E}(t)\mathbf{x}(t) - \mathbf{y}(t))^T \mathbf{R}(t)^{-1} (\mathbf{E}(t)\mathbf{x}(t) - \mathbf{y}(t)) \\ + \sum_{t=0}^{t_f - 1} \mathbf{w}(t)^T \mathbf{Q}(t)^{-1} \mathbf{w}(t), \tag{7.2.5}$$

subject to $\widehat{\mathbf{x}}(t)$ being a solution to the evolving model (7.2.1). The covariance matrices $\mathbf{R}(t)$, $\mathbf{Q}(t)$ of observation and model errors, respectively, are commonly chosen as the weight matrices in (7.2.5), so that the least-squares determination is also the minimum-variance estimate; other, nonstatistical weights are possible. Typically there is an estimate of the initial conditions, $\widehat{\mathbf{x}}(0)$, with some statement of uncertainty, $\mathbf{P}(0) \equiv \langle (\widehat{\mathbf{x}}(0) - \mathbf{x}(0))(\widehat{\mathbf{x}}(0) - \mathbf{x}(0))^T \rangle$, and (7.2.5) is often augmented with another term of form $(\widehat{\mathbf{x}}(0) - \mathbf{x}(0))^T \mathbf{P}(0)^{-1} (\widehat{\mathbf{x}}(0) - \mathbf{x}(0))$. Estimates of the statevector so derived yield optimally mapped fields and serve as a basis for optimal estimates of other physical quantities, such as vorticity, frequency/wavenumber spectra, and so forth. No term involving the covariance of $\mathbf{x}(t)$ is included in (7.2.5), as it is intended that the model and $\mathbf{P}(0)$ should be a complete specification of any prior statistical information about the statevector. The reader will appreciate that there are formidable practical problems involved in carrying out these calculations. The work of Chiu and Desaubies (1987), previously described, involved a particularly simple version of this problem, in which $\mathbf{x}(t)$ remains constant over time.

Many variations on the problem posed are of interest. For example, in the "model-identification" problem,[4] one seeks an estimate of model parameters like eddy coefficients that will give the best fit to the observations. The "initialization" problem seeks to produce an improved estimate of the initial conditions from the later observations and the model dynamics. The formalism of modeling in the tomographic problem is not restricted to the elements of $\mathbf{x}(t)$. To the extent that the error elements of the solution, $\mathbf{n}(t)$, are subject to known dynamic evolution equations, they too can be modeled. For example, moored instruments move through time according to the physics governing mooring dynamics (recall chapter 5), often dominated by tidal currents. Some noise elements (*e.g.*, the unchanging errors in anchor positions) satisfy a particularly simple evolution equation, $n_{\text{anc}}(t+1) = n_{\text{anc}}(t)$.

In a purely mathematical sense, the determination of the minimum of (7.2.5) subject to the model is the same problem as already discussed in chapter 6 – the

[4]In chapter 5, we referred to "ray or mode identification," a different usage from parameter estimation in a model.

introduction of time merely being a bookkeeping index for keeping track of the subvectors of $\mathbf{x}$, $\mathbf{y}$. One introduces the model constraints either with Lagrange multipliers or in the minimum-square sense analogous to (6.3.7) and forms and solves the appropriate normal equations. In practice, the large number of time steps typically involved in running GCMs renders the problem dimension so vast as to defeat simple simultaneous-equations solution methods. Much of the remainder of this chapter can be regarded as a technical discussion of the way in which a time-evolving model relates sub-elements $\mathbf{x}(t)$, $\mathbf{x}(t+1)$ of the full statevector $\mathbf{x}$ and how the resulting structures can be exploited to produce efficient algorithms for finding the appropriate minimum of (7.2.5).

7.3. State Estimation: Practice

For the initial discussion, it is assumed that the model, (7.2.1), is either linear or linearizable to the form

$$\mathbf{x}(t+1) = \mathbf{A}(t)\,\mathbf{x}(t) + \mathbf{B}(t)\,\mathbf{q}(t) + \mathbf{\Gamma}(t)\,\mathbf{w}(t)\,. \tag{7.3.1}$$

$\mathbf{A}(t)$ is an $N \times N$ matrix that may depend on time, and the product $\mathbf{B}(t)\,\mathbf{q}(t)$ is a very general form for writing forces/sources/sinks/boundary conditions (*e.g.*, Brogan, 1985). Typically, $\mathbf{q}(t)$ represents the independent degrees of freedom in the boundary conditions and external sources, and $\mathbf{B}(t)$ distributes $\mathbf{q}(t)$ over the model grid. For example, if the northern boundary of the model has boundary values of $\mathbf{x}(t)$ given by $q_1(t)$, and the southern boundary has values $q_2(t)$, then

$$\mathbf{B}(t)\,\mathbf{q}(t) = \begin{bmatrix} 0 & 0 \\ 0 & 0 \\ \vdots & 0 \\ 1 & 0 \\ \vdots & \vdots \\ 0 & 1 \\ 0 & 1 \\ 1 & \vdots \\ \vdots & 0 \\ 0 & 0 \end{bmatrix} \begin{bmatrix} q_1(t) \\ q_2(t) \end{bmatrix},$$

where the nonzero elements in the first column coincide with the northern boundary grid elements, and those in the second column correspond with those of the southern boundary. These representations are not unique. A time-dependent $\mathbf{B}$ could represent regions of varying ice cover, where the effects of wind stress are imposed intermittently by resetting the elements of $\mathbf{B}(t)$ to zero or unity, as required.

In similar fashion, $\boldsymbol{\Gamma}(t)\mathbf{w}(t)$ represents the unknown elements of the forces/boundary conditions, where $\boldsymbol{\Gamma}(t)$ is known, and the covariance $\mathbf{Q}(t)$ of $\mathbf{w}(t)$ is specified. If $\mathbf{Q}(t)$ is finite, it is a statement that something is known of the magnitudes of the unknown elements.

The forecast problem. The model has been run from some estimated initial conditions at time $t = 0$, with best estimates of forcing, and so forth, up to time t, when the observations begin. Define $\tilde{\mathbf{x}}(t, -)$ as the model estimate; no data have yet been used. Suppose (in a manner to be discussed later) an estimate of the uncertainty of this "state estimate," denoted $\mathbf{P}(t, -)$, has been produced. Observations $\mathbf{y}(t)$ with error covariance $\mathbf{R}(t)$ are available. How can these observations be used to improve on the model forecast $\tilde{\mathbf{x}}(t, -)$?

This problem is one of "nowcasting," or, in control terminology, of "filtering." First consider a simple test: Are the observations consistent with the model? To answer this question, substitute $\tilde{\mathbf{x}}(t, -)$ into (7.2.4) and compare the difference

$$\mathbf{y}(t) - \mathbf{E}(t)\,\tilde{\mathbf{x}}(t, -) \tag{7.3.2}$$

between observations and model estimate; if this difference equaled zero, there would be no reason to change, $\tilde{\mathbf{x}}(t) = \tilde{\mathbf{x}}(t, -)$, but the agreement between model and observation would lead to a reduced uncertainty estimate. Such perfect agreement does not occur, and one wants to use the difference (7.3.2) to make corrections. The recursive solution (6.7.5)–(6.7.7) shows how this is done: make a weighted average of the two estimates in the form

$$\tilde{\mathbf{x}}(t) = \tilde{\mathbf{x}}(t, -) + \mathbf{K}(t)\,(\mathbf{y}(t) - \mathbf{E}(t)\,\tilde{\mathbf{x}}(t, -)), \tag{7.3.3a}$$

with

$$\mathbf{K}(t) = \mathbf{P}(t, -)\mathbf{E}^{T}(t)[\mathbf{E}(t)\,\mathbf{P}(t, -)\mathbf{E}^{T}(t) + \mathbf{R}(t)]^{-1}, \tag{7.3.3b}$$

and new uncertainty

$$\mathbf{P}(t) = \mathbf{P}(t, -) - \mathbf{K}(t)\,\mathbf{E}(t)\,\mathbf{P}(t, -)\,. \tag{7.3.4}$$

We have arrived at what is usually known as the "Kalman filter" [various algebraically equivalent expressions for (7.3.3) and (7.3.4) can be derived]. The forecasting problem is now solved by using the model (7.3.1) with $\mathbf{w}(t) = \mathbf{0}$ to compute $\tilde{\mathbf{x}}(t+1, -)$, which is the optimum estimate from both model dynamics and observations up to time t.

Evaluation of $\widehat{\mathbf{x}}(t)$ from (7.3.3) requires knowledge of $\mathbf{P}(t, -)$. Let $\widehat{\mathbf{x}}(t-1) = \mathbf{x}(t-1) + \gamma(t-1)$, where $\langle \gamma(t-1)\gamma^T(t-1) \rangle = \mathbf{P}(t-1)$. For a linear model,

$$\gamma(t) = \mathbf{A}(t-1)\gamma(t-1) + \mathbf{\Gamma}(t-1)\mathbf{w}(t-1)\,, \qquad (7.3.5)$$

that is, the state error propagates in exactly the same way as the statevector, but with an added component owing to the unknown forcing/source/control terms. The covariance of this new error is

$$\begin{aligned}\mathbf{P}(t,-) &= \Big\langle \big[\mathbf{A}(t-1)\,\gamma(t-1) + \mathbf{\Gamma}(t-1)\,\mathbf{w}(t-1)\big]\big[\mathbf{A}(t-1)\,\gamma(t-1) \\ &\qquad + \mathbf{\Gamma}(t-1)\,\mathbf{w}(t-1)\big]^T \Big\rangle \\ &= \mathbf{A}(t-1)\,\mathbf{P}(t-1)\,\mathbf{A}^T(t-1) + \mathbf{\Gamma}(t-1)\,\mathbf{Q}(t-1)\,\mathbf{\Gamma}^T(t-1)\,. \end{aligned} \qquad (7.3.6)$$

Depending on the details of the model, any error structure present at t can grow, decay, or remain bounded as time progresses.

Equations (7.3.3), (7.3.4), and (7.3.6) now lead to an obvious recursion, employing the data as they come in, forecasting the statevector at the time of the next observations, comparing and combining as in (7.3.3*a*). At the time the forecast step is taken, one has used all extant data and all of one's dynamic and statistical insight by way of the model and the covariance matrices. This optimality and sequential data use have led to widespread real-time applications (*e.g.*, in the controls for landing airplanes). A large literature exists on both the theory and practice of Kalman filtering (Sorenson, 1985). There have been some applications to oceanographic problems, as reviewed by Ghil and Malanotte-Rizzoli (1991), but only the most tentative use has so far been attempted with tomographic data. Howe *et al.* (1987) replaced the dynamic model with a rule that the perturbation statevector decays to zero with a 10-day time constant, thus asserting that the ocean tends to its climatology with time. The model uncertainty covariance $\mathbf{Q}(t)$ was specified to grow with time to the estimated covariance of mesoscale eddies, reaching its maximum value after 10 days. The only simpler "dynamics" would be one of pure persistence, the assertion $\mathbf{x}(t+1) = \mathbf{x}(t)$, in which case the Kalman filter reduces to recursive least-squares. For very large model uncertainty, (7.3.3) reduces to (6.5.10) with $\mathbf{S}^{-1} = \mathbf{0}$. That is, if the model skill is very poor, the best estimate is based on the static inversion of the observations available at time t, with prior estimates carrying no information about the present. In this way, the model-oriented result reduces neatly to the data-oriented result, depending on the relative uncertainties of model and data.[5]

Smoothing. It will be apparent that we have not solved the optimization problem stated in the objective function (7.2.5): the data future to time t, which supposedly have been stored, have not been used to calculate $\hat{\mathbf{x}}(t)$, nor has anything been said about estimating the control variables $\mathbf{w}(t)$. Algorithms that address these problems are known as "smoothers." Suppose, to be specific, that the Kalman filter procedure has been carried out over the full time span $0 \le t \le t_f$ and that the state estimate $\hat{\mathbf{x}}(t)$ and its uncertainty $\mathbf{P}(t)$ have been stored. We now wish to step *backward* from t_f systematically improving the previous estimates. The estimate that will minimize the mean square error is (Bryson and Ho, 1975)

$$\hat{\mathbf{x}}(t,+) = \hat{\mathbf{x}}(t) + \mathbf{L}(t)\,[\hat{\mathbf{x}}(t+1) - \mathbf{A}(t)\,\hat{\mathbf{x}}(t) - \mathbf{B}(t)\,\mathbf{q}(t)], \tag{7.3.7}$$

with new uncertainty

$$\mathbf{P}(t,+) = \mathbf{P}(t) + \mathbf{L}(t)\,[\mathbf{P}(t+1,+) - \mathbf{P}(t+1,-)]\,\mathbf{L}^T(t), \tag{7.3.8a}$$

where

$$\mathbf{L}(t) = \mathbf{P}(t)\,\mathbf{A}^T(t)\mathbf{P}^{-1}(t+1,-)\,. \tag{7.3.8b}$$

$\hat{\mathbf{x}}(t,+)$ is now a third estimate of the state at time t, along with $\hat{\mathbf{x}}(t,-)$, $\hat{\mathbf{x}}(t)$. Note that the form of (7.3.7) is again a weighted average of two quantities, in this case of the prior best estimate of $\hat{\mathbf{x}}(t)$ with the difference between what the model predicted the new best estimate of $\hat{\mathbf{x}}(t+1)$ should have been and what it actually turned out to be.

An estimate can also be made of the unknown control terms, through the knowledge of how the system evolved in the future. The best estimate is

$$\hat{\mathbf{w}}(t-1,+) = \mathbf{M}(t-1)\{\hat{\mathbf{x}}(t,+) - \hat{\mathbf{x}}(t,-)\}, \tag{7.3.9}$$

with uncertainty

$$\mathbf{Q}(t-1,+) = \mathbf{Q}(t-1) + \mathbf{M}(t-1)\{\mathbf{P}(t,+) - \mathbf{P}(t,-)\}\mathbf{M}^T(t-1)\,, \tag{7.3.10}$$

where $\mathbf{M}(t-1) \equiv \mathbf{Q}(t-1)\mathbf{\Gamma}^T\mathbf{P}^{-1}(t,-)$. The corrections to $\hat{\mathbf{x}}(t)$, $\hat{\mathbf{w}}(t-1)$ are proportional to their respective uncertainties $\mathbf{P}(t)$, $\mathbf{Q}(t-1)$.

[5] "Large" uncertainty is measured in terms of any one of several available matrix norms. Obtaining the static inversion limit depends on there being no nullspace in $\mathbf{E}(t)$. If a nullspace is present, the model, no matter how great its uncertainties, will always contain information about the statevector, and the static limit cannot be achieved.

The observations are used only in the forward, filtering step (7.3.3*a*); the backward recursive sequence (7.3.7)–(7.3.10) involves only the model and the previous computation. As with the Kalman filter, the smoother reduces to the static, data-oriented estimate when the model uncertainty greatly exceeds the observational noise – if $\mathbf{R}(t)$ has no nullspace.

The recursion (7.3.7)–(7.3.10) is usually known as the "RTS smoother," for Rauch, Tung, and Streibel (1965). Other equivalent forms are known that place less demand on computer storage than does the RTS method, but are less transparent. A full discussion of this interesting subject is left to the many available textbooks. Applications of smoothing to altimetry data are given by Gaspar and Wunsch (1989) and Fukumori *et al.* (1992). The second paper is specifically directed at the problem of reducing the computing burden by solving the optimization problem approximately, rather than rigorously. No tomographic application of a smoothing algorithm with real data has been published.[6]

The nonlinear-model problem. Two distinct nonlinearities occur in practice: model nonlinearity and nonlinearity in the observation equation (7.2.4). The latter nonlinearity has already been encountered in the static inverse problem of chapter 6 (travel times, *etc.*, being nonlinear in S and u). It is usually handled by linearizing the equation about a reference state.

Consider model nonlinearity. The model and data are estimating the true state trajectory $\mathbf{x}(t),\ 0 \le t \le t_f$. If there exists a reasonably accurate prior estimate, $\widehat{\mathbf{x}}_0(t), 0 \le t \le t_f$, we linearize the model calculation about this state as before:

$$\mathbf{x}(t) = \widehat{\mathbf{x}}_0(t) + \Delta\mathbf{x}(t), \qquad \mathbf{f} = \mathbf{f}_0(t) + \Delta\mathbf{f}(t)\,.$$

The model in (7.2.1) is linearized as

$$\mathbf{x}_0(t+1) + \Delta\mathbf{x}(t+1) = \mathbf{L}(\mathbf{x}_0(t),\, \mathbf{f}_0(t),\, t)$$

$$+\frac{\partial \mathbf{L}\,(\mathbf{x}_0(t),\, \mathbf{f}_0(t),\, t)}{\partial \mathbf{x}_0(t)}\,\Delta\mathbf{x}(t) + \frac{\partial \mathbf{L}\,(\mathbf{x}_0(t),\, \mathbf{f}_0(t),\, t)}{\partial \mathbf{f}_0(t)}\,\Delta\mathbf{f}(t)\,,$$

or

$$\Delta\mathbf{x}(t+1) = \frac{\partial \mathbf{L}}{\partial \mathbf{x}_0(t)}\,\Delta\mathbf{x}(t) + \mathbf{B}_1(t)\mathbf{q}_1(t), \quad \mathbf{B}_1(t)\mathbf{q}_1(t) = \frac{\partial \mathbf{L}}{\partial \mathbf{f}_0(t)}\,\Delta\mathbf{f}(t)\,.$$

[6] Spiesberger and Metzger (1991*c*) and others have employed a form of objective mapping in which the time dimension is treated in the same way as the spatial dimensions – using a temporal covariance function, rather than the dynamic model that appears in sequential estimators based on Kalman-like filtering. Such objective mapping schemes are special cases of so-called Wiener filtering and smoothing algorithms.

Use of a linearized model defines the "linearized Kalman filter" (LKF). A smoother may be linearized in the same way.

Another possibility for linearization is to use the latest state estimate, $\mathbf{x}_0(t) \equiv \widehat{\mathbf{x}}(t)$, as the reference. A systematic procedure for this computation is called the "extended Kalman filter" (EKF), again with equivalent application to the smoothing problem. The EKF might appear a better choice than the LKF, but there are practical problems of stability, both numerical and statistical. We leave the discussion to the huge literature on the subject (Gelb, 1974; Sorenson, 1985).

Example: A Rossby-wave ocean. For illustration, we return to the model-oriented inversion of a single meridional slice through a Rossby-wave ocean (fig. 4.4), which served as an example for the static inverse (section 6.5). Here we make use of our explicit knowledge of the physics governing the time evolution.

The field of sound-slowness is supposed to be well described by the evolution equation (4.2.15), with only a single vertical mode present. For specificity, it is further assumed that the time series of the measured ray travel times has been Fourier-analyzed and that we are dealing with a single frequency, $\frac{1}{3}$ cycle per year. It is useful to put this model of the time evolution into the standard form (7.3.1). We follow the formulation of Gaspar and Wunsch (1989), allowing for the fact that here we are using many wavenumbers at a single frequency. The statevector consists of the in-phase and out-of-phase coefficients, as given by (4.2.17). For time $t = \Delta t$, we have $\mathbf{x}(\Delta t) = \mathbf{A}\,\mathbf{x}(0)$, where

$$\mathbf{A} = \left\{ \begin{matrix} \mathbf{A}_0 & \mathbf{0} & \mathbf{0} & \cdot & \mathbf{0} & \cdot \\ \mathbf{0} & \mathbf{A}_0 & \mathbf{0} & \cdot & \cdot & \mathbf{0} \\ \cdot & \cdot & \cdot & \cdot & \cdot & \cdot \\ \mathbf{0} & \mathbf{0} & \cdot & \cdot & \mathbf{0} & \mathbf{A}_0 \end{matrix} \right\},$$

with

$$\mathbf{A}_0 = \left\{ \begin{matrix} \cos(\omega\Delta t) & -\sin(\omega\Delta t) \\ \sin(\omega\Delta t) & \cos(\omega\Delta t) \end{matrix} \right\},$$

so that $\mathbf{A}$ is block-diagonal. Setting $\Delta t = 1$ gives the general form

$$\mathbf{x}(t+1) = \mathbf{A}\,\mathbf{x}(t)\,. \tag{7.3.11}$$

Equation (7.3.11) has the requisite canonical form for a model without forcing. The travel-time observations are given in standard form by (4.2.7) [ℓ is defined in terms of the statevector elements by (4.2.6*b*)] and the addition of a noise term.

To make an interesting example, assume that there is an estimate of the initial state, $\widehat{\mathbf{x}}(0)$, with uniform, large uncertainty of 100 m rms at all wavenumbers.

Observations yield in-phase and out-of-phase amplitudes of the travel-time variation over the 3-year cycle. The model's time step is taken as $\frac{1}{6}$ year, but with observations available only at the irregular times $t = 0.5,\ 1.2,\ 1.8$, and 2.5 years. The problem is simplified by assuming that the model is perfect [*i.e.*, that $\mathbf{w}(t) = \mathbf{0}$ in (7.3.1)]. [Known controls, $\mathbf{q}(t)$, could easily be accommodated, but would provide no additional insight.]

We have computed the Kalman-filtered and smoothed estimates of the statevector and its uncertainty. In the treatment of the static problem (section 6.5) we spoke of the difficulty of deriving information about 380 unknowns from five noisy pieces of information. There we had no information about the out-of-phase component. Now there is some information, but the problem is still heavily underdetermined, serving as a useful illustration of the time-dependent methodology.

Fig. 7.3a deals with estimates of $x_1(t)$, corresponding to the amplitude of the wavenumber at point A in fig. 4.3. The filtered estimate, which is a model prediction without data until $t = 0.5$ year, initially decreases in magnitude owing to the model physics (the nearly perfect agreement with the true value at 0.5 year, prior to use of the data, is coincidence). As time evolves, the intermittently available data give rise to adjustments in $\hat{x}_1(t)$ that tend to bring the estimates closer to the true value – estimates are almost everywhere consistent within one standard error. The smoothed estimate, $\hat{x}_1(t, +)$, is much "smoother," as its name implies; it carries information about the "future" observations backward in time and is a radically improved estimate of the original, very poorly known initial condition. The smoothed and filtered estimates hardly differ toward the end of the observation period, enough information having been accumulated to remove the very large initial error; but with the very noisy observations we postulated, new useful information accumulates slowly.

Fig. 7.3b shows the behavior of the time evolution of the uncertainty of $\hat{x}_1(t)$, $\hat{x}_1(t, +)$, that is, of the elements depicted in fig. 7.3a. The uncertainties are asymptotic – as more data arrive, they are only slowly improving at the end of the computation. The uncertainty has a natural periodicity, in the absence of data, of one-half the wave period, because (7.3.6) is quadratic in $\mathbf{A}$. That periodicity competes with the irregular interval postulated for the observation availability. Use of the smoother (*i.e.*, of future observations) reduces the uncertainty for early times. The structure of the uncertainties and the solutions is very similar to that for the other elements of the statevector for which information is available (fig. 7.3c).

The diagonal elements of $\mathbf{P}(0, +)$ are shown in fig. 7.3d. Recall that the initial uncertainty of all elements was $10^{-2}\ \mathrm{km}^2 = 10^4\ \mathrm{m}^2$. Most of the elements retain that uncertainty, despite all the physics and the measurements. It is no

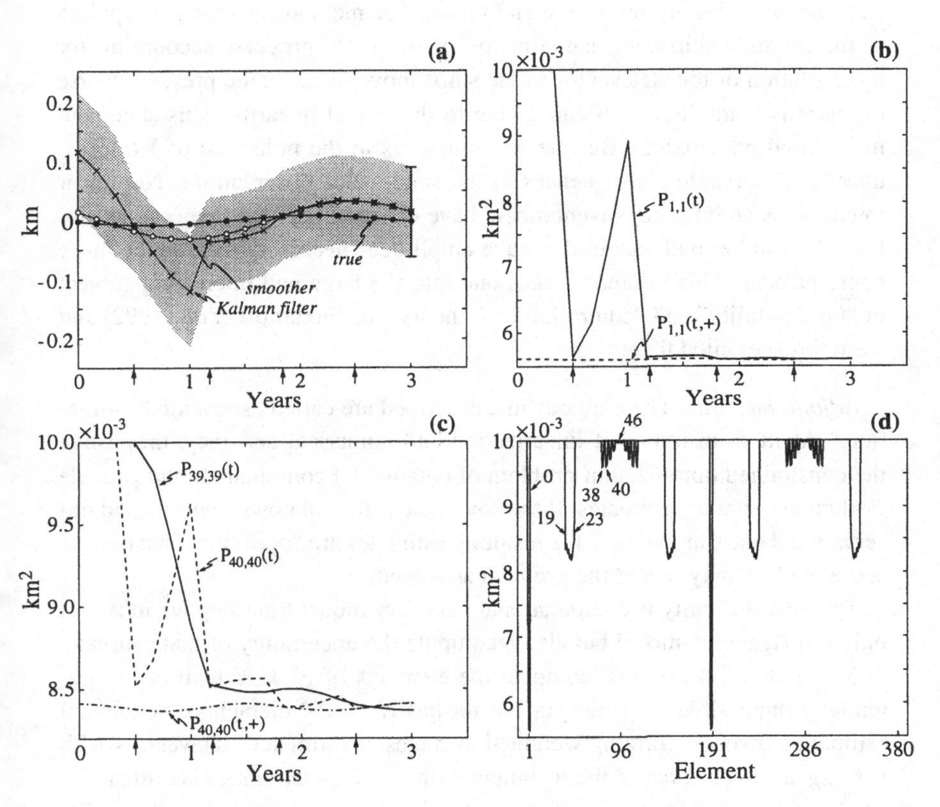

Fig. 7.3. (a) Evolution of the element $x_1(t)$ of the statevector for the Rossby-wave model. The Kalman-filtered estimates $\widehat{x}_1(t)$ (crosses) start with large error at $t = 0$, but converge toward the "true" values (dots) as data become available. The smoothed estimate $\widehat{x}_1(t, +)$ (open circles) gives better values at early times. Times of observations are shown. (b) The uncertainties $P_{1,1}(t)$ of the filtered estimate $\widehat{x}_1(t)$ and $P_{1,1}(t, +)$ of the smoothed estimate $\widehat{x}_1(t, +)$. (c) Uncertainties $P_{39,39}(t)$ and $P_{40,40}(t)$ of the filtered in-phase and out-of-phase components, respectively, and $P_{40,40}(t, +)$ of the smoothed out-of-phase component. (d) Uncertainty of $P_{i,i}(0, +)$ of the smoothed estimate $\widehat{x}(t, +)$ at t=0, as a function of the element number i.

surprise that a reduction in uncertainty occurs only for zero wavenumber (i = 1, 2, 189, 190) and for those wavenumbers that are in resonance with the ray loops (i = 39, 40, 41, 42, ..., 47, 48, *etc.*) and the first ray harmonic (i = 77, 78, 81, 82, ..., 93, 94, *etc.*), as discussed in chapters 4 and 6.

What has been gained by the extra effort of using a dynamic model? We have obtained information concerning the out-of-phase components, which were invisible to the static inversion. Further, each new time step has improved the state estimates for all times, past and future. The method can easily be applied to the multiple-ship coverage situation of fig. 6.16, properly accounting for the evolution of the statevector as the ships move. Finally, the present inverse problem is made more difficult owing to the model linearity. This statement may sound paradoxical. But the wavenumbers in the nullspace of $\mathbf{E}(t)$ produce no observable consequences in the single-slice observations. Nonlinear interactions among the wavenumbers have observable consequences, and inferences can be made about the wave amplitudes necessary to produce those consequences. This inference leads one into the large and interesting subject of "observability" and "controllability" theory; see Fukumori *et al.* (1992) and the references cited there.

Adjoint methods. The methods just described are called "sequential estimation." Apart from issues of linearization and numerical accuracy, they solve the constrained optimization problem of data/model combination and provide explicit uncertainty estimates. If the covariances for unknown controls and observational noise are known, the resulting estimates are "optimum" in a precise sense, and we may regard the problem as solved.

The only difficulty is computational. At every model time step we must not only propagate the model but also recompute the uncertainty of that estimate, (7.3.6) and (7.3.8), that is, compute the elements of $N \times N$ matrices (a potentially impossible job if the statevector has $N = 10^8$ elements). Sequential estimation involves forming weighted averages of estimated statevectors with varying accuracies; hence the recomputation of the uncertainties is critical.

This computing burden on sequential estimation is so forbidding that a different approach has recently attracted attention (*e.g.*, Talagrand and Courtier, 1987; Thacker and Long, 1988; Wunsch, 1988; Tziperman and Thacker, 1989). In the oceanographic and meteorological literatures this method has come to be called the "adjoint method." It has a much longer history in control and estimation theory where it is commonly known as the "Pontryagin principle."

The problem is identical with that already considered: find the minimum, by least-squares, of J in (7.2.5), subject to the dynamic model (7.2.1) with partially known and unknown sources, boundary conditions, and controls. The algorithm for finding the minimum proceeds somewhat differently. Instead of solving sequentially in time, the model is introduced explicitly into the objective function of (7.2.5), using a set of vector Lagrange multipliers, $\boldsymbol{\mu}(t)$, as in (6.3.13):

$$J'' = J' - 2\sum_{t=0}^{t_f-1} \boldsymbol{\mu}^T(t+1)[\mathbf{x}(t+1) - \mathbf{L}(\mathbf{x}(t), \mathbf{r}, t, \mathbf{q}(t), \mathbf{w}(t))] . \quad (7.3.12)$$

Differentiation of (7.3.12) with respect to $\mathbf{x}(t)$, $\mathbf{w}(t)$, $\boldsymbol{\mu}(t)$ is straightforward and leads to a set of normal equations for the unknowns. The solution to this (large) set of equations is the required estimate. The number of unknowns has been increased by the dimension of $\boldsymbol{\mu}(t)$ times the number of time steps. These equations are written out in a number of places (*e.g.*, Wunsch, in press) and fully determine all of the problem unknowns, including $\mathbf{x}(t)$, $\boldsymbol{\mu}(t)$, and $\mathbf{w}(t)$. The time evolution of the Lagrange multipliers is governed by $(\partial \mathbf{L}(t)/\partial \mathbf{x}(t))^T$, defining what is usually called the "adjoint model"; $\boldsymbol{\mu}(t)$ are the "adjoint solution." With linear models, $(\partial \mathbf{L}(t)/\partial \mathbf{x}(t))^T = \mathbf{A}^T$. For a linear model, an exact solution to the normal equations is possible (Brogan, 1985; Wunsch, 1988). More generally (*e.g.*, Thacker and Long, 1988; Ghil and Malanotte-Rizzoli, 1991), the equations are solved iteratively, exploiting the special relationships of the Lagrange multipliers to the statevector and control unknowns.

This Pontryagin-principle method seeks the solution over the entire time span at once, finding the solution by iteration, and not sequentially as in filter/smoother methods. Because of its use of all of the equations simultaneously, no weighted averaging is done, and one need not compute the solution covariances – resulting in a very large computational savings. Whether or not the solution without an uncertainty estimate is useful is up to the investigator. In any event, it is possible to prove formally (Bryson and Ho, 1975; Thacker, 1986) that the state estimates from the adjoint method and from smoothing are the same for linear problems, as indeed they must be if the same objective function and models are employed. Applied to the problem depicted in fig. 7.3a, $\widehat{\mathbf{x}}(t, +)$ would result. As with the sequential methods, we are as yet unaware of any direct use of adjoint methods with tomographic data and shall leave the subject, at this stage, as an important tool that will surely be central to future tomographic experiments.

7.4. Extensions: Control, Identification, and Adaptive Methods

Most models contain parameterizations of physics that are not completely understood. Common examples are mixing coefficients, which constitute an attempt to reduce extremely complex small-scale processes to a few simple parameters. The "model-identification" problem refers to the question of what values of these model parameters will give the best fit to the observations (keeping in mind the possibility that none of the values will provide an acceptable fit). Thus model identification is essentially the same as some of the inverse problems described in earlier chapters.

A somewhat different model-identification problem separately treats elements of the model, such as sources, sinks, and boundary conditions, as formal unknowns of the system. Consider the linear Rossby-wave equation (2.18.27) in an open-ocean domain (*i.e.*, with boundaries at least partially lying in the open sea). As a practical matter, the values of the dynamic variables on these open boundaries will be very uncertain. If tomography provides us with copious information about the interior fields, it becomes possible to calculate the boundary conditions that gave rise to the observed fields. Alternatively, the wind field blowing over the ocean will be partially or wholly uncertain. Again, with adequate interior tomographic or other observations, one may wish to calculate an estimate of the wind field. Precisely this calculation was described by Schröter and Wunsch (1986) (fig. 7.4) for a steady, nonlinear circulation model. Both these and related problems are usually denoted "control" problems or sometimes "boundary control" problems.[7] One is essentially asking what boundary conditions are required to drive the system to the observed state or sequence of states.

The distinction between these types of problems is somewhat arbitrary; to the extent that the wind curl is regarded as part of the model parameters, we have an identification problem; otherwise it is a boundary-control variable. In practice, procedures for solving such problems are the same, except that boundary-control problems can be linear, whereas the identification problems are very often nonlinear.

A general procedure for handling the model-identification problem is "state vector augmentation." Consider the linear barotropic Rossby-wave equation (2.18.27), but for a homogeneous fluid, written in terms of the pressure p, and with an additional friction term:

$$\frac{\partial \nabla^2 p}{\partial t} + \beta \frac{\partial p}{\partial x} - \varepsilon p = \widehat{\mathbf{k}} \cdot \nabla \times \boldsymbol{\tau}, \tag{7.4.1}$$

where $\boldsymbol{\tau}$ is the wind stress, and ε is an unknown coefficient of bottom friction. The tomographic observations are as before in (7.2.4). Suppose that (7.4.1) is discretized, so that the statevector in p evaluated at grid point j is

$$\mathbf{x}(t) = [p_j(t)], \tag{7.4.2}$$

satisfying a linear equation of form (7.3.1). Now augment the state to

$$\mathbf{x}'(t) = \begin{bmatrix} \mathbf{x}(t) \\ \varepsilon \end{bmatrix}.$$

[7] Strictly speaking, because we are using a partial differential system, the full terminology would be "distributed-system boundary-control problem." More conventional control problems are described by a system of ordinary differential equations and consequently are of much lower dimensionality.

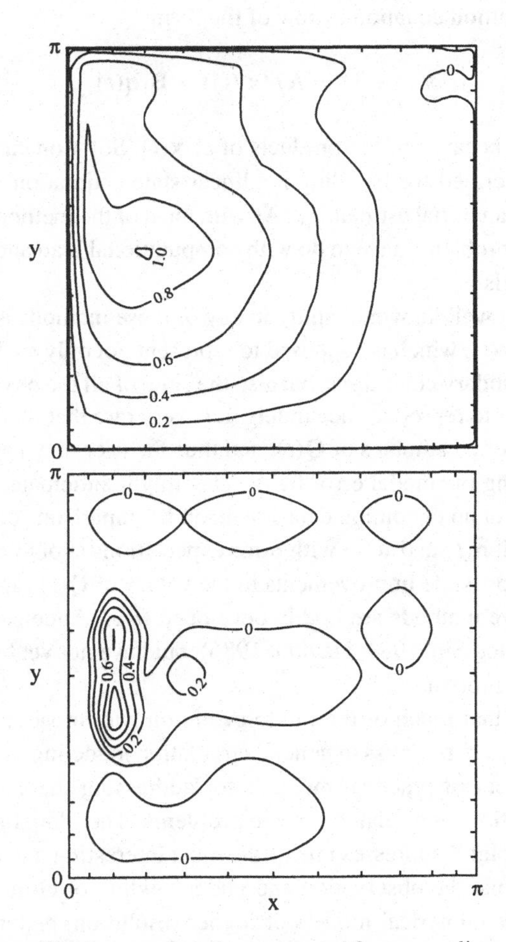

Fig. 7.4. Top: Normalized stream function computed from a nonlinear, steady GCM by Schröter and Wunsch (1986). In the computation, the wind field was treated as partially uncertain, and it was presumed that noisy tomographic measurements of the interior circulation (vorticity) were available. In addition, a sea-surface elevation was supplied in approximate form along a single zonal line near the basin center, mimicking an altimetric measurement. The solution was chosen to maximize the mass transport of the western-boundary current subject to the dynamics and the noisy observations. Bottom: Sensitivity of the maximum transport in the solution in the top panel to perturbations in the tomographic estimates of circulation. Such calculations show the data flow in the constrained model and are also extremely important in deciding where to deploy resources at sea. The sensitivity is proportional to the Lagrange multipliers (adjoint solution) of the objective function.

The state evolution equation is now of the form

$$\mathbf{x}'(t+1) = \mathbf{A}_L(\mathbf{x}'(t)) + \mathbf{B}_1\mathbf{q}(t)$$

and is nonlinear because of the products of ε, $\mathbf{x}(t)$. Solution can be by any of the methods described for handling nonlinear state estimation, including linearization about an initial estimate ε_0. As with most of the methods discussed in this chapter, the problems have to do with computational load and the existence of realistic models.

Often the least well known quantity in any of these methods is $\mathbf{Q}(t)$, the covariance of the $\mathbf{w}(t)$, which is supposed to represent not only such quantities as misspecified boundary conditions, but also the suite of all the physics the model fails to include or to reproduce accurately. It is very rare that modelers produce quantitatively useful estimates of $\mathbf{Q}(t)$, and thus the user may have grave difficulty in specifying the model error fields. One might anticipate, however, that the combination of an oncoming data stream and comparisons of the statistical behaviors of both $\widehat{\mathbf{n}}(t)$ and $\widehat{\mathbf{w}}(t)$ with prior expectations should carry information concerning possible improvements in the values of $\mathbf{Q}(t)$ [and $\mathbf{R}(t)$]. Such so-called adaptive methods are highly developed (*e.g.*, Anderson and Moore, 1979; Goodwin and Sin, 1984; Haykin, 1986), but have not yet been applied to the tomographic problem.

We anticipate that much of the future evolution and impact of tomography will be closely tied to progress in general circulation modeling of the ocean and in the development of practical methods for addressing the state-estimation, model-identification, and related inverse problems. The global-scale problem discussed in chapter 8 addresses one of the most interesting and difficult of all scientific problems: the observation and understanding of climate change. Its solution demands numerical models of higher resolution, better physics, and faster computation than models available today. Better models, in turn, require more and better data of the type that only tomography and a few other known technologies can provide.

CHAPTER 8

THE BASIN SCALE

In this chapter we discuss problems peculiar to very long range ocean acoustic transmissions, from sub-basin scales (a few megameters) to antipodal transmissions (20 Mm). Spatial averages on a basin scale are desirable for the study of ocean climate. Traditional methods have not given the data required for an understanding of ocean variability on this scale, largely because of the effects of the intense mesoscale variability. For example, local measurements at 1 km depth show a month-to-month mesoscale variability of order 1° C, suppressing any evidence of climate variability. A typical mesoscale coherence distance is 100 km. Horizontal averaging over 5–10 Mm reduces the mesoscale "noise" variance by a factor of order 100. Simulated records of axial travel times at ranges of order 10 Mm are shown in fig. 8.1. The climatic trend is taken as 0.2 s/year and is discernible over the mesoscale fluctuations after a few years.

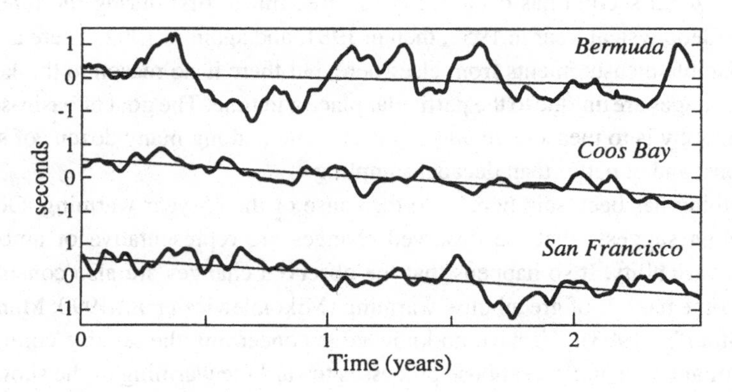

Fig. 8.1. Computer simulation of travel-time variations along three ray paths from Heard Island to the designated areas (fig. 8.7). The mesoscale-induced fluctuations based on the model of Semtner and Chervin (1988) are superimposed on the greenhouse-induced trend in 1990 according to the model of Bryan *et al.* (1988). There is more fluctuation for the Bermuda path than for the two eastward paths, as a result of passing through the intense eddy region in the area of the Agulhas Current retroflection.

Thus, ranges of 5–10 Mm are gyre-resolving and mesoscale-suppressing; they are the preferred dimension for acoustic thermometry of the ocean climate. The speed of sound (3000 knots) makes the measurements instantaneous for climate purposes. The inherent time scales are decadal, and this poses severe demands on the observing system.

Any conceivable global acoustic grid will necessarily be coarse. Our view is that the tomographic observations cannot and should not stand alone;[1] the emphasis here is on combining tomographic and other data with ocean general circulation models. This application differs from the earlier uses of ocean acoustic tomography, namely, the inversion of a relatively dense acoustic network toward the construction of fields of $C(x, y, z, t)$ and $u(x, y, z, t)$. In our interpretation of the inverse problem as a continuous transition, from dealing with data-intensive experiments (with some dynamic constraints) to data-scarce experiments (with massive model constraints), this chapter is an illustration of the latter category.

8.1. Climate Variability

Fig. 8.2 shows a warming at sound-channel depth of the order of 0.2°C in 35 years along an east–west section through the subtropical gyre of the Atlantic Ocean (Roemmich and Wunsch, 1984; Parrilla *et al.*, 1994). The associated change in acoustic travel time is roughly −4 s. The conclusion is that decadal variability in the oceans can produce easily observable changes in acoustic travel time.

This zonal section has been occupied three times: first during the International Geophysical Year in 1957, then in 1981, and again in 1992. There are no comparable measurements from elsewhere, but there is no reason to think that such changes are unique to the particular place and time. The goal of basin-scale tomography is to measure changes in heat content along many dozens of such sections, and at better than decadal sampling.

Nothing has been said here as to the cause of the 35-year warming. Ocean modeling suggests that the observed changes are representative of ambient ocean variability. It so happens that the observed changes are also consistent with some models of greenhouse warming (Mikolajewicz *et al.*, 1990; Manabe and Stouffer, 1993). We have no knowledge concerning the relative contributions of ambient and greenhouse processes toward the warming of the subtropical North Atlantic.

[1] In an essay on observing the oceans in the 1990s, Munk and Wunsch (1982*a*) develop the theme that the acoustic observations nicely complement satellite altimetry measurements.

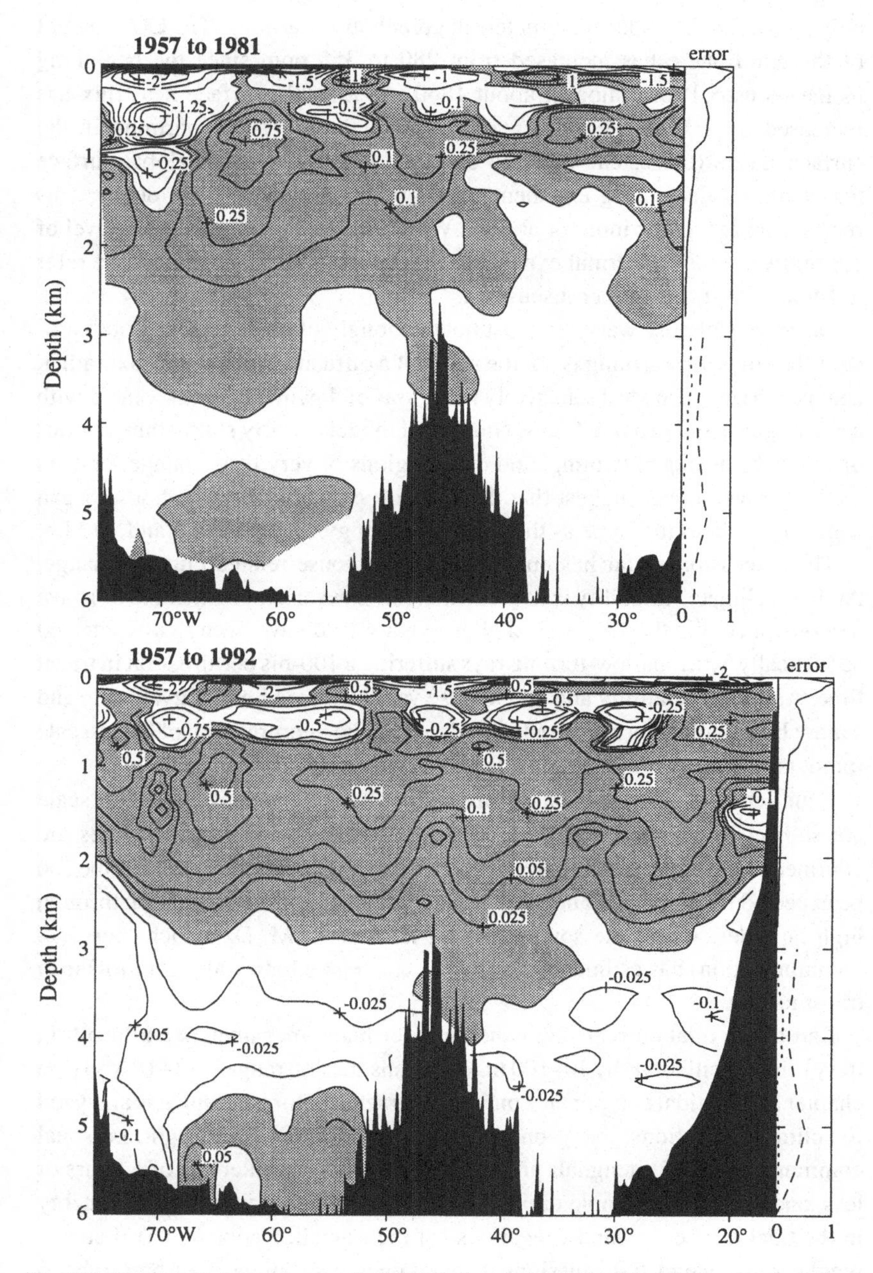

Fig. 8.2. Changes in termperature (in °C) relative to the International Geophysical Year (IGY) section in 1957 (Roemmich and Wunsch, 1984; Parrilla *et al.*, 1994). The section is across the North Atlantic Ocean along 24°N.

Next we consider some parameters of greenhouse warming. The CO_2 content of the atmosphere has increased from 280 to 355 ppm since the beginning of the industrial revolution in about 1860. As a result, surface heat flux has increased by 2 W/m^2. For orientation, take 0.02°C/year of warming in the surface layers (consistent with the observed change in mean global surface temperature), decreasing exponentially to 0.005°C/year at 1 km depth. This requires an extra heat input of about 2 W/m^2 and leads to a rise in sea level of 1.8 mm/year due to thermal expansion. These are acceptable values. We refer to Munk (1990) for further discussion.

Ocean greenhouse warming must not be thought of as a uniform global process. The interior warming is not the result of a diffusive downward flux; rather, heat is carried downward selectively in regions of downwelling associated with a convergence of horizontal flow. Numerical models clearly show some regions of twice the average warming, and other regions of very little change, or even cooling. The models suggest that the scale of variability in greenhouse ocean warming is about the same as the scale of ocean gyres, between 5 and 10 Mm.

The discussion so far has emphasized greenhouse-related climate change. Ambient climate variability is of great interest. Lawson and Palmer (1984) have demonstrated that the 1982–83 El Niño event would have been easily detected acoustically, with shallow-turning rays suffering a 100-ms perturbation in travel time. Ambient variability and greenhouse variability are both of gyre scale and cannot be separated by simple spatial filtering. The interpretation of basin-scale tomographic array measurements is a model-oriented inverse problem.

Time series of climate-induced travel-time fluctuations on a 5–10-Mm scale are subject to seasonal and tidal "noise." Seasonal temperature changes are confined to the upper few hundred meters. Seasonal variations in travel time can be expected for steep rays and high modes, particularly for propagation through high latitudes, where the sound channel is shallow. M. Dzieciuch (personal communication) has estimated $\tau_{\text{winter}} - \tau_{\text{summer}} = 3$ s for a California-to-Japan transmission.

Barotropic tidal currents are coherent over many megameters and result in travel-time oscillations by 10–100 ms over transmission ranges of 1–10 Mm (see chapter 3). The tidal components may be subtracted before the signal is analyzed for climate variations, using one of three methods (D. Cartwright, personal communication): (i) If signals are transmitted at regular intervals of 6 hours or less, one may design simple digital filters that will discriminate against energy in the tidal frequency bands, regardless of their detailed content. (ii) If such a regular sequence of transmissions is maintained over 29 days, or preferably a year, one may apply classic methods of analysis to extract "tidal constants" for the path and subsequently subtract the predicted tidal components. (iii) Tidal constants along the acoustic path may be inferred *a priori* from existing global models.

8.2. Some Experimental Considerations

Early experiments depended exclusively on explosive sources. Here the frequencies are typically very low, the bandwidth large, and the intensities high, as desired. The frequency is determined by the bubble pulse and decreases with increasing intensity and decreasing depth. The sources, both in the explosion phase and in the subsequent bubble pulses, tend to be nonlinear and not very repeatable. It is difficult to conceive of a phase-coherent analysis yielding adequate time resolution. Recent work and future plans depend on the use of coded, electrically driven sources. The requirement for decadal time series makes it almost mandatory that these be cable-connected to shore. This raises some issues concerning modification of the transmitted signal by bottom interaction near the source. With regard to receivers, these can be cable-connected to shore or autonomous, moored in deep water, intermittently monitored, and occasionally replaced without disturbing the continuity of the observational time series.

The general observational requirements for the non-explosive sources are discussed in chapter 5. Here we review a few considerations peculiar to very long-range transmission. Attenuation is now a major consideration. The working frequency is set by a balance between two conflicting requirements: resolution and signal-to-noise ratio (SNR). The former calls for a large bandwidth (5.2.9) and hence a high center frequency. The latter requirement calls for a low attenuation and hence low frequency (5.1.4). Cost favors the high frequencies.

Table 8.1 gives some possible parameters, following a procedure identical with that used in constructing table 5.1. The Atlantic attenuation values used here are about twice those in the Pacific; spreading losses are taken as spherical on a flat Earth; SNR is for a single hydrophone. For each of these reasons, one can consider circumstances that will lead to an improvement in the overall performance, as compared with the values in the table. But it is not easy to attain simultaneously a favorable resolution and a favorable SNR!

Internal waves at 5 Mm range are expected to produce a bias and a spread both of order 100 ms (4.4.3). The bias is of the order of the yearly climatic changes and need not concern us if the internal wave activity is statistically steady. Spread will interfere with the resolution of adjoining ray arrivals and is a matter of concern at long ranges.

Table 8.1. *Representative parameters for very long range transmissions*

Range (Mm)	1[a]	5	10
Frequency (Hz)	250	100	70
Source level (dB re 1 μPa at 1 m)	192	195	195
Attenuation (dB)	8	6	6
Noise level (dB re 1 μPa/$\sqrt{\text{Hz}}$)	68	70	75
Bandwidth (Hz)	83	33	23
Integration times (s)	196	1000	1200
SNR single hydrophone (dB)	19	15	15

[a]The first column corresponds to the values in table 5.1.

8.3. A Brief Historical Review

We summarize the very limited experience with basin-scale transmissions. The history of very long-range acoustic transmissions goes back almost 40 years. Initially these transmissions were limited to single explosive events [with the notable exceptions of Gordon Hamilton's explosive time series SCAVE 1961–3 and similar work by Johnson (1969) in the Pacific]. We first give a brief historical account of these events; they provide a useful reference for some of the phenomena discussed in this chapter.

Surprisingly, the earliest experiment was also the loudest. In 1955, during a period when numerous nuclear explosions were being set off to measure various effects in different environments, a 30-kton bomb known as WIGWAM was detonated at 650 m depth in deep water off the coast of California (Sheehy and Halley, 1957). The explosion produced acoustic echoes from many islands, seamounts, and other topographic features throughout the entire Pacific Ocean (fig. 8.3) (see section 8.8 for discussion). So far as we know, there have been no other deep nuclear explosions, so WIGWAM constituted a singular event, and one hopes it will remain so.

The earliest antipodal transmission (March, 1960) was an almost casual add-on to a geophysical survey. The R/V *Vema* and HMAS *Diamantina* used pressure detonators to fire 300-pound amatol charges near the sound axis off Perth, Australia. The detonations were clearly recorded by axial hydrophones off Bermuda, nearly halfway around the Earth (fig. 8.4). This fulfilled a 1944 prediction by Ewing and Worzel (1948), immediately following the discovery of the SOFAR channel (see Epilogue), that it would not be surprising if transmissions could

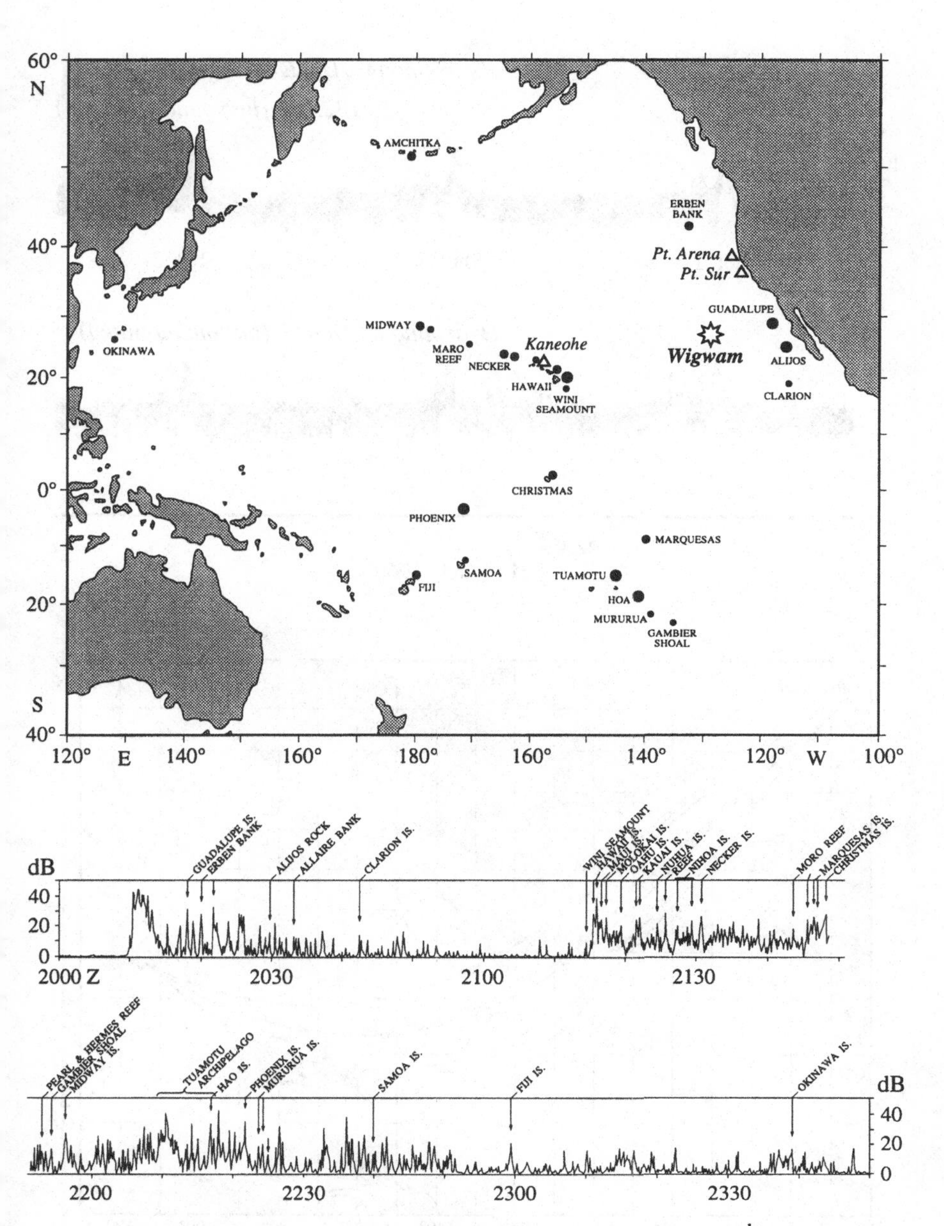

Fig. 8.3. The signal from the WIGWAM explosion on 1955 May 14^{h}.0000000 Z, recorded at Pt. Sur. The signal was also recorded at Kaneohe and Pt. Arena. The identified scatterers were located by triangulation from the three receiving stations and are designated by dots, with size of dot indicating intensity relative to the direct arrival: −4 to −14 dB, −14 to −24 dB, < −24 dB. (Adapted from Sheehy and Halley, 1957.)

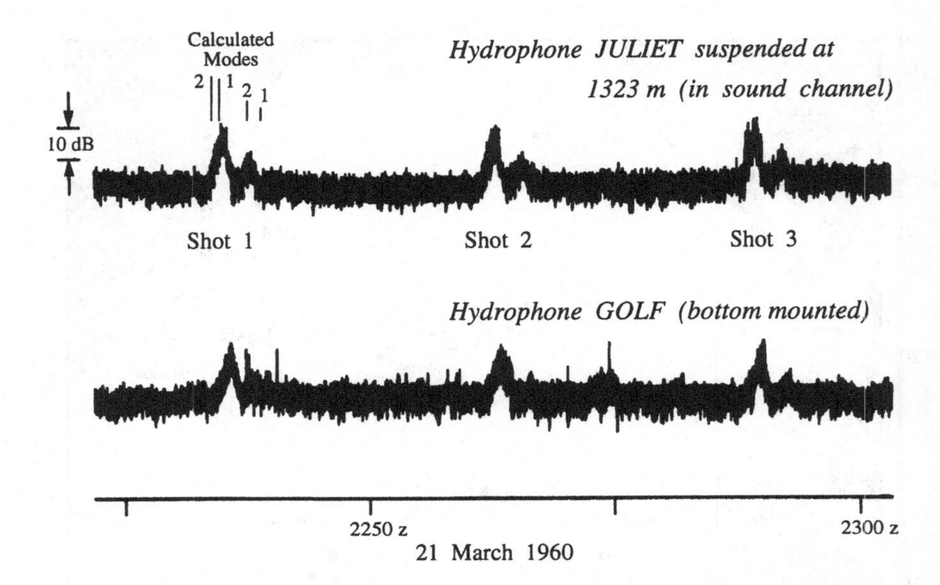

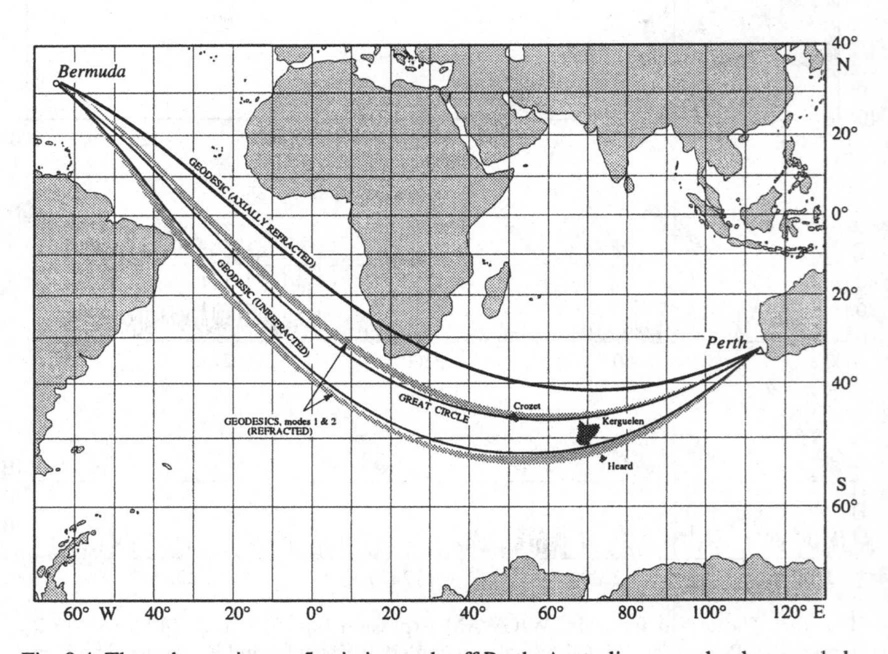

Fig. 8.4. Three detonations at 5-min intervals off Perth, Australia, were clearly recorded at Bermuda, halfway around the globe. The computed arrival times for modes 1 and 2 are indicated in the upper left corner. (Adapted from Shockley *et al.*, 1982.)

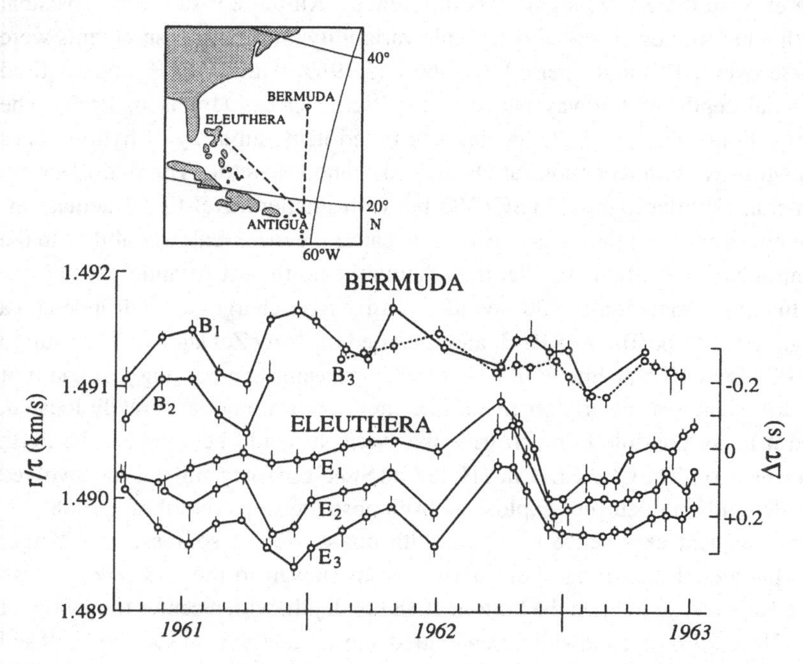

Fig. 8.5. Project SCAVE, showing variations in the mean axial sound-speed and the corresponding perturbation in travel time (relative scale), according to Hamilton (1977). Error bars give full spread of data. The source is at Antigua, the receivers at Bermuda (B) and Eleuthera (E).

be detected at ranges of 10,000 nautical miles. The Bermuda hydrophones were tended by Gordon Hamilton of the Lamont Geological Observatory of Columbia University. In 1961, 1 year following the Perth transmission, Hamilton (1977) initiated the SCAVE (Sound Channel Axis Velocity Experiment) transmission from Antigua to both Bermuda and Eleuthera (fig. 8.5). Ranges were of the order of 2 Mm. Precisely located and timed SOFAR explosive charges were fired at axial depth off Antigua, using the hydrophone array of the Atlantic Missile Range. From the axial cutoff of the received signal at the Bermuda and Eleuthera field stations, Hamilton was able to ascertain travel times to within 30 ms. Over a period of 27 months, the rms travel-time variations were estimated at 200 ms, with time scales of a few months; the maximum change was by 500 ms in 3 months. Hamilton remarked on the lack of correlation between the fluctuations for the Antigua–Bermuda and Antigua–Eleuthera paths (Bermuda and Eleuthera are separated by more than 1 Mm). Fluctuations at the three individual Eleuthera hydrophones (60-km separations) were clearly correlated, but even so there were significant differences. All these results are consistent with what we now know of mesoscale variability. Similar measurements were made over a 17-month period in 1966 and 1967, with SOFAR charges fired at axial depth off Midway Island in the Pacific Ocean (Johnson, 1969). The detonations were precisely located and timed using an array of hydrophones off Midway, with receptions at Oahu, Wake, and Eniwetok. The findings were generally similar to those in SCAVE, but with smaller travel-time fluctuations. We now know that this is as expected, because the mesoscale variability in the central North Pacific is smaller than that in the northwest Atlantic.

In April 1964, eighteen 50-pound explosive SUS charges were dropped on a flight from Cape Town to Perth and recorded off New Zealand at ranges up to 10,000 km (Kibblewhite *et al.*, 1966). One transmission passing just south of Heard Island was clearly received. The shot points are not accurately located, and it is not possible to reconstruct the events as could be done for the Perth transmission. The CHASE (Cut a Hole And Sink 'Em) program in 1966 involved the demolition of surplus explosives from obsolete ships (Northrop, 1968).[2]

Subsequent experience has been with non-explosive sources. Spiesberger has pioneered long-range transmission from Hawaii to the U.S. West Coast. The Kaneohe source, on the bottom at 183 m depth, with a center frequency of 133 Hz and 17-Hz bandwidth, transmitted intermittently from 1983 to 1989 and was received at a number of stations at ranges of 3–4 Mm (Spiesberger *et al.*, 1989*a,b*, 1992; Headrick *et al.*, 1993). In addition, a moored source from the 1987 Reciprocal Tomography Experiment (RTE87) (Worcester *et al.*, 1991*b*) was recorded at a range of about 3 Mm from May to September 1987 (fig. 8.6). This was the longest distance over which measured acoustic arrivals have been identified with specific rays (Spiesberger *et al.*, 1994). Spiesberger and Metzger (1991*c*) interpreted the diminishing travel times in terms of an increased range-averaged depth-integrated heat content by 4.2×10^8 J/m^2 over the 4-month period. [Dushaw *et al.* (1993*c*) estimate warming within the RTE87 triangle to be about four times larger.]

The Heard Island Feasibility Test (section 8.7) demonstrated transmissions by non-explosive sources up to antipodal ranges (Munk and Forbes, 1989; Baggeroer and Munk, 1992; Munk *et al.*, 1994). The Heard Island transmissions were from a moving source ship and lasted for only 5 days, so they have no bearing on measuring a variable ocean climate.

[2]We mention here another explosive "experiment" that achieved long-range acoustic detection. In 1883 the Krakatau volcano (in the Sunda Strait between Java and Sumatra) violently erupted and nearly destroyed itself. The atmospheric blast wave circumnavigated the globe *at least* three times. The pressure signature was recorded on ink charts at gasworks and on barometers worldwide (Simkin and Fiske, 1983).

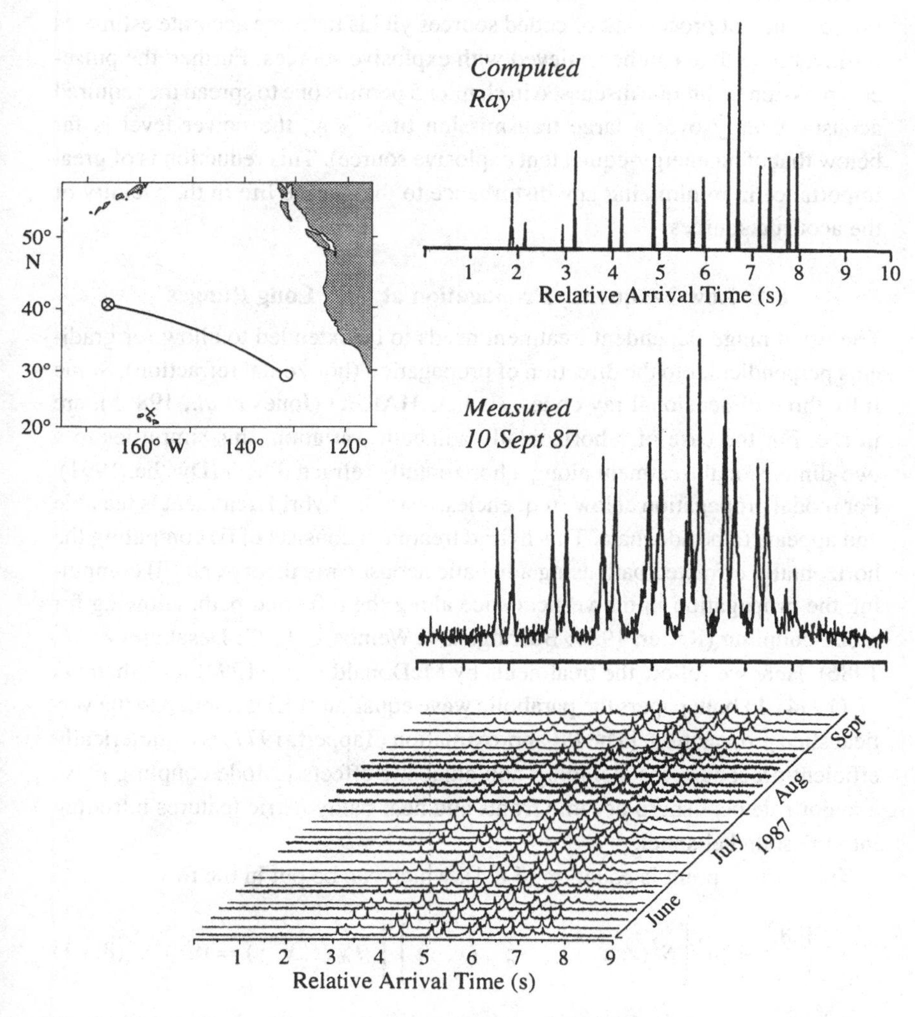

Fig. 8.6. Transmission from an RTE87 transceiver ⊗ (Worcester *et al.*, 1991*b*) to a receiver ○ at about 3 Mm range. (The Kaneohe source × is also shown.) The RTE87 source (250 Hz, 62.5-Hz bandwidth) was moored at the axis (860 m depth) in 5.5 km of water. The zero on the travel-time axis is about 30 min. A daily averaged record for 10 September compares favorably to the computed ray arrival pattern; see Spiesberger *et al.* (1994) for further details. The lower panel shows daily averages recorded every third day from 1 June until 10 September 1987. (From Spiesberger and Metzger, 1991*c*, with permission.)

There are some inherent advantages to the use of non-explosive sources. Phase-coherent processing of coded sources yields far more accurate estimates of travel time than can be achieved with explosive sources. Further, the pulse-compression technique discussed in chapter 5 permits one to spread the required acoustic energy over a large transmission time (*e.g.*, the power level is far below that of an energy-equivalent explosive source). This reduction is of great importance in minimizing any disturbance to the marine life in the vicinity of the acoustic sources.

8.4. Low-Frequency Propagation at Very Long Ranges

The usual range-dependent treatment needs to be extended to allow for gradients perpendicular to the direction of propagation (horizontal refraction). Some fully three-dimensional ray codes, such as HARPO (Jones *et al.*, 1986*b*), are in use. For the case of a horizontally adiabatic variation, this simplifies to a two-dimensional treatment along a horizontally refracted path (Dysthe, 1991). For modal propagation at low frequencies, a similar hybrid treatment is feasible and appears to be adequate. This hybrid treatment consists of (i) computing the horizontally refracted path using adiabatic acoustic ray theory and (ii) computing the propagation in the vertical slice along the refracted path, allowing for mode coupling (Keller, 1958; Burridge and Weinberg, 1977; Desaubies *et al.*, 1986). Here we follow the treatments by McDonald *et al.* (1994) and Shang *et al.* (1994). In both papers the parabolic wave equation (PE) is applied to the vertical slice propagation. [The PE approximation (Tappert, 1977) is a numerically efficient procedure for including non-adiabatic effects.] Mode coupling plays a major role near sharp oceanic fronts and near bathymetric features intruding into the sound channel.

The starting point is equation (4.6.4), which can be put in the form

$$\left\{\frac{\partial^2}{\partial z^2} + \omega^2 \left[S^2(z; x, y) - s^2_{p,m}(x, y) \right] \right\} P_m(z; x, y) = 0, \tag{8.4.1}$$

where $P_m(z; x, y)$ is the *local* mode function. Here, $s_{p,m}$ is the local phase slowness of mode m at point x, y. Boundary conditions are $P_m(0; x, y) = 0$ (pressure-release surface) and $P_m \to 0$ as $z \to -\infty$. McDonald *et al.* (1994) represent the Earth beneath the ocean bottom at $z = -h(x, y)$ by a homogeneous "bottom fluid" with density 1.5 times that of water, and with sound-speed 100 m/s above that of the bottom water. In general, the first 30 or 40 modes are trapped in the ocean sound channel. Higher modes penetrate significantly into the underlying fluid. McDonald and associates dissipate the bottom-penetrating energy by 0.1 dB per wavelength. Shang and co-workers follow a similar procedure, but with a somewhat different density and sound-speed for the bottom fluid. Solutions to (8.4.1) permit the mapping of the eigenvalue $s_{p,m}(x, y)$ over the Earth's surface and the construction of horizontally refracted ray paths for each mode at each frequency.

Let $\widetilde{x}$ be the distance from the source along the horizontally refracted path. The solution (4.6.2) within the vertical slice $(\widetilde{x}, z)$ is now written

$$p(\widetilde{x}, z, t) = \sum_m Q_m(\widetilde{x})\ P_m(z; \widetilde{x})\ e^{i\omega t}\ . \tag{8.4.2}$$

For a range-independent environment, this is simply

$$p(\widetilde{x}, z, t) = \sum_m P_m(z)\ e^{-i(k_m\widetilde{x} - \omega t)}, \tag{8.4.3}$$

with constant $k_m = \omega s_{p,m}$. In the adiabatic approximation, $P_m(z) \to P_m(z; \widetilde{x})$, and

$$k_m(\widetilde{x}) = \frac{\omega}{\widetilde{x}} \int_0^{\widetilde{x}} d\widetilde{x}\ s_{p,m}(\widetilde{x})\ . \tag{8.4.4}$$

To obtain a measure of mode scattering, Shang and co-workers and McDonald and associates employ the PE solution

$$p_{\mathrm{PE}}\,(\widetilde{x}, z, t) = \widetilde{x}^{-\frac{1}{2}}\ \Psi(\widetilde{x}, z)\ e^{-i(k_0\widetilde{x} - \omega t)} \tag{8.4.5}$$

where $k_0 = \omega S_0$ is a reference wavenumber, and Ψ obeys the equation

$$2i\, k_0\, \frac{\partial \Psi}{\partial \widetilde{x}} + \frac{\partial^2 \Psi}{\partial z^2} + \omega^2 \left[S^2(\widetilde{x}, z) - S_0^2 \right] \Psi = 0\ . \tag{8.4.6}$$

Writing $p(\widetilde{x}, z, t) = p(\widetilde{x}, z) e^{i\omega t}$, the orthogonality of $P_m(z; \widetilde{x})$ permits the evaluation of the range-dependent modal amplitudes $Q_m(\widetilde{x})$ according to

$$Q_m(\widetilde{x}) = \int dz\ p_{\mathrm{PE}}(\widetilde{x}, z)\, P_m(z; \widetilde{x})\ . \tag{8.4.7}$$

We refer to Shang *et al.* (1994) and McDonald *et al.* (1994) for further discussion.

8.5. Refracted Geodesics

For an ellipsoidal Earth with equatorial radius R_{EQ} and polar radius $R_{\text{PO}} = R_{\text{EQ}}\sqrt{1-e^2}$, the range for the Perth–Bermuda transmission (Shockley *et al.*, 1982; Munk *et al.*, 1988) is 19,820.7 km, as compared with 19,822.1 for a sphere with the appropriately weighted mean radius, using $e^2 = 0.006694605$ for the ellipticity, or equivalently $f = \frac{1}{2}e^2 = 1/298.75$ for the "flattening."

The geodesic (minimum path) is thus shorter by 1.4 km. But the important point is not this slight range difference, but the fact that the geodesic path lies significantly to the south of the great circle (fig. 8.4). The offset is the result of the shortest path being displaced toward polar latitudes, where the Earth's radius is smaller. The conclusion is that for long-range transmissions we cannot ignore Earth flattening, and great-circle geometry fails catastrophically for near-antipodal ranges.

In the adiabatic limit, the propagation on an ellipsoidal Earth is given by the equations (Munk *et al.*, 1988; Heaney *et al.*, 1991)

$$\frac{d\phi}{ds} = \frac{\cos\alpha}{\mu(\phi)}, \qquad \frac{d\lambda}{ds} = \frac{\sin\alpha}{\nu(\phi)\,\cos\phi},$$
$$\frac{d\alpha}{ds} = \frac{\sin\alpha\,\tan\phi}{\nu(\phi)} - \frac{1}{s_p}\left(\frac{\sin\alpha}{\mu(\phi)}\frac{\partial}{\partial\phi} - \frac{\cos\alpha}{\nu(\phi)\,\cos\phi}\frac{\partial}{\partial\lambda}\right)s_p, \tag{8.5.1}$$

$$\mu(\phi) = \frac{R_{\text{EQ}}(1-e^2)}{(1-e^2\sin^2\phi)^{\frac{3}{2}}}, \qquad \nu(\phi) = \frac{R_{\text{EQ}}}{(1-e^2\sin^2\phi)^{\frac{1}{2}}} \tag{8.5.2}$$

for both modal and ray representations, where s is arc length along the ray, λ is east longitude, ϕ is north latitude (the angle between the *local* surface normal and the equatorial plane), α is the local ray direction clockwise from north, and $s_p = \widetilde{S}$ is the phase slowness associated with ray n or mode m and equals the sound-slowness at the turning point (2.10.7); s_p is derived from (8.4.3), as previously discussed. The second term in the expression for $d\alpha/ds$ is essentially the gradient of s_p normal to the geodesic, and this is elegantly computed by Heaney *et al.* (1991) by taking horizontal derivatives of the equations governing s_p. The interpretation is that the ray turns away from the local geodesic at the same rate that it would turn away from a straight line if the surface were locally flat.

For the case that the sound-speed is a function of latitude only, the equation for $d\alpha/ds$ is equivalent to

$$H = \nu(\phi)\,S(\phi)\,\cos\phi\,\sin\alpha = \text{constant}, \tag{8.5.3a}$$

which is Snell's law on an ellipsoid. When S is constant,

$$H = \nu(\phi)\, \cos\phi\, \sin\alpha = \text{constant}, \tag{8.5.3b}$$

yields the equation for the geodesic, which reduces to the great-circle equation

$$\cos\phi\, \sin\alpha = \text{constant}, \tag{8.5.3c}$$

on a sphere.

The Heard Island Feasibility Test afforded an opportunity to compare the computed geodesics with observations. The source ship was steered as closely as possible to a straight and steady course, necessarily directed into the prevailing wind and sea during each hour-long transmission. The ship position was monitored every 10 s using the Global Positioning System (GPS). The acoustic signal was measured at Ascension Island, approximately 10,000 km to the northwest (fig. 8.7). If the ship's heading were directly along the geodesic launch azimuth, we would expect the measured Doppler to be consistent with the

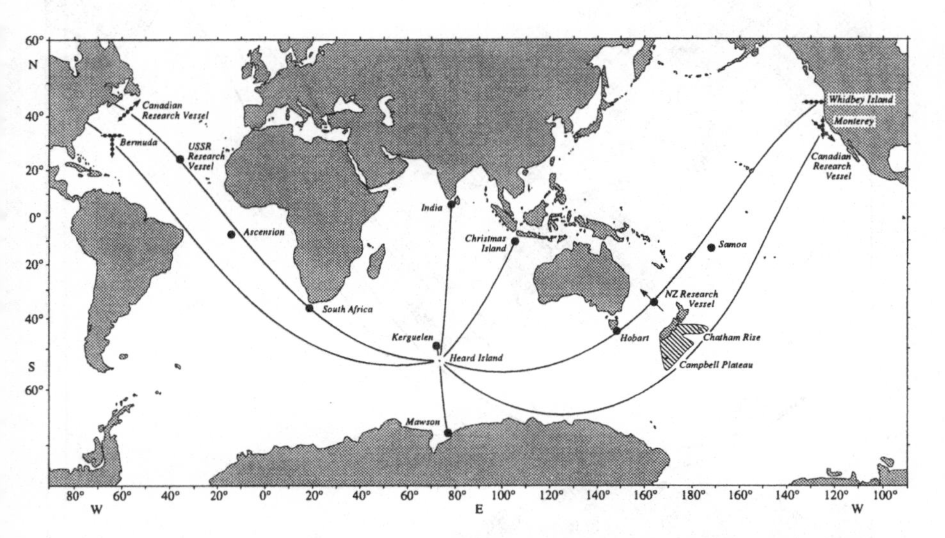

Fig. 8.7. Heard Island Feasibility Test. The sources were suspended from the center well of the R/V *Cory Chouest* 50 km southeast of Heard Island. Black circles indicate receiver sites. Horizontal lines represent horizontal receiver arrays off the American West Coast and off Bermuda. Vertical lines designate vertical arrays off Monterey and Bermuda. Lines with arrows off California and Newfoundland indicate Canadian towed arrays. Ray paths from the source to receivers are along refracted geodesics, which would be great circles but for the Earth's nonspherical shape and the ocean's horizontal sound-speed gradients. Signals were received at all sites except the vertical array at Bermuda, which sank, and the Japanese station off Samoa.

ship's total speed; if it were at a right angle to the launch azimuth, we would expect zero Doppler. A comparison of the GPS positions and the measured Doppler thus provides a measure of the acoustic launch angle at the source. Forbes and Munk (1994) derived a launch azimuth from Heard Island to Ascension Island of 268.1 ±0.1°, as compared with 268.05° for the refracted geodesic launch azimuth of mode 1 at 57 Hz. The corresponding angle for the axially refracted geodesic is 268.2°, and for the nonrefracted geodesic it is 265.9°.

Forbes (1994) found two geodesic paths from Heard Island to Whidbey Island (fig. 8.8). The measured Doppler yielded a launch azimuth of 130°,

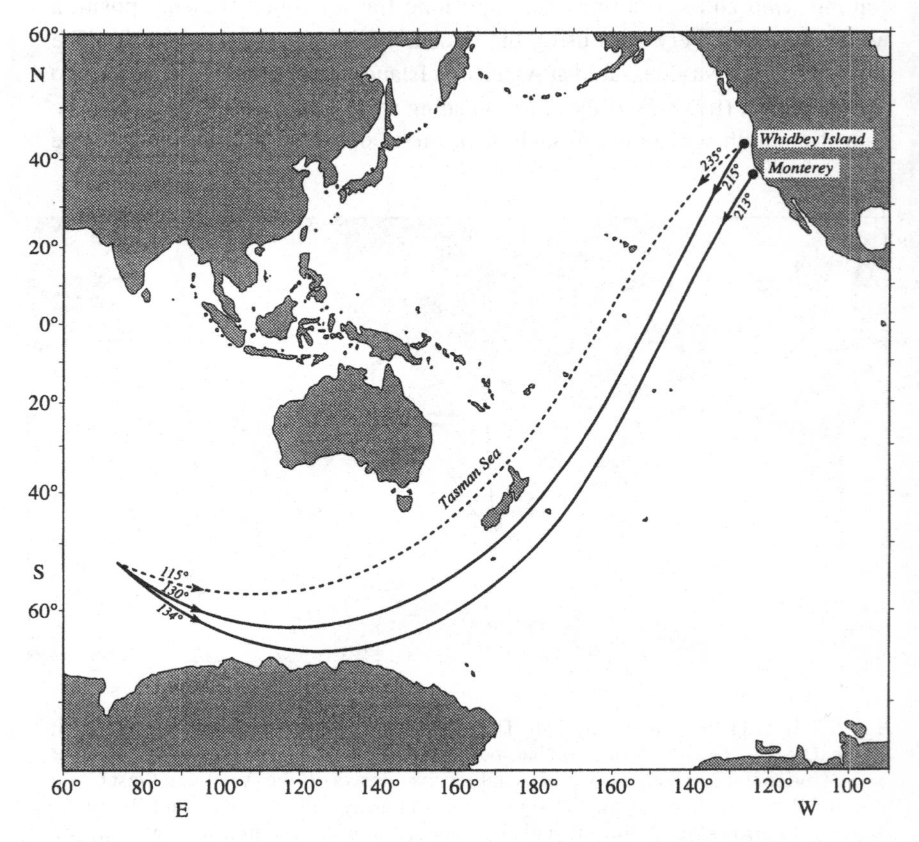

Fig. 8.8. Refracted geodesics from Heard Island to the American West Coast. The dashed path was apparently not successful. The path with a launch azimuth of 130° and a receiver azimuth of 215° was drawn with some artistic license. (From Forbes, 1994, with permission.)

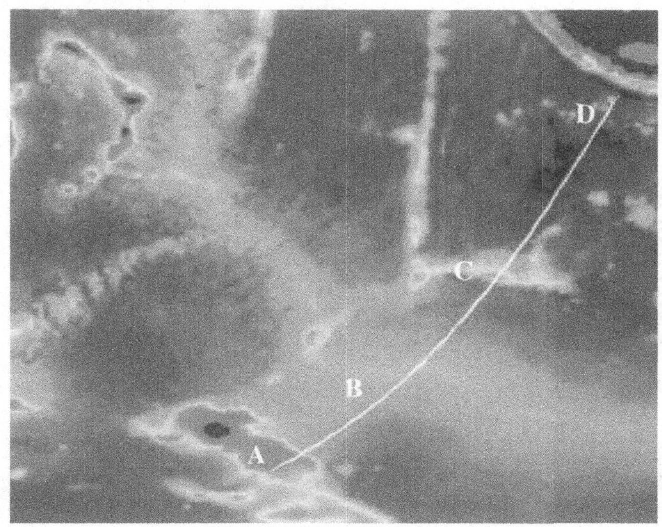

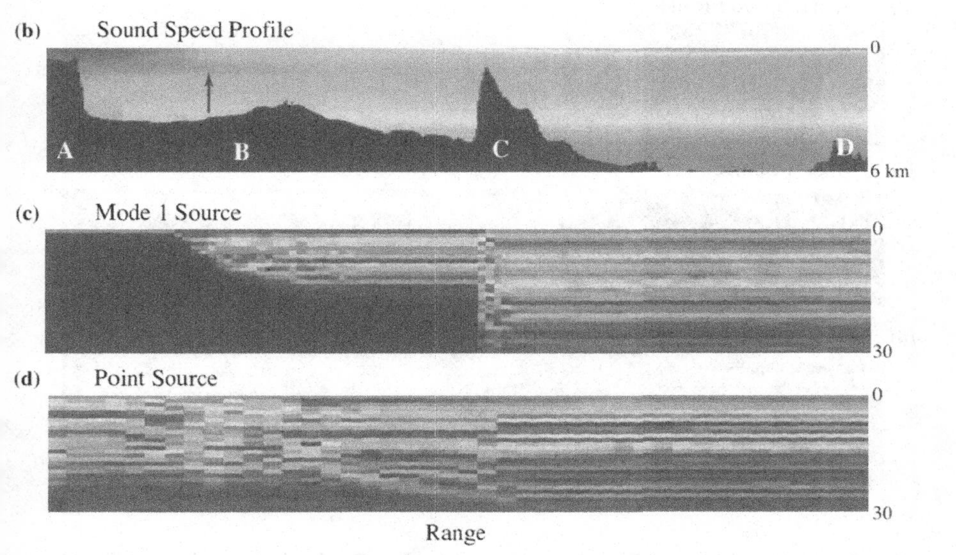

Fig. 8.10. Computed mode scatter for transmission from Heard Island to Christmas Island: (a) refracted geodesic for mode 1, (b) bottom bathymetry and the Levitus climatological sound speed field along the refracted path, increasing from blue (1440 m/s) through green through yellow to red (1550 m/s), (c) relative intensities of modes 1 (top) to 30 (bottom), increasing by 10 dB from blue to red, for mode 1 excitation, and (d) for multimode excitation by a point source at 175 m. The water at the source and receiver is sufficiently deep that there are no significant bathymetric terminal interactions. The transmission path crosses the Antarctic Circumpolar Front (ACF) to the left of point B (arrow), where the sound speed profile changes from a polar duct to an interior channel. At point C the path crosses Broken Ridge. (From McDonald *et al.*, 1994, with permission.)

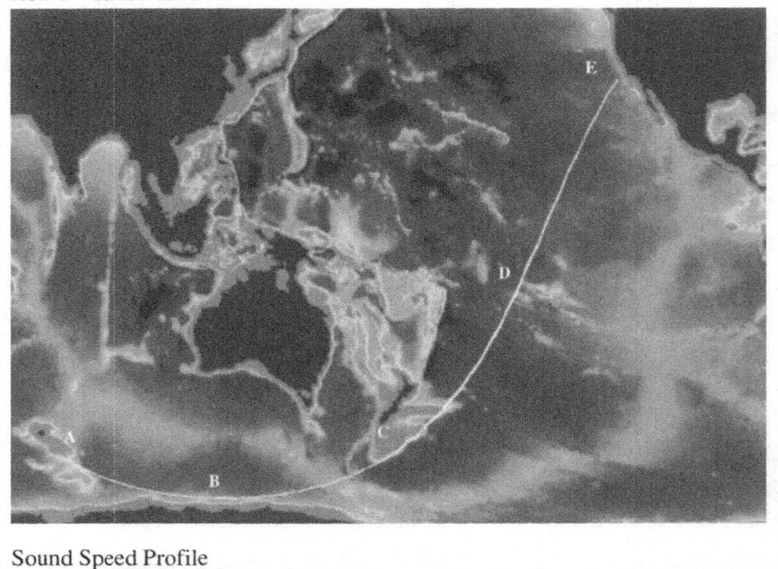

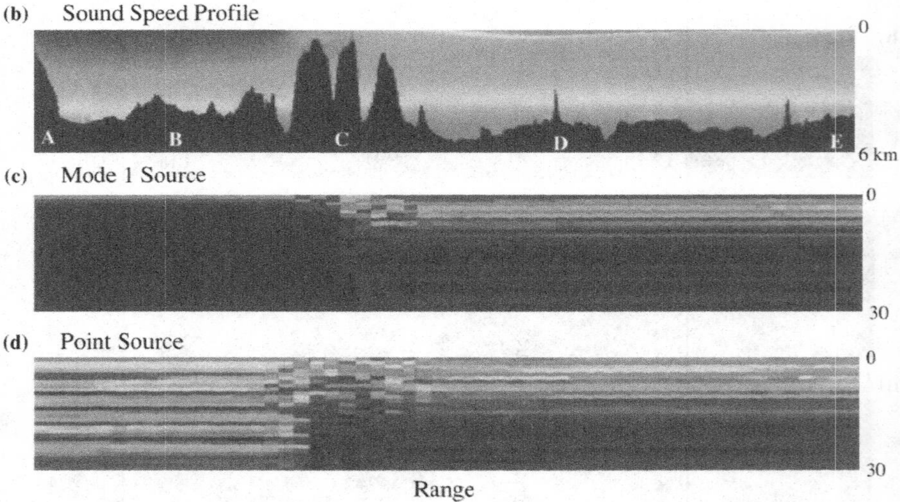

Fig. 8.11. As in fig. 8.10, but for the Heard Island to California transmission. The path crosses the ACF near the location of the three shoals centered at C (Campbell Plateau, Chatham Rise).

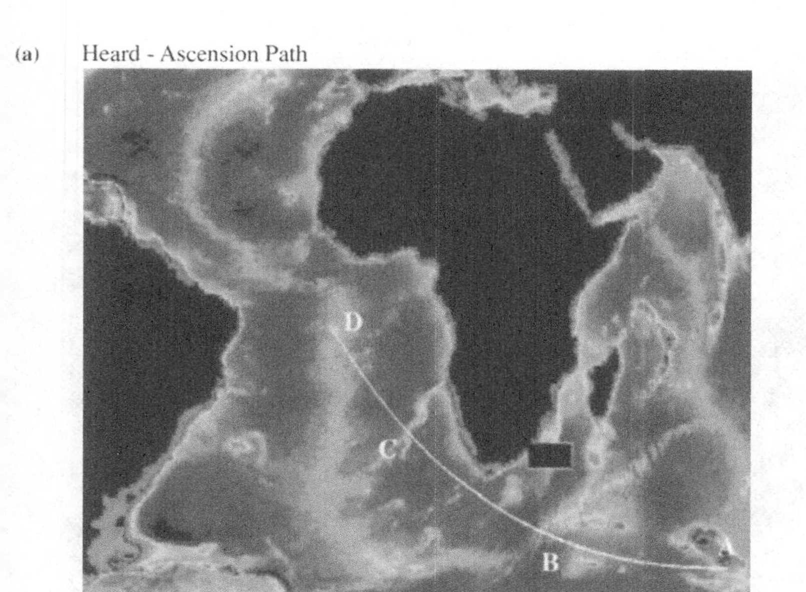

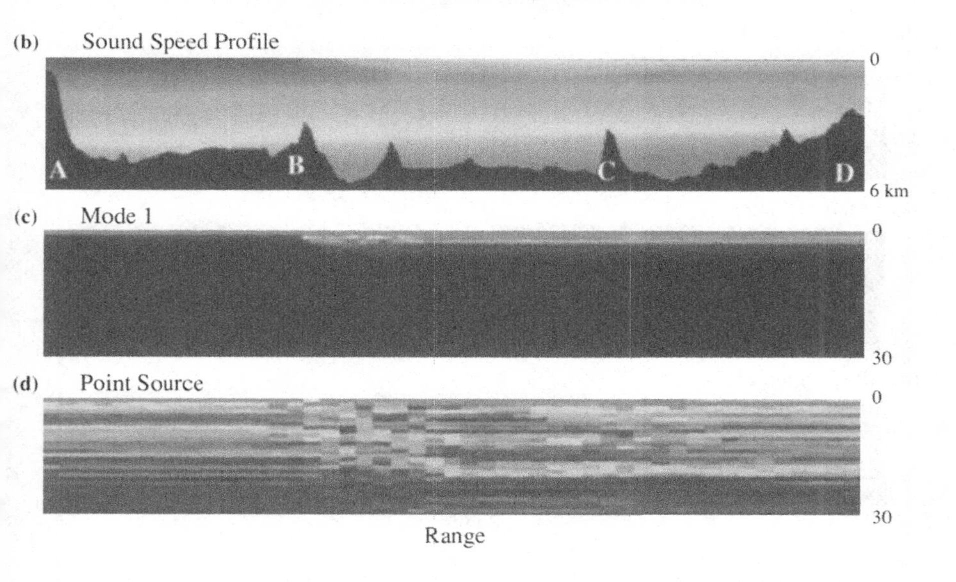

Fig. 8.12. Heard Island to Ascension Island. The Antarctic circumpolar convergence (AAC) is crossed near point B.

Fig. 8.13. Point Sur, California, to New Zealand. A glitch in the sound speed field near New Zealand (to the far right) is responsible for the apparent terminal mode scatter.

consistent with a path to the *east* of New Zealand. At Whidbey Island, multiple hydrophones could be accessed for horizontal beam-forming, giving an arrival azimuth of 215° from east of New Zealand. Thus the measured launch and arrival azimuths preclude a passage through the Tasman Sea. A series of transverse ridges lying between northern New Zealand and New Caledonia and Fiji appears to have blocked the transmission from exiting the Tasman Sea.

Thirty years prior to the Heard Island Feasibility Test, explosive charges detonated in the sound channel off Perth, Australia, were clearly recorded at Bermuda. As first reported by John Ewing (Maurice Ewing's brother) in "Notes and personalia" (Ewing, 1960), the recording was considered particularly remarkable inasmuch as the great circle is blocked by Kerguelen Bank and the Crozet Islands. Problems of bathymetric blockage and departures from great circles occupy much of this chapter. A discussion of the Perth–Bermuda transmission followed the original note by 22 years (Shockley *et al.*, 1982). Munk *et al.* (1988) revisited the problem in an attempt to account for the bathymetric leakage and the unexplained double pulse at Bermuda (fig. 8.4). Using the geodesic equations (8.5.1) in the limit of low modes of high frequency, $s_p \to S_{AX}$, they found that the southward displacement of the geodesic relative to the great circle avoided the blockage by Kerguelen Bank and the Crozet Islands, but when refraction was taken into account, the refracted geodesic was displaced far northward and collided with the tip of South Africa. Bermuda lies in the shadow of the Cape of Good Hope. Theoretical attempts to account for the Bermuda detection by diffraction and bottom scatter near the cape were unsuccessful.

For the explosive source in the Perth-to-Bermuda transmission, the center frequency is estimated at 15 Hz, and the high-frequency limit $s_p = S_{AX}$ is not a useful approximation. By allowing for horizontal modal refraction at 15 Hz (rather than for $f \to \infty$), Heaney *et al.* (1991) were able to establish an unblocked transmission path (fig. 8.4).

We have not yet taken into account the refraction by bathymetry (section 4.7). In general, rays are repelled from regions of high temperature (high sound-speed), small depths, and high latitudes. All of these effects depend on mode number and frequency, or on ray number. Heaney *et al.* (1991) found that there were two groups of Perth–Bermuda eigenrays: group A passed just south of the Cape of Good Hope; group B passed almost 1 Mm to the south of the cape and eventually interacted slightly with bathymetry off Brazil (fig. 8.4). Travel times were

$$A: \; 13,364 \pm 5 \text{ s observed}, \qquad B: \; 13,394 \pm 5 \text{ s observed},$$
$$\quad\; 13,354 \pm 5 \text{ s computed}, \qquad \quad\; 13,403 \pm 9 \text{ s computed}.$$

This accounts satisfactorily for the double pulses measured at two Bermuda hydrophones for each of three shots. Munk *et al.* (1988) incorrectly attributed the double pulses to reflections from the steep southeastern flank of the island to the offshore hydrophones.

8.6. Spheroidal Caustics

The foregoing numerical path construction allowed for volumetric, bathymetric, and spheroidal refractions. The spheroidal effect can be treated analytically, and it is interesting to inquire whether or not it can account for the observed double pulses. Longuet-Higgins (1990) has considered the ray paths near the antipode of a slightly oblate ellipsoid of mean radius R and flattening f. A ray caustic in the form of an asteroid or 4-star centered at the antipode S′ at latitude ϕ, with outer radius

$$r_{\text{outer}} = \pi \; f \; R \; \cos^2\phi,$$

and inner radius $\frac{1}{2}\, r_{\text{outer}}$ (fig. 8.9), separates the region near the antipode from the rest of the globe. At any point P within the 4-star there are 4 ray arrivals A, B, C, D; outside there are 2 ray arrivals. We consider only ranges of less than half the Earth's circumference, in which case the number of rays is reduced to 2 inside and 1 outside the 4-star.

The antipodal travel time is $\tau_{\text{AP}} = \pi\; RS \approx 13,400$ s; differences between the two ray arrivals within the 4-star are of order $f\, \tau_{\text{AP}} = 45$ s, where $f \approx \frac{1}{300}$

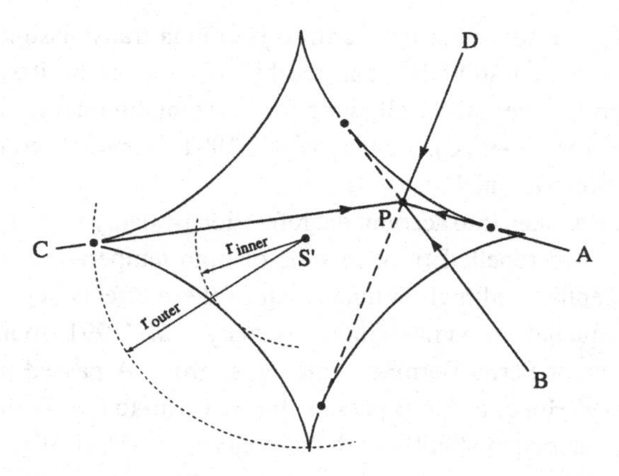

Fig. 8.9. Ray caustics near the source antipode S′ on an ellipsoidal Earth (Longuet-Higgins, 1990). At any point P within the 4-star there are 4 rays, each tangent to one of the caustics. Outside the 4-star there are only 2 rays. As P moves to the northeast toward the nearest caustic, the angle between rays A and B diminishes.

is the Earth flattening, as compared with the measured 30 s for the Perth-to-Bermuda transmission (fig. 8.4). However, using $\phi = 32°$ yields $r_{\text{outer}} = 50$ km, as compared with a distance of 180 km from Bermuda to the antipode. We conclude that ellipsoidal refraction plays a significant role in determining the two-pulse arrival, but that other factors (such as continental repulsion off Brazil) may also be involved.

Nothing has been said about intensities. We expect the intensities along the caustics to be high, and at any of the cusps to be even higher. These are increasing orders of singularity in catastrophe theory. In the real ocean, internal waves and other irregularities cause the acoustic wave fronts to become increasingly wiggly, and under the expected saturated conditions the caustic features must become increasingly diffuse; perhaps the two arrival peaks at Bermuda that were numerically accounted for by Heaney *et al.* (1991) were largely associated with the caustic cusps spreading into the area outside the 4-star.

With diminishing oblateness, the 4-star caustic collapses to a single point at the antipode, the most severe singularity. The suggestion that waves propagating on a sphere converge at the antipode goes back to Marconi in 1922; Gerson *et al.* (1969) conducted an experiment between Perth and Bermuda (!) to test the hypothesis of an antipodal focus for high-frequency radio waves. They found the reception at the antipode to be superior to that at two other stations, each located about 1700 km closer to the transmitter. We do not have any observational evidence concerning the amplification of sound waves at antipodal receptions.

8.7. Mode Stripping and Repopulation

In any basin-scale transmission, we are bound to encounter significant variations along the path. We make a distinction between two cases: (i) *topological* variations, such as the splitting of the sound channel into two channels, or the transition from a polar (RSR) duct to a temperate (RR) sound channel, and (ii) mesoscale and other statistical variability. The topological case can be avoided, or at least minimized, by the choice of path; there is not much one can do about avoiding mesoscale variability.

It has long been recognized that CW transmissions across fronts can be associated with dramatic intensity changes; in fact, this has become part of the dogma of anti-submarine warfare. Brekhovskikh and Lysanov (1991, fig. 1.11) give an example of a 10-dB intensity change across the Gulf Stream. Akulichev (1989) has measured a 17-dB drop associated with crossing the Kuroshio current. The receiver was at 100 m depth off Kamchatka; the source was towed at 100 m depth from subtropical to subarctic waters, keeping a fixed range of

2100 km. These are not surprising results. They are roughly what one expects from adiabatic propagation theory. But they illustrate the difficulty of doing tomographic work under such extreme conditions.

Fig. 8.10 shows the evolution of the 57-Hz PE solution along the mode-1 path from Heard Island to Christmas Island, as reported by McDonald *et al.* (1994). The second panel shows an interior sound minimum to the right, and a polar duct to the left. The transition near point B (arrow) is associated with the Antarctic Circumpolar Front (ACF). At this point, for an initial mode-1 excitation, the energy is scattered into successively higher modes, until the first 15 modes are significantly populated. At point C the transmission encounters the Broken Ridge, with strong bathymetric scattering into higher modes. There is no further significant interaction, volumetric or bathymetric, until Christmas Island.

The long transmission to California (fig. 8.11) crosses the ACF at a glancing angle near the location of the three shoals centered at C. Thus, volumetric and bathymetric scatterings are not well separated. There is a gradual transition of mode 1 to adjoining higher modes (third panel), and at the same time a sharp stripping of modes above $m = 10$ (fourth panel). On arrival in California, the first 8 modes are well populated, in agreement with vertical array measurements by Baggeroer *et al.* (1994).

The Atlantic path to Ascension (fig. 8.12) crosses the ACF near point B, leading to some limited scatter of mode 1 into neighboring modes. For future reference, we include a proposed transmission path from Pt. Sur, California, to New Zealand (fig. 8.13). This takes place in a temperate profile with no shoaling, and there is little scattering among the lowest 30 modes. The California point source (left in lower three panels) is on the axis at 600 m depth in 3000 m of water. (Under anticipated actual conditions, the source is close to the bottom, and considerable sea floor interaction is expected.)

Returning to the Heard Island–Christmas Island transmission illustrated in fig. 8.10, McDonald *et al.* (1994) computed the arrival pattern in the time domain. After the path integration at 21 frequencies (equally spaced between 52 and 62 Hz) for each of the 30 excited modes, the resulting spectral amplitudes at the receiver were added and Fourier-transformed to the time domain. We refer to McDonald *et al.* (1994) for a discussion of the procedure. Fig. 8.14 gives a comparison of the measured and computed arrival patterns. The patterns show comparable spreads and complexities, and a vague resemblance in overall structure. A closer agreement is not to be expected. The $C(z; x)$ field used in the computation is based on the climatological mean, which must differ significantly from the instantaneous field at the time of the transmission. In

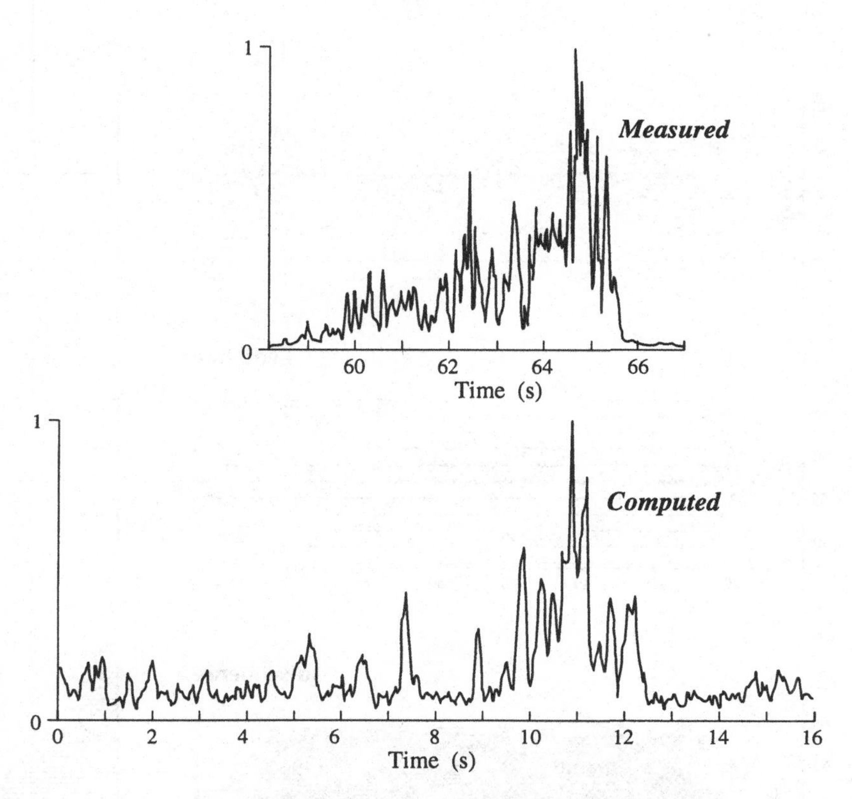

Fig. 8.14. Measured (top) and computed (bottom) arrival patterns at Christmas Island.

fact, details in the measured patterns are not correlated between successive transmissions.

8.8. Basin Reverberation

Basin-scale experiments are accompanied by a sustained reverberation following the direct arrival. In the case of WIGWAM (fig. 8.3), the direct arrival at Pt. Sur 10 min after the explosion was followed by more than 1.5 hours of recordings 10 dB above background. The explosion produced acoustic echoes from many islands, seamounts, and other topographic features throughout the Pacific Ocean. As indicated in the figure, all of the recorded peaks in the echo signals represent what is nearly backscatter.

In May 1966, CHASE V was detonated off Cape Mendocino, California, at a depth of 1125 m, with a yield of about 1 kton. This involved the demolition of surplus explosives from obsolete ships. The signal was recorded off Hawaii

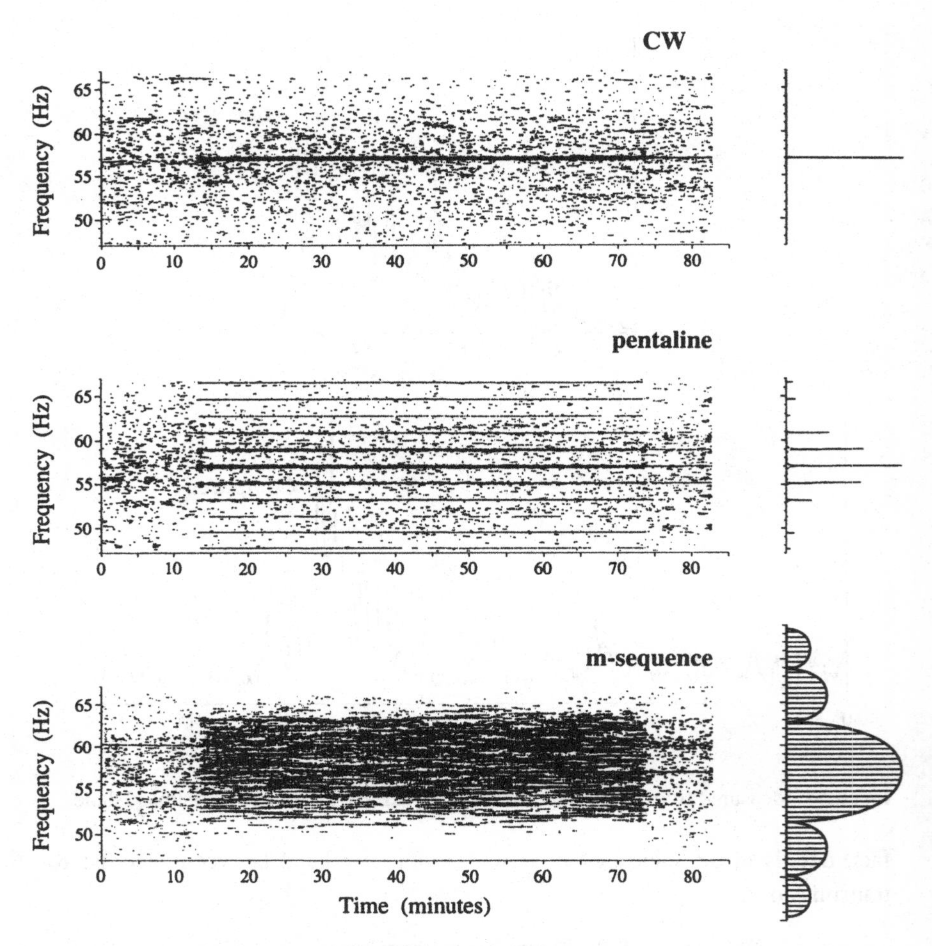

Fig. 8.15. Heard Island Feasibility Test. Three signal modulations and their spectra as recorded at Ascension Island. Top: 57-Hz continuous-wave tone. Middle: pentaline code. Bottom: m-sequence pseudo-random code. These are sonograms, or time–frequency plots. Some 60-Hz noise is evident before and after the transmissions. The persistence of the lines for several minutes after the 1-hour transmission is attributed to scattering and multiple horizontal paths.

by hydrophones suspended from R/P *FLIP*. Reverberations lasted 2.5 hours beyond the direct arrival. The earliest reflections came from the continental slope and were −3 dB relative to the direct arrival; subsequent reflections from the Hawaiian arc were of higher intensity, up to −20 dB. Northrop (1968) remarked that the higher signal level of the Hawaiian reflections may have been due to steeper slopes and lack of sedimentary cover. Munk and Zachariasen (1991) attempted to interpret the scattered intensity in terms of the principles discussed at the end of chapter 4.

For the Heard Island Feasibility Test there is a distinct "afterglow" following the 1-hour coded transmission (fig. 8.15). We refer particularly to the 57-Hz carrier (which contains half the transmitted power) in the m-sequence. (The 60-Hz line that is present prior to and following the transmission is due to ship electric noise.)

Basin reverberation complicates the analysis of long-range acoustic experiments. But if it is possible to identify individual scatterers in the arrival sequence, this might provide useful tomographic information along the scattered paths.

8.9. The Future of Basin-Scale Tomography

Clearly, we know very little about basin-scale tomography. At this time we can only state some underlying principles and report on some observed phenomena. Considerations of the SNR may need to take into account the signal-connected noise (reverberation). We have yet to learn whether or not there are identifiable and stable ray arrivals at the 5–10-Mm scale. The inversion formalism may have to include consideration of volumetric and bathymetric mode-to-mode scatter.

EPILOGUE

THE SCIENCE OF OCEAN ACOUSTIC TOMOGRAPHY

We conclude with a short discussion of the scientific capabilities that have been demonstrated over the past 15 years and the resulting insights into ocean processes.

This book is being written toward the end of a period of intensive technical developments of tomography, and our emphasis is necessarily on the machinery involved in making the method work, with much less emphasis on the science. Although the reader may have the impression that acoustic tomography is unusual in having such an extended period of development, almost every technology now widely used at sea has undergone a similar 15–20-year gestation period (Wunsch, 1989).[1]

It is convenient to divide the findings into a number of overlapping themes.[2]

Mesoscale mapping. The idea for ocean acoustic tomography emerged in the wake of the clear realization by oceanographers in the early and middle 1970s that the ocean was turbulent – filled with mesoscale eddies and other variabilities on space scales of $O(100 \text{ km})$ and on time scales of months. The experience in the field programs conducted in the 1970s – to describe and understand the mesoscale – was that shipboard and moored technologies were inadequate to sample such fields. Programs like the Mid-Ocean Dynamics Experiment (*e.g.*, MODE Group, 1978) and its successors (*e.g.*, Fu *et al.*, 1982; Mercier and Colin de Verdière, 1985; Hua *et al.*, 1986) yielded distorted maps of mesoscale fields, either because ship surveys were employed to map rapidly evolving fields or

[1] Recall that even such a now-familiar and reliable technology as moored current-meter observations was developed during a 20-year period of turmoil (Heinmiller, 1983), with early years of failed and missing moorings, inoperable tape recorders, and nasty, late surprises such as the discovery that measured current speeds depended strongly on mooring type (Gould and Sambuco, 1975), or that measured current directions were influenced by the instrument depth (Hendry and Hartling, 1979).

[2] We use the established acronyms to identify experiments. Table A.1 gives references and summarizes results.

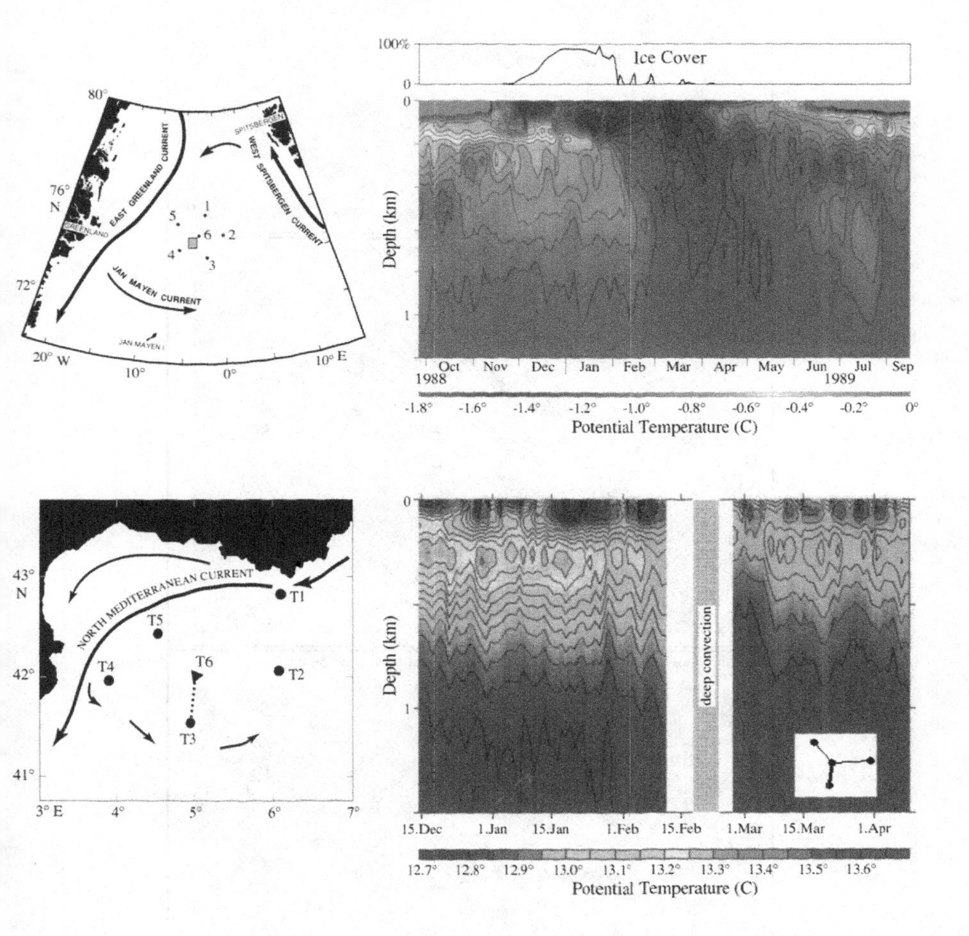

Fig. E.1. Time–depth evolution of potential temperature in the Greenland Sea (top) and the western Mediterranean Sea (bottom). The Greenland Sea time history is an average over the indicated region at the center of the array (Morawitz *et al.*, 1994). The Mediterranean evolution (in 0.5°C contours) is a range average between moorings T3 and T6 (THETIS Group, 1994; Send *et al.*, in press). The Mediterranean instruments were not working during the deep convection phase.

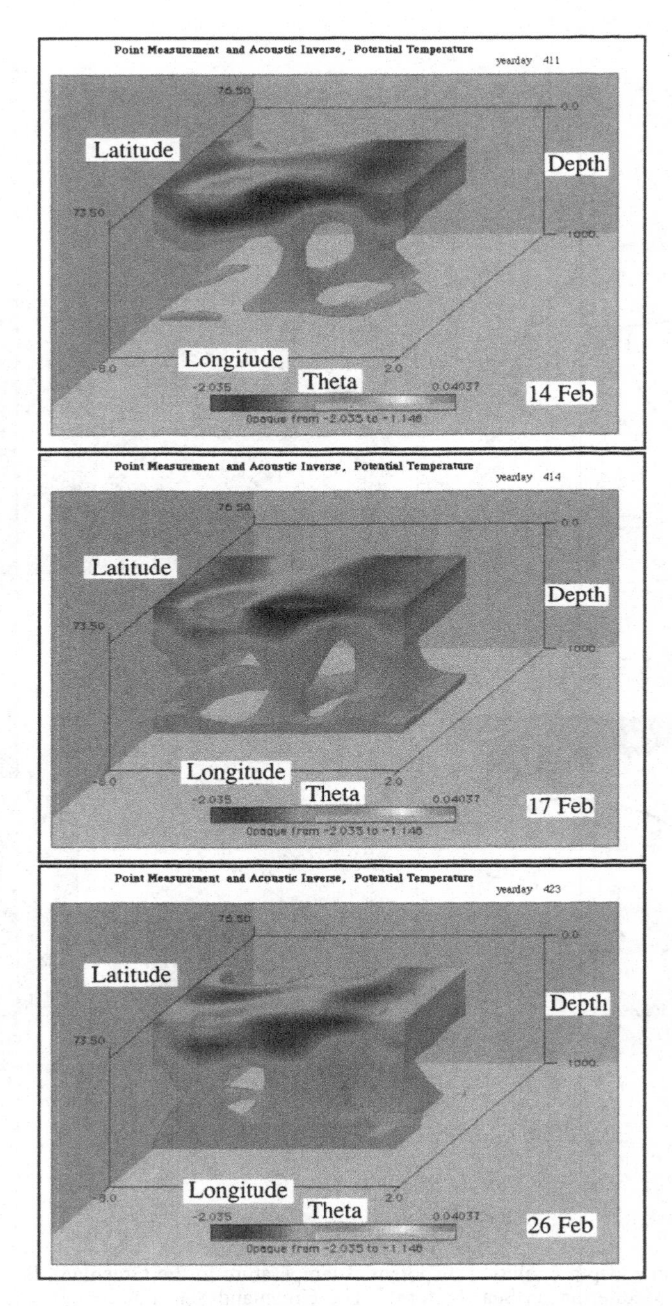

Fig. E.2. Formation and evolution of convective chimney structures in the Greenland Sea. Water with potential temperature less than –1.14°C is contained within the colored blobs. In the period from 14 to 26 February 1989, the subsurface maximum virtually disappears. (From Morawitz *et al.*, 1994, with permission.)

because there were too few *in situ* instruments to depict the detailed structures that governed the time and space evolutions.

The very first three-dimensional tomography experiment (Ocean Tomography Group, 1982) had as its central goal the production of three-dimensional time-evolving maps of the ocean using a modest number of acoustic instrument moorings. Such maps would be directly comparable to those produced during the MODE program, which involved a very large number of moorings combined with extensive shipboard measurements. That use of tomography will be recognized as most like its use in medicine. The 1981 experiment was described earlier (see figs. 6.17 and 6.18), and it succeeded in making synoptic maps of the ocean, despite the somewhat primitive nature of the acoustic sources then available. Ten years later, the AMODE experiment, using improved sources and augmented by shipborne receivers, demonstrated that tomography was indeed capable of providing detailed and accurate mapping over most of the water column (see fig. 1.6). But in the interval since the 1981 demonstration, intensive mesoscale mapping exercises have fallen somewhat out of favor in physical oceanography, and our view of the best use of tomography for understanding the mesoscale has evolved. Particularly, as discussed in chapters 7 and 8, the availability of general circulation models capable of resolving mesoscale eddies leads one to employ the tomographic measurements as integral constraints on the models – with the models used to study the detailed eddy interactions.

Should tomographic instrumentation ever become sufficiently inexpensive, one would expect to see renewed interest in detailed mapping of special regions, with emphasis on rapidly evolving features. That particular application, however, does not take advantage of the unique tomographic capability to provide integral (rather than differential) information.

Convection. One of the most interesting applications of tomography has been to study the mechanisms governing oceanic convection. Convective processes are believed to be the chief mechanisms by which the properties of the surface ocean and deep ocean are connected, with important consequences for ocean circulation and climate. Convection occurs only in extremely limited oceanic areas, and within those areas it is temporally highly intermittent and spatially extremely compact, thus posing an enormously demanding sampling problem.

Tomographic instruments were deployed as essential components in two recent field programs to study deep convection (fig. E.1). In the Greenland Sea experiment of 1988–9 (Worcester *et al.*, 1993; Sutton *et al.*, 1994; Pawlowicz *et al.*, in press), six acoustic transceivers were deployed to study deep-water formation and the response of the gyre to variations in wind stress and ice cover, in conjunction with other measurements made as part of the international

Greenland Sea Project. A region encompassing the area of a convective chimney was extracted from a three-dimensional inverse (fig. E.2) (Morawitz *et al.*, 1994), and its time history is shown in the upper panel of fig. E.1. Near the end of February, a subsurface temperature maximum disappeared over a large area of the central Greenland Sea. Whereas the water column was modified to about 1000 m depth over much of the gyre, the surface remained colder than the deeper water, contrary to what might be expected from simple models of convective renewal. Sutton (1993), using normal mode data to improve near-surface resolution, reported that the findings were consistent with a scenario for deep mixing presented by Rudels (1990) in which the warm, salty subsurface layer plays a crucial role: The atmosphere cools the free surface, ice is formed, and the injection of salt into the cold surface waters leads to convective overturn. Ice is then melted by the heat brought to the surface from below, and the process is repeated. Sutton also reported that a number of theories of deep mixing were inconsistent with the findings.

Gaillard (1994) and Morawitz *et al.* (1994), using three-dimensional inversions, have shown the formation and evolution of convective chimney structures with spatial scales of 30–40 km (fig. E.2). Morawitz *et al.* (1994) estimated an annual average water-mass renewal rate of about 0.1 Sv, consistent with tracer measurements.

In the winter of 1991–2, a tomography array, again consisting of six moorings, was deployed in the western Mediterranean Gulf of Lions (THETIS Group, 1994; Send *et al.*, in press). The near-surface layer was well sampled by the acoustics, which showed cooling and subsequent entrainment of the warmer Levantine Intermediate Water from below, in agreement with mixed-layer modeling. Send *et al.* (in press) found general agreement of the total heat loss with modeled surface heat fluxes, implying a confinement of water by the local circulation, which should be an important factor in setting the location and extent of the deep convection region.[3] The deep convection event was followed by rapid "capping" in the near-surface region and subsequent restratification. These authors estimated an annual mean deep water replenishment of 0.3 Sv.

Vorticity. As discussed in chapter 3, vorticity, $\omega = \nabla \times \mathbf{v}$, can be regarded as the primary dynamic field for describing large-scale, rotating stratified fluids. Much of the theory of the ocean circulation and its variability is the story of injection of vorticity into the ocean at the sea surface (directly through the wind-stress curl and indirectly through buoyancy exchange with the atmosphere)

[3] That was not the case for the Greenland Sea, where horizontal advection processes were important at times (Pawlowicz *et al.*, in press).

and its subsequent propagation and transformation in the oceanic interior. In a rotating stratified fluid, a derived quantity, usually called "Ertel's potential vorticity," emerges as the most readily useful variable. In one form it is

$$\zeta_{\text{pot}} = \rho^{-1}(\boldsymbol{\omega} + 2\boldsymbol{\Omega}) \cdot \boldsymbol{\nabla}\rho\,, \tag{E.1}$$

where $\boldsymbol{\Omega}$ is the Earth's rotation vector, and ρ is the fluid density.

The circulation around any horizontal closed path is

$$\hat{\mathbf{k}} \cdot \int\!\!\int \boldsymbol{\omega}\, dA = \oint \mathbf{v} \cdot d\boldsymbol{\ell}$$

and is readily measured by tomography, producing areal-average values of $\boldsymbol{\omega} \cdot \hat{\mathbf{k}}$ with very high accuracy. Because one can use the same instruments to determine the density field, the potential vorticity (E.1) can be readily obtained, permitting the direct testing of oceanographic theory in a way that is difficult using current meters or other point measurements.

We now refer to two tomographic experiments in very different ocean environments: the RTE87 triangle array north of the Hawaiian Islands on a spatial scale of 1000 km (Dushaw *et al.*, 1994) (fig. E.3), and the SYNOP tomographic array on the southern edge of the Gulf Stream on a scale of 100 km (Chester *et al.*, 1994) (fig. E.4). We note that the measured relative vorticities are in the expected ratio of order 1 : 1000, $10^{-8}\ \text{s}^{-1}$ in the open Pacific, as compared with $10^{-5}\ \text{s}^{-1}$ in the western-boundary current.

The physical regime considered by Dushaw *et al.* (1994) was that of the low-frequency limit of the barotropic vorticity equation (a special limiting case governing the evolution of ζ_{pot}), which is approximately[4]

$$H\mathbf{u} \cdot \nabla_h(f/H) = \hat{\mathbf{k}} \cdot \nabla \times (\boldsymbol{\tau}/\rho H), \tag{E.2}$$

where f is the Coriolis parameter, $H(x, y)$ the bottom topography, $\boldsymbol{\tau}$ the wind stress, and $\mathbf{u} = [u, v]$ the depth-averaged currents. This expression relates the current $\mathbf{u}(x, y)$ to the *local* wind stress $\boldsymbol{\tau}(x, y)$. The approximation omits the time derivative of the relative vorticity, a term that Dushaw *et al.* (1994) estimated from the vorticity time series and found to be negligible. Theory suggests that this equation should govern oceanographic motions at midlatitudes varying on time scales of 1 year and longer, and with spatial scales of 1000 km and larger.

[4]It is not appropriate, although common practice, to refer to equation (E.2) as the "Sverdrup balance," a relationship applying only to stratified flows (with a mid-depth level of zero motion) at zero frequency.

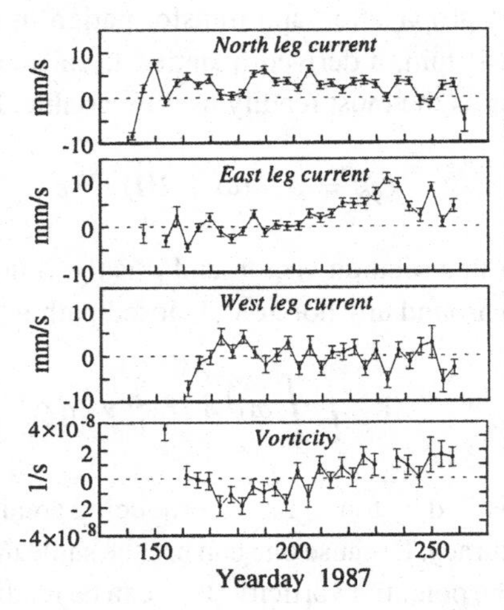

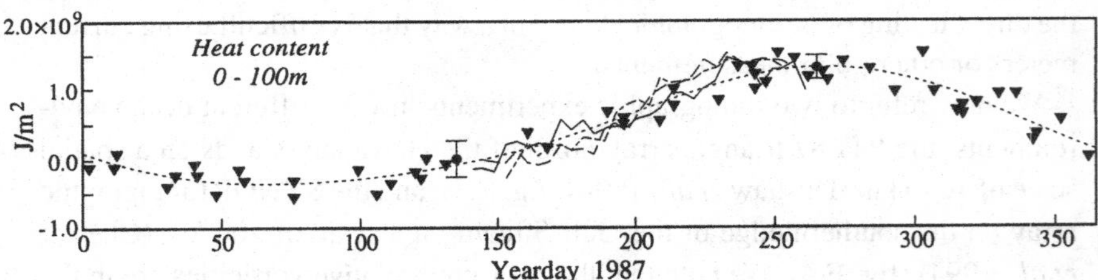

Fig. E.3. Reciprocal 1987 transmissions along three legs of the megameter Pacific triangle shown in fig. 6.28 (Worcester *et al.*, 1990, 1991*b*; Dushaw *et al.*, 1993*c*, 1994). The low-frequency depth-averaged currents are shown for each of the three legs. The area averaged vorticity is obtained by integrating the currents around the experiment triangle. The acoustically derived heat content (solid is north, dashed is east, dash-dot is west) increases in accord with the annual cycle (dotted) computed from XBT data (triangles). Vertical bars indicate the tomographic uncertainty limits.

Using estimates of the wind-stress curl from meteorological analyses, Dushaw *et al.* (1994) demonstrated that the observed megameter-scale variability was an order of magnitude *larger* than the variability computed from (E.2). Those authors concluded that the currents and vorticity were dominated by non-local forcing, with the results stated rigorously as averages over the triangle (with the uncertainty in the meteorological forcing the dominant error source).

Chester *et al.* (1994) deployed a tomographic array on the southern edge of the Gulf Stream, in the recirculation region (fig. E.4). The availability of several triangles made it possible to define the vorticity as a function of position relative to the stream, and in principle to determine its gradients. In practice, what

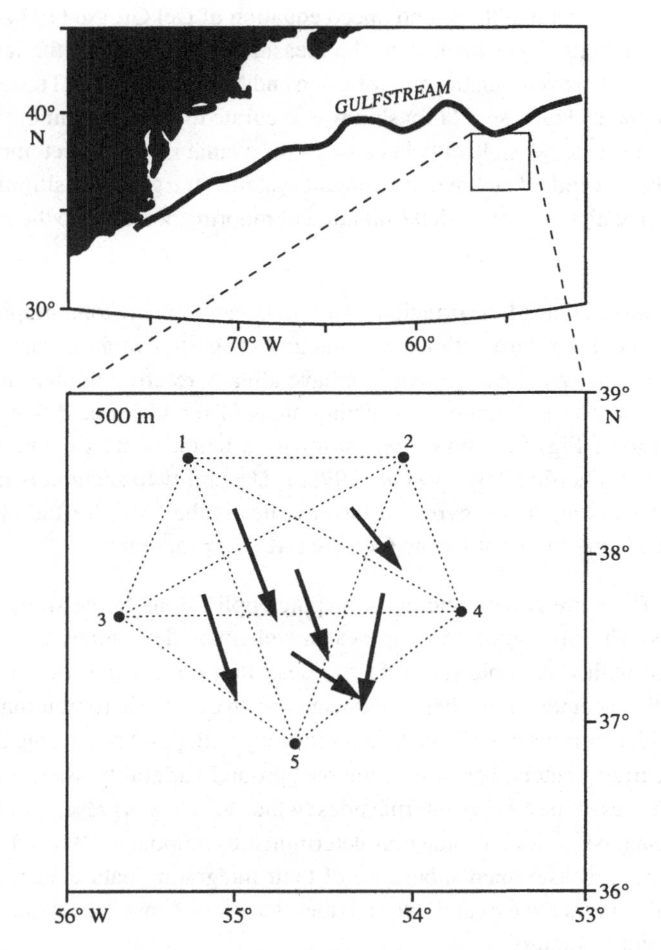

Fig. E.4. Tomographic array near the southern edge of the Gulf Stream described by Chester *et al.* (1994). Arrows indicate the Eliassen-Palm vorticity flux away from the Gulf Stream.

Plumb (1986) calls a generalized Eliassen-Palm flux was used to demonstrate the radiation of wave-like energy *away* from the meandering current.

Equation of state. The equations of state for sea water are fundamental relationships $\rho(T, Sa, p)$ used in most dynamic calculations, with the corresponding relation for sound-speed $C(T, Sa, p)$ being a special case. One of the more surprising results of the experiment north of Hawaii (Worcester *et al.*, 1991*b*; Spiesberger and Metzger, 1991*a,b*; Spiesberger, 1993; Dushaw *et al.*, 1993*b*) was the conclusion that the sound-speed equation of Del Grosso (1974) was in significantly better agreement with the measurements than was the later, and presumably improved, relationship of Chen and Millero (1977).[5] The ability to show that the Del Grosso relationship was accurate to within about 0.05 m/s at 4000 m was made possible only because of the availability of both tomography and another recently developed instrument system – the global positioning system (GPS), which permitted determination of mooring positions to the requisite precision.

Heat content. One of the principal assets of ocean acoustic tomography is the ability to provide robust estimates of integral properties of an ocean volume, such as vorticity and heat content. We have already referred to measurements of heat content in the limited convecting areas of the Greenland Sea and the Gulf of Lions. Fig. E.3 shows the seasonal variation of heat content in the upper eastern Pacific (Dushaw *et al.*, 1993*c*). The ability to accurately measure variable heat contents on gyre and basin scales is the basis for the proposed Acoustic Thermometry of Ocean Climate (ATOC) program.

Tides. There are two distinct tomographic applications to the study of tidal processes. The difference in reciprocal travel times determines the currents associated with barotropic (or surface) tides. Sum travel times determine the vertical displacement of isotherms associated with baroclinic (or internal) tides. Surface tidal currents have been difficult to study with point measuring devices, such as current meters, because of the background variability associated with internal waves. In particular, internal tides (which are always present) impose on the tidal analysis a troublesome non-deterministic component (Wunsch, 1975). Tomographic measurements, because of their integrating nature, suppress the short-scale internal waves and internal tides, leaving a conspicuous open-ocean tidal-current signature.

As an example, we quote the result for the eastward component of the tidal current as obtained on the north leg of the RTE87 Pacific triangle (Dushaw *et al.*, 1994, in press) (see table 3.1). For the M_2 constituent, the amplitude and Greenwich phase were 1.31 ± 0.03 cm/s and $223 \pm 1^\circ$, respectively. Cartwright *et al.* (1992) obtained 1.42 cm/s and 218° from Schwiderski global tidal elevations, with an uncertainty of 10% in amplitude and 6° in phase. With the increasing precision of satellite altimetry the error bars of the elevation-based tidal constants will go down, and it remains to be seen which of the methods will give the more accurate results.

[5]This difference does not affect the equations of state for sea water within the standard error of the equation (5×10^{-6} cm^3/g in the specific volume) (Millero and Li, 1994).

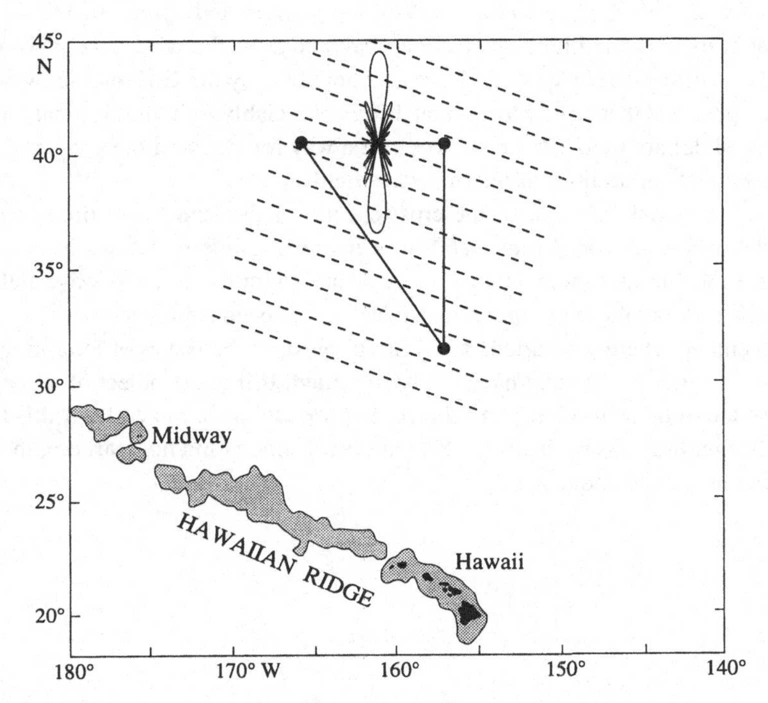

Fig. E.5. Beam pattern (in dB relative to maximum) of a line array of length 750 km for 160-km incident wavelength. The antenna has the greatest response for propagation at right angles. The tomographic array consists of three such line arrays with different orientations. (From Dushaw *et al.*, in press, with permission.)

The high quality of the measurements made it possible to separate a baroclinic M_2 constituent that was phase-locked to the surface tides over the Hawaiian ridge 2000 km to the south. The direction of propagation could be determined with the high angular resolution of the tomographic array (fig. E.5). Dushaw *et al.* (in press) inferred a northward-propagating internal tide with a flux of about 180 W/m. This accounts for a small fraction of the global tidal dissipation of 3×10^{12} W.

Internal waves. In the forward problem, a number of statistical properties of the ocean internal wave field have been related to statistical properties in the acoustic signature. The RTE83 (Flatté and Stoughton, 1986; Stoughton *et al.*, 1986) demonstrated the feasibility of an inverse solution, namely, that significant limits can be imposed on the range-averaged internal wave intensity based on the wander of ray-like arrivals. Similarly, for the SLICE89 experiment

(Duda *et al.*, 1992; Colosi *et al.*, 1994), the wander and spread of the ray-like arrivals, and the blunting of the arrival wedge of the mode-like arrivals (section 4.4), can all be quantitatively accounted for by the GM internal wave model, provided that energy levels that differ appreciably from the nominal (but are plausible) are used. This result opens the way for studying the geographic and temporal variabilities of internal wave fields.

In a reciprocal experiment, the cross-spectra of the sum travel times $s(t)$ and the difference travel times $d(t)$ have interesting interpretations in terms of the heat flux and momentum flux associated with the internal wave field (section 3.2), but those interpretations have not yet been exploited.

We end up where we started: Many, if not most, of the issues of long-range ocean acoustic tomography have not been settled. If it is the object of a book to give a definitive account of a subject, then we are much too early; if it is to help further progress by sharing what has been learned (and not learned), then the timing may be about right.

APPENDIX A

A PERSONAL CHRONICLE

The idea for acoustic tomography arose abruptly.[1] Because we can mark a clear conceptual start, the following brief chronicle of the development of ocean acoustic tomography, given from our perspective as participants, may be of interest.

Some hopes for "Monitoring the oceans acoustically" were voiced at the thirtieth anniversary of the founding of the Office of Naval Research (ONR) (Munk and Worcester, 1976). A "Preliminary report on ocean acoustic monitoring" was prepared during the JASON[2] Summer Study (JSN-77-8) by Garwin, Munk, and Wunsch. That work was expanded into "Ocean acoustic tomography: a scheme for large scale monitoring" (Munk and Wunsch, 1979), which examined the acoustic and inverse theoretical requirements for mapping the oceans with mesoscale resolution. It concluded that "an acoustic tomographic system appears to be both practical and useful." The name "ocean acoustic tomography" was deliberately chosen to arouse the reader's curiosity as to what it is all about. Response by the oceanographic community was varied; those with a

[1] There were some precursors, as always. LaCasce and Beckerle (1975) suggested studying Rossby waves by their effects on acoustic signals. The relationship between temperature and sound-speed is so close that this type of possibility must have occurred to many acousticians. For example, H. N. Opland, who had spent his career on problems of anti-submarine warfare, approached Athelton Spilhaus in 1975 for suggestions concerning "global geophysical projects with some realistic benefit to mankind" (personal communication with Walter Munk, 1985). Spilhaus, then at NOAA, reminded Opland that the bathythermograph had been invented in 1937 for measuring temperature profiles so that "sound transmission paths could be predicted." About 20 years later it had occurred to Spilhaus that the process might be reversed, and he suggested this idea as a topic for Opland, for the purpose of providing "synoptic inputs to climate models." Opland moved to Washington in 1976 and started a company (called Loki Associates, after the Norwegian god of mischief) to put Spilhaus's suggestion to practice. For the next year Opland contacted U.S. government agencies for support, but receiving no encouragement he returned to private industry. We are unaware of any previous published estimates of the methodology and practice for inferring oceanic fields from acoustics measurements.

[2] JASON, comprising a group of university professors, has advised the U.S. government on defense and other problems since 1960.

background in inverse theory regarded the inverse problem as trivial, but found the acoustic applications to be of interest, whereas the marine acousticians were interested in the inverse problem.

A.1. Overture

Advances in several fields were prerequisites for the development of ocean acoustic tomography: an understanding of underwater sound propagation, a statistical description of oceanic processes (especially internal waves and other fine structure), and the availability of inverse methods for inference from measurements. In this section we focus on the history of the acoustic developments crucial to tomography. A description of the development of the required inverse methods could also be written. Readers interested in the development of inverse methods for oceanographic applications should consult Wunsch (in press). Several events stand out:

SOFAR channel. In 1944, Ewing and Worzel (1948) departed Woods Hole aboard the *Saluda* to test a theory of an ocean acoustic waveguide. A deep receiving hydrophone was hung from *Saluda*. A second ship dropped 4-pound charges at distances up to 900 miles. Detonations and the hydrophone were both in the sound channel. Ewing and Worzel heard, for the first time, the characteristic signature of a SOFAR (*s*ound *f*ixing *a*nd *r*anging) transmission building up to its climax (fig. A.1):

bump bump bump bump **bump**

In the words of those authors, "the end of the sound channel transmission was so sharp that it was impossible for the most unskilled observer to miss it." What is more, that was one of the rare cases in which an oceanographic experiment

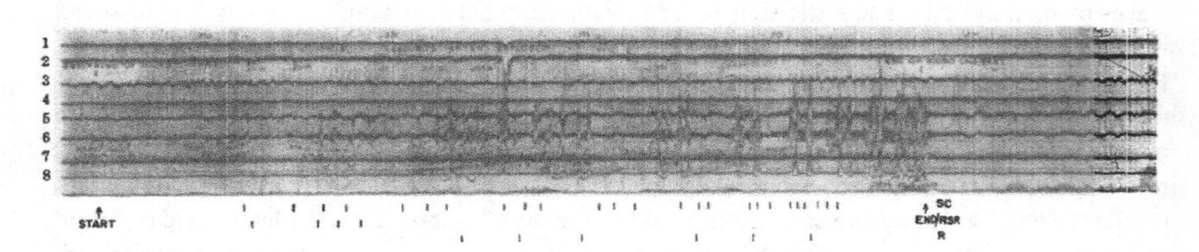

Fig. A.1. Discovery of SOFAR channel: shot 43 recorded aboard the *Saluda* on April 3, 1944. Charges were exploded at 4000 feet depth and a range of 320 nautical miles. Times are labeled for 370, 371, . . . , 374 s following the explosion. Channels are identified as follows: 1, time break; 2, rectified; 3, high-frequency; 4, low-frequency; 5, low-frequency; 6, high-frequency; 7, rectified; 8, rectified. (Adapted from Ewing and Worzel, 1948.)

confirmed a previously developed theory. Later in the same year, Ewing and Worzel established a bottom-mounted hydrophone connected to a field station on Eleuthera Island in the Bahamas. They spoke even then of transmissions over 10,000 miles.[3]

Shortly thereafter, Brekhovskikh and his colleagues independently discovered the surface-ducted sound channel (Brekhovskikh, 1949; Rozenberg, 1949). According to Brekhovskikh (personal communication, 1989), some work had been planned for 1946 in the Sea of Japan, but the equipment was not ready. Rather than lose ship time, it was decided to make some spontaneous measurements of sound transmission. Charges were dropped from a vessel, detonating at 100 m depth, with a suspended hydrophone drifting at 100 m, not unlike the *Saluda* experiment. Brekhovskikh wrote that

> Something very strange was observed in the course of the experiment. Peak amplitude decreased markedly only for the first 30 nm, whereas at greater distances the decrease was hardly noticeable. The acoustic signal form was also drastically different at different distances. At small ones it resembled shock waves, whereas at long distances the signal started very weakly, then increased with time resembling a thunder in the final stage before coming to an abrupt end. It was my duty to treat these results. It appeared that the only way to explain them was to take into account the existence of an acoustic wave guide with its axis at a depth of about 150 m.... The picture fitted very well the summer hydrological conditions in the Sea of Japan. Since this work could have military applications its publication was delayed till 1948.... Due to the weakened international scientific relations we did not know of Ewing's paper until later.

It is surprising that Ewing, the experimentalist, was testing a theory, whereas Brekhovskikh, the theorist, made his discovery looking at data.

The next few years saw a rapid sequence of discoveries, including the anomalous absorption of sound in sea water, convergence-zone focusing, and the biological origin of certain ambient noises. A growing acoustic community under U.S. Navy sponsorship set out to exploit the sound channel for submarine detection. Most of the work was classified, and when it was eventually reported in the open literature some 15 years later, the authorship bore little or no relation to the people who had done the pioneering work.[4] In the meantime, the acoustic

[3]The realization that upward refraction by the pressure gradient should be associated with extraordinarily long sound listening ranges goes back to the U-boat operation during World War I (Lichte, 1919).

[4]An authoritative review can be found in the "green book" produced by the Committee of Undersea Warfare's Panel on Underwater Acoustics, under the auspices of the National Research Council, NRC 1950. See also Bell (1962), Wood (1965), Klein (1968), and Lasky (1973, 1974).

and oceanographic communities drifted apart, separated by a veil of secrecy, to the detriment of both.[5]

Ocean fine structure. By 1960, the acoustic community had become greatly concerned with the extreme variability of acoustic signals transmitted through the oceans. Fade-outs and phase jumps are the rule, not the exception, in long-range CW sound transmissions, implying an ocean variability of small scale and high frequency. In an attempt to account for the variability, the acoustic community invented their own oceans. Sometimes the ocean was modeled as an imperfect transmission line, the imperfections consisting of an ocean fine structure with a $-\frac{5}{3}$ wavenumber spectrum characteristic of homogeneous isotropic turbulence in the inertial subrange. We now know that the ocean fine structure is neither homogeneous nor isotropic and has little to do with turbulence in the usual sense. In fact, the fine structure is dominated by the variable straining due to the ever-present internal waves. The space and time scales of internal waves can be reasonably fitted by an empirical spectrum (Garrett and Munk, 1972; Munk, 1981).

In 1973, Clark and Kronengold (1974) transmitted continuously for many months along the 1250-km path from Eleuthera to Bermuda. This was a continuation of the pioneering MIMI transmission (named for the Miami-Michigan collaboration) across the Straits of Florida (Steinberg and Birdsall, 1966). The essential new feature was that the explosive sources of earlier experiments had been replaced by a piezoelectric transducer. In essence, the experiment consisted of recording the relative phase and intensity of the received signal using an oscillator that was perfectly synchronized with the 406-Hz transmitter (fig. A.2).

The long-term CW transmissions gave time series for the acoustic phase $\phi(t)$ and phase rate $\dot{\phi}(t)$ of the received signal, which consisted of a superposition of many ray paths. The interpolation of phase is ambiguous across deep-intensity fade-outs, giving rise to a random walk of the "extended phase" over many

[5]Problems of classification and security are part of the oceanographic scene, and especially in ocean acoustics. In the development of acoustic tomography, a very deliberate effort was made to avoid a repetition of what had happened in the early fifties. The development has depended on collaboration with the navy in two ways: for research support from ONR, and for the use of the navy listening facilities. The navy receivers have served to augment the sparse arrays of moored tomographic receivers. It is also prudent to monitor source performance in real time, as has been demonstrated dramatically on a number of occasions. Most of us have clearances to visit the required facilities, and the monitoring in real time is arranged on a personal basis and does not involve our institutions. We are pleased to report that these relatively informal arrangements have worked satisfactorily to the present. As this is being written, multiple use of the facilities for security and research (including seismological and biological studies, among others) is under active consideration.

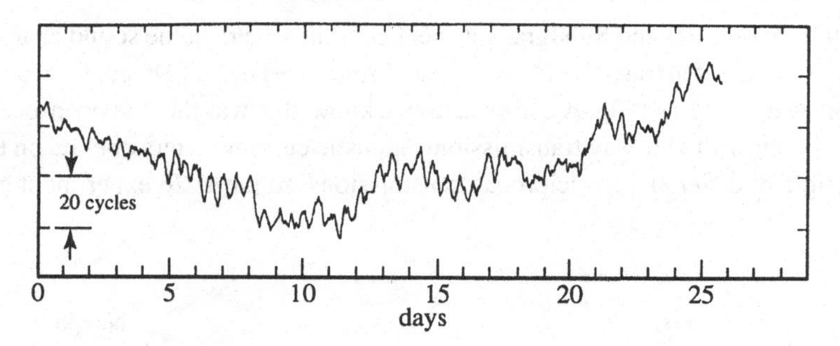

Fig. A.2. Time series of phase for the 406-Hz CW transmission from Eleuthera to Bermuda (1250 km). This pioneering experiment was conducted during September and October 1973 by Birdsall, Clark, Kronengold, and Steinberg (Dyson *et al.*, 1976).

cycles that is of statistical rather than geophysical origin.[6] By taking account of the statistical random walk, the phase $\phi_i(t)$ and phase rate $\dot{\phi}_i(t)$ for a single ray path i can be inferred from the CW phase and phase rate (Dyson *et al.*, 1976). The result is $\langle\dot{\phi}_i^2\rangle = 1.6 \times 10^{-5}$ s^{-2}. This parameter depends on the fluctuations of sound-speed along the transmission path and can be calculated using the GM internal wave spectrum, leading to the result $\langle\dot{\phi}_i^2\rangle = 2.7 \times 10^{-5}$ s^{-2} (Munk and Zachariasen, 1976). There are no loose parameters here, and the data sets are entirely independent – one acoustic, the other based on traditional oceanographic measurements. The agreement gives an oceanographic interpretation of the measured "decorrelation time" of a few minutes. It was this rough agreement that convinced Munk that acoustic fluctuations and internal waves had something to do with one another. This work has been greatly expanded by members of the JASON group. Dashen (1979) introduced path integrals as a convenient formalism for calculating the propagation of waves in random media, leading to the important distinction between *saturated* and *unsaturated* paths. That led to a monograph by Flatté *et al.* (1979, see also Flatté, 1983), with emphasis on acoustic fluctuation statistics produced by internal waves. But there are other sources of ocean fine structure, such as that due to interleaving of different water masses, and the problem is by no means settled (section 4.4).

[6]The ambiguity is avoided by transmitting a broadband coded signal and measuring arrival time by correlating the received signal with a replica of the transmitted signal, as discussed in chapter 5. A broadband signal encoded using a binary m-sequence was transmitted in the very first MIMI transmissions across the Straits of Florida (Birdsall, 1965), but the results were discarded because no one knew what to do with all of the data.

A.2. Reciprocal Transmission Experiment of 1976

In 1976, Worcester and Snodgrass suspended transceivers in the sound channel from two ships separated by 25 km, located southwest of San Diego, California (Worcester, 1977*a,b*) (fig. A.3). As far as we know, that was the first open-ocean experiment with two-way transmission. Acoustic current meters operate on the principle of differencing reciprocal transmissions, so the 1976 experiment can

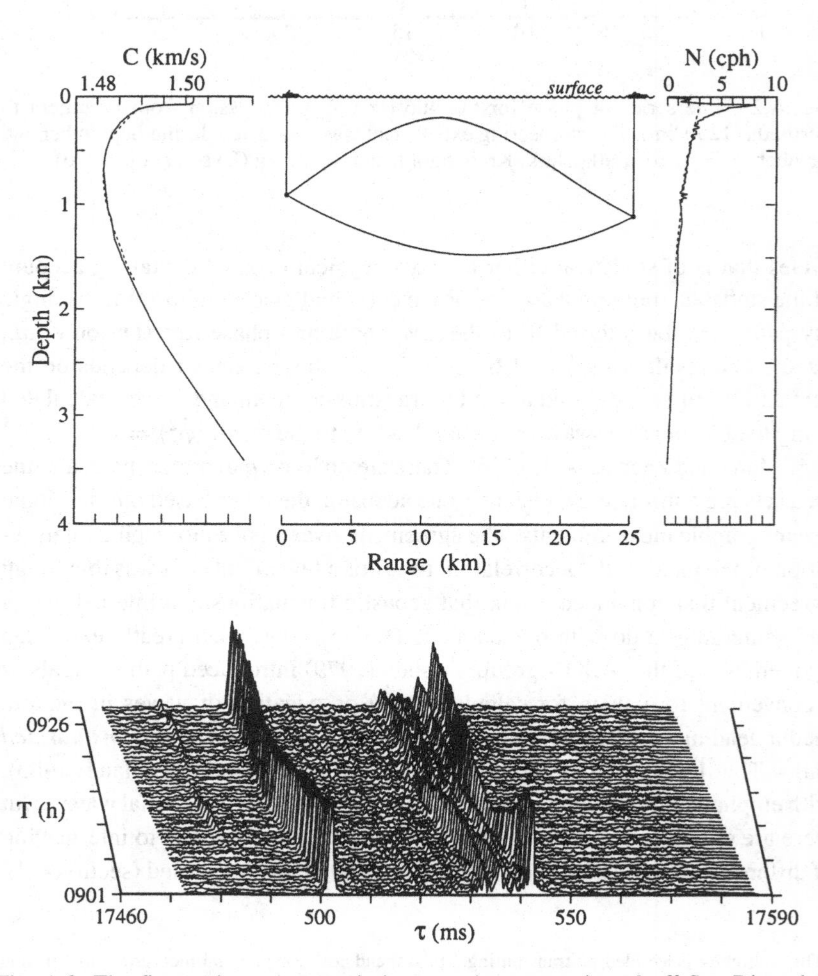

Fig. A.3. The first reciprocal transmission experiment, conducted off San Diego by Worcester (1977*a,b*) in April 1976. Top: Two ray paths for the 25-km range. Bottom: Arrival pattern at 30-s intervals clearly separates the early lower path from the later and far more complex shallow arrival.

be regarded as the sea trial of a 25-km current meter. The broadband acoustic transmitters were centered at 2250 Hz. The geometry was selected to give two purely refracted ray paths with travel times differing by about 40 ms. Fig. A.3 clearly shows the distinction between the unsaturated lower path and the partially saturated upper path. The phase-coded signal yielded 20-μs travel-time precision, which has not been matched in subsequent experiments because at longer ranges the frequency and bandwidth, and hence precision, are necessarily lower. The experiment, which would now be considered a vertical slice tomography experiment, and which served as the basis for what might be considered the first tomography dissertation (Worcester, 1977*a*), yielded a useful estimate of the 3-cm/s current component along the upper path relative to that along the lower path. It convinced us of the feasibility of measuring ocean currents using acoustic transmissions.

A.3. Resolution, Identification, Stability, 1977–1980

By 1977, the first tomography group was being formed. Spindel and Porter, from Woods Hole, brought their experience in underwater acoustics, in autonomous instrumentation, in setting deep-sea acoustic moorings, and in precise acoustic position-keeping. Webb, from Woods Hole, had built acoustic sources for SOFAR floats. Wunsch, from MIT, provided the leadership in tomographic data analysis. Birdsall and Metzger, from the University of Michigan, provided expertise in signal processing. Munk and Worcester, from Scripps, participated in both theoretical and experimental aspects of the work. Support came from the ONR under the heading "Variability of Sound Transmission through the Ocean Interior," featuring experiments for testing predicted fluctuations, rather than for tomography *per se*. The formation of the group was originally something of a shotgun wedding under the persuasion of H. Bezdek of the ONR, who wanted to avoid duplicate efforts; yet as we write this book, many members of the original group are still working together on the Acoustic Thermometry of Ocean Climate (ATOC) project.

In the fall of 1978, a phase-coded signal was transmitted at 10-min intervals for 48 days from a moored source southwest of Bermuda westward to "a bottom receiver[7] at ~900-km range" (Spiesberger *et al.*, 1980). The resolution of the

[7]This refers to a navy SOSUS array (*so*und *su*rveillance *s*ystem). The tomographic community is greatly indebted to the U.S. Navy for having permitted the use of these remarkable facilities, starting in 1978. The foregoing quotation was deliberately vague to honor an agreement with the U.S. Navy to give ranges only to the nearest 100 km, and absolute travel times to the nearest 100 s (there is no restriction on *relative* travel times). The editor of the *Journal of Physical Oceanography* originally declined publication unless precise coordinates were published. We took the view that precise coordinates should not be required, as they were not needed to interpret the results.

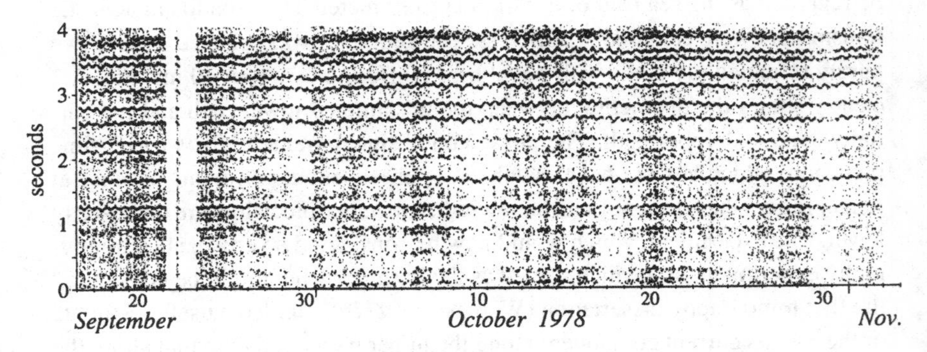

Fig. A.4. Arrival pattern (in *relative* units) for a 900-km transmission from a moored Webb source (220 Hz) southwest of Bermuda. Arrivals were recorded at 10-min intervals over a 48-day period. (From Spiesberger *et al.*, 1980, with permission.)

220±5-Hz Webb source proved adequate to track 14 multipaths unambiguously for the duration of the experiment (fig. A.4). The position of each arrival peak exceeding a prescribed SNR is marked by a dot whose diameter is proportional to the SNR. The eye naturally finds a pattern by incoherently averaging the dots. The *resolution*, *identification*, and *stability* were adequate to measure mesoscale-induced travel-time variability over a 900-km path. The dot plot also shows semi-diurnal wiggles of roughly 8 ms amplitude; these can be accounted for by a 2-cm/s barotropic tidal current, in accordance with co-tidal charts (Munk *et al.*, 1981*b*).

This work came just in time to save the tomography effort from an early demise. A proposal had been submitted to the National Science Foundation to augment the existing ONR support. One of the reviewers wrote that "travel times along ray paths are meaningless in a saturated environment." He continued with a statement that individual ray arrivals could not be resolved, and even if resolved could not be identified, and even if identified would not be stable. The proposal was declined in September 1979. We responded by sending the dot plot shown in fig. A.4, with the statement that"we have resolved, identified and tracked 13 rays for over 2 months, see enclosed figure." The proposal was subsequently accepted.

Preparations for a 1981 demonstration experiment then got under way. In April 1979, a 300-km transmission north of Bermuda in an early test of the WHOI equipment showed a 0.7-s travel-time increase over a 19-day period, accompanied by large variations in the amplitudes and separations of ray arrivals. The results are related to a southern meander of the Gulf Stream (Spindel and Spiesberger, 1981).

In May 1980, off California, vertical arrival angles were measured with a four-element vertical array with 1.5 wavelengths between elements during a test of the SIO instrumentation. Worcester (1981) found excellent agreement between measured and predicted arrival patterns with respect to both time and angle. A vertical array has been part of the Scripps receiving equipment ever since. It is useful in three ways: to confirm ray identification, to separate upward and downward rays arriving at nearly the same time, and to increase the SNR.

In summary, the 4 years starting in 1977 were spent in developing the tools, theoretical and experimental, for mesoscale tomographic mapping.

A.4. The 1981 Demonstration Experiment

The first three-dimensional test of ocean acoustic tomography was performed during 1981 in a 300-km × 300-km square to the southwest of Bermuda (fig. 6.18). After 2 years of discussion, we agreed on a rather odd mooring configuration that placed four acoustic sources to the west and five receivers to the east of the square. Both the experiment location and scale were deliberately chosen to resemble the 1976 MODE experiment. The purpose of MODE had been to map the recently discovered mesoscale variability using current meters and CTDs. We wanted to demonstrate what could be done with acoustic techniques. Working in a 300-km square made it possible to achieve mesoscale resolution with just eight or nine moorings. The ranges were compatible with existing source and receiver technology (Spindel *et al.*, 1982). The sources were extensions of the technology employed in neutrally buoyant SOFAR floats. They consisted of open-end resonant tubes approximately one-quarter wavelength long, driven at one end by a piezoelectric transducer. Source calibrations showed that the transmitting voltage response was quite complicated (fig. A.5). Nonetheless, the sources were adequate (barely) to transmit phase-coded m-sequences with digits containing 14 cycles of the 224-Hz carrier (62.5 ms digit duration).[8]

The Gauss-Markov inversion procedure described in chapter 6 yielded a time series of tomographic maps at 3-day intervals; CTD surveys were conducted

[8]The first source mooring was installed on 3 February, and by 6 February all four sources were in place. We had made arrangements with the U.S. Navy to monitor the sources on a SOSUS array as they were being installed. It is difficult to convey the satisfaction of hearing the signal arrive at a remote site at the expected moment. It is also difficult to convey the sense of disappointment when it does not, leading to an immediate and total adjustment in the lives of the investigators. On 21 February, source 2 was not heard, and this was confirmed on the next transmission day. Spindel and his associates had just set the last receiver mooring and were on their way home. They immediately flew to Nassau to rejoin the R/V *Oceanus* and repair the faulty source. This is described in some detail to give the reader a flavor of the early (and some later) experiments.

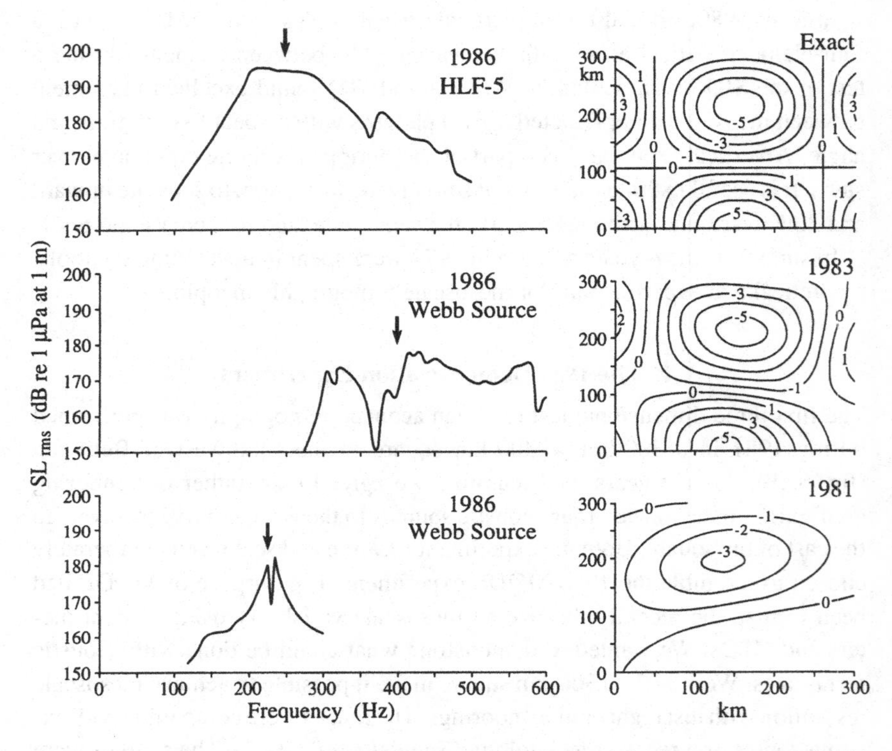

Fig. A.5. Development of tomographic sources. The left panels give rms source levels as a function of frequency, with center frequencies indicated by arrows, showing the evolution toward greater bandwidth, simpler source functions, and increased power. The right top panel shows an assumed sound speed anomaly (m/s). Computer simulations (right lower panels) show a dramatic improvement resulting from replacement of the 1981 source (5 ms rms error) with the 1983 source (2 ms rms error). (From Cornuelle *et al.*, 1985, with permision.)

at the beginning and end (fig. 6.18). The ocean appears to undergo significant changes over periods as short as 6 days (see year days 100 and 106). The acoustic maps show westward movement, whereas from the CTD data we can only infer it.

Even allowing for the evolution of the field during the 3 weeks required for each CTD survey, the comparison in fig. 6.18 between the tomographically derived maps and the CTD maps is disappointing. We obtained smoothed but qualitatively correct patterns, albeit with large error. It was subsequently determined (Cornuelle *et al.*, 1985, app. B) that the travel-time error was 5 ms rms, significantly larger than the expected 2 ms rms, due to interference between closely spaced (though formally resolved) ray arrivals. Inversions with 5 ms

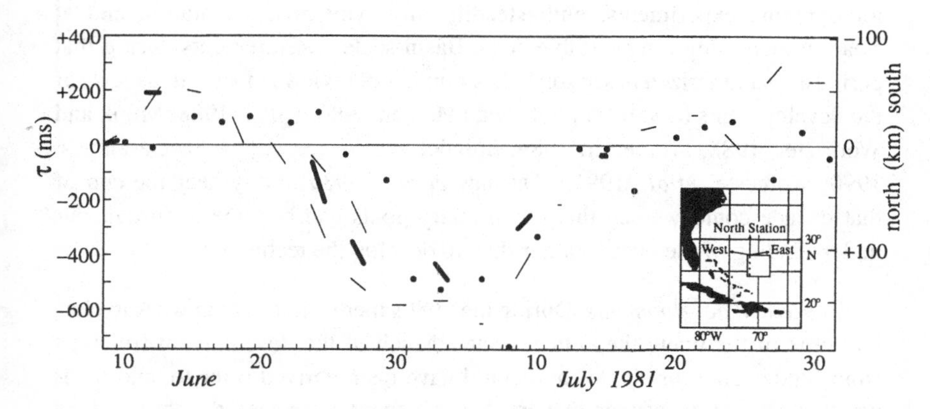

Fig. A.6. Propagation across the Gulf Stream (source locations shown in inset). Line segments denote intermittently measured travel-time perturbations (left scale). Dots mark position of the north wall of the Gulf Stream inferred from surface temperatures measured by NOAA-6 satellite (right scale). The two scales are related by a simple Gulf Stream model. (From Spiesberger *et al.*, 1983, with permission.)

travel-time error have significantly less resolution than those with 2 ms error (fig. A.5). The interference is virtually eliminated by going to sources with larger bandwidth. The fight for bandwidth is a major theme underlying the first 10 years of experimental work.

The acoustic signals were also monitored at long ranges (1000–2000 km) using SOSUS arrays. Fig. A.6 shows a 50-day cycle in travel times, with a total range of almost 1 s for transmissions northward across the Gulf Stream (Spiesberger *et al.*, 1983). Simultaneous measurements of sea-surface temperature taken by the NOAA-6 satellite indicated a meandering of the Gulf Stream in the region intercepted by the acoustic transmission path, with a latitudinal displacement of the north wall of the Gulf Stream by 180 km. As expected, travel time is shortest when the displacement is farthest north, with the largest fraction of warm Sargasso Sea water along the path. There is agreement between estimated and measured magnitudes.

A.5. A Decade of Development

Much to our chagrin, and in spite of the fact that we had called the 1981 experiment a "demonstration," most of the 1980s decade was spent in a struggle to demonstrate the validity of the tomographic measurements, to develop autonomous instrumentation adequate for megameter ranges, and to understand the capabilities and limitations of the technique. We conducted a series of tomographic experiments, with steadily improving instrumentation and at steadily increasing ranges (table A.1). Basin-scale measurements during that period are summarized in section 8.3. A number of reviews of various aspects of the developments have been published (Mercer, 1986; Knox, 1988; Munk and Worcester, 1988; Worcester, 1989; Spindel and Worcester, 1990; Desaubies, 1990; Worcester *et al.*, 1991*a*; Dushaw *et al.*, 1993*a*). Only near the end of that decade could we say that our primary goals had become to further our understanding of the ocean, rather than to develop the technique.

Theoretical developments. During the 1980s there was ongoing work to clarify some of the many theoretical issues. Much of the development followed from fundamental principles and could have been derived prior to, and independently of, the experimental work; but as is so often the case, the observations frequently motivated the theoretical analyses. All of the important results have been discussed in earlier chapters. The goal in this section is to highlight some of the more significant developments and to describe how they occurred.

Following the 1981 experiment, Munk and Wunsch took sabbatical leave and shared an office at the University of Cambridge, providing an opportunity to continue their collaboration. That led to a series of papers on various aspects of the tomographic problem, including the combination of satellite altimetry with tomography (Munk and Wunsch, 1982*a*), up/down ambiguity in midlatitudes (Munk and Wunsch, 1982*b*), and ray/mode duality in the ocean environment (Munk and Wunsch, 1983). Later in the decade, theoretical studies and numerical simulations by a number of investigators led to greater understanding of the effects of nonlinear bias, the oceanographic information available in a vertical slice, the horizontal geometries required to map the ocean mesoscale field with a specified precision, and the properties of tomographic reconstructions of the two-dimensional vector field of current.

To a surprising extent, fundamental advances in understanding the sampling properties of acoustic tomography were the results of modeling the ocean using a Fourier basis in the horizontal, so that the results could be easily interpreted in wavenumber space. Range-dependent inversions of the sum travel times from the 1983 Reciprocal Transmission Experiment using a Fourier basis yielded estimates of the range-dependent temperature field that were in good agreement with independent XBT measurements (Howe *et al.*, 1987) (fig. A.7). Even though the uncertainty of the range-dependent component of the field was large, the amount of range-dependent resolution led to an effort by Cornuelle and Howe (1987) to understand the sampling properties in a vertical slice (sections 4.2 and 6.9). Simulation of moving-ship tomography geometries by Cornuelle *et al.* (1989) using two-dimensional Fourier expansions led

Table A.1. *Significant ocean tomography experiments*

Year	Experiment	Institutions[a]	Range/Location/Comments
1976	Two-Way Accoustic Transmission (TWATE)	SIO	25 km/Offshore San Diego/Measured currents relative to ship drift (Worcester, 1977*a,b*, 1979).
1978	900-km Propagation Test	WHOI/SIO/UM	900 km/Northwest Atlantic/Demonstrated multipath resolution, stability, and identification (Spiesberger, 1980; Spiesberger *et al.*, 1980; Brown *et al.*, 1980; Munk *et al.*, 1981*b*; Spiesberger and Worcester, 1981; Legters *et al.*, 1983).
1981	Tomography Demonstration Experiment	WHOI/SIO/UM/MIT	300 km × 300 km/Northwest Atlantic/Demonstrated mesoscale sound-speed pattern recognition (Ocean Tomography Group, 1982; Spindel, 1982; Spindel *et al.*, 1982; Worcester and Cornuelle, 1982; Cornuelle, 1983; Metzger, 1983; Spiesberger *et al.*, 1983; Spiesberger and Worcester, 1983; Cornuelle *et al.*, 1985; Chiu, 1985; Chiu and Desaubies, 1987; Gaillard and Cornuelle, 1987; Spiesberger, 1989).
1983	RTE83	WHOI/SIO/UM	300 km/Northwest Atlantic/Demonstrated current measurement by reciprocal transmissions; demonstrated single-slice tomography (Worcester *et al.*, 1985*b*; Howe, 1986; Flatté and Stoughton, 1986; Stoughton *et al.*, 1986; Howe *et al.*, 1987).
1983	Florida Straits	UMiami	20- and 45-km triangles/Florida Straits/Demonstrated measurement of areal average relative vorticity; monitored Florida Current transport and meanders (Palmer *et al.*, 1985; DeFerrari and Nguyen, 1986; Monjo, 1987; Ko, 1987; Ko *et al.*, 1989; Chester, 1989; Chester *et al.*, 1991).
1983–9	Pacific Basin, Hawaii to mainland	WHOI/UM	3000–4000 km/North Pacific/Found reproducible features in arrival pattern at 4000 km (Bushong, 1987; Spiesberger *et al.*, 1989*a,b*, 1992; Spiesberger and Metzger, 1992; Headrick *et al.*, 1993).
1984	Marginal Ice Zone Tomography Experiment	WHOI	53 and 161 km/Fram Strait/Demonstrated feasibility of tomography in the marginal ice zone (MIZ); suggested feasibility of surface-wave tomography (Lynch *et al.*, 1987, 1989; Miller, 1987; Romm, 1987; Chiu *et al.*, 1987; Miller *et al.*, 1989).

(continued)

Table A.1 (*cont.*)

Year	Experiment	Institutions[a]	Range/Location/Comments
1984	Bottom-mounted Gulf Stream	WHOI/MIT	19–51 km/Northwest Atlantic/Demonstrated feasibility of using surface- and bottom-reflected ray paths from bottom-mounted instruments (Malanotte-Rizzoli *et al.*, 1982, 1985; Spiesberger and Spindel, 1985; Spiesberger *et al.*, 1985; Cornuelle and Malanotte-Rizzoli, 1986; Agnon *et al.*, 1989).
1987	RTE87	SIO/WHOI/UM	750, 1000, and 1275 km/Central North Pacific/Measured heat content, tidal currents, and vorticity; showed that barotropic currents and vorticity were much greater than expected from basic Sverdrup dynamics (Worcester *et al.*, 1985*a*, 1990, 1991*b*; Spindel and Worcester, 1986; Jin and Worcester, 1989; Spiesberger and Metzger, 1991*a,b,c*; Dushaw, 1992; Dushaw *et al.*, 1993*b,c*, 1994, in press; Spiesberger, 1993; Spiesberger *et al.*, 1994).
1988–9	GSP88	SIO/WHOI/UM/UW	200-km pentagon/Greenland Sea/Measured temperature, heat content, barotropic currents, and tides in an Arctic environment; measured the evolution of deep mixing in the Greenland Sea during winter; conducted engineering tests of moving-ship tomography (Jin and Wadhams, 1989; Greenland Sea Project Group, 1990; Peckham *et al.*, 1990; Worcester *et al.*, 1993; Sutton, 1993; Sutton *et al.*, 1993, 1994; Jin *et al.*, 1993, in press; Lynch *et al.*, 1993*a,b*; Pawlowicz 1994; Pawlowicz *et al.*, in press; Morawitz *et al.*, 1994).
1988–9	Gulf Stream (SYNOP)	WHOI/MIT/IFREMER	200-km pentagon/Gulf Stream/Measured barotropic and baroclinic currents and vorticity in the Gulf Stream recirculation region; estimated eddy statistics (Chester, 1993; Chester *et al.*, 1994; Chester and Malanotte-Rizzoli, in press).
1988	Monterey Canyon	WHOI/NPGS	54 km/Monterey Canyon/Demonstrated surface-wave tomography (Hippenstiel *et al.*, 1992; Miller *et al.*, 1993; Westreich *et al.*, in press).
1989	SLICE89	SIO/UW	1000 km/Northeast Pacific/Receptions on a 3000-m long vertical receiving array tested the horizontal sampling properties of single slice tomography; internal wave scattering found to be more significant than expected (Howe *et al.*, 1991; Duda *et al.*, 1992; Cornuelle *et al.*, 1992, 1993; Colosi, 1993; Worcester *et al.*, 1994; Colosi *et al.*, 1994).
1990	ATE90	UW/SIO/UM	1000–2000 km/Northwest Atlantic/Demonstrated the utility of incorporating tomographic

1990	GASTOM	IFREMER/ SHOM	300-km pentagon/Bay of Biscay/Studied mesoscale variability associated with Mediterranean water in the Bay of Biscay (Piquet-Pellorce *et al.*, 1992).
1991	Heard Island Feasibility Test (HIFT)	SIO/UW/UM/ CSIRO	5–18 Mm/Indian, Atlantic, and Pacific oceans/Demonstrated the feasibility of using global scale acoustic transmissions to measure ocean warming (Munk and Forbes, 1989; Munk, 1990, 1991; Semtner and Chervin, 1990; Baggeroer and Munk, 1992; Munk *et al.*, 1994; Birdsall *et al.*, 1994*a,b*; McDonald *et al.*, 1994; Shang *et al.*, 1994; Chiu *et al.*, 1994*b*; Heard and Chapman, 1994; Dzieciuch and Munk, 1994; Forbes and Munk, 1994; Forbes, 1994; Palmer *et al.*, 1994; Georges *et al.*, 1994; Fraser and Morash, 1994; Burenkov *et al.*, 1994; Brundrit and Krige, 1994; Bowles *et al.*, 1994).
1991–2	AMODE/MST	SIO/UW/UM	350–670 km/Northwest Atlantic/Measurement of gyre-scale general circulation and heat transport; test assimilation of tomographic data in numerical models; test moving ship tomography (Cornuelle *et al.*, 1989; Howe *et al.*, 1989*a,b*; AMODE-MST Group, 1994).
1991–2	THETIS-1	IFREMER/ IFM/IACM	Irregular 200-km pentagon/Gulf of Lions (Mediterranean)/Measured winter deep water formation and associated circulation (THETIS Group, 1994; Send *et al.*, in press; Gaillard, 1994).
1992	Barents Sea Tomography Experiment	NPGS/WHOI	35 km/Barents Sea/Used hybrid ray–mode inversions to study the dynamics of the Polar Front in shallow water (Chiu *et al.*, 1994*a*).
1993	ATE-93	NIO	270 km/Arabian Sea/Tested applicability of tomographic techniques in Arabian Sea.
1993–4	THETIS-2	IFREMER/IFM/ IACM/WHOI	221–605 km/Western Mediterranean/Monitored variability of thermal structure and heat content in the western Mediterranean.

[a] CSIRO, Commonwealth Scientific and Industrial Research Organization (Hobart, Australia); IACM, Institute for Applied and Computational Mathematics (Heraklion, Greece); IFM, Institut für Meereskunde (University of Kiel, Kiel, Germany): IFREMER, Institut Français de Recherche pour l'Exploration de la Mer (Brest, France); MIT, Massachusetts Institute of Technology (Cambridge, Massachusetts); NIO, National Institute of Oceanography (Goa, India); NPGS, Naval Postgraduate School (Monterey, California); SHOM, Service Hydrographique et Oceanographique de la Marine (Brest, France); SIO, Scripps Institution of Oceanography (La Jolla, California); UM, University of Michigan (Ann Arbor, Michigan); UMiami, University of Miami (Miami, Florida); UW, University of Washington (Seattle, Washington); WHOI, Woods Hole Oceanographic Institution (Woods Hole, Massachusetts).

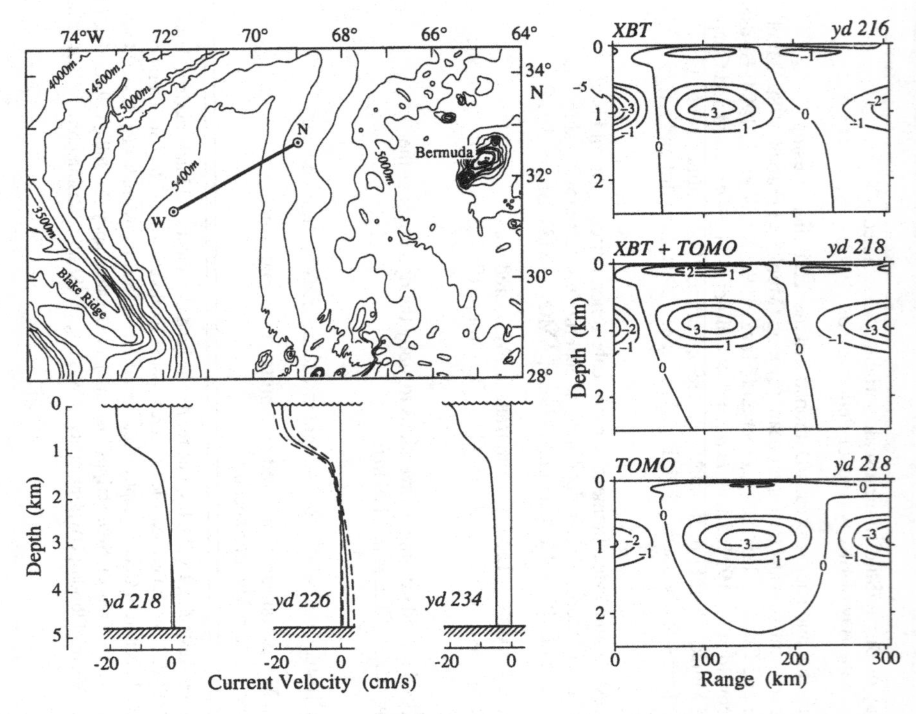

Fig. A.7. 1983 Reciprocal Transmission Experiment. Top left: Plan view of the experiment. Tomography transceivers were moored at N and W at 1300 m depth. Bottom left: Range-averaged current profiles for year days 218, 226, and 234, with associated uncertainties. Right: Sound-speed perturbation (m/s) as a function of range. The upper panel, for year day 216, is from an inversion using only XBT data. The lower panel is from a tomographic inversion on day 218 immediately following the XBT survey. The center panel is from an inversion combining the XBT and tomographic data. (Adapted from Howe *et al.*, 1987.)

to an improved understanding of horizontal sampling properties (sections 1.4 and 6.8). We should have appreciated the horizontal sampling properties much sooner, because they are just an application of the projection-slice theorem, but we failed to do so because we were focusing on geometries with only a few moored instruments.

Finally, the conduct of tomographic experiments in many different locations during the 1980s led us to appreciate more fully the sensitivity of the vertical resolution to the local sound-speed profile. The analytic temperate and polar profiles discussed extensively in chapter 2 help in understanding basic principles, but are of limited utility in detailed experiment planning.

Observations. In 1981 the tomography group faced a *Catch-22* situation: The resolution of broadband sources was needed to demonstrate feasibility, and feasibility had to be demonstrated to gain support for acquiring broadband sources! There was nothing particularly difficult about the tomographic requirements; it just happened that the combination of low frequency, broad bandwidth, and high pressure required was not available, as it had not been needed in naval or other applications.

While analysis of the data from the 1981 experiment was still in progress, the instrumentation used in that experiment was already being modified to test the feasibility of measuring ocean currents and vorticity using reciprocal transmissions over 300-km ranges (Worcester *et al.*, 1985*b*). The 224-Hz sources and receivers used in 1981 were too narrowband to provide the $O(1\text{-ms})$ travel-time precision needed to measure the expected $O(10\text{-ms})$ differential travel-time signal. They were replaced by similar 400-Hz sources of larger diameter, in a not entirely successful attempt to increase bandwidth (fig. A.5). The SIO and WHOI receiver electronics were basically those used in 1981, but modified to accommodate the higher sampling rates.

An attempt in 1983 to measure ocean currents and vorticity in a 300-km triangular array of 400-Hz transceivers yielded estimates of velocity and temperature, but not of the circulation around the triangle, as had been hoped (Howe *et al.*, 1987) (fig. A.7). The high-frequency (> 0.5-cpd) travel-time fluctuations of the oppositely traveling signals were found to be highly correlated (section 3.6), implying that the oppositely directed ray paths were sufficiently close spatially and temporally to see nearly the same internal wave field and that the measured differential travel times would therefore yield meaningful estimates of the large-scale current field.[9]

In parallel with the efforts at WHOI and SIO to make reciprocal measurements at 300 km range in midocean, DeFerrari independently developed 400-Hz tomographic instrumentation to make reciprocal measurements with 20- and 45-km transceiver triangles in the Florida Straits during 1983 (DeFerrari and Nguyen, 1986; Ko *et al.*, 1989; Chester, 1989; Chester *et al.*, 1991). That region is oceanographically intensely interesting, but also extraordinarily complex. The depth changes from 100 m to nearly 800 m in a distance of 50 km, so that virtually all rays intersect the rapidly changing bottom. Under those conditions it was impossible to resolve and identify individual ray arrivals from bottom-interacting paths (Palmer *et al.*, 1985). Travel times were instead

[9]In an ironic twist, the current meters deployed on the transceiver moorings failed, making it impossible to compare the tomographically derived currents with conventional point measurements.

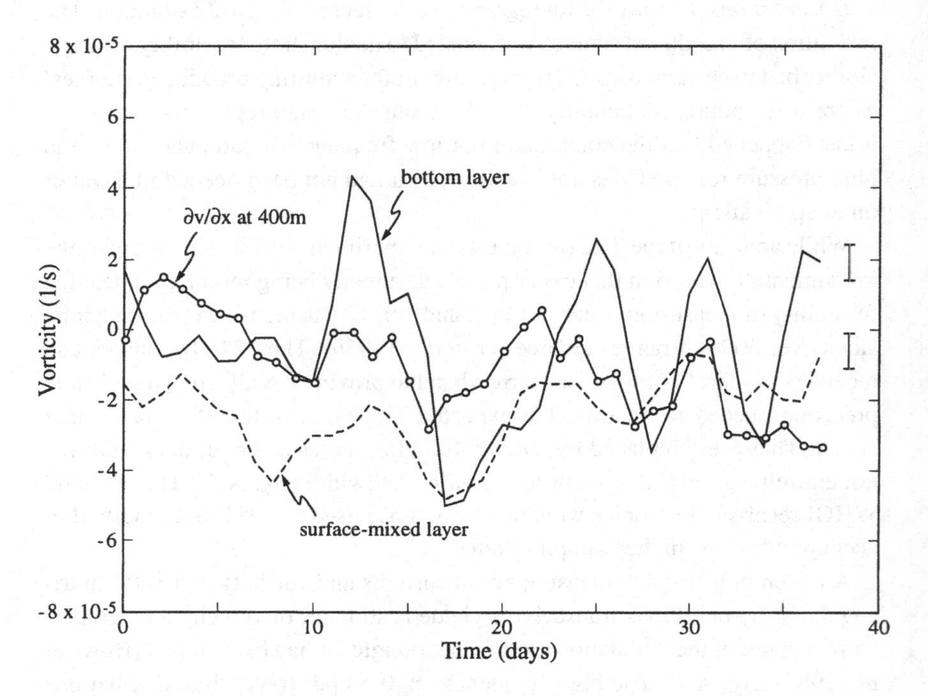

Fig. A.8. Estimated areal-average vertical component of the relative vorticity, $\partial v/\partial x - \partial u/\partial y$, from reciprocal tomographic measurements in the Florida Current. Estimates are for a bottom layer of about 200 m thickness (the bottom depth varies from about 600 to 750 m), and for a surface-mixed layer of about 100 m thickness. The value of $\partial v/\partial x$ (one element of the relative vorticity) from two current meters at 400 m depth provides a rough check on the tomographic magnitudes. (Adapted from Chester *et al.*, 1991.)

computed from the envelope of the multipath arrival *groups*, averaged over many transmissions. The mean flow (the Gulf Stream) is extremely strong, with large shears. The axis of the jet shifts laterally across the topography on time scales of days to months. Despite what must be regarded as an unfavorable environment for a determination of the oceanographic fields by tomographic methods, Ko *et al.* (1989) and Chester *et al.* (1991) succeeded in making estimates of vorticity in several range-independent layers (fig. A.8).

By the beginning of 1984 it had become clear that the time was ripe to develop a new generation of tomographic instrumentation. The instrumentation for the Woods Hole digital-buoy system (DIBOS) was an outgrowth of earlier work by Spindel and Porter (1977; Spindel *et al.*, 1978; Spindel, 1979) using moored and freely drifting buoys to measure phase fluctuations. The Scripps system was housed in spherical pressure capsules developed by Snodgrass (1968) in the late 1960s to measure deep-sea tides. Microprocessors were primitive, and programs were written in assembly language. The existing transceiver electronics were strained to the limit in terms of computer processing power, memory, and data storage capacity. The need for assembly language programming made changing the software to accommodate changing experimental needs difficult and error-prone. The sources were also incapable of supporting experiments at longer ranges. Two largely parallel developments of second-generation autonomous transceivers ensued. One development was aimed at extending the technology to reciprocal tomography at 1000 km range, and the second was aimed at producing a cost-effective commercial instrument capable of performing regional tomographic experiments at ranges of a few hundred kilometers.

The key development required to make long-range reciprocal transmissions practical was an improved source at lower frequency, with greater bandwidth, and with higher output power. Existing hydraulically driven designs were modified to provide a battery-powered source with a 100-Hz bandwidth centered at 250 Hz and a sound pressure level of 193 dB re 1 μPa at 1 m (fig. A.5). The 100-Hz bandwidth enabled the generation of three-cycle (12-ms) digits.

In parallel with the development of the HLF-5 source, SIO scientists undertook the development of new transceiver electronics.[10] The system was configured with a powerful central microprocessor controlling a number of single-chip computers (Worcester *et al.*, 1985*a*). Those receivers were equipped with vertical arrays of four to eight hydrophones for arrival-angle discrimination. Data storage, which was based on small hard-disk technology, was no longer limiting (Peckham *et al.*, 1990). The acoustic long-baseline system used to track mooring motion was an integral part of the receiver, as was the rubidium frequency standard used to maintain time with 1-ms precision. The new transceiver system, initially developed for deployment in the RTE87 1000-km triangle in the central North Pacific, has been the mainstay of subsequent SIO experiments.

More or less in parallel, Webb Research Corporation (WRC), working closely with WHOI and MIT scientists, developed a commercially available 400-Hz transceiver suitable for shorter-range experiments. The system employed an improved version of the 400-Hz organ-pipe source, with a source level of 180 dB re 1 μPa at 1 m (Boutin *et al.*, 1989), a computer-compensated clock developed by P. Tillier (to reduce energy requirements), and a single-channel receiver.

[10]The new system was named AVATAR (*a*dvanced *v*ertical *a*rray *t*omographic *a*coustic *r*eceiver). In the Hindu religion, an avatar is a god come down to Earth in material form, but the French word *avatar*, meaning vicissitudes, is perhaps more appropriate in the tomographic context.

That development was a major step toward making tomographic techniques available to the wider community of oceanographers.

Both new instruments were sufficiently reliable to make possible long-term experiments, lasting up to a year. The emphasis in the late 1980s then shifted from experiments largely designed to develop the technique to longer term experiments largely designed to learn about the ocean. The results from those experiments, including the 1987 Reciprocal Tomography Experiment, the 1988–9 Gulf Stream (SYNOP) Tomography Experiment, and the 1988–9 Greenland Sea Tomography Experiment, were summarized in the Epilogue.

To meet a requirement for better horizontal resolution, development of moving-ship tomography (MST) was initiated late in the decade. Moored AVATAR/HLF-5 autonomous transceivers were used to transmit to a ship-suspended vertical receiving array lowered at many different locations. The principal new requirement was for precise positioning of the sources and receivers so that the absolute ranges would be known to better than 5–10 m (Cornuelle *et al.*, 1989). That required a combination of differential GPS positioning to locate the ship and acoustic techniques to locate the subsurface hydrophones relative to the ship, as discussed in section 5.7 (Howe *et al.*, 1989*a,b*). Preliminary equipment tests were done in conjunction with the Greenland Sea Tomography Experiment deployment and recovery cruises during summer 1988 and 1989 (with a GPS reference station at Ny Ålesund, Spitzbergen!). Data obtained during the 1991–2 Acoustic Mid-Ocean Dynamics Experiment (AMODE) were consistent with position accuracies to a few meters halfway between Bermuda and Puerto Rico. That experiment was the most ambitious tomographic mapping experiment to date (fig. 1.6).

This section has so far focused on *autonomous* and *ship-suspended* instrumentation, best suited for measurements of limited duration. Those have provided great flexibility for sub-basin transmissions. For basin and global experiments, with their requirements of high power and long duration, the decision tends to favor shore-connected instruments. The east Pacific Basin transmissions of Spiesberger and Metzger used a shore-connected U.S. Navy source off Kaneohe, Hawaii, and all of the basin-scale experiments conducted to date have relied heavily on existing U.S. Navy SOSUS receivers that are cable-connected to shore.

A.6. Testing

Much effort during the 1980s was devoted to testing the accuracy of tomographic inversions. The issue is difficult because of the very different natures of integrating acoustic measurements and conventional point measurements.

The obvious approach is to compare inversions for the sound-speed field using acoustic data only with inversions for the sound-speed field using traditional point measurements only (*e.g.*, the top and bottom panels in fig. A.7). A second and better approach, suggested in section 6.9, is to test for consistency of the acoustic and other measurements by performing inversions that combine all of the data and then testing the data residuals to be sure they behave as expected (*e.g.*, the middle panel in fig. A.7). Such a combined inversion provides the best possible estimate of the ocean, given the available data, and so should be the normal analysis approach. It is not always easy in such a combined inversion to tell to what extent the various data types are redundant, and so provide direct checks on one another, and to what extent they are independent, but consistent with one another and with the ocean model.

It is important to realize that at the most fundamental level, what needs testing is the acoustic *forward problem*, not the inverse methods. The linear inverse methods described in chapters 6 and 7 are well-known mathematical techniques that provide rigorous estimates of the unknown ocean model parameters together with the uncertainties of the estimates. Examination of the data residuals provides built-in checks on the procedure. Provided that the forward problem has been accurately modeled, the solution and its uncertainty directly follow. This leads to a third (and most direct) approach: Starting with traditional point measurements, use standard inverse methods to map the sound-speed field, *together with the associated uncertainties*, and then use the mapped fields to compute the travel times, *together with their uncertainties*, for comparison with the measured travel times. This *validity* test is quite different from an evaluation of the *utility* of tomography, which requires a comparison of the resolution, accuracy, and required effort as compared with other methods.

Tests that were conducted during the 1980s included all three types of comparisons mentioned. They showed that tomographic techniques can provide measurements of thermal structure, heat content, and velocity on scales from hundreds of kilometers to 1000 km that are consistent with independent measurements within the estimated uncertainties.

Sound-speed. The SLICE89 experiment (sections 2.16 and 6.9) provided the best test to date of the acoustic forward problem (Howe *et al.*, 1991; Duda *et al.*, 1992; Cornuelle *et al.*, 1993; Worcester *et al.*, 1994; Colosi *et al.*, 1994). The experiment consisted of 1000-km transmissions from a near-axial HLF-5 source to a 3000-m-long vertical receiving array suspended from R/P *FLIP* at 1000 km range, together with nearly 300 XBT, AXBT, and CTD casts along the propagation path while the transmissions were in progress. All of the XBT, AXBT, and CTD data were objectively mapped, and the resulting range-dependent

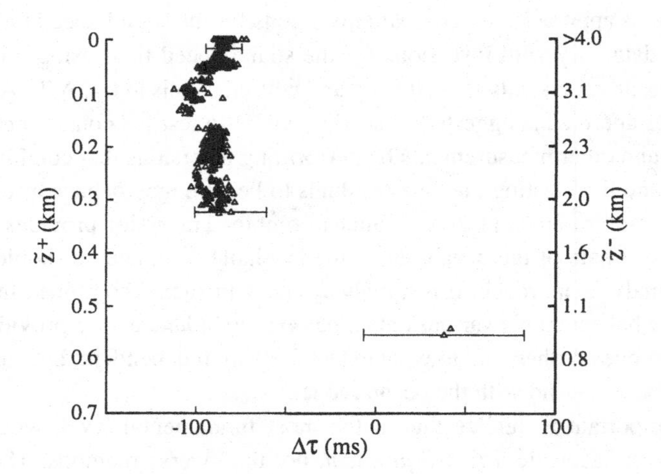

Fig. A.9. Measured minus predicted travel times as functions of ray turning depths from the SLICE89 experiment. (Adapted from Worcester *et al.*, 1994.)

sound-speed field was used to compute the predicted travel times of the ray-like arrivals in that experiment (fig. 2.21). The predicted travel times were then subtracted from the time-mean of the measured travel times to find the discrepancies (fig. A.9). The plotted error bars are combinations of the uncertainties in the predicted and measured travel times, with the uncertainty in the predicted travel times due to the uncertainty in the sound-speed field estimates from the direct measurements dominating.[11] The measured travel times are consistent with predictions for the early arrivals with upper turning depths above 350 m and are marginally consistent for the near-axial final cutoff rays.

Although the rather dry comparison between predicted and measured travel times in fig. A.9 provides the most direct test of the validity of the tomographic method, the travel-time data obtained in the SLICE89 experiment were also inverted, as described in section 6.9, to obtain the range-dependent sound-speed field, for comparison with direct measurements (fig. 6.27). Cornuelle *et al.* (1993) found that the range average of the sound-speed field constructed from travel times differed from that constructed from the XBT, AXBT, and CTD data alone by less than about 0.3 m/s (roughly 0.07° C), except near the surface (fig. A.10). The negative anomaly between 500 and 1200 m is due to the

[11]The relatively large travel-time anomalies of about -100 ms for rays with upper turning depths between the surface and about 350 m are most likely due to a 150-m error in the estimated range (which is not reflected in the error bars), although a bias in the direct estimates of sound-speed of 0.2 m/s would have the same effect.

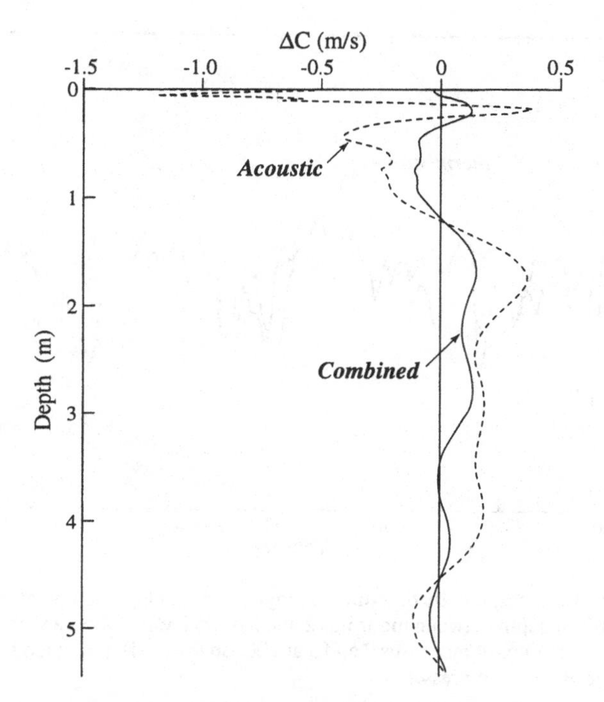

Fig. A.10. Test of the range-averaged sound-speed profile from the SLICE89 experiment. The dashed curve is the profile determined from acoustic travel times alone minus that determined from the CTD, XBT, and AXBT data. The solid curve is the profile determined from the combined acoustic, CTD, XBT, and AXBT data minus that determined from the CTD, XBT, and AXBT data. (Adapted from Cornuelle *et al.*, 1993.)

late-arriving, near-axial final cutoffs in fig. A.9. The 0.3-m/s discrepancy between 500 and 1200 m is marginally consistent with the estimated uncertainties, as it must be because the axial travel-time perturbations in fig. A.9 were marginally consistent with the travel times estimated from the direct measurements. The residuals from combined inversions of the acoustic travel times and the direct measurements are consistent with their expected error levels.

Such agreement was considered a benchmark in the development of ocean acoustic tomography, permitting future emphasis to be placed on the scientific problems, not the method.

SLICE89 lasted only 9 days, giving a comparison at essentially one point in time, because the mesoscale sound-speed field changed little during the experiment. The Gulf Stream (SYNOP) Tomography Experiment, in contrast, lasted nearly 300 days, giving a long-term comparison between moored temperature

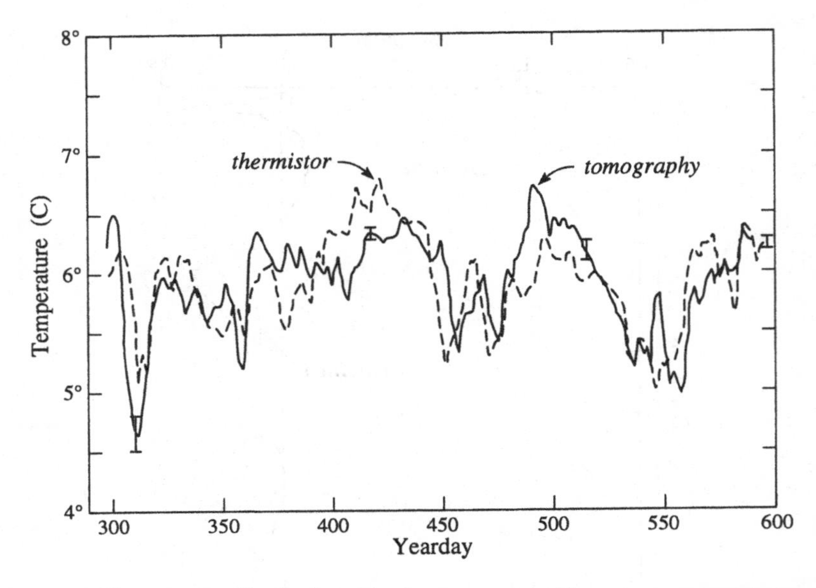

Fig. A.11. The tomographically determined range-averaged layer temperatures between 1000 and 1400 m depth between moorings 2 and 3 of SYNOP (solid), and the average of two thermistor readings at moorings 2 and 3 at 1000 m (dashed). (Adapted from Chester and Malanotte-Rizzoli, in press.)

measurements and tomographically determined temperatures (Chester, 1993; Chester *et al.*, 1994; Chester and Malanotte-Rizzoli, in press). The range-averaged temperature in a layer between 1000 and 1400 m computed from the acoustic travel-time data for a 203-km path was compared with the average of the temperatures from thermistors at 1000 m depth on the two moorings (fig. A.11). The comparison is generally favorable, but is limited by the accuracy of the conventionally determined range-averaged temperature.

Velocity. Most of the efforts to validate tomographic determinations of sound-speed and heat content have used XBT, AXBT, and CTD profiles to map the temperature and sound-speed fields in the plane containing the source and receiver. The problem of obtaining profiles of current at sufficiently close spacing to map the current field is more difficult, because the technology for making absolute current measurements with instruments lowered from shipboard does not exist, so that sparse data from moored current meters must be used for the comparisons. Dushaw *et al.* (1993*a*) have presented the available evidence.

The most convincing comparisons have been for the barotropic tides. The tides provide a convenient large-scale signal with which to verify the ability of differential travel times to measure gyre-scale barotropic currents. The 1987 Reciprocal Tomography Experiment yielded tidal harmonic constants determined from the acoustically derived barotropic currents in excellent agreement with those determined from current meter measurements and from empirical numerical tidal models, as discussed in section 3.6 (table 3.1).

The best test of the ability of tomographic experiments to measure low-frequency ($\leq$ 1 cpd) currents was made in the Gulf Stream (SYNOP) Tomography Experiment, because of its combination of short transmission ranges (100–200 km) and extensive moored current meter measurements. The acoustically determined, range-averaged current at 500 m depth compares favorably with the average of three current meter measurements at 500 m depth from instruments at the ends and midpoint of the acoustic path (Chester *et al.*, 1994) (fig. A.12). The zonal path was about 170 km long, so the current meters were separated by 85 km, comparable to the O(100-km) correlation scale. Spectra of the current show that the acoustic measurements suppress energy for periods of less than about 20 days, as might be expected given the range averaging inherent in the measurements. Tomographic and current meter estimates of the statistical properties of this energetic region are also in agreement (Chester, 1993) (table A.2).

The 1987 Reciprocal Tomography Experiment (Dushaw *et al.*, 1994) and the 1988–9 Greenland Sea Tomography Experiment also yielded comparisons between tomographically determined, low-frequency barotropic currents and independent estimates (fig. A.12). The independent measurements were made at a single location in both experiments, although the paths were many mesoscale correlation lengths long. The tomographic estimates appear consistent with the independent measurements, but stronger statements are not possible given the limited conventional data sets. The degree to which the tomographic estimates suppress variability with 10–20-day periods is more striking in both cases than for the Gulf Stream experiment.

A.7. Coda

Tomography is one of the tools now available to physical oceanographers to study open-ocean processes with spatial scales of 100–1000 km. Groups in the United States, Germany, France, India, and Japan have regional-scale tomographic programs. Recent concerns about the possible effects of low-frequency acoustic transmissions on marine mammals and other marine life (Potter, 1994) make research on this issue essential to the future development of acoustic techniques for monitoring the ocean.

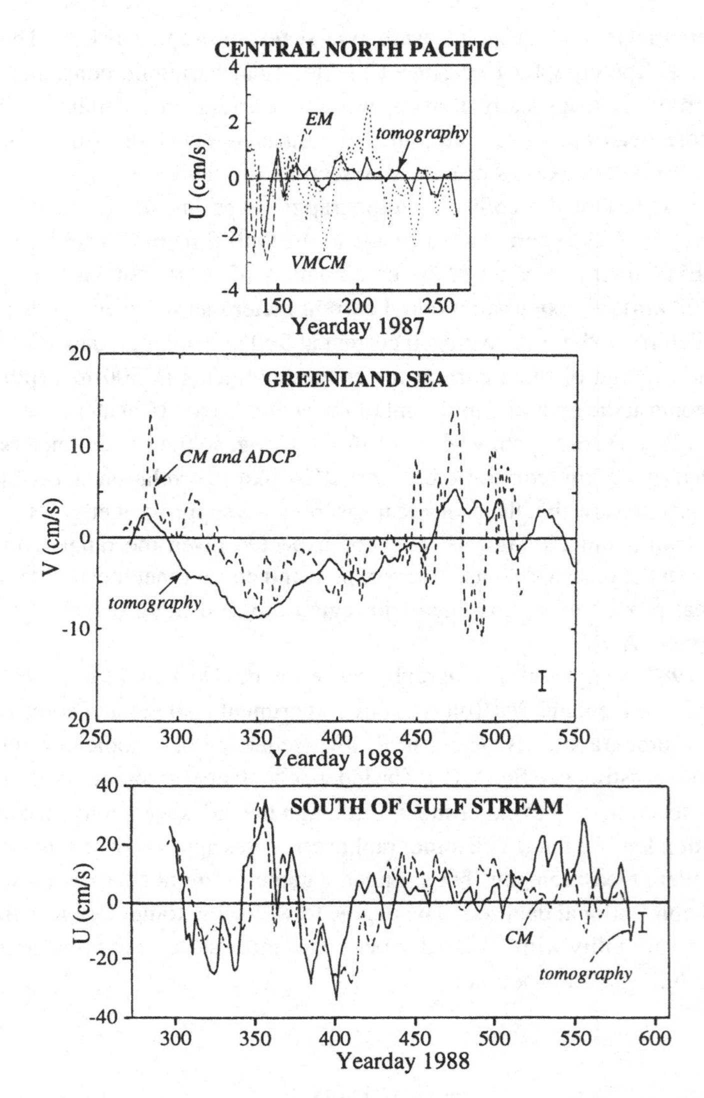

Fig. A.12. Comparison of low-frequency (<1 cpd), range-averaged, tomographic currents (solid) with point current measurements (dashed). The uncertainty estimates are for the tomographically derived currents. Note the different velocity scales. RTE87: zonal, barotropic current compared with barotropic current determined from three moored vector measuring current meters (VMCM), and with electromagnetic field measurements (EM), both located near the middle of the acoustic path. (Adapted from Dushaw *et al.*, 1994.) Greenland Sea: Meridional barotropic current compared with current determined from current meter and acoustic Doppler current profiler (ADCP) data (W. Morawitz, personal communication). SYNOP: zonal current at 500 m compared with the average of three current meters at 500 m depth located on the acoustic path. (Adapted from Chester *et al.*, 1994.)

Table A.2. *Eddy statistics at 500 m from SYNOP*

		Tomography	Current Meter
$\frac{1}{2}(\bar{u}^2 + \bar{v}^2)$	cm^2/s^2	12.3 ± 26.1	33.1 ± 17.8
$\frac{1}{2}\overline{(u'^2 + v'^2)}$	cm^2/s^2	225.4 ± 123.4	382.4 ± 54.9
$\overline{T'^2}$	$°C^2$	1.1 ± 2.7	2.5 ± 1.7
$\overline{u'v'}$	cm^2/s^2	5.0 ± 76.1	-12.6 ± 54.6
$\overline{u'T'}$	$°C \cdot cm/s$	-2.3 ± 4.5	-3.3 ± 4.7
$\overline{v'T'}$	$°C \cdot cm/s$	-4.6 ± 6.2	-6.3 ± 5.0
ζ	$10^{-6}s^{-1}$	-2.1 ± 4.3	-2.1 ± 3.9
$\overline{\zeta'^2}$	$10^{-12}s^{-2}$	0.25 ± 1.2	

Note: u and v are the current components, T is temperature, and ζ is relative vorticity. A bar indicates a temporal mean, and a prime indicates perturbation from the mean.

Source: (Data from Chester, 1993.)

Perhaps the key outstanding question is whether or not tomographic techniques can be extended to ocean regimes in which acoustic propagation is more complex (*e.g.*, longer ranges, regions with weak or double sound-speed channels, and shallow water). Much work is yet required to evaluate the information content of acoustic observables other than ray travel times and modal group delays. These questions are, of course, related. Extending tomography to new regimes or geometries may well involve the need to use acoustic observables other than travel time.

APPENDIX B

OCEAN ACOUSTIC PROPAGATION ATLAS

Climatological sound-speed profiles and predicted acoustic arrival patterns for selected locations worldwide (fig. B.1) are displayed in this appendix. The locations are from a regular grid spaced at 15° increments in latitude and 20° increments in longitude. [Worcester and Ma (in press) provide results for all grid locations exceeding 2000 m depth.] The atlas is organized by location, beginning at 75°N and proceeding southward. At each latitude, results are presented in order of longitude, proceeding westward from the prime meridian. The interpretation of the plots presented here is discussed at length in section 2.16. Each panel is described briefly, proceeding counterclockwise from the bottom left.

The sound-speed profiles (bottom left) were computed from annual-average climatological temperature and salinity data due to Levitus (1982), using the Del Grosso (1974) sound-speed equation. The Levitus climatology is a horizontally smoothed picture of the ocean, so the results do not properly represent the behavior to be expected in frontal regions.

Acoustic normal mode functions 1 and 7 computed for 70 Hz are displayed at bottom center. The amplitude normalization is arbitrary. The group velocity for each mode is given immediately below the mode function.

Time fronts in τ, z-space for a fixed range of 500 km (bottom right) show the arrival structure for a source on the sound-channel axis, when one exceeding 100 m depth exists, or for a source at 100 m depth. The source depth used is indicated by a small arrow and is listed on each figure. A minimum depth of 100 m was chosen in part to reflect experimental realities and in part to give simpler time fronts, because sources located in shallow surface ducts can give complex ray arrival patterns. Normal mode predictions are more appropriate for propagation in a shallow surface duct. The interpretation of time fronts is discussed in section 2.4.

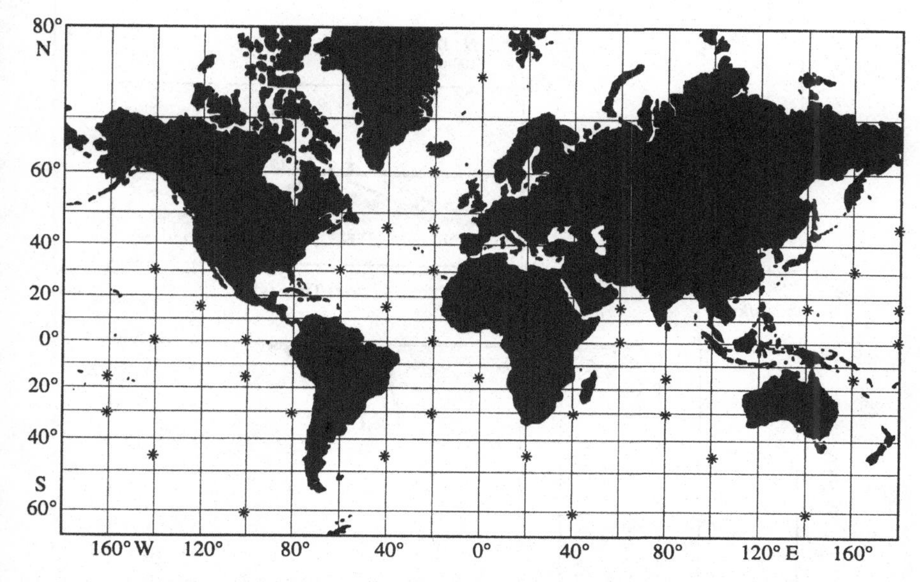

Fig. B.1. Selected locations (*) for which sound-speed profiles and predicted acoustic arrival patterns are included in this atlas.

The WKBJ approximation (Brown, 1981) was used to predict the arrival pattern at 500 km range (top right) for a pulse with a center frequency of 250 Hz and a pulse length of 0.012 s (three cycles of carrier). The predicted waveform was complex demodulated, and only the arrival amplitude is displayed. The source and receiver were both selected to be at the same depth as the source used to construct the time-front diagram. The arrival spacing therefore corresponds to that predicted from a horizontal slice through the time front at the source depth (arrow). Geometric arrivals are labeled, where practical, with their ray identifier, $\pm p$, where $+(-)$ indicates a ray that initially travels upward (downward) at the source and has a total of p upper plus lower turning points between the source and receiver.

Finally, equation (2.5.2) was used to compute the action A (top right) from the sound-speed profile.

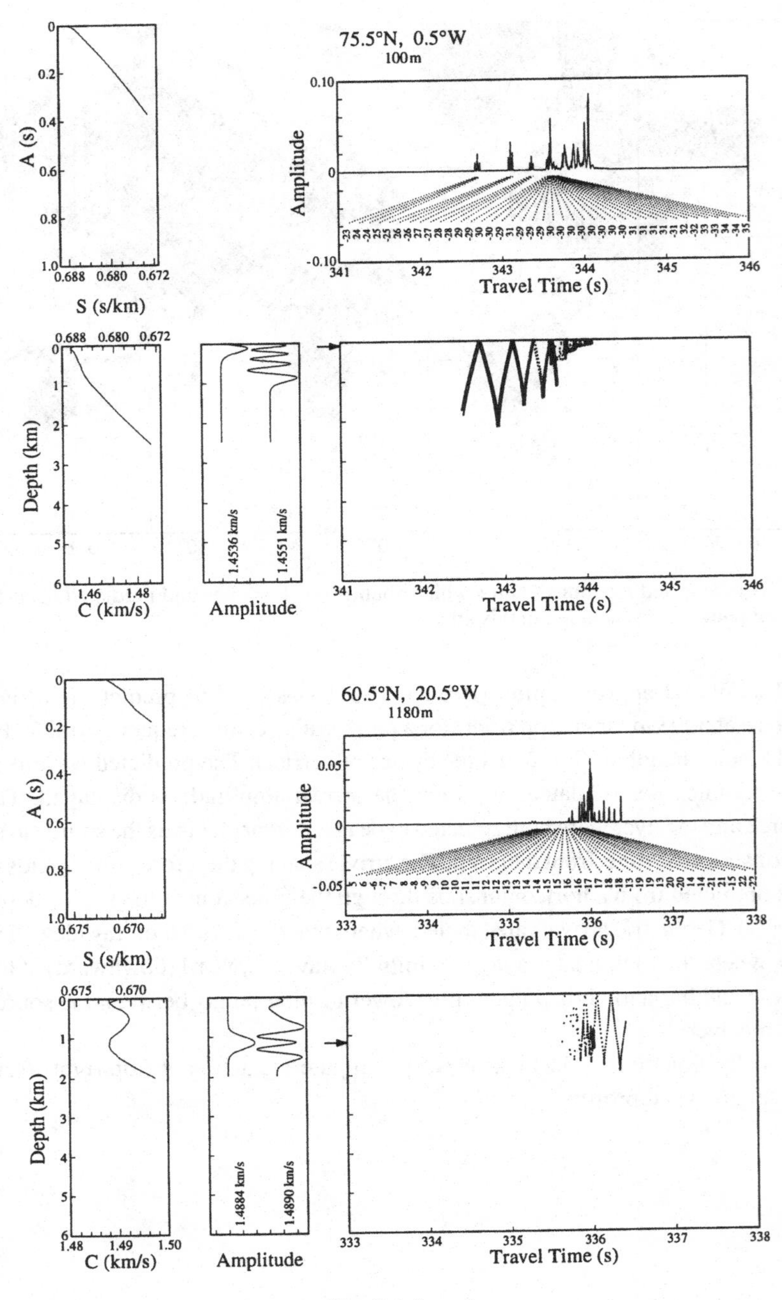
75.5°N, 0.5°W
100m
Amplitude
Travel Time (s)
A (s)
S (s/km)
Depth (km)
C (km/s)
1.4536 km/s
1.4551 km/s
Amplitude
Travel Time (s)
60.5°N, 20.5°W
1180m
Amplitude
Travel Time (s)
A (s)
S (s/km)
Depth (km)
C (km/s)
1.4884 km/s
1.4890 km/s
Amplitude
Travel Time (s)

Fig. B.1 (*cont.*)

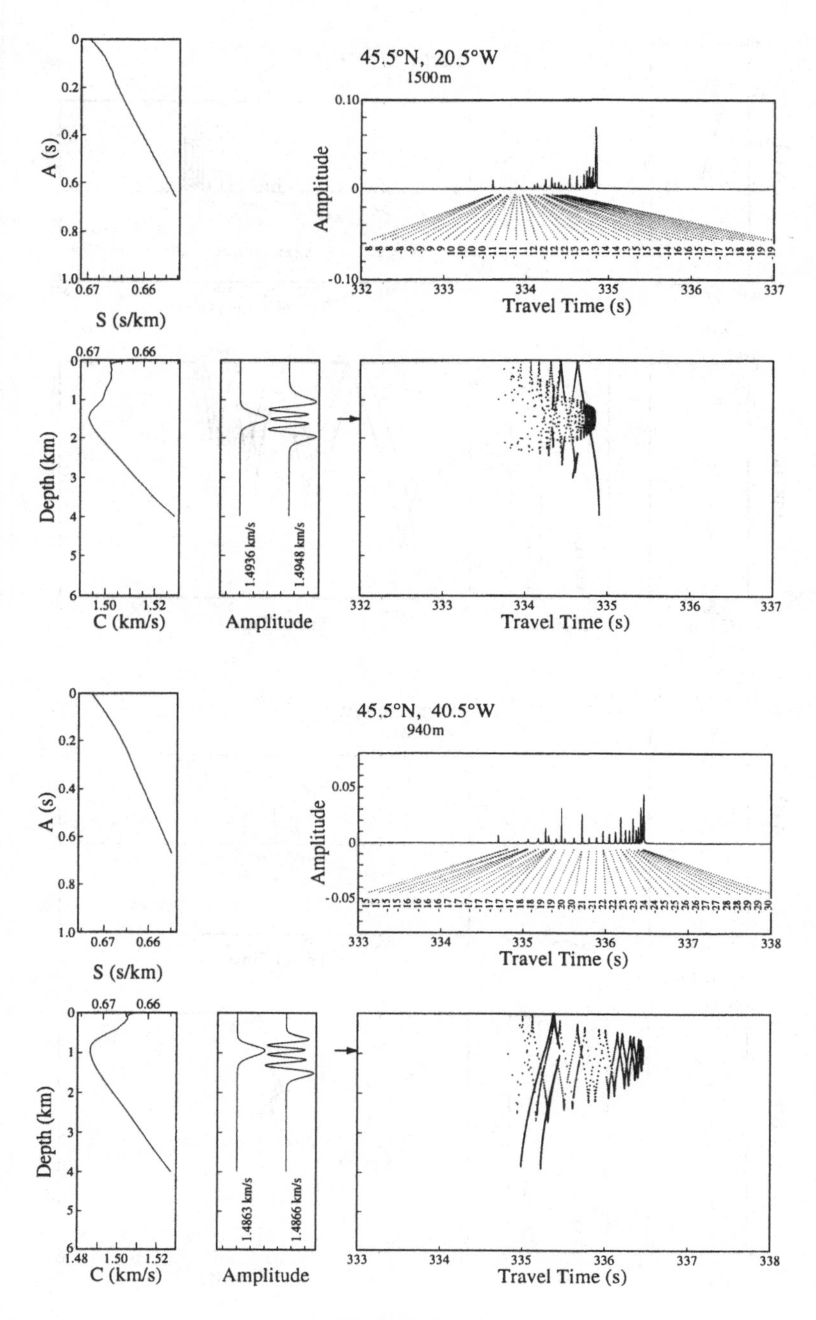
45.5°N, 20.5°W
1500m
A (s)
S (s/km)
Amplitude
Travel Time (s)
Depth (km)
C (km/s)
1.4936 km/s
1.4948 km/s
45.5°N, 40.5°W
940m
1.4863 km/s
1.4866 km/s

Fig. B.1 (*cont.*)

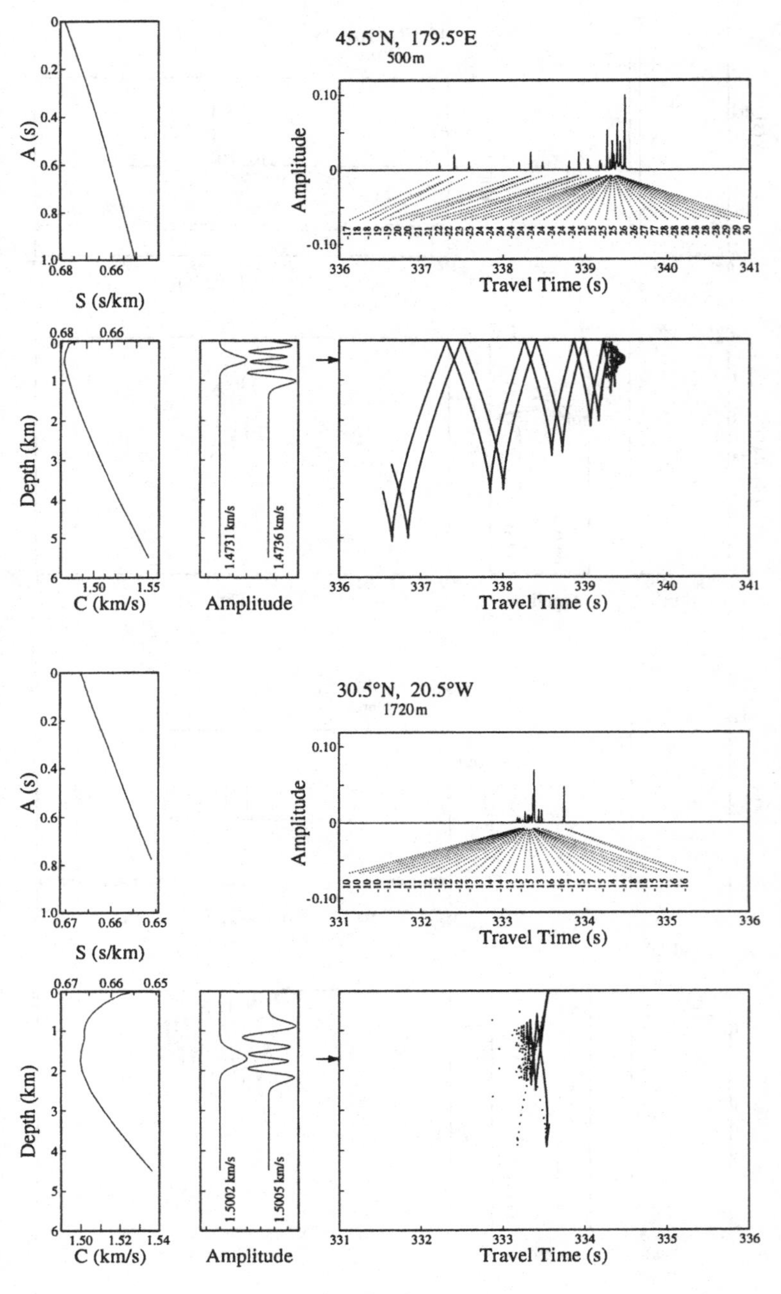
45.5°N, 179.5°E
500m
Amplitude
Travel Time (s)
A (s)
S (s/km)
Depth (km)
C (km/s)
1.4731 km/s
1.4736 km/s
30.5°N, 20.5°W
1720m
1.5002 km/s
1.5005 km/s

Fig. B.1 (*cont.*)

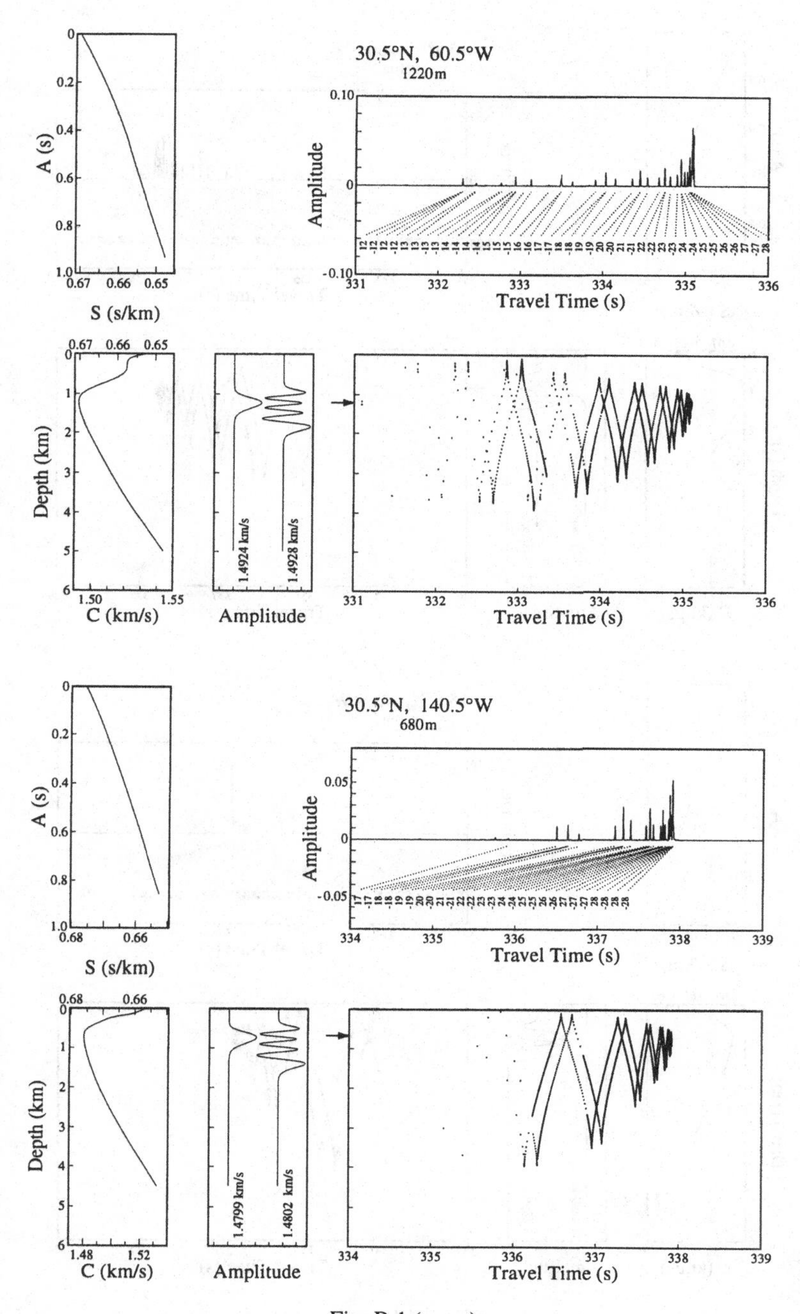
30.5°N, 60.5°W
1220m
A (s)
S (s/km)
Amplitude
Travel Time (s)
Depth (km)
C (km/s)
1.4924 km/s
1.4928 km/s
30.5°N, 140.5°W
680m
1.4799 km/s
1.4802 km/s

Fig. B.1 (*cont.*)

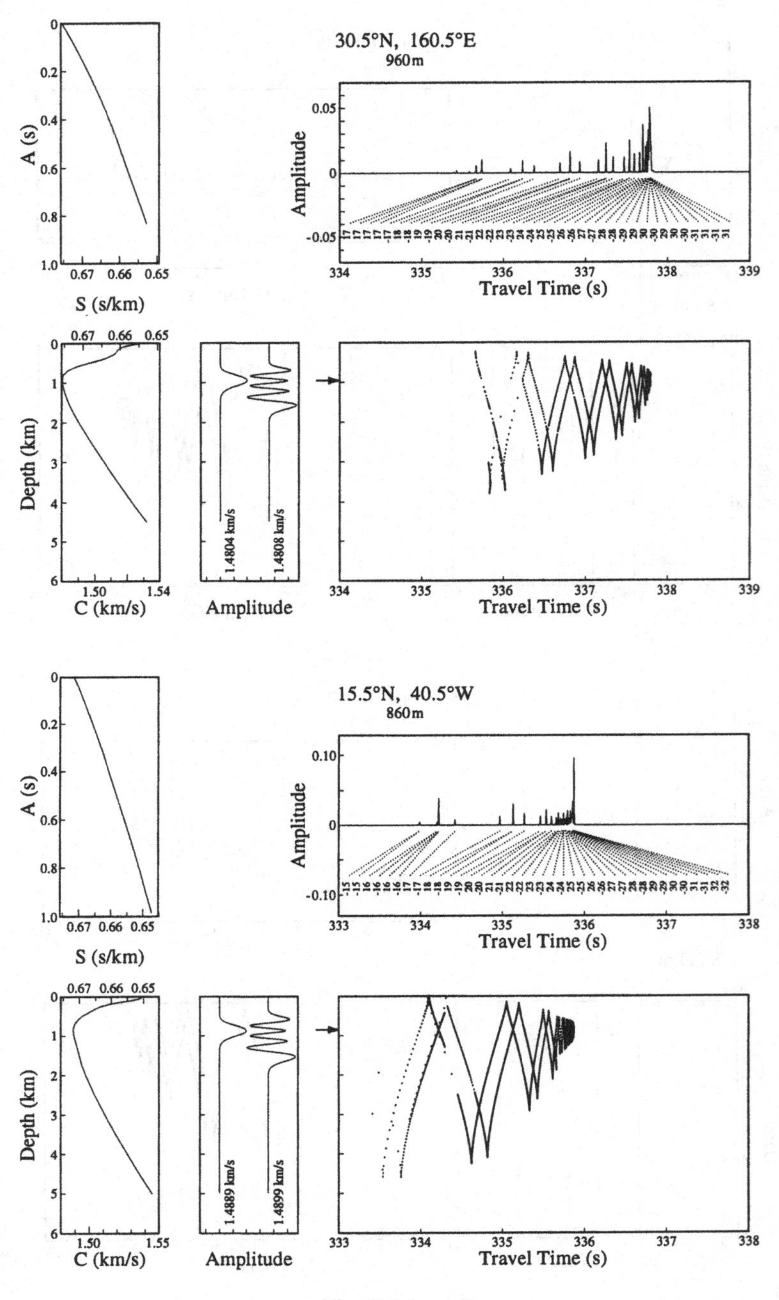
30.5°N, 160.5°E
960m
Amplitude
Travel Time (s)
A (s)
S (s/km)
Depth (km)
C (km/s)
1.4804 km/s
1.4808 km/s
15.5°N, 40.5°W
860m
1.4889 km/s
1.4899 km/s

Fig. B.1 (*cont.*)

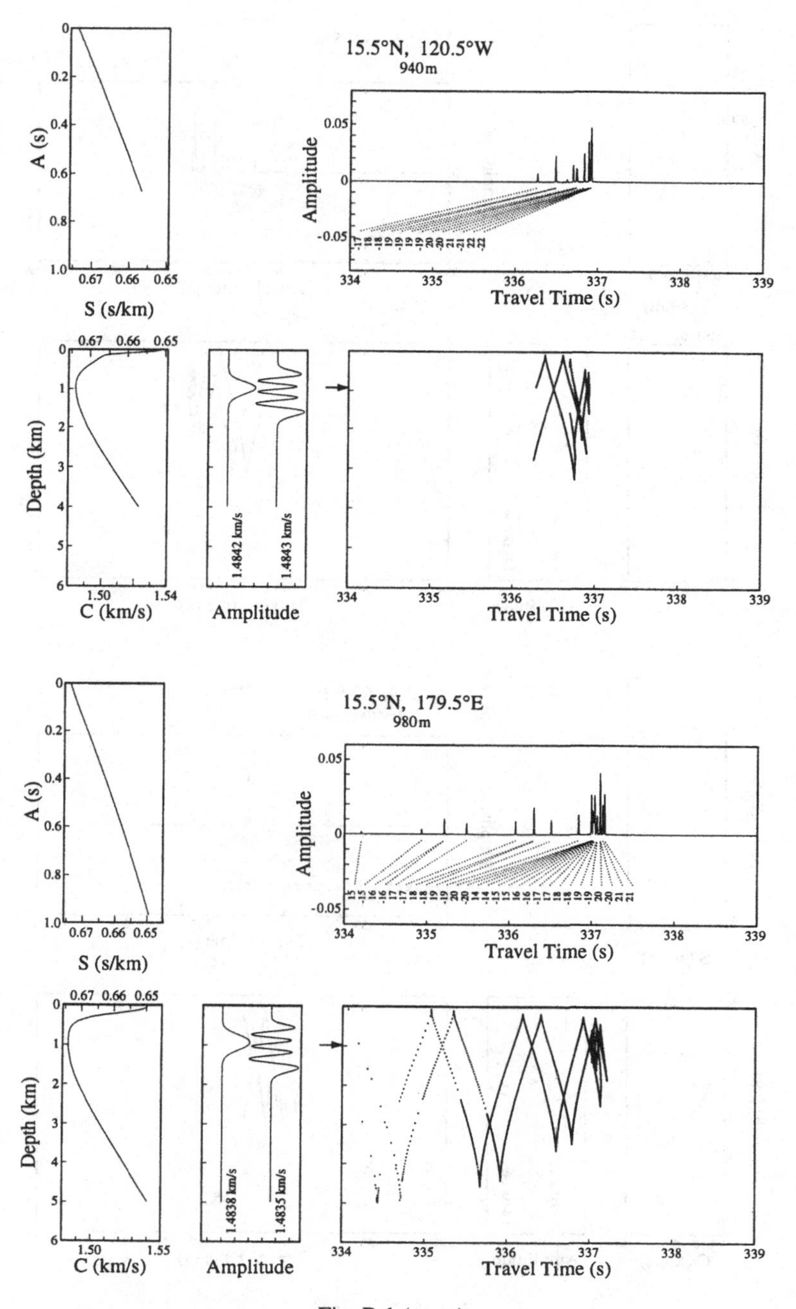

15.5°N, 120.5°W
940m
Amplitude
Travel Time (s)
A (s)
S (s/km)
Depth (km)
C (km/s)
1.4842 km/s
1.4843 km/s
Amplitude
Travel Time (s)
15.5°N, 179.5°E
980m
Amplitude
Travel Time (s)
A (s)
S (s/km)
Depth (km)
C (km/s)
1.4838 km/s
1.4835 km/s
Amplitude
Travel Time (s)

Fig. B.1 (*cont.*)

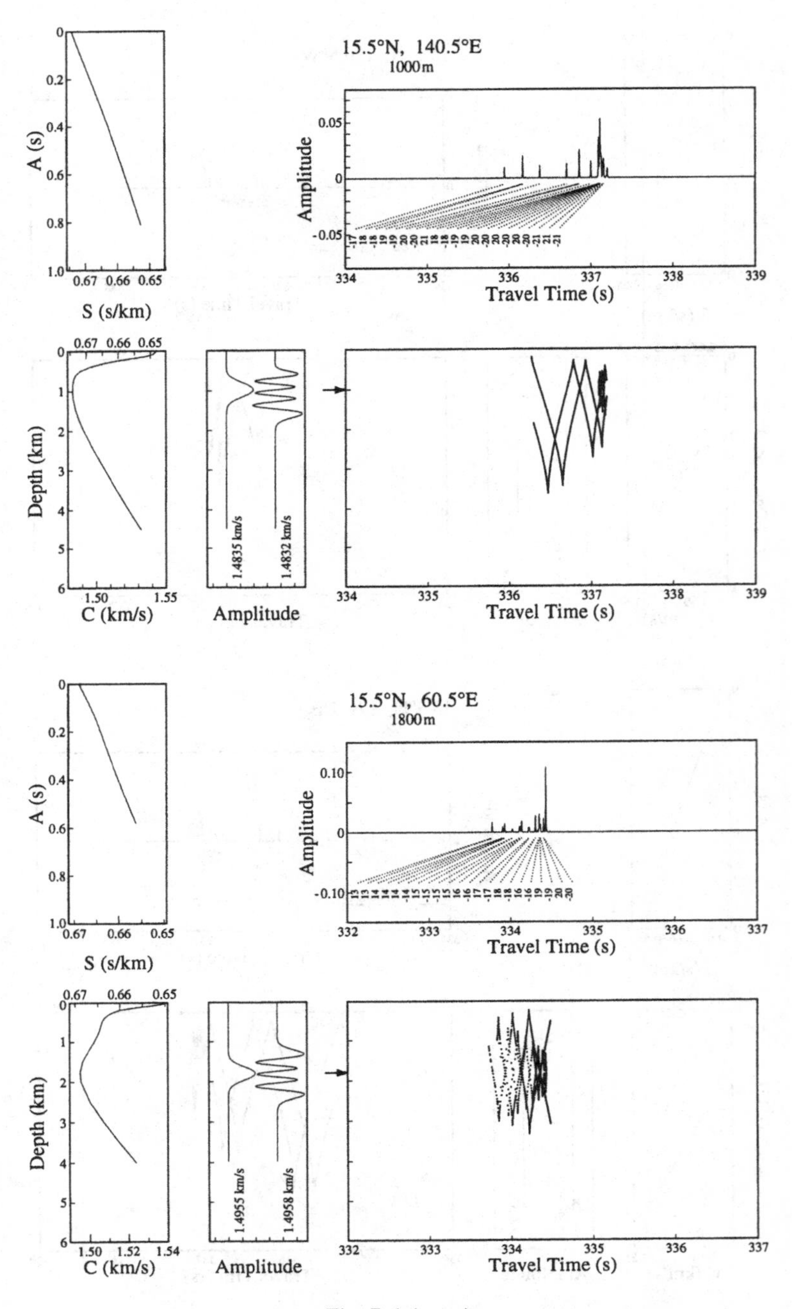
15.5°N, 140.5°E
1000m
A (s)
S (s/km)
Amplitude
Travel Time (s)
Depth (km)
C (km/s)
1.4835 km/s
1.4832 km/s
15.5°N, 60.5°E
1800m
1.4955 km/s
1.4958 km/s

Fig. B.1 (*cont.*)

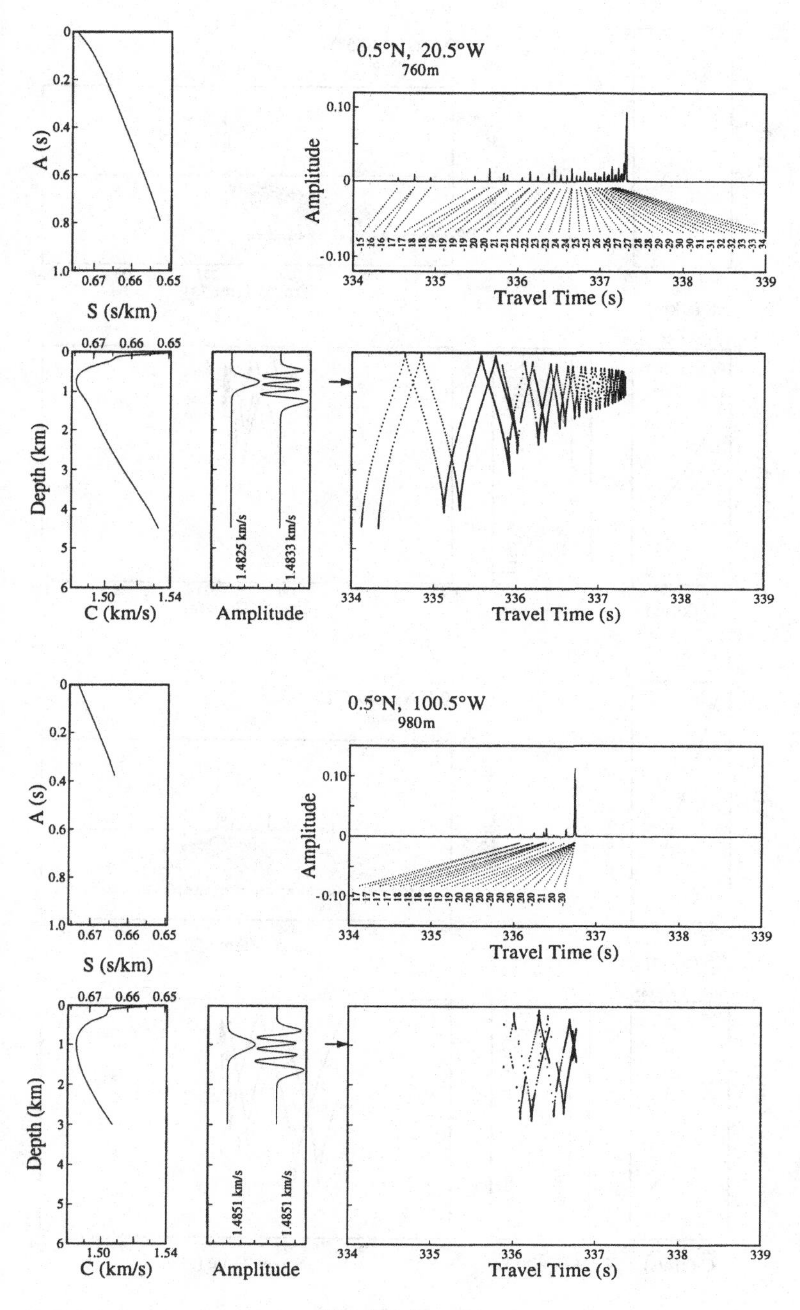
0.5°N, 20.5°W
760m
A (s)
S (s/km)
Amplitude
Travel Time (s)
Depth (km)
C (km/s)
1.4825 km/s
1.4833 km/s
0.5°N, 100.5°W
980m
1.4851 km/s
1.4851 km/s

Fig. B.1 (*cont.*)

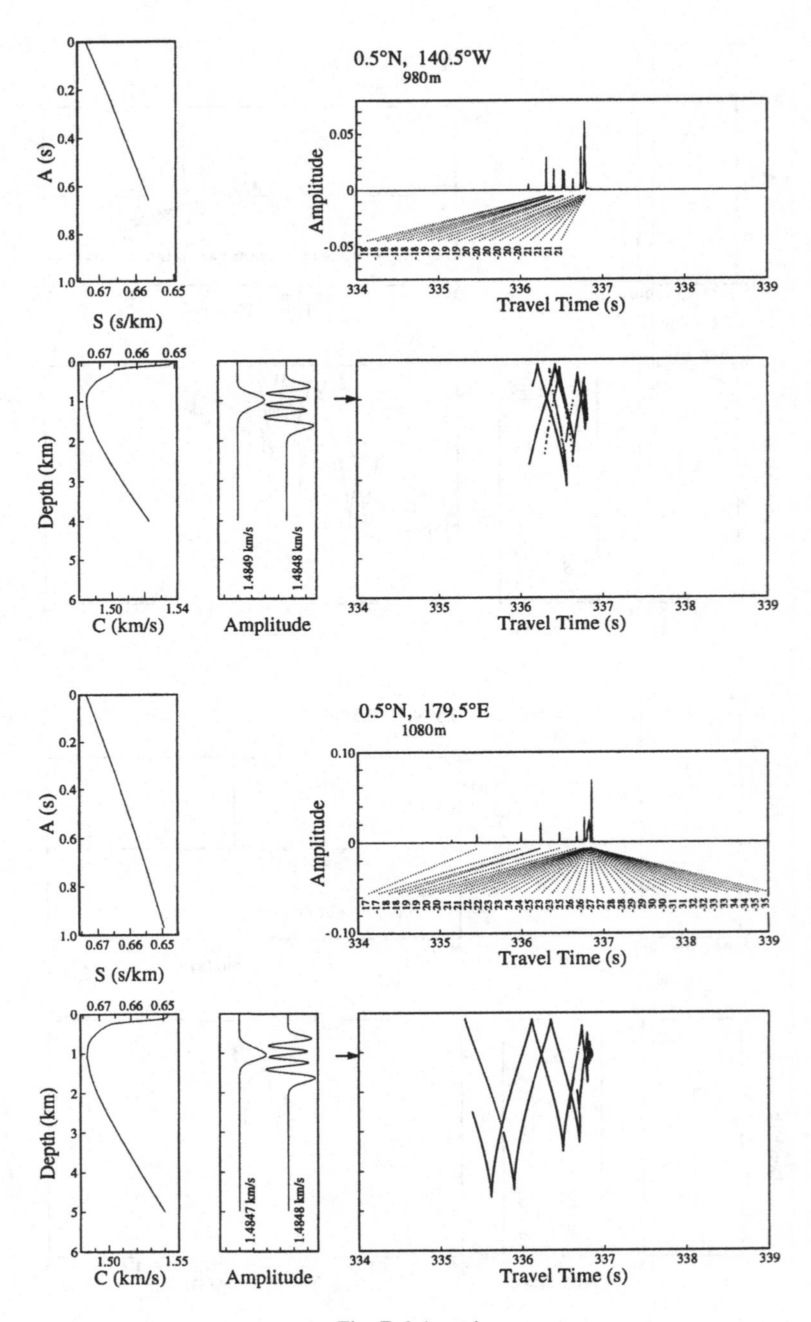
0.5°N, 140.5°W
980m
A (s)
S (s/km)
Amplitude
Travel Time (s)
Depth (km)
C (km/s)
1.4849 km/s
1.4848 km/s
Amplitude
Travel Time (s)
0.5°N, 179.5°E
1080m
A (s)
S (s/km)
Amplitude
Travel Time (s)
Depth (km)
C (km/s)
1.4847 km/s
1.4848 km/s
Amplitude
Travel Time (s)

Fig. B.1 (*cont.*)

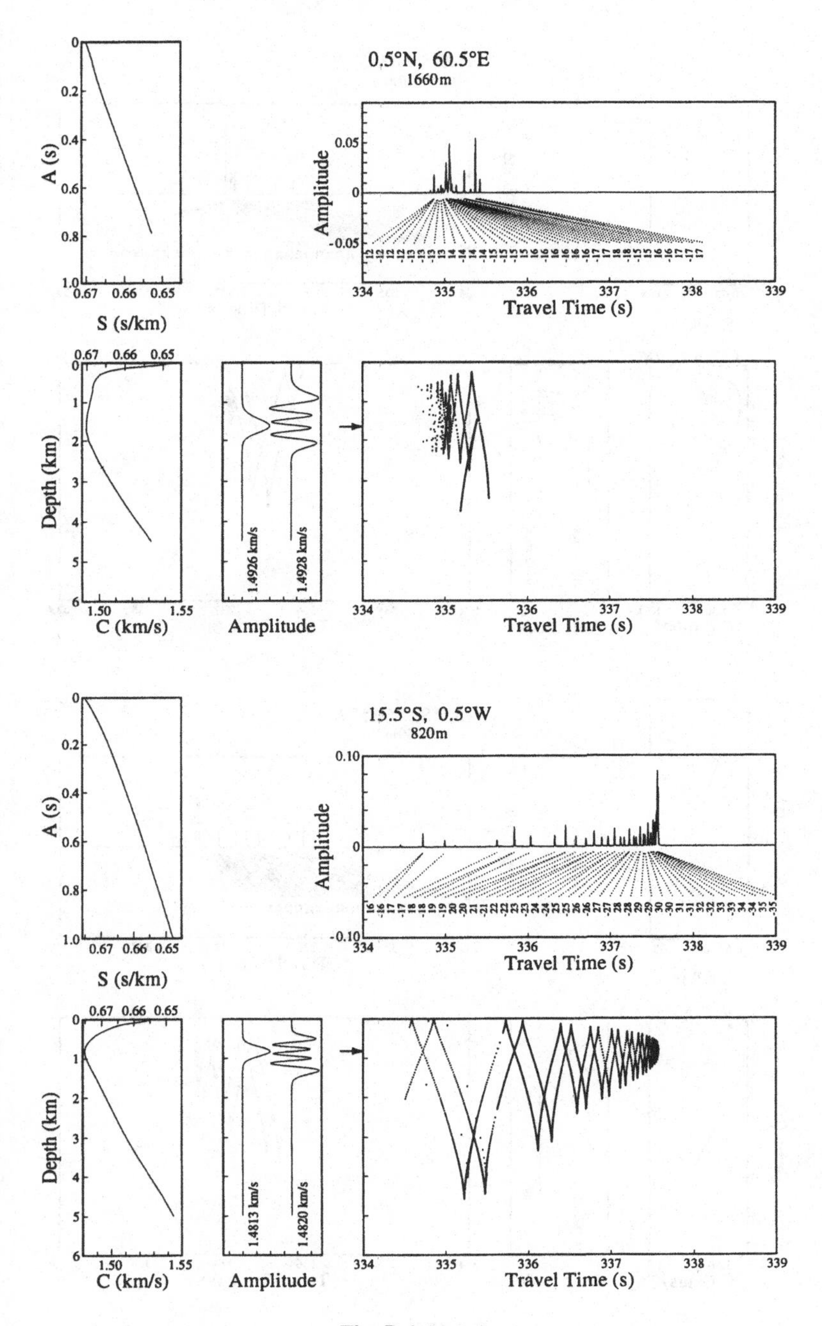

0.5°N, 60.5°E
1660m
A (s)
S (s/km)
Amplitude
Travel Time (s)
Depth (km)
C (km/s)
1.4926 km/s
1.4928 km/s
15.5°S, 0.5°W
820m
1.4813 km/s
1.4820 km/s

Fig. B.1 (*cont.*)

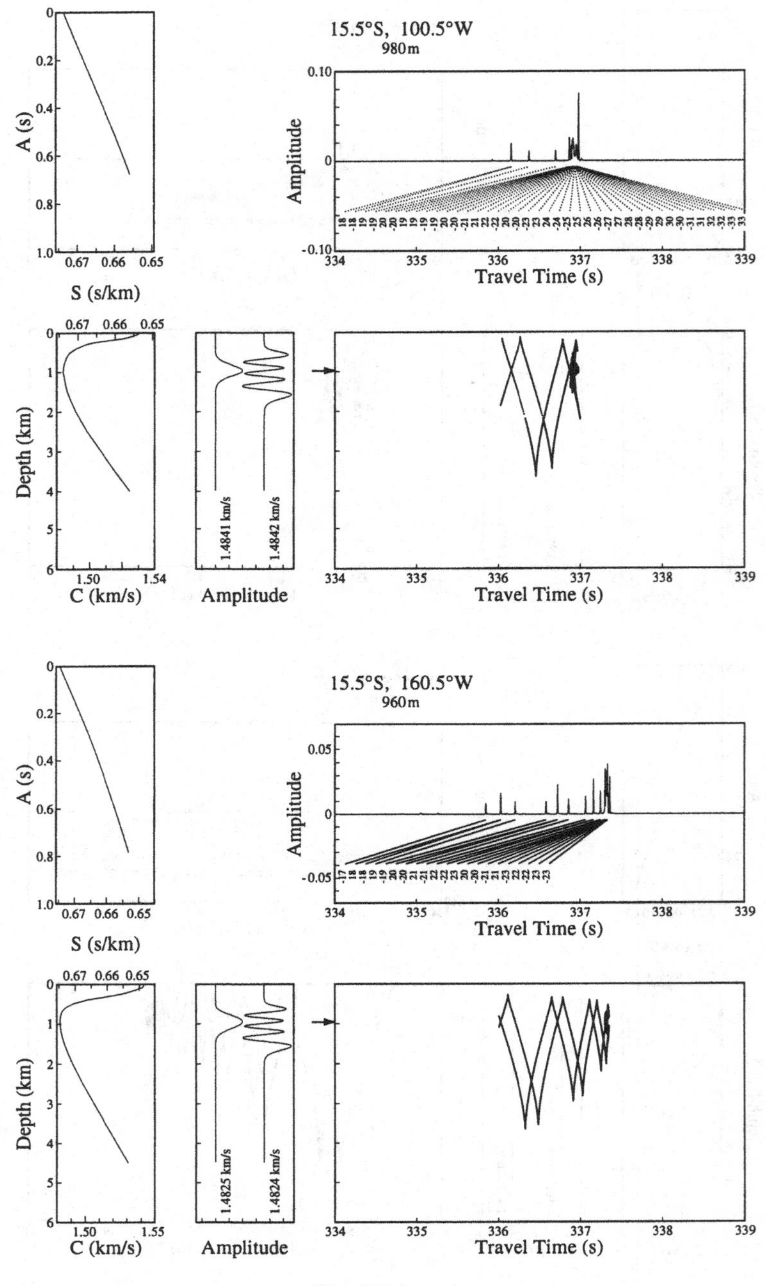
15.5°S, 100.5°W
980m
A (s)
S (s/km)
Amplitude
Travel Time (s)
Depth (km)
C (km/s)
1.4841 km/s
1.4842 km/s
15.5°S, 160.5°W
960m
1.4825 km/s
1.4824 km/s

Fig. B.1 (*cont.*)

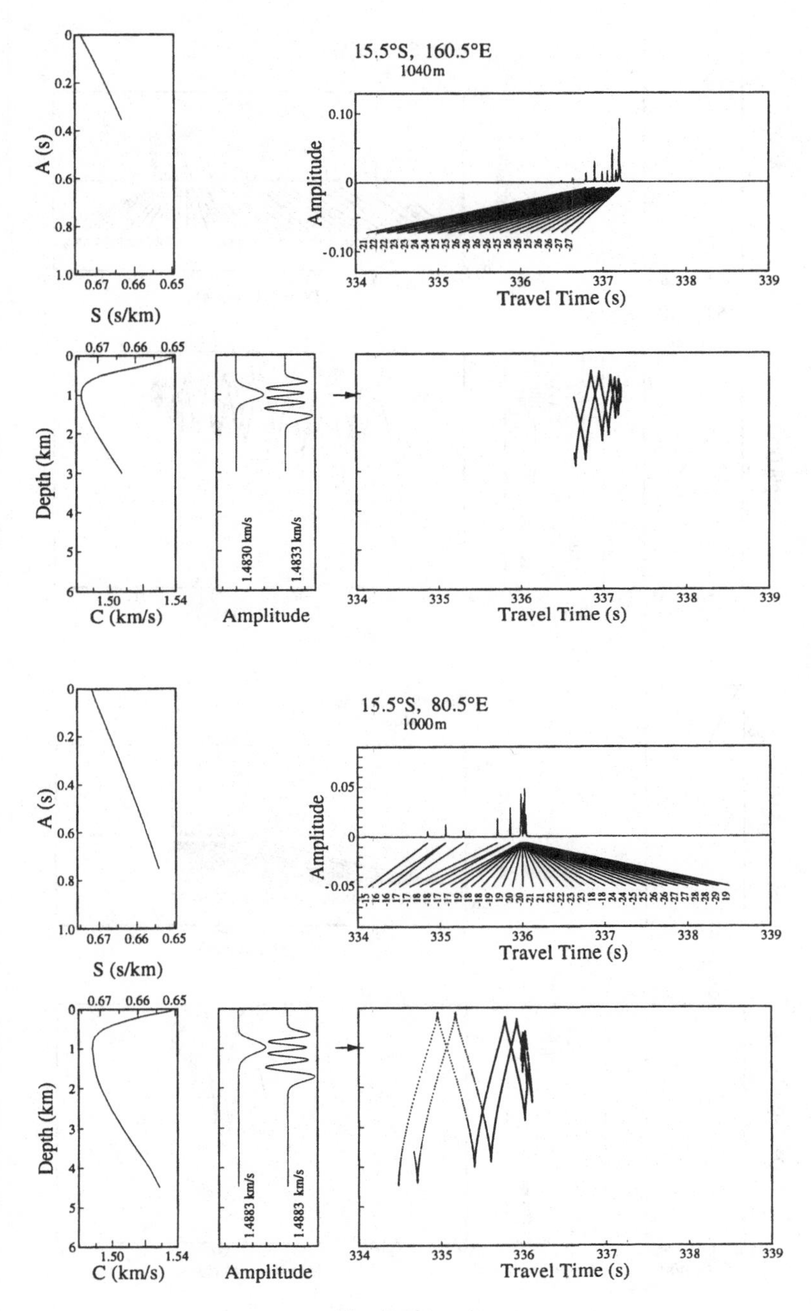
15.5°S, 160.5°E
1040m
A (s)
S (s/km)
Amplitude
Travel Time (s)
Depth (km)
C (km/s)
1.4830 km/s
1.4833 km/s
15.5°S, 80.5°E
1000m
1.4883 km/s
1.4883 km/s

Fig. B.1 (*cont.*)

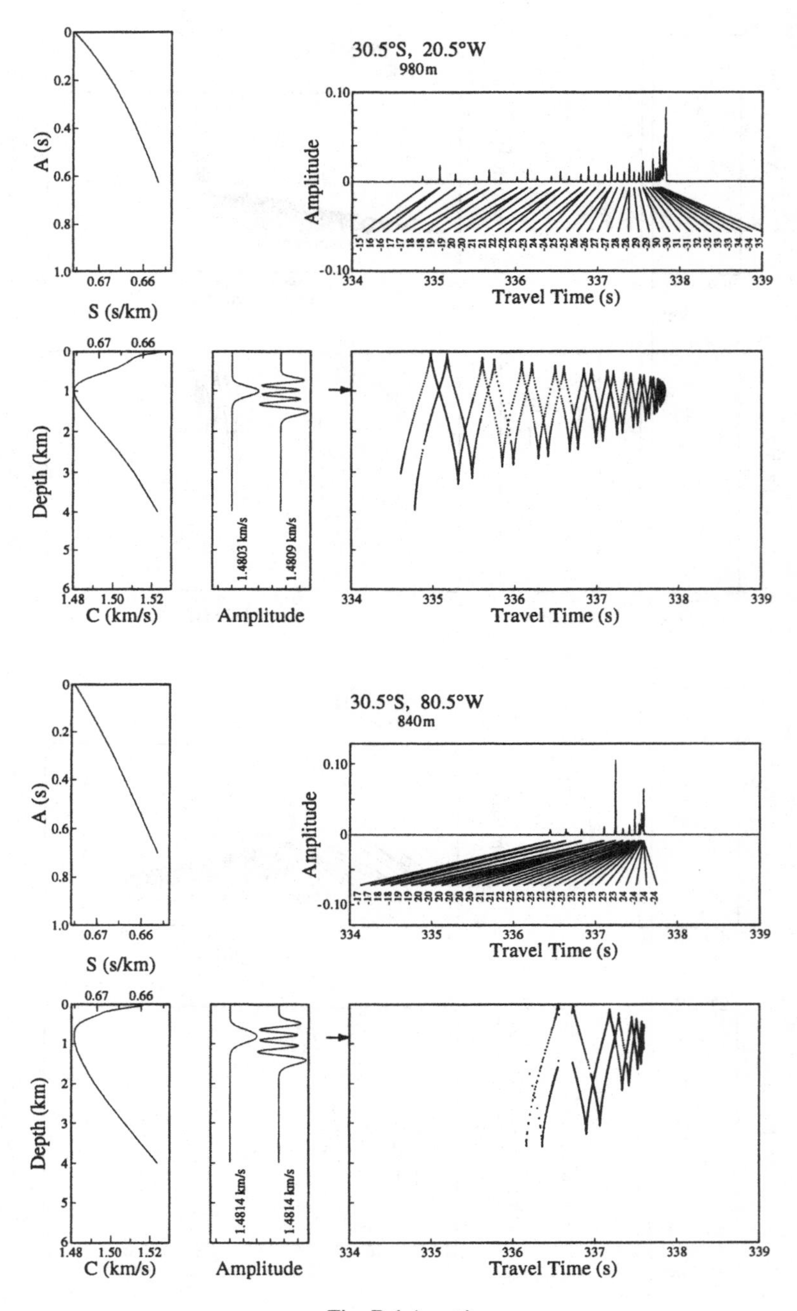
30.5°S, 20.5°W
980m
A (s)
S (s/km)
Amplitude
Travel Time (s)
Depth (km)
C (km/s)
1.4803 km/s
1.4809 km/s
Amplitude
Travel Time (s)
30.5°S, 80.5°W
840m
A (s)
S (s/km)
Amplitude
Travel Time (s)
Depth (km)
C (km/s)
1.4814 km/s
1.4814 km/s
Amplitude
Travel Time (s)

Fig. B.1 (*cont.*)

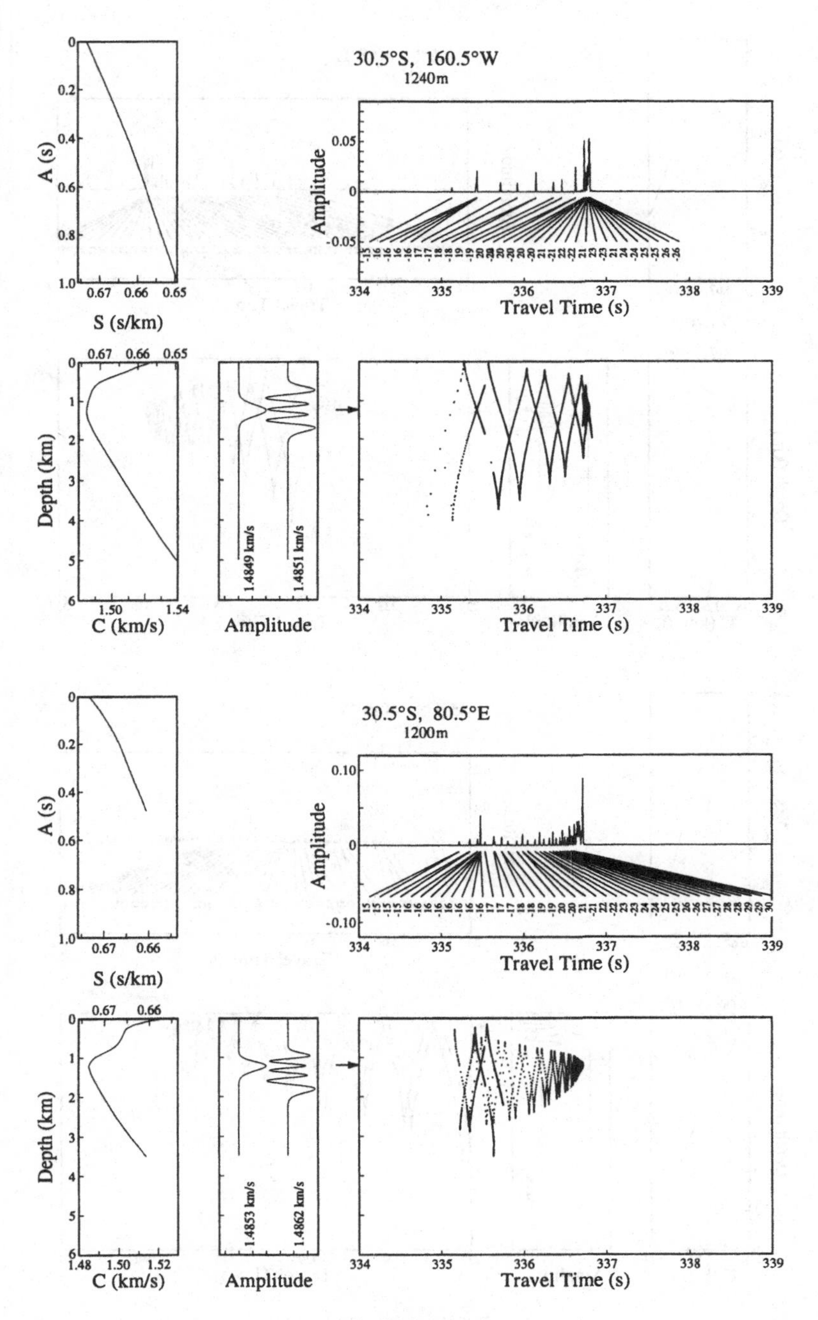
30.5°S, 160.5°W
1240m
A (s)
S (s/km)
Amplitude
Travel Time (s)
Depth (km)
C (km/s)
1.4849 km/s
1.4851 km/s
30.5°S, 80.5°E
1200m
1.4853 km/s
1.4862 km/s

Fig. B.1 (*cont.*)

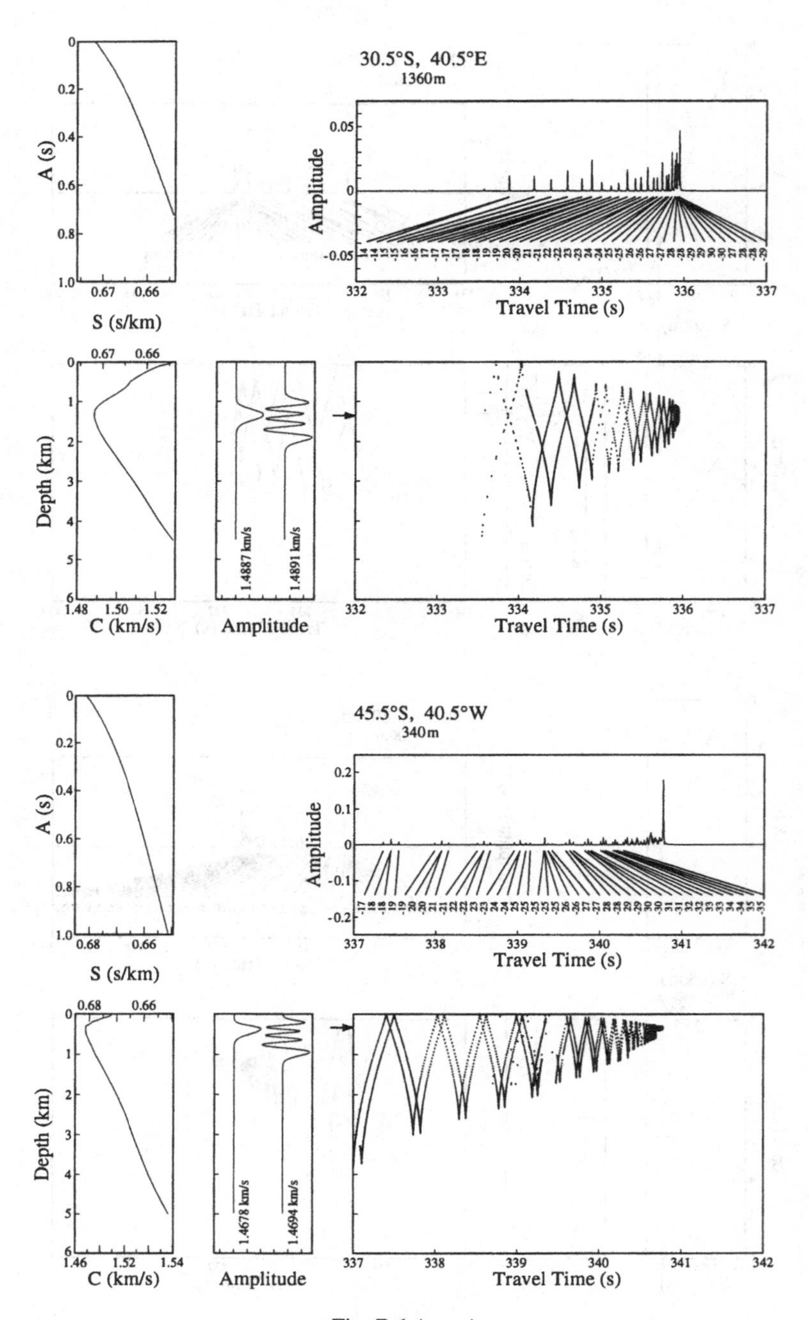
30.5°S, 40.5°E
1360m
A (s)
S (s/km)
Amplitude
Travel Time (s)
Depth (km)
C (km/s)
1.4887 km/s
1.4891 km/s
45.5°S, 40.5°W
340m
1.4678 km/s
1.4694 km/s

Fig. B.1 (*cont.*)

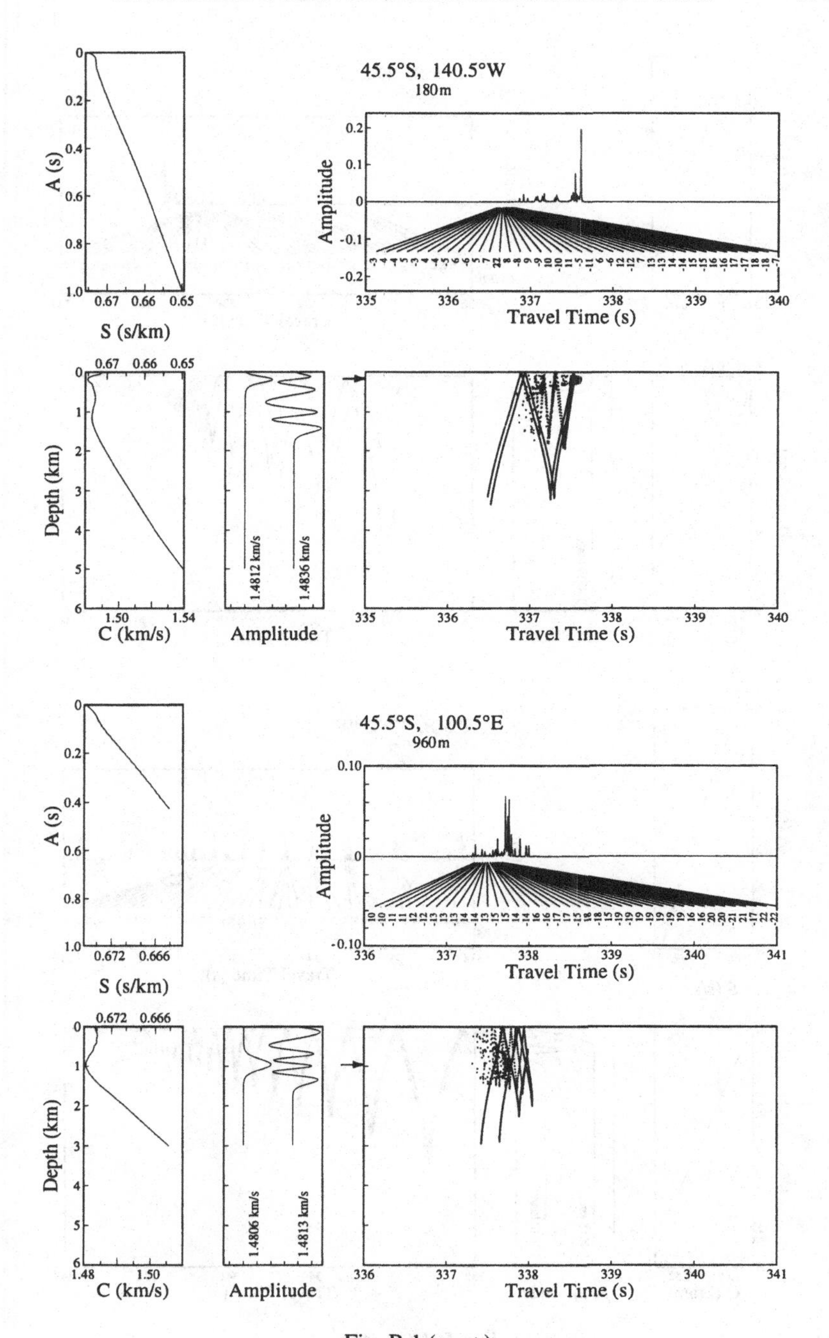
45.5°S, 140.5°W
180m
A (s)
S (s/km)
Amplitude
Travel Time (s)
Depth (km)
C (km/s)
1.4812 km/s
1.4836 km/s
45.5°S, 100.5°E
960m
1.4806 km/s
1.4813 km/s

Fig. B.1 (*cont.*)

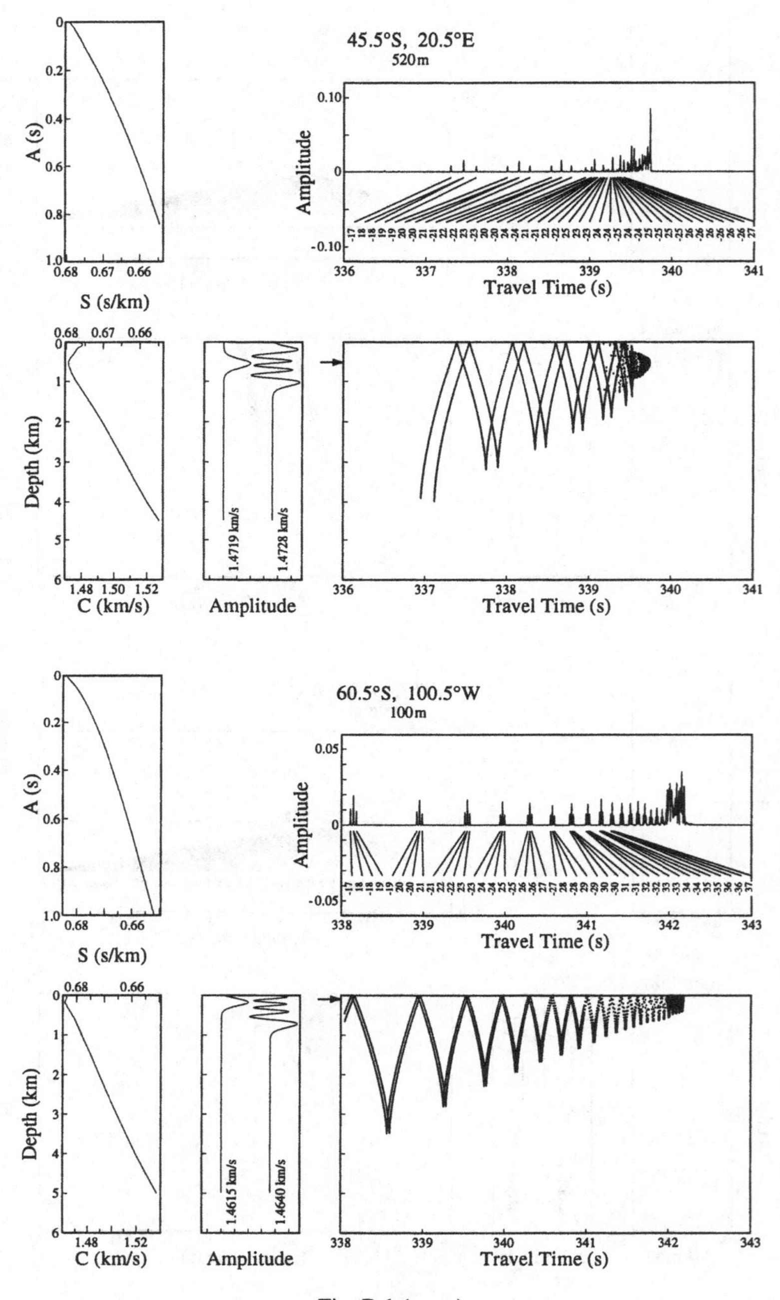
45.5°S, 20.5°E
520m
A (s)
S (s/km)
Amplitude
Travel Time (s)
Depth (km)
C (km/s)
1.4719 km/s
1.4728 km/s
60.5°S, 100.5°W
100m
1.4615 km/s
1.4640 km/s

Fig. B.1 (*cont.*)

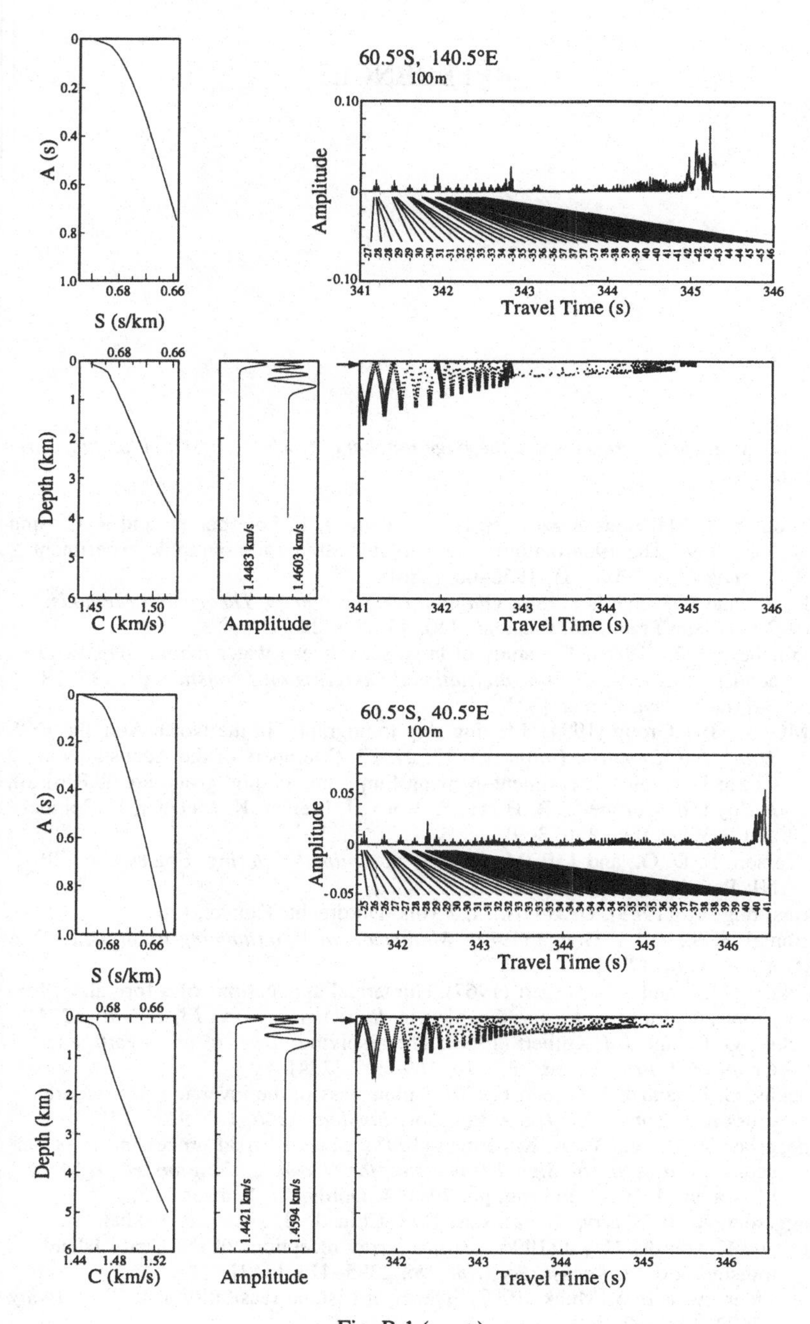
60.5°S, 140.5°E
100m
Amplitude
Travel Time (s)
A (s)
S (s/km)
Depth (km)
C (km/s)
Amplitude
1.4483 km/s
1.4603 km/s
60.5°S, 40.5°E
100m
1.4421 km/s
1.4594 km/s

Fig. B.1 (*cont.*)

REFERENCES

Numbers in square brackets are the page numbers on which these publications are cited.

Agnon, Y., P. Malanotte-Rizzoli, B. D. Cornuelle, J. L. Spiesberger, and R. C. Spindel (1989). The 1984 bottom-mounted Gulf Stream tomographic experiment. *J. Acoust. Soc. Am., 85*, 1958–66. [386]

Aki, K., and P. Richards (1980). *Quantitative Seismology, Theory and Methods*, 2 vols. San Francisco: Freeman. [40, 45, 260. 283]

Akulichev, V. A. (1989). The study of large-scale ocean water inhomogeneities by acoustic methods. In *13th International Congress on Acoustics*, pp. 117–28. Belgrade: Sava Center. [341]

AMODE-MST Group (1994). Moving ship tomography in the North Atlantic. *EOS, Trans. Am. Geophys. Union, 75*, 17, 21, 23. (Members of the Acoustic Mid-Ocean Dynamics Experiment–Moving Ship Tomography group are T. Birdsall, J. Boyd, B. Cornuelle, B. Howe, R. Knox, J. Mercer, K. Metzger, R. Spindel, and P. Worcester.) [26, 369]

Anderson, B. D. O., and J. B. Moore (1979). *Optimal Filtering*. Englewood Cliffs, NJ: Prentice-Hall. [322]

Armstrong, M. (1989). *Geostatistics*, 2 vols. Dordrecht: Kluwer. [263]

Arthnari, T. S., and Y. Dodge (1981). *Mathematical Programming in Statistics*. New York: Wiley. [276]

Backus, G. E., and J. F. Gilbert (1967). Numerical applications of a formalism for geophysical inverse theory. *Geophys. J. Roy. Astron. Soc., 13*, 247–76. [224]

Backus, G. E., and J. F. Gilbert (1968). The resolving power of gross earth data. *Geophys. J. Roy. Astron. Soc., 16*, 169–205. [224]

Backus, G. E., and J. F. Gilbert (1970). Uniqueness in the inversion of inaccurate gross earth data. *Phil. Trans. Roy. Soc. London, A266*, 123–92. [224]

Baggeroer, A. B., and W. A. Kuperman (1993). Matched field processing in ocean acoustics. In *Acoustic Signal Processing for Ocean Exploration*, ed. J. M. F. Moura and I. M. G. Lourtie, pp. 79–114. Dordrecht: Kluwer. [302]

Baggeroer, A., B. Sperry, K. Lashkari, C.-S. Chiu, J. H. Miller, P. Mikhalevsky, and K. von der Heydt (1994). Vertical array receptions of the Heard Island transmissions. *J. Acoust. Soc. Am., 96*, 2395–413. [342]

Baggeroer, A., and W. Munk (1992). The Heard Island feasibility test. *Phys. Today, 45*, 22–30. [332, 369]

Barth, N. H., and C. Wunsch (1989). Oceanographic experiment design by simulated annealing. *J. Phys. Oceanogr., 20*, 1249–63. [287]

Bell, T. G. (1962). Sonar and submarine detection. USN Underwater Sound Laboratory Report 545. [357]

Bengtsson, L., M. Ghil, and E. Källen (eds.) (1981). *Dynamic Meteorology Data Assimilation Methods*. Berlin: Springer-Verlag. [306]

Bennett, A. F. (1992). *Inverse Methods in Physical Oceanography*. Cambridge University Press. [224, 275]

Birdsall, T. G. (1965). MIMI multipath measurements. *J. Acoust. Soc. Am., 38*, 919. [359]

Birdsall, T. G. (1976). On understanding the matched filter in the frequency domain. *IEEE Trans. Educ., 19*, 168–9. [192]

Birdsall, T. G., and K. Metzger (1986). Factor inverse matched filtering. *J. Acoust. Soc. Am., 79*, 91–9. [191, 192, 221]

Birdsall, T. G., K. Metzger, and M. A. Dzieciuch (1994*a*). Signals, signal processing, and general results. *J. Acoust. Soc. Am., 96*, 2343–52. [369]

Birdsall, T. G., K. Metzger, M. A. Dzieciuch, and J. Spiesberger (1994*b*). Integrated autocorrelation phase at 1 period lag. *J. Acoust. Soc. Am., 96*, 2353–6. [369]

Blokhintsev, D. I. (1946). The propagation of sound in an inhomogeneous and moving medium. I. *J. Acoust. Soc. Am., 18*, 322–8. [117]

Blokhintsev, D. I. (1952). *The Acoustics of an Inhomogeneous Moving Medium* (trans. R. T. Beyer and D. Mintzer). Providence: Brown University Research Analysis Group. [117]

Blokhintsev, D. I. (1956). Acoustics of a nonhomogenous moving medium. National Advisory Committee on Aeronautics Technical Memorandum 1399. [117]

Boden, L., J. B. Bowlin, and J. L. Spiesberger (1991). Time domain analysis of normal mode, parabolic, and ray solutions of the wave equation. *J. Acoust. Soc. Am., 90*, 954–8. [38]

Bomford, G. (1980). *Geodesy*. Oxford University Press. [40]

Borish, J., and J. B. Angell (1983). An efficient algorithm for measuring the impulse response using pseudorandom noise. *J. Audio Eng. Soc., 31*, 478–88. [195]

Boutin, P. B., J. Kemp, S. Liberatore, J. Lynch, N. Witzell, K. Metzger, and D. Webb (1989). Results of the Lake Seneca directivity, source level, and pulse response tests of the MIT 400 Hz Webb tomography sources. Woods Hole Oceanographic Institution Technical Memorandum WHOI-1-89. [373]

Bouyoucos, J. V. (1975). Hydroacoustic transduction. *J. Acoust. Soc. Am., 57*, 1341–51. [182]

Bowditch, N. (1984). *American Practical Navigator*. Washington: Defense Mapping Agency Hydrographic Center, U.S. Government Printing Office, United States Hydrographic Office, publication no. 9. [178]

Bowles, A. E., M. Smultea, B. Würsig, D. P. DeMaster, and D. Palka (1994). Relative abundance and behavior of marine mammals exposed to transmissions from the Heard Island Feasibility Test. *J. Acoust. Soc. Am., 96*, 2469–84. [369]

Boyles, C. A. (1965). Theory of focusing plane waves by spherical, liquid lenses. *J. Acoust. Soc. Am., 38*, 393–405. [166]

Bracewell, R. N. (1956). Strip integration in radio astronomy. *Aust. J. Phys., 9*, 198–217. [285]

Brekhovskikh, L. M. (1949). Concerning the propagation of sound in an underwater acoustic channel. *Dokl. Akad. Nauk SSSR, 69*, 157–60. [357]

Brekhovskikh, L. M., and Y. Lysanov (1991). *Fundamentals of Ocean Acoustics*, 2nd ed. Berlin: Springer-Verlag. [33, 51, 61, 62, 64, 66, 137, 158, 159, 168, 171, 341]

Bretherton, F. P., R. E. Davis, and C. Fandry (1976). A technique for objective analysis and design of oceanographic instruments applied to MODE-73. *Deep-Sea Res., 23*, 559–82. [262]

Brogan, W. L. (1985). *Modern Control Theory*, 2nd ed. Englewood Cliffs, NJ: Prentice-Hall/Quantum. [279, 310, 319]

Brown, M. G. (1981). Application of the WKBJ Green's function to acoustic propagation in horizontally stratified oceans. *J. Acoust. Soc. Am., 71*, 1427–32. [10, 42, 51, 96, 101, 216, 383]

Brown, M. G. (1982). Inverting for the ocean sound speed structure. Ph.D. thesis, University of California, San Diego. [42, 51, 96]

Brown, M. G. (1984). Linearized travel time, intensity, and waveform inversions in the ocean sound channel – a comparison. *J. Acoust. Soc. Am., 75*, 1451–61. [302]

Brown, M. G., and F. D. Tappert (1987). Catastrophe theory, caustics and traveltime diagrams in seismology. *Geophys. J. Roy. Astron. Soc., 88*, 217–29. [51]

Brown, M. G., W. H. Munk, J. L. Spiesberger, and P. F. Worcester (1980). Long-range acoustics in the northwest Atlantic. *J. Geophys. Res., 85*, 2699–703. [302, 367]

Brundrit, G. B., and L. Krige (1994). Heard Island signals through the Agulhas retroflection region. *J. Acoust. Soc. Am., 96*, 2464–8. [369]

Bryan, K., S. Manabe, and M. J. Spelman (1988). Interhemispheric asymmetry in the transient response of a coupled ocean-atmosphere model to a CO_2 forcing. *J. Phys. Oceanogr., 18*, 851–67. [323]

Bryson, A. E., and Y.-C. Ho (1975). *Applied Optimal Control*, rev. ed. New York: Hemisphere. [313, 319]

Budden, K. G. (1961). *The Wave-guide Mode Theory of Wave Propagation*. Englewood Cliffs, NJ: Prentice-Hall. [67]

Burenkov, S. V., A. N. Gavrilov, A. Y. Uporin, A. V. Furduev, and N. N. Andregev (1994). Long range sound transmission from Heard Island to Krylov Underwater Mountain. *J. Acoust. Soc. Am., 96*, 2458–63. [369]

Burridge, R., and H. Weinberg (1977). Horizontal rays and vertical modes. In *Wave Propagation and Underwater Acoustics*, ed. J. B. Keller and J. S. Papadakis, pp. 86–152. Berlin: Springer-Verlag. [334]

Bushong, P. J. (1987). Tomographic measurements of barotropic motions. M.S. thesis, Massachusetts Institute of Technology–Woods Hole Oceanographic Institution Joint Program in Oceanography. [367]

Cartwright, D. E., R. D. Ray, and B. V. Sanchez (1992). A computer program for predicting oceanic tidal currents. NASA Technical Memorandum 104578, Goddard Space Flight Center, Greenbelt, MD. [131, 352]

Chapman, M. F., P. D. Ward, and D. D. Ellis (1989). The effective depth of a Pekeris ocean waveguide, including shear wave effects. *J. Acoust. Soc. Am., 85*, 648–53. [171]

Chen, C.-T., and F. J. Millero (1977). Speed of sound in seawater at high pressures. *J. Acoust. Soc. Am., 62*, 1129–35. [352]

Chester, D. B. (1989). Acoustic tomography in the Straits of Florida. M.S. thesis, Massachusetts Institute of Technology–Woods Hole Oceanographic Institution Joint Program in Oceanography. [367, 371]

Chester, D. B. (1993). A tomographic view of the Gulf Stream Southern Recirculation Gyre at 38°N, 55°W. Ph.D. thesis, Massachusetts Institute of Technology–Woods Hole Oceanographic Institution Joint Program in Oceanography, WHOI-93-28. [368, 378, 379, 381]

Chester, D. B., and P. Malanotte-Rizzoli (in press). A tomographic view of the Gulf Stream Southern Recirculation Gyre at 38°N, 55°W. *J. Geophys. Res.* [368, 378]

Chester, D. B., P. Malanotte-Rizzoli, and H. A. DeFerrari (1991). Acoustic tomography in the Straits of Florida. *J. Geophys. Res., 96*, 7023–48. [367, 371, 372]

Chester, D., P. Malanotte-Rizzoli, J. Lynch, and C. Wunsch (1994). The eddy radiation field of the Gulf Stream as measured by ocean acoustic tomography. *Geophys. Res. Lett., 21*, 181–4. [349, 350, 351, 368, 378, 379, 380]

Chiu, C.-S. (1985). Estimation of planetary wave parameters from the data of the 1981 Ocean Acoustic Tomography Experiment. Ph.D. thesis, Massachusetts Institute of Technology–Woods Hole Oceanographic Institution Joint Program in Oceanography. [367]

Chiu, C.-S., and Y. Desaubies (1987). A planetary wave analysis using the acoustic and conventional arrays in the 1981 Ocean Tomography Experiment. *J. Phys. Oceanogr., 17*, 1270–87. [291, 292, 309, 367]

Chiu, C.-S., J. F. Lynch, and O. M. Johannessen (1987). Tomographic resolution of mesoscale eddies in the marginal ice zone: a preliminary study. *J. Geophys. Res., 92*, 6886–902. [367]

Chiu, C.-S., J. H. Miller, R. H. Bourke, J. F. Lynch, and R. D. Muench (1994*a*). Acoustic images of the Barents Sea Polar Front. *EOS, Trans. Am. Geophys. Union, 75*, 118. [369]

Chiu, C.-S., A. J. Semtner, C. M. Ort, J. H. Miller, and L. L. Ehret (1994*b*). A ray variability analysis of sound transmission from Heard Island to California. *J. Acoust. Soc. Am., 96*, 2380–8. [369]

Clark, J. G., and M. Kronengold (1974). Long-period fluctuations of CW signals in deep and shallow water. *J. Acoust. Soc. Am., 56*, 1071–83. 148, 358]

Cohn, M., and A. Lempel (1977). On fast *m*-sequence transforms. *IEEE Trans. Inform. Theory, IT-23*, 135–7. [195]

Collins, M. D., and W. A. Kuperman (1994). Overcoming ray chaos. *J. Acoust. Soc. Am., 95*, 3167–70. [157]

Colosi, J. A. (1993). The nature of wavefront fluctuations induced by internal gravity waves in long-range oceanic acoustic pulse transmissions. Ph.D. thesis, University of California, Santa Cruz. [368]

Colosi, J. A., S. M. Flatté, and C. Bracher (1994). Internal-wave effects on 1000-km oceanic acoustic pulse propagation: simulation and comparison with experiment. *J. Acoust. Soc. Am., 96*, 452–68. [101, 148, 149, 153, 154, 354, 367, 368, 375]

Cornuelle, B. D. (1983). Inverse methods and results from the 1981 Ocean Acoustic Tomography Experiment. Ph.D. thesis, Massachusetts Institute of Technology–Woods Hole Oceanographic Institution Joint Program in Oceanography. [209, 213, 367]

Cornuelle, B. D. (1985). Simulations of acoustic tomography array performance with untracked or drifting sources and receivers. *J. Geophys. Res., 90*, 9079–88. [209, 213]

Cornuelle, B. D., and B. M. Howe (1987). High spatial resolution in vertical slice ocean acoustic tomography. *J. Geophys. Res., 92*, 11680–92. [136, 139, 140, 141, 142, 366]

Cornuelle, B. D., and P. Malanotte-Rizzoli (1986). A maximum-gradient inverse for the Gulf Stream system. *J. Geophys. Res., 91*, 10566–80. [275, 368]

Cornuelle, B. D, W. H. Munk, and P. F. Worcester (1989). Ocean acoustic tomography from ships. *J. Geophys. Res., 94*, 6232–50. [23, 24, 26, 89, 213, 215, 285, 286, 366, 369, 374]

Cornuelle, B. D., C. Wunsch, D. Behringer, T. G. Birdsall, M. G. Brown, R. Heinmiller, R. A. Knox, K. Metzger, W. H. Munk, J. L. Spiesberger, R. C. Spindel, D. C. Webb, and P. F. Worcester (1985). Tomographic maps of the ocean mesoscale. 1: Pure acoustics. *J. Phys. Oceanogr., 15*, 133–52. [201, 288, 289, 290, 291, 292, 364, 367]

Cornuelle, B. D., P. F. Worcester, J. A. Hildebrand, W. S. Hodgkiss, Jr., T. F. Duda, B. M. Howe, J. A. Mercer, and R. C. Spindel (1992). Vertical slice ocean acoustic tomography at 1000-km range in the North Pacific Ocean. Scripps Institution of Oceanography Reference Series, 92-17, University of California, San Diego, La Jolla, CA. [99, 368]

Cornuelle, B. D., P. F. Worcester, J. A. Hildebrand, W. S. Hodgkiss, Jr., T. F. Duda, J. Boyd, B. M. Howe, J. A. Mercer, and R. C. Spindel (1993). Ocean acoustic tomography at 1000-km range using wavefronts measured with a large-aperture vertical array. *J. Geophys. Res., 98*, 16365–77. [99, 281, 282, 292, 293, 294, 368, 375, 376, 377]

Crawford, G. B., R. J. Lataitis, and S. F. Clifford (1990). Remote sensing of ocean flows by spatial filtering of acoustic scintillations: theory. *J. Acoust. Soc. Am., 88*, 442–54. [125]

Creager, K. C., and L. M. Dorman (1982). Location of instruments on the seafloor by joint adjustment of instrument and ship positions. *J. Geophys. Res., 87*, 8379–88. [212]

Daley, R. (1991). *Atmospheric Data Analysis*. Cambridge University Press. [306]

Dashen, R. (1979). Path integrals for waves in random media. *J. Math. Phys., 20*, 894–920. [359]

Dashen, R., and W. Munk (1984). Three models of global ocean noise. *J. Acoust. Soc. Am., 76*, 540–54. [137, 181, 182]

Daubechies, I. (1992). *Ten Lectures on Wavelets*. Philadelpha: SIAM. [225]

Davis. R. E. (1985). Objective mapping by least squares fitting. *J. Geophys. Res., 90*, 4773–7. [275]

Decarpigny, J.-N., B. Hamonic, and O. B. Wilson, Jr. (1991). The design of low-frequency underwater acoustic projectors: present status and future trends. *IEEE J. Oceanic Eng., 16*, 107–22. [185]

DeFerrari, H. A., and H. B. Nguyen (1986). Acoustic reciprocal transmission experiments, Florida Straits. *J. Acoust. Soc. Am., 79*, 299–315. [367, 371]

Del Grosso, V. A. (1974). New equation for the speed of sound in natural waters (with comparisons to other equations). *J. Acoust. Soc. Am., 56*, 1084–91. [34, 352, 382]

Desaubies, Y. (1990). Ocean acoustic tomography. In *Oceanographic and Geophysical Tomography: Proc. 50th Les Houches Ecole d'Eté de Physique Théorique and NATO ASI*, ed. Y. Desaubies, A. Tarantola, and J. Zinn-Justin, pp. 159–202. Amsterdam: Elsevier. [366]

Desaubies, Y., C.-S. Chiu, and J. H. Miller (1986). Acoustic mode propagation in a range-dependent ocean. *J. Acoust. Soc. Am., 80*, 1148–60. [158, 334]

Deutsch, R. (1965). *Estimation Theory*. Englewood Cliffs, NJ: Prentice-Hall. [259]

Doolittle, R., A. Tolstoy, and M. Buckingham (1988). Experimental confirmation of horizontal refraction of CW acoustic radiation from a point source in a wedge-shaped ocean environment. *J. Acoust. Soc. Am., 83*, 2117–25. [168, 169]

Dowling, A. P., and J. E. Ffowcs Williams (1983). *Sound and Sources of Sound*. Chichester: Ellis Horwood. [38]

Dozier, L. B., and F. D. Tappert (1978*a*). Statistics of normal mode amplitudes in a random ocean. I. Theory. *J. Acoust. Soc. Am., 63*, 353–65. [162]

Dozier, L. B., and F. D. Tappert (1978*b*). Statistics of normal mode amplitudes in a random ocean. II. Computations. *J. Acoust. Soc. Am., 64*, 533–47. [162]

Duda, T. F., S. M. Flatté, J. A. Colosi, B. D. Cornuelle, J. A. Hildebrand, W. S. Hodgkiss, Jr., P. F. Worcester, B. M. Howe, J. A. Mercer, and R. C. Spindel (1992). Measured wave-front fluctuations in 1000-km pulse propagation in the Pacific Ocean. *J. Acoust. Soc. Am., 92*, 939–55. [98, 354, 368, 375]

Dushaw, B. D. (1992). The 1987 Gyre Scale Reciprocal Acoustic Tomography Experiment. Ph.D. thesis, Scripps Institution of Oceanography, University of California, San Diego, La Jolla, CA. [131, 368]

Dushaw, B. D., D. B. Chester, and P. F. Worcester (1993*a*). A review of ocean current and vorticity measurements using long-range reciprocal acoustic transmissions. In *OCEANS '93: Engineering in Harmony with the Ocean, pp. I-298 to I-305*. New York: IEEE. [129, 134, 366, 378]

Dushaw, B. D., P. F. Worcester, B. D. Cornuelle, and B. M. Howe (1993*b*). On equations for the speed of sound in seawater. *J. Acoust. Soc. Am., 93*, 255–75. [34, 351, 368]

Dushaw, B. D., P. F. Worcester, B. D. Cornuelle, and B. M. Howe (1993*c*). Variability of heat content in the central North Pacific in summer 1987 determined from long-range acoustic transmissions. *J. Phys. Oceanogr., 23*, 2650–66. [297, 298, 332, 350, 352, 368]

Dushaw, B. D., P. F. Worcester, B. D. Cornuelle, and B. M. Howe (1994). Barotropic currents and vorticity in the central North Pacific Ocean during summer 1987 determined from long-range reciprocal acoustic transmissions. *J. Geophys. Res., 99*, 3263–72. [129, 131, 133, 349, 350, 352, 368, 379, 380]

Dushaw, B. D., B. D. Cornuelle, P. F. Worcester, B. M. Howe, and D. S. Luther (in press). Barotropic and baroclinic tides in the central North Pacific Ocean determined from long-range reciprocal acoustic transmissions. *J. Phys. Oceanogr.* [131, 148, 352, 353, 368]

Dyson, F., W. Munk, and B. Zetler (1976). Interpretation of multipath scintillations Eleuthera to Bermuda in terms of internal waves and tides. *J. Acoust. Soc. Am., 59*, 1121–33. [359]

Dysthe, K. B. (1991). Note on averaged horizontal refraction for long distance propagation in an ocean sound channel. *J. Acoust. Soc. Am., 91*, 1369–74. [164, 334]

Dzieciuch, M., and W. Munk (1994). Differential Doppler as a diagnostic. *J. Acoust. Soc. Am., 96*, 2414–24. [369]

Eckart, C., and G. Young (1939). A principal axis transformation for non-Hermitian matrices. *Bull. Am. Math. Soc., 45*, 118–21. [247]

Ehrenberg, J. E., T. E. Ewart, and R. D. Morris (1978). Signal-processing techniques for resolving individual pulses in a multipath environment. *J. Acoust. Soc. Am., 63*, 1801–8. [201]

Eisler, T. J., R. New, and D. Calderone (1982). Resolution and variance in acoustic tomography. *J. Acoust. Soc. Am., 72*, 1965–77. [224]

Eisler, T. J., and D. A. Stevenson (1986). Performance bounds for acoustic tomography in a vertical ocean slice. *IEEE J. Oceanic Eng., OE-11*, 72–8. [224]

Ewart, T. E., J. E. Ehrenberg, and S. A. Reynolds (1978). Observations of the phase and amplitude of individual Fermat paths in a multipath environment. *J. Acoust. Soc. Am., 63*, 1801–8. [201]

Ewing, J. (1960). Notes and personalia. *Trans. Am. Geophys. Union, 41*, 670. [339]

Ewing, M., and J. L. Worzel (1948). Long-range sound transmission. *Geol. Soc. Am. Memoir, 27*, part III, 1–35. [328, 356]

Farmer, D. M., and S. F. Clifford (1986). Space-time acoustic scintillation analysis: a new technique for probing ocean flows. *IEEE J. Oceanic Eng., OE-11*, 42–50. [125]

Farmer, D. M., and G. B. Crawford (1991). Remote sensing of ocean flows by spatial filtering of acoustic scintillations: observations. *J. Acoust. Soc. Am., 90*, 1582–91. [125]

Fisher, F. H., and V. P. Simmons (1977). Sound absorption in sea water. *J. Acoust. Soc. Am., 62*, 558–64. [177]

Flatté, S. M. (1983). Wave propagation through random media: contributions from ocean acoustics. *Proc. IEEE, 71*, 1267–94. [133, 148, 149, 359]

Flatté, S. M., R. Dashen, W. Munk, K. Watson, and F. Zachariasen (1979). *Sound Transmission Through a Fluctuating Ocean*. Cambridge University Press. [12, 32, 33, 36, 133, 136, 148, 149, 151, 203, 359]

Flatté, S. M., and R. B. Stoughton (1986). Theory of acoustic measurement of internal wave strength as a function of depth, horizontal position and time. *J. Geophys. Res., 91*, 7709–20. [133, 136, 149, 353, 367]

Flatté, S. M., and R. B. Stoughton (1988). Predictions of internal-wave effects on ocean acoustic coherence, travel-time variance, and intensity moments for very long-range propagation. *J. Acoust. Soc. Am., 84*, 1414–24. [133, 149, 150, 196, 206]

Fletcher, R., and M. J. D. Powell (1963). A rapidly convergent descent method for minimization. *Computer J., 6*, 163–8. [291]

Forbes, A. M. G. (1994). The Tasman Blockage – an acoustic sink for the Heard Island Feasibility Test? *J. Acoust. Soc. Am., 96*, 2428–31. [338, 369]

Forbes, A. M. G., and W. Munk (1994). Doppler-inferred launch angles of global acoustic ray paths. *J. Acoust. Soc. Am., 96*, 2425–7. [338, 369]

Franchi, E. R., and M. J. Jacobson (1972). Ray propagation in a channel with depth-variable sound speed and current. *J. Acoust. Soc. Am., 52*, 316–31. [119]

Franchi, E. R., and M. J. Jacobson (1973*a*). An environmental-acoustics model for sound propagation in a geostrophic flow. *J. Acoust. Soc. Am., 53*, 835–47. [119]

Franchi, E. R., and M. J. Jacobson (1973*b*). Effect of hydrodynamic variations on sound transmission across a geostrophic flow. *J. Acoust. Soc. Am., 54*, 1302–11. [119]

Francois, R. E., and G. R. Garrison (1982*a*). Sound absorption based on ocean measurements. Part I: Pure water and magnesium sulfate contributions. *J. Acoust. Soc. Am., 72*, 896–907. [177]

Francois, R. E., and G. R. Garrison (1982*b*). Sound absorption based on ocean measurements. Part II. Boric acid contribution and equation for total absorption. *J. Acoust. Soc. Am., 72*, 1879–90. [177]

Fraser, I. A., and P. D. Morash (1994). Observation of the Heard Island signals near the Gulf Stream. *J. Acoust. Soc. Am., 96*, 2448–57. [369]

Fu, L., T. Keffer, P. P. Niiler, and C. Wunsch (1982). Observations of mesoscale variability in the western North Atlantic: a comparative study. *J. Marine Res., 40*, 809–48. [264, 346]

Fukumori, I., J. Benveniste, C. Wunsch, and D. B. Haidvogel (1992). Assimilation of sea surface topography into an ocean circulation model using a steady-state smoother. *J. Phys. Oceanogr., 23*, 1831–55. [314, 318]

Gaillard, F. (1985). Ocean acoustic tomography with moving sources or receivers. *J. Geophys. Res., 90*, 11891–8. [213, 261]

Gaillard, F. (1994). Monitoring convection in the Gulf of Lion with tomography: a 3-D view. *EOS, Trans. Am. Geophys. Union, 75*, 118. [348]

Gaillard, F., and B. D. Cornuelle (1987). Improvement of tomographic maps by using surface-reflected rays. *J. Phys. Oceanogr., 17*, 1458–67. [291, 367, 369]

Garmany, J. (1979). On the inversion of travel times. *Geophys. Res. Lett., 6*, 277–9. [283]

Garrett, C., and W. Munk (1972). Space-time scales of internal waves. *Geoph. Fl. Dyn., 3*, 225–64. [148, 358]

Garrison, G. R., R. E. Francois, E. W. Early, and T. W. Wen (1983). Sound absorption measurements at 10–650 kHz in arctic waters. *J. Acoust. Soc. Am., 73*, 492–501. [177]

Gaspar, P., and C. Wunsch (1989). Estimates from altimeter data of barotropic Rossby waves in the northwestern Atlantic ocean. *J. Phys. Oceanogr., 19*, 1821–44. [314, 315]

Gelb, A. (ed.) (1974). *Applied Optimal Estimation*. Cambridge, MA: MIT Press. [315]

Georges, T. M., L. R. Boden, and D. R. Palmer (1994). Features of the Heard Island signals received at Ascension. *J. Acoust. Soc. Am., 96*, 2441–7. [369]

Gerson, N. C., J. G. Hengen, R. M. Pipp, and J. B. Webster (1969). Radio-wave propagation to the antipode. *Canadian J. Physics, 47*, 2143–59. [341]

Ghil, M., and P. Malanotte-Rizzoli (1991). Data assimilation in meteorology and oceanography. *Adv. Geophys., 33*, 141–266. [306, 312, 319]

Gill, A. E. (1982). *Atmosphere–Ocean Dynamics*. New York: Academic Press. [3]

Gill, P. E., W. Murray, and M. H. Wright (1981). *Practical Optimization*. New York: Academic Press. [280]

Golomb, S. W. (1982). *Shift Register Sequences*, rev. ed. Laguna Hills, CA: Aegean Park Press. [193, 194, 218]

Golub, G. H., and C. F. Van Loan (1989). *Matrix Computation*, 2nd ed. Baltimore: Johns Hopkins University Press. [261]

Goncharov, V. V., and A. G. Voronovich (1993). An experiment on matched-field acoustic tomography with continuous wave signals in the Norway Sea. *J. Acoust. Soc. Am., 93*, 1873–81. [302]

Goodwin, G. C., and K. S. Sin (1984). *Adaptive Filtering Prediction and Control*. Englewood CLiffs, NJ: Prentice-Hall. [322]

Gould, W. J., and E. Sambuco (1975). The effect of mooring type on measured values of ocean currents. *Deep-Sea Res., 22*, 55–62. [346]

Grace, O. D., and S. P. Pitt (1970). Sampling and interpolation of bandlimited signals by quadrature methods. *J. Acoust. Soc. Am., 48*, 1311–18. [185]

Greenland Sea Project Group (1990). Greenland Sea Project – a venture toward improved understanding of the oceans' role in climate. *EOS, Trans. Am. Geophys. Union, 71*, 750–1, 754–5. [101, 368]

Hamilton, G. R. (1977). Time variations of sound speed over long paths in the ocean. In *International Workshop on Low-Frequency Propagation and Noise* (Woods Hole, MA, 14–19 October 1974), pp. 7–30. Washington, DC: Department of the Navy. [331]

Hamilton, K. G., W. L. Siegmann, and M. J. Jacobson (1977). Combined influence of spatially uniform currents and tidally varying sound speed on acoustic propagation in the deep ocean. *J. Acoust. Soc. Am., 62*, 53–62. [119, 331]

Hamilton, K. G., W. L. Siegmann, and M. J. Jacobson (1980). Simplified calculation of ray-phase perturbations due to ocean-environmental variations. *J. Acoust. Soc. Am., 67*, 1193–206. [119]

Haykin, S. (1986). *Adaptive Filter Theory*. Englewood Cliffs, NJ: Prentice-Hall. [322]

Hayre, H. S., and I. D. Tripathi (1967). Ray path in a linearly moving, inhomogeneous, layered ocean model. *J. Acoust. Soc. Am., 41*, 1373–4. [119]

Headrick, R. H., J. L. Spiesberger, and P. J. Bushong (1993). Tidal signals in basin-scale acoustic transmissions. *J. Acoust. Soc. Am., 93*, 790–802. [332, 367]

Heaney, K. D., W. A. Kuperman, and B. E. McDonald (1991). Perth-Bermuda sound propagation 1960: adiabatic mode interpretation. *J. Acoust. Soc. Am., 90*, 2586–94. [336, 339, 341]

Heard, G. J., and N. R. Chapman (1994). Heard Island Feasibility Test: analysis of Pacific path data obtained with a horizontal line array. *J. Acoust. Soc. Am., 96*, 2389–94. [369]

Heinmiller, R. D. (1983). Instruments and methods. In *Eddies in Marine Science*, ed. A. R. Robinson, pp. 542–67. Berlin: Springer-Verlag. [346]

Heller, G. S. (1953). Propagation of acoustic discontinuities in an inhomogeneous moving liquid medium. *J. Acoust. Soc. Am., 25*, 950–1. [117]

Helstrom, C. W. (1968). *Statistical Theory of Signal Detection*, 2nd ed. London: Pergamon Press. [186, 188, 190, 197, 206]

Hendry, R. M., and A. J. Hartling (1979). A pressure-induced direction error in nickel-coated Aanderaa current meters. *Deep-Sea Res., 26*, 327–35. [346]

Herman, G. T. (ed.) (1979). *Image Reconstruction from Projections: Implementation and Applications*. Berlin: Springer-Verlag. [287]

Herman, G. T. (1980). *Image Reconstruction from Projections: The Fundamentals of Computerized Tomography*. New York: Academic Press. [287]

Hippenstiel, R., E. Chaulk, and J. H. Miller (1992). An adaptive tracker for partially resolved acoustic arrivals with application to ocean acoustic tomography. *J. Acoust. Soc. Am., 92*, 1759–62. [218, 368]

Hoerl, A. E., and R. W. Kennard (1970*a*). Ridge regression: biased estimation for non-orthogonal problems. *Technometrics, 12*, 55–67. [234]

Hoerl, A. E., and R. W. Kennard (1970*b*). Ridge regression: applications to non-orthogonal problems. *Technometrics, 12*, 69-82. [234]

Horvat, D. C. M., J. S. Bird, and M. M. Goulding (1992). True time-delay bandpass beamforming. *IEEE J. Oceanic Eng., 17*, 185–92. [185, 202]

Howe, B. M. (1986). Ocean acoustic tomography: mesoscale velocity. Ph.D. thesis, University of California, San Diego. [135, 367]

Howe, B. M. (1987). Multiple receivers in single vertical slice ocean acoustic tomography experiments. *J. Geophys. Res., 92*, 9479–86. [10]

Howe, B. M., P. F. Worcester, and R. C. Spindel (1987). Ocean acoustic tomography: mesoscale velocity. *J. Geophys. Res., 92*, 3785–805. [129, 131, 202, 312, 366, 367, 370, 371]

Howe, B. M., J. A. Mercer, and R. C. Spindel (1989*a*). A floating acoustic-satellite tracking (FAST) range. In *Proceedings of Marine Data Systems '89* (New Orleans, LA, 26–28 April, 1989), pp. 225–30. Stennis Space Center, MS: Marine Technology Society, Gulf Coast Section. [214, 369, 374]

Howe, B. M., J. A. Mercer, R. C. Spindel, and P. F. Worcester (1989*b*). Accurate positioning for moving ship tomography. In *Oceans '89* (18–21 Sept. 1989, Seattle, WA), pp. 880–6. New York: IEEE. [214, 369, 374]

Howe, B. M., J. A. Mercer, R. C. Spindel, P. F. Worcester, J. A. Hildebrand, W. S. Hodgkiss, Jr., T. F. Duda, and S. M. Flatté (1991). SLICE89: a single slice tomography experiment. In *Ocean Variability and Acoustic Propagation*, ed. J. Potter and A. Warn-Varnas, pp. 81–6. Dordrecht: Kluwer. [98, 368, 375]

Hua, B. L., J. C. McWilliams, and W. B. Owens (1986). An objective analysis of the Polymode Local Dynamics Experiment. Part II: Streamfunction and potential vorticity fields during the intensive period. *J. Phys. Oceanogr.*, *16*, 506–22. [346]

Huang, T. S. (ed.) (1979). *Picture Processing and Digital Filtering*, 2nd ed. Berlin: Springer-Verlag. [225]

Itzikowitz, S., M. J. Jacobson, and W. L. Siegmann (1982*a*). Short-range acoustic transmissions through cyclonic eddies between a submerged source and receiver. *J. Acoust. Soc. Am.*, *71*, 1131–44. [166]

Itzikowitz, S., M. J. Jacobson, and W. L. Siegmann (1982*b*). Modelling of long-range acoustic transmissions through cyclonic and anticyclonic eddies. *J. Acoust. Soc. Am.*, *73*, 1556–66. [166]

Jensen, F. B., W. A. Kuperman, M. B. Porter, and H. Schmidt (1994). *Computational Ocean Acoustics*. New York: AIP Press. [38]

Jin, G., and P. Wadhams (1989). Travel time changes in a tomography array caused by a sea ice cover. *Prog. Oceanogr.*, *22*, 249–75. [368]

Jin, G., and P. F. Worcester (1989). The feasibility of measuring ocean pH by long range acoustics. *J. Geophys. Res.*, *94*, 4749–56. [201, 368]

Jin, G., J. F. Lynch, R. Pawlowicz, and P. Wadhams (1993). Effects of sea ice cover on acoustic ray travel times, with applications to the Greenland Sea Tomography Experiment. *J. Acoust. Soc. Am.*, *94*, 1044–57. [368]

Jin, G., J. F. Lynch, R. Pawlowicz, and P. F. Worcester (in press). Acoustic scattering losses in the Greenland Sea marginal ice zone during the 1988–89 tomography experiment. *J. Acoust. Soc. Am.*, *96*. [368]

Johnson, R. H. (1969). Synthesis of point data and path data in estimating SOFAR speed. *J. Geophys. Res.*, *74*, 4559–70. [328, 332]

Jolliffe, I. T. (1986). *Principal Component Analysis*. Berlin: Springer-Verlag. [263]

Jones, R. M., and T. M. Georges (1994). Nonperturbative ocean acoustic tomography inversion. *J. Acoust. Soc. Am.*, *96*, 439–51. [283]

Jones, R. M., T. M. Georges, L. Nesbitt, R. Tallamraju, and A. Weickmann (1990). Vertical-slice ocean-acoustic tomography – extending the Abel inversion to non-axial sources and receivers. In *Premier Congres Français d'acoustique* (10–13 April 1990, Lyon), ed. P. Filippi and M. Zakharia, pp. 1013–16. Les Ulis, France: Editions de physique. [46]

Jones, R. M., T. M. Georges, and J. P. Riley (1986*a*). Inverting vertical-slice tomography measurements for asymmetric ocean sound-speed profiles. *Deep-Sea Res.*, *33*, 601–19. [283]

Jones, R. M., J. P. Riley, and T. M. Georges (1986*b*). *HARPO*. NOAA, Wave Propagation Laboratory, Boulder, CO. [334]

Jones, R. M., E. C. Shang, and T. M. Georges (1993). Nonperturbative modal tomography inversion. Part I. Theory. *J. Acoust. Soc. Am.*, *94*, 2296–302. [283]

Kamel, A., and L. B. Felsen (1982). On the ray equivalent of a group of modes. *J. Acoust. Soc. Am.*, *71*, 1445–52. [59]

Keller, J. B. (1954). Geometrical acoustics. I. The theory of weak shock waves. *J. Appl. Phys.*, *25*, 938–47. [117]

Keller, J. B. (1958). Surface waves on water of nonuniform depth. *J. Fluid. Mech.*, *4*, 607. [334]

Kibblewhite, A. C., R. N. Denham, and P. H. Barker (1966). Long-range sound propagation study in the Southern Ocean – Project Neptune. *J. Acous. Soc. Am.*, *38*, 629–43. [332]

Kirkpatrick, S., C. D. Gelatt, and M. P. Vecchi (1983). Optimization by simulated annealing. *Science*, *220*, 671–80. [280]

Klein, E. (1969). Underwater sound research and applications before 1939. *J. Acoust. Soc. Am., 43*, 931. [357]

Knox, R. A. (1988). Ocean acoustic tomography: a primer. In *Oceanic Circulation Models: Combining Data and Dynamics*, ed. D. L. T. Anderson and J. Willebrand, pp. 141–88. Dordrecht: Kluwer. [366]

Ko, D. S. (1987). Inversion methods and results from the 1983 Straits of Florida Acoustic Reciprocal Transmission Experiment. Ph.D. thesis, University of Miami, Miami, FL. [367]

Ko, D. S., H. A. DeFerrari, and P. Malanotte-Rizzoli (1989). Acoustic tomography in the Florida Strait: temperature, current and vorticity measurements. *J. Geophys. Res., 94*, 6197–211. [367, 371, 372]

Kornhauser, E. T. (1953). Ray theory for moving fluids. *J. Acoust. Soc. Am., 25*, 945–9. [117]

Koza, J. R. (1992). *Genetic Programming: On the Programming of Computers by Means of Natural Selection.* Cambridge, MA: MIT Press. [280, 301]

LaCasce, E. O., Jr., and J. C. Beckerle (1975). Preliminary experiment to measure periodicities and large-scale ocean movements with acoustic signals. *J. Acoust. Soc. Am., 57*, 966–7. [355]

Lanczos, C. (1961). *Linear Differential Operators*. New York: Van Nostrand. [247]

Lapwood, E. R. (1975). The effect of discontinuities in density and rigidity on torsional eigenfrequencies of the earth. *Geophys. J. Roy. Astron. Soc., 40*, 453–64. [88]

Lasky, M. (1973). Review of World War I acoustic technology USN. *J. Underwater Acoust., 24*, 363. [357]

Lasky, M. (1974). A historical review of underwater acoustic technology 1916–1939. *J. Underwater Acoust., 24*, 597. [357]

Lawson, C. L., and R. J. Hanson (1974). *Solving Least Squares Problems*. Englewood Cliffs, NJ: Prentice-Hall. [234, 301]

Lawson, L. M., and D. R. Palmer (1984). Acoustic ray-path fluctuations induced by El Niño. *J. Acoust. Soc. Am., 75*, 1343–5. [326]

Legters, G. R., N. L. Weinberg, and J. G. Clark (1983). Long-range Atlantic acoustic multipath identification. *J. Acoust. Soc. Am., 73*, 1571–80. [367]

Levitus, S. (1982). *Climatological Atlas of the World Ocean*. NOAA professional paper 13, U.S. Department of Commerce. [93, 382]

Lichte, H. (1919). On the influence of horizontal temperature layers in sea water on the range of underwater sound signals. *Physikalische Zeitschrift, 17*, 385–9. [357]

Lidl, R., and H. Niederreiter (1986). *Introduction to Finite Fields and Their Applications*. Cambridge University Press. [193]

Liebelt, P. B. (1967). *An Introduction to Optimal Estimation*. Reading, MA: Addison-Wesley. [238, 260]

Longuet-Higgins, M. (1982). On triangular tomography. *Dynam. Atmos. Oceans, 7*, 33–46. [123]

Longuet-Higgins, M. (1990). Ray paths and caustics on a slightly oblate ellipsoid. *Proc. Roy. Soc. Lond., A428*, 283–90. [340]

Lorenc, A. C. (1986). Analysis methods for numerical weather prediction. *Q. J. Roy. Met. Soc., 112*, 1177–94. [306]

Lovett, J R. (1980). Geographic variation of low-frequency sound absorption in the Atlantic, Indian, and Pacific Oceans. *J. Acoust. Soc. Am., 67*, 338–40. [177]

Luenberger, D. G. (1984). *Linear and Non-Linear Programming*, 2nd ed. Reading, MA: Addison-Wesley. [277, 278, 280]

Luther, D. S., J. H. Filloux, and A. D. Chave (1991). Low-frequency, motionally induced electromagnetic fields in the ocean. 2. Electric field and Eulerian current comparison. *J. Geophys. Res., 96*, 12797–814. [131]

Lynch, J. F., J. H. Miller, and C.-S. Chiu (1989). Phase and travel-time variability of adiabatic acoustic normal modes due to scattering from a rough sea surface, with applications to propagation in shallow-water and high-latitude regions. *J. Acoust. Soc. Am., 85*, 83–9. [367]

Lynch, J. F., S. D. Rajan, and G. V. Frisk (1991). A comparison of broadband and narrow-band modal inversions for bottom geoacoustic properties at a site near Corpus Christi, Texas. *J. Acoust. Soc. Am., 89*, 648–65. [74, 75, 76, 78]

Lynch, J. F., R. C. Spindel, C.-S. Chiu, J. H. Miller, and T. G. Birdsall (1987). Results from the 1984 Marginal Ice Zone Experiment preliminary tomography transmissions: implications for marginal ice zone, Arctic, and surface wave tomography. *J. Geophys. Res., 92*, 6869–85. [367]

Lynch, J. F., H. X. Wu, P. Wadhams, and P. F. Worcester (1993*a*). Ice edge noise observations from the 1988–89 Greenland Sea Tomography Experiment. In *Proceedings of the European Conference on Underwater Acoustics*, ed. M. Weydert, pp. 611–19. Amsterdam: Elsevier. [368]

Lynch, J. F., H. X. Wu, R. Pawlowicz, P. F. Worcester, R. E. Keenan, H. C. Graber, O. M. Johannessen, P. Wadhams, and R. A. Shuchman (1993*b*). Ambient noise measurements in the 200–300-Hz band from the Greenland Sea Tomography Experiment. *J. Acoust. Soc. Am., 94*, 1015–33. [368]

McDonald, B. E., M. D. Collins, W. A. Kuperman, and K. D. Heaney (1994). Comparison of data and model predictions for Heard Island acoustic transmissions. *J. Acoust. Soc. Am., 96*, 2357–70. [334, 335, 342, 369]

MacKenzie, K. V. (1981). Nine-term equation for sound speed in the oceans. *J. Acoust. Soc. Am., 70*, 807–12. [33]

Magnus, J. R., and H. Neudecker (1988). *Matrix Differential Calculus with Applications in Statistics and Econometrics*. New York: Wiley. [257]

Malanotte-Rizzoli, P., B. D. Cornuelle, and D. B. Haidvogel (1982). Gulf Stream acoustic tomography: modelling simulations. *Ocean Modelling, 46*, 10–19. [368]

Malanotte-Rizzoli, P., and W. R. Holland (1986). Data constraints applied to models of the ocean general circulation. Part I: The steady case. *J. Phys. Oceanogr., 16*, 1665–87. [268]

Malanotte-Rizzoli, P., J. Spiesberger, and M. Chajes (1985). Gulf Stream variability for acoustic tomography. *Deep-Sea Res., 32*, 237–50. [368]

Manabe, S., and R. J. Stouffer (1993). Century scale effect of increased atmospheric CO_2 on the ocean-atmosphere system. *Nature, 364*, 215–18. [324]

Menemenlis, D., and D. Farmer (1992). Acoustical measurement of current and vorticity beneath ice. *J. Atmos. Oceanic Tech., 9*, 827–49. [185]

Menke, W. (1989). *Geophysical Data Analysis: Discrete Inverse Theory*, 2nd ed. New York: Academic. [249]

Mercer, J. A. (1986). Acoustic oceanography by remote sensing. *IEEE J. Oceanic Eng., OE-11*, 51–7. [366]

Mercer, J. A. (1988). Non-reciprocity of simulated long-range acoustic transmissions. *J. Acoust. Soc. Am., 84*, 999–1006. [126]

Mercer, J. A., and J. R. Booker (1983). Long-range propagation of sound through oceanic mesoscale structures. *J. Geophys. Res., 88*, 689–99. [56, 282]

Mercier, H., and A. Colin de Verdière (1985). Space and time scales of mesoscale motions in the eastern North Atlantic. *J. Phys. Oceanogr., 15*, 171–83. [264, 346]

Metzger, K. M., Jr. (1983). Signal processing equipment and techniques for use in measuring ocean acoustic multipath structures. Ph.D. thesis, University of Michigan, Ann Arbor. [185, 192, 202, 219, 221, 367]

Metzger, K. M., Jr., and R. J. Bowens (1972). An ordered table of primitive polynomials over GF(2) of degrees 2 through 19 for use with linear maximal sequence generators. Cooley Electronics Laboratory Technical Memorandum no. 107, University of Michigan. [218]

Mikolajewicz, U., B. D. Santer, and E. Maier-Reimer (1990). Ocean response to greenhouse warming. *Nature, 345*, 589–93. [324]

Milder, D. M. (1969). Ray and wave invariants for SOFAR channel propagation. *J. Acoust. Soc. Am., 46*, 1259-63. [137, 158]

Miller, J. C. (1982). Ocean acoustic rays in the deep six sound channel. *J. Acoust. Soc. Am., 71*, 859–62. [45]

Miller, J. C. (1986). Hamiltonian perturbation theory for acoustic rays in a range dependent sound channel. *J. Acoust. Soc. Am., 79*, 338–46. [44]

Miller, J. H. (1987). Estimation of sea surface wave spectra using acoustic tomography. Ph.D. thesis, Woods Hole Oceanographic Institution–Massachusetts Institute of Technology Joint Program in Oceanography. [367]

Miller, J. H., J. F. Lynch, and C.-S. Chiu (1989). Estimation of sea surface spectra using acoustic tomography. *J. Acoust. Soc. Am., 86*, 326–45. [367]

Miller, J. H., J. F. Lynch, C.-S. Chiu, E. L. Westreich, J. S. Gerber, R. Hippenstiel, and E. Chaulk (1993). Acoustic measurements of surface gravity wave spectra in Monterey Bay using mode travel time fluctuations. *J. Acoust. Soc. Am., 94*, 954–74. [368]

Millero, F., and X. Li (1994). Comments on "On equations for the speed of sound in sea water." *J. Acoust. Soc. Am., 95*, 2757–9. [352]

Milne, P. H. (1983). *Underwater Acoustic Positioning Systems.* New York: E & F. N. Spon. [212]

MODE Group (1978). The Mid-ocean Dynamics Experiment. *Deep-Sea Res., 25*, 859–910. [346]

Monjo, C. L. (1987). Modeling of acoustic transmission in the Straits of Florida Acoustic Reciprocal Transmission Experiment. Ph.D. thesis, University of Miami, Miami, FL. [367]

Morawitz, W. M. L., P. Sutton, P. F. Worcester, B. D. Cornuelle, J. Lynch, and R. Pawlowicz (1994). Evolution of the 3-dimensional temperature structure and heat content in the Greenland Sea during 1988–89 from tomographic measurements. *EOS, Trans. Am. Geophys. Union, 75*, 118. [348, 368]

Morse, P. M., and H. Feshbach (1953). *Methods of Theoretical Physics*, 2 vols. New York: McGraw-Hill. [62]

Munk, W. (1974). Sound channel in an exponentially stratified ocean, with application to SOFAR. *J. Acoust. Soc. Am., 55*, 220–6. [36, 106, 107, 108, 109, 110, 111, 112, 113, 114]

Munk, W. (1980). Horizontal deflection of acoustic paths by mesoscale eddies. *J. Phys. Oceanogr., 10*, 596–604. [164, 165, 166, 167]

Munk, W. (1981). Internal waves and small scale processes. In *Evolution of Physical Oceanography – Scientific Surveys in Honor of Henry Stommel*, ed. B. A. Warren and C. Wunsch, pp. 264-91. Cambridge, MA: MIT Press. [148, 358]

Munk, W. (1986). Acoustic monitoring of ocean gyres. *J. Fluid Mech., 173*, 43–53. [121]

Munk, W. (1990). The Heard Island Experiment. *Naval Research Reviews, 42*, 2–22. [326, 369]

Munk, W. (1991). Refraction of acoustic modes in very long-range transmissions. In *Ocean Variability and Acoustic Propagation*, ed. J. Potter and A. Warn-Varnas, pp. 539–43. Dordrecht: Kluwer. [169, 369]

Munk, W., and A. M. G. Forbes (1989). Global ocean warming: an acoustic measure? *J. Phys. Oceanogr., 19*, 1765–78. [332, 369]

Munk, W., W. C. O'Reilly, and J. L. Reid (1988). Australia–Bermuda sound transmission experiment (1960) revisited. *J. Phys. Oceanogr., 18*, 1876–98. [336, 339, 340]

Munk, W., R. Spindel, A. Baggeroer, and T. Birdsall (1994). The Heard Island feasibility test. *J. Acoust. Soc. Am., 96*, 2330–42. [332, 369]

Munk, W., and P. F. Worcester (1976). Monitoring the ocean acoustically. In *Science, Technology, and the Modern Navy, Thirtieth Anniversary, 1946–1976*, ed. E. I. Salkovitz, pp. 497–508. Office of Naval Research, Arlington, VA (ONR-37). Also appears as: (1977). Weather and climate under the sea – the navy's habitat. In *Science and the Future Navy – A Symposium, 30th Anniversary Volume ONR*, pp. 42–52. Washington, DC: National Academy of Sciences. [355]

Munk, W., and P. F. Worcester (1988). Ocean acoustic tomography. *Oceanography, 1*, 8–10. [366]

Munk, W., P. F. Worcester, and F. Zachariasen (1981*a*). Scattering of sound by internal wave currents: the relation to vertical momentum flux. *J. Phys. Oceanogr., 11*, 442–54. [121]

Munk, W., and C. Wunsch (1979). Ocean acoustic tomography: a scheme for large scale monitoring. *Deep-Sea Res., 26*, 123–61. [9, 28, 42, 107, 225, 284, 355]

Munk, W., and C. Wunsch (1982*a*). Observing the ocean in the 1990's. *Phil. Trans. Roy. Soc., A307*, 439–64. [28, 121, 124, 125, 207, 210, 285, 324, 366]

Munk, W., and C. Wunsch (1982*b*). Up/down resolution in ocean acoustic tomography. *Deep-Sea Res., 29*, 415–36. [15, 51, 86, 255, 257, 264, 267, 299, 366]

Munk, W., and C. Wunsch (1983). Ocean acoustic tomography: rays and modes. *Rev. Geophys. Space Phys., 21*, 777–93. [46, 61, 66, 74, 283, 366]

Munk, W., and C. Wunsch (1985). Biases and caustics in long-range acoustic tomography. *Deep-Sea Res., 32*, 1317–46. [56, 107]

Munk, W., and C. Wunsch (1987). Bias in acoustic travel time through an ocean with adiabatic range-dependence. *Geophys. Astrophys. Fluid Dyn., 39*, 1–24. [56, 107, 137]

Munk, W., and F. Zachariasen (1976). Sound propagation through a fluctuating stratified ocean: theory and observation. *J. Acoust. Soc. Am., 59*, 818–38. [148, 359]

Munk, W., and F. Zachariasen (1991). Refraction of sound by islands and seamounts. *J. Atmos. Oceanic Technol., 8*, 554–74. [170, 171, 345]

Munk, W., B. Zetler, J. G. Clark, D. Porter, J. Spiesberger, and R. Spindel (1981*b*). Tidal effects on long-range sound transmission. *J. Geophys. Res., 86*, 6399–410. [132, 362, 367]

Newhall, B. K., M. J. Jacobson, and W. L. Siegmann (1977). Effect of a random ocean current on acoustic transmission in an isospeed channel. *J. Acoust. Soc. Am., 62*, 1165–75. [119]

Noble, B., and J. W. Daniel (1977). *Applied Linear Algebra*, 2nd ed. Englewood Cliffs, NJ: Prentice-Hall. [247]

Northrop, J. (1968). Submarine topographic echoes from CHASE V. *J. Geophys. Res., 73*, 3909–16. [332, 344]

Norton, S. J. (1988). Tomographic reconstruction of two-dimensional vector fields: application to flow imaging. *Geophys. J. Roy. Astron. Soc., 97*, 162–8. [125, 126]

Ocean Tomography Group (1982). A demonstration of ocean acoustic tomography. *Nature, 299*, 121–5. [151, 288, 347, 367]

Officer, C. B. (1958). *Introduction to the Theory of Sound Transmission*. New York: McGraw-Hill. [61]

Palmer, D. R., T. M. Georges, J. J. Wilson, L. D. Weiner, J. A. Paisley, R. Mathiesen, R. R. Pleshek, and R. R. Mabe (1994). Reception at Ascension of the Heard Island Feasibility Test transmissions. *J. Acoust. Soc. Am., 96*, 2432–40. [369]

Palmer, D. R., L. M. Lawson, D. A. Seem, and Y. H. Daneshzadeh (1985). Ray path identification and acoustic tomography in the Straits of Florida. *J. Geophys. Res., 90*, 4977–89. [367, 371]

Papoulis, A. (1977). *Signal Analysis*. New York: McGraw-Hill. [186, 187]

Parker, R. L. (1972). Understanding inverse theory with grossly inadequate data. *Geophys. J. Roy. Astron. Soc., 29*, 123–38. [276]

Parker, R. L. (1977). Understanding inverse theory. *Ann. Revs. Earth and Planet. Seism., 5*, 35–64. [224]

Parker, R. L. (1994). *Geophysical Inverse Theory*. Princeton University Press. [224]

Parrilla, G., A. Lavin, H. Bryden, M. Garcia, and R. Millard (1994). Rising temperatures in the subtropical North Atlantic Ocean over the past 35 years. *Nature, 369*, 48–51. [324, 325]

Pawlowicz, R. (1994). Tomographic observations of deep convection and the thermal evolution of the Greenland Sea Gyre 1988–89. Ph.D. thesis, Massachusetts Institute of Technology–Woods Hole Oceanographic Institution Joint Program in Oceanography/Oceanographic Engineering. [368]

Pawlowicz, R., J. F. Lynch, W. B. Owens, P. F. Worcester, W. M. L. Morawitz, and P. J. Sutton (in press). Thermal evolution of the Greenland Sea Gyre in 1988–89. *J. Geophys. Res.* [347, 348, 368]

Peckham, D. A., D. Horwitt, and K. R. Hardy (1990). Application of SCSI hard disk drives in marine instrumentation. In *Marine Instrumentation '90* (27 Feb.–1 March 1990, San Diego), pp. 165–7. Spring Valley, CA: West Star Publications. [368, 373]

Pedlosky, J. (1987). *Geophysical Fluid Dynamics*, 2nd ed. Berlin: Springer-Verlag. [112, 122, 268]

Pierce, A. D. (1989). *Acoustics, An Introduction to Its Physical Principles and Applications*. Woodbury, NY: Acoustical Society of America/AIP. [12, 116, 117, 128]

Piquet-Pellorce, F., F. Martin-Lauzer, and F. Evennou (1992). Données de tomographie acoustique recueillies pendant la campagne GASTOM 90. *Report D'Etudes No. 9/92/CMO/EO*, Etablissement Principal du Service Hydrographique et Océanographique de la Marine–CMO, Brest, France. [369]

Plumb, R. A. (1986). Three-dimensional propagation of transient quasi-geostrophic eddies and its relationship with the eddy forcing of the time-mean flow. *J. Atmos. Sci., 43*, 1657–78. [351]

Pond, S., and G. L. Pickard (1983). *Introductory Physical Oceanography*, 2nd ed. London: Pergamon Press. [35]

Porter, R. P. (1973). Dispersion of axial SOFAR propagation in the western Mediterranean. *J. Acoust. Soc. Am., 53*, 181–91. [70]

Potter, J. R. (1994). ATOC: sound policy or enviro-vandalism? Aspects of a modern media-fueled policy issue. *J. Environ. Devel., 3*, 47–62. [379]

Press, W. H., S. A. Teukolsky, W. T. Vetterling, and B. P. Flannery (1992*a*). *Numerical Recipes in C*, 2nd ed. Cambridge University Press. [301]

Press, W. H., S. A. Teukolsky, W. T. Vetterling, and B. P. Flannery (1992*b*). *Numerical Recipes in Fortran*, 2nd ed. Cambridge University Press. [301]

Pridham, R. G., and R. A. Mucci (1979). Shifted sideband beamformer. *IEEE Trans. Acoust., Speech, Signal Processing, ASSP-27*, 713–22. [185, 202]

Radon, J. (1917). Ueber die Bestimmung von Funktionen durch ihre Integralwerte längs gewisser Mannigfaltigkeiten, *Ber. Saechs. Akademie der Wissenschaften, Leipzig, Mathematisch-Physikalische Klasse, 69*, 262–77. [285]

Rauch, H. E., F. Tung, and C. T. Streibel (1965). Maximum likelihood estimates of linear dynamic systems. *AIAA Journal, 3*, 1445–50 (reprinted in Sorenson, 1985). [314]

Richman, J. G., C. Wunsch, and N. G. Hogg (1977). Space and time scales of mesoscale motion in the western North Atlantic. *Rev. Geophys. Space Phys., 15*, 385–420. [264, 288, 289]

Rihaczek, A. (1969). *Principles of High-Resolution Radar*. New York: McGraw-Hill. [203, 205]

Ripley, B. D. (1981). *Spatial Statistics*. New York: Wiley. [263, 275, 301]

Rodi, W. L., P. Glover, T. M. C. Li, and S. S. Alexander (1975). A fast, accurate method for computing group-velocity partial derivatives for Rayleigh and Love modes. *Bull. Seism. Soc. Am., 65*, 1105–14. [76]

Roemmich, D., and C. Wunsch (1984). Apparent changes in the climate state of the deep North Atlantic Ocean. *Nature, 307*, 447–50. [324, 325]

Romm, J. J. (1987). Applications of normal mode analysis to ocean acoustic tomography. Ph.D. thesis, Massachusetts Institute of Technology. [74, 76, 367]

Rossby, T. (1975). An oceanic vorticity meter. *J. Marine Res., 33*, 213–22. [122]

Rowland, S. W. (1979). Computer implementation of image reconstruction formulas. In *Image Reconstruction from Projections. Implementation and Applications*, ed. G. T. Herman, pp. 9–80. Berlin: Springer-Verlag. [285]

Rozenberg, L. L. (1949). Concerning one new phenomenon in hydroacoustics. *Dokl. Akad. Nauk SSSR, 69*, 175–6. [357]

Rudels, B. (1990). Haline convection in the Greenland Sea. *Deep-Sea Res., 36*, 1491–511. [348]

Sanford, T B. (1974). Observations of strong current shears in the deep ocean and some implications on sound rays. *J. Acoust. Soc. Am., 56*, 1118–21. [126, 127, 200]

Sasaki, Y. (1970). Some basic formalisms in numerical variational analysis. *Mon. Wea. Rev., 98*, 875–83. [234]

Scales, L. E. (1985). *Introduction to Non-Linear Optimization*. Berlin: Springer-Verlag. [280]

Schröter, J., and C. Wunsch (1986). Solution of nonlinear finite difference ocean models by optimization methods with sensitivity and observational strategy analysis. *J. Phys. Oceanogr., 16*, 1855–74. [268, 303, 320, 321]

Schwiderski, E. W. (1980). Ocean tides. *Mar. Geodesy, 3*, 161–255. [131]

Seber, G. A. F. (1977). *Linear Regression Analysis*. New York: Wiley. [268]

Seber, G. A. F., and C. J. Wild (1989). *Nonlinear Regression*. New York: Wiley. [280, 281]

Semtner, A. J., and R. M. Chervin (1988). A simulation of the global ocean circulation with resolved eddies. *J. Geophys. Res., 93*, 15502–22. [323]

Semtner, A. J., and R. M. Chervin (1990). Environmental effects on acoustic measures of global ocean warming. *J. Geophys. Res., 95*, 12973–82. [369]

Semtner, A. J., and R. M. Chervin (1992). Ocean general circulation from a global eddy-resolving model. *J. Geophys. Res., 97*, 5493–550. [304, 305]

Send, U. (in press). Peak tracking by simultaneous inversion. *Geophys. Res. Lett.* [218, 369]

Send, U., F. Schott, F. Gaillard, and Y. Desaubies (in press). Observation of a deep convection regime with acoustic tomography. *J. Geophys. Res.* [348]

Shang, E. C. (1989). Ocean acoustic tomography based on adiabatic mode theory. *J. Acoust. Soc. Am., 85*, 1531–7. [75]

Shang, E. C., and Y. Y. Wang (1991). On the calculation of modal travel time perturbation. *Sov. Phys. Acoust., 37*, 411–13. [75]

Shang, E. C., and Y. Y. Wang (1992). On the possibilities of monitoring El Niño by using modal ocean acoustic tomography. *J. Acoust. Soc. Am., 91*, 136–40. [75]

Shang, E. C., and Y. Y. Wang (1993*a*). The nonlinearity of modal travel time perturbation. In *Computational Acoustics*, vol. 1, ed. R. Lau, D. Lee, and A. Robinson, pp. 385–97. Amsterdam: Elsevier. [159]

Shang, E. C., and Y. Y. Wang (1993*b*). Acoustic travel time computation based on PE solution. *J. Computational Acoust., 1*, 91–100. [159]

Shang, E. C., Y. Y. Wang, and T. M. Georges (1994). Dispersion and repopulation of Heard–Ascension modes. *J. Acoust. Soc. Am., 96*, 2371–9. [334, 335, 369]

Sheehy, M. J., and R. Halley (1957). Measurement of attenuation of low-frequency underwater sound. *J. Acoust. Soc. Am., 29*, 464–9. [328, 329]

Shockley, R. C., J. Northrop, P. G. Hansen, and C. Hartdegen (1982). SOFAR propagation paths from Australia to Bermuda: comparison of signal speed algorithms and experiments. *J. Acoust. Soc. Am., 71*, 51–60. [329, 336, 339]

Simkin, T., and R. Fiske (1983). *Krakatau 1883 – The Volcanic Eruption and Its Effects*. Washington, DC: Smithsonian Institution Press. [332]

SIZEX Group (1989). SIZEX experiment report. The Nansen Remote Sensing Center technical report 23. [101]

Slichter, L. B. (1932). The theory of the interpretation of seismic travel-time curves in horizontal structures. *Physics, 3*, 273–95. [50]

Smith, K. B., M. G. Brown, and F. D. Tappert (1992*a*). Ray chaos in underwater acoustics. *J. Acoust. Soc. Am., 91*, 1939–49. [139, 155]

Smith, K. B., M. G. Brown, and F. D. Tappert (1992*b*). Acoustic ray chaos induced by mesoscale ocean structure. *J. Acoust. Soc. Am., 91*, 1950–9. [155, 156, 157]

Sneddon, I. N. (1972). *The Use of Integral Transforms*. New York: McGraw-Hill. [45]

Snodgrass, F. E. (1968). Deep sea instrument capsule. *Science, 162*, 78–87. [373]

Sorenson, H. W. (ed.) (1985). *Kalman Filtering: Theory and Application*. New York: IEEE Press. [312, 315]

Spencer, C., and D. Gubbins (1980). Travel time inversion for simultaneous earthquake location and velocity structure determination in laterally varying media. *Geophys. J. Roy. Astron. Soc., 63*, 95–116. [16]

Spiesberger, J. L. (1980). Stability of long range ocean acoustic multipaths. Ph.D. thesis, Scripps Institution of Oceanography, University of California, San Diego, La Jolla, CA. [367]

Spiesberger, J. L. (1985*a*). Gyre-scale acoustic tomography: biases, iterated inversions, and numerical methods. *J. Geophys. Res., 90*, 11869–76. [56]

Spiesberger, J. L. (1985*b*). Ocean acoustic tomography: travel time biases. *J. Acoust. Soc. Am., 77*, 83–100. [56]

Spiesberger, J. L. (1989). Remote sensing of western boundary currents using acoustic tomography. *J. Acoust. Soc. Am., 86*, 346–51. [367]

Spiesberger, J. L. (1993). Is Del Grosso's sound-speed algorithm correct? *J. Acoust. Soc. Am., 93*, 2235–7. [351, 368]

Spiesberger, J. L., T. G. Birdsall, K. Metzger, Jr., R. A. Knox, C. W. Spofford, and R. C. Spindel (1983). Measurements of Gulf Stream meandering and evidence of seasonal thermocline development using long range acoustic transmissions. *J. Phys. Oceanogr.*, *13*, 1836–46. [365, 367]

Spiesberger, J. L., P. J. Bushong, K. Metzger, Jr., and T. G. Birdsall (1989*a*). Ocean acoustic tomography: estimating the acoustic travel time with phase. *IEEE J. Oceanic Eng.*, *14*, 108–19. [332, 367]

Spiesberger, J. L., P. J. Bushong, K. Metzger, Jr., and T. G. Birdsall (1989*b*). Basin-scale tomography: synoptic measurements of a 4000-km length section in the Pacific. *J. Phys. Oceanogr.*, *19*, 1073–90. [332, 367]

Spiesberger, J. L., P. Malanotte-Rizzoli, and E. B. Welsh (1985). Travel time and geometry of steep acoustic rays subject to Gulf Stream variability. *J. Acoust. Soc. Am.*, *78*, 260–3. [368]

Spiesberger, J. L., and K. Metzger, Jr. (1991*a*). New estimates of sound speed in water. *J. Acoust. Soc. Am.*, *89*, 1697–700. [351, 368]

Spiesberger, J. L., and K. Metzger, Jr. (1991*b*). A new algorithm for sound speed in seawater. *J. Acoust. Soc. Am.*, *89*, 2677–88. [351, 368]

Spiesberger, J. L., and K. Metzger, Jr. (1991*c*). Basin-scale tomography: a new tool for studying weather and climate. *J. Geophys. Res.*, *96*, 4869–89. [314, 332, 333, 368]

Spiesberger, J. L., and K. Metzger, Jr. (1992). Basin-scale ocean monitoring with acoustic thermometers. *Oceanography*, *5*, 92–8. [367]

Spiesberger, J. L., K. Metzger, and J. A. Furgerson (1992). Listening for climatic temperature change in the northeast Pacific: 1983–1989. *J. Acoust. Soc. Am.*, *92*, 384–96. [367]

Spiesberger, J. L., and R. C. Spindel (1985). Gulf Stream tomography: preliminary results from an experiment. In *Proceedings of the Gulf Stream Workshop*, ed. R. Watts, pp. 479–94. Exeter: University of Rhode Island. [368]

Spiesberger, J. L., R. C. Spindel, and K. Metzger (1980). Stability and identification of ocean acoustic multipaths. *J. Acoust. Soc. Am.*, *67*, 2011–17. [361, 362, 367]

Spiesberger, J. L., E. Terray, and K. Prada (1994). Successful ray modeling of acoustic multipaths over a 3000 km section in the Pacific with rays. *J. Acoust. Soc. Am.*, *95*, 3654–7. [332, 333, 368]

Spiesberger, J. L., and P. F. Worcester (1981). Fluctuations of resolved acoustic multipaths at long range in the ocean. *J. Acoust. Soc. Am.*, *70*, 565–76. [367]

Spiesberger, J. L., and P. F. Worcester (1983). Perturbations in travel time and ray geometry due to mesoscale disturbances: a comparison of exact and approximate calculations. *J. Acoust. Soc. Am.*, *74*, 219–25. [56, 367]

Spindel, R. C. (1979). An underwater acoustic pulse compression system. *IEEE Trans. Acoust., Speech, Signal Processing*, *ASSP-27*, 723–8. [372]

Spindel, R. C. (1982). Ocean acoustic tomography: a new measuring tool. *Oceanus*, Woods Hole Oceanographic Institution, *25*, 12–21. [367]

Spindel, R. C. (1985), Signal processing in ocean tomography. In *Adaptive Methods in Underwater Acoustics*, ed. H. G. Urban, pp. 687–710. Dordrecht: Reidel. [175]

Spindel, R. C., and R. P. Porter (1977). A mobile coherent, low frequency acoustic range. *IEEE J. Oceanic Eng.*, *OE-2*, 331–7. [372]

Spindel, R. C., and J. L. Spiesberger (1981). Multipath variability due to the Gulf Stream. *J. Acoust. Soc. Am.*, *69*, 982–8. [362]

Spindel, R. C., and P. F. Worcester (1986). Technology in ocean acoustic tomography. *Mar. Tech. Soc. J.*, *20*, 68–72. [182, 368]

Spindel, R. C., and P. F. Worcester (1990). Ocean acoustic tomography programs: accomplishments and plans. In *OCEANS '90* (24–26 Sept. 1990, Washington, DC), pp. 1–10. New York: IEEE. [366]

Spindel, R. C., K. R. Peal, and D. E. Koelsch (1978). A microprocessor acoustic data buoy. *Proc. IEEE Oceans '78*, 527–31. [372]

Spindel, R. C., P. F. Worcester, D. C. Webb, P. Boutin, K. Peal, and A. Bradley (1982). Instrumentation for ocean acoustic tomography. In *OCEANS '82* (20–22 Sept. 1982, Washington, DC), pp. 92–9. New York: IEEE. [208, 210, 212, 363, 367]

Spofford, C. W., and A. P. Stokes (1984). An iterative perturbation approach for ocean acoustic tomography. *J. Acoust. Soc. Am., 75*, 1443–50. [281]

Stallworth, L. A. (1973). A new method for measuring ocean and tidal currents. In *Oceans '73* (proceedings of the IEEE International Conference on Engineering in the Ocean Environment, 25–28 Sept. 1973, Seattle, WA), pp. 55–8. New York: IEEE. [121]

Stallworth, L. A., and M. J. Jacobson (1970). Acoustic propagation in an isospeed channel with uniform tidal current and depth change. *J. Acoust. Soc. Am., 48*, 382–91. [119]

Stallworth, L. A., and M. J. Jacobson (1972*a*). Sound transmission in an isospeed ocean channel with depth-dependent current. *J. Acoust. Soc. Am., 51*, 1738–50. [119]

Stallworth, L. A., and M. J. Jacobson (1972*b*). Acoustic propagation in a uniformly moving ocean channel with depth-dependent sound speed. *J. Acoust. Soc. Am., 52*, 344–55. [119]

Steinberg, J. C., and T. G. Birdsall (1966). Underwater sound propagation in the Straits of Florida. *J. Acoust. Soc. Am., 39*, 301–15. [148]

Sternberg, R. L. (1987). Beamforming with acoustic lenses and filter plates. In *Progress in Underwater Acoustics*, ed. H. M. Merklinger, pp. 651–5. New York: Plenum Press. [166]

Stoughton, R. B., S. M. Flatté, and B. M. Howe (1986). Acoustic measurements of internal wave rms displacement and rms horizontal current off Bermuda in late 1983. *J. Geophys. Res., 91*, 7721–32. [134, 135, 200, 201, 353, 367]

Strang, G. (1986). *Introduction to Applied Mathematics*. Wellesley, MA: Wellesley-Cambridge Press. [247, 277]

Sutton, P. J. (1993). The upper ocean in the Greenland Sea during 1988–89 from modal analyses of tomographic data. Ph.D. thesis, Scripps Institution of Oceanography, University of California, San Diego, La Jolla, CA. [296, 348, 368]

Sutton, P. J., W. M. L. Morawitz, B. D. Cornuelle, G. Masters, and P. F. Worcester (1994). Incorporation of acoustic normal mode data into tomographic inversions in the Greenland Sea. *J. Geophys. Res., 99*, 12487–502. [101, 297, 347, 368]

Sutton, P. J., P. F. Worcester, G. Masters, B. D. Cornuelle, and J. F. Lynch (1993). Ocean mixed layers and acoustic pulse propagation in the Greenland Sea. *J. Acoust. Soc. Am., 94*, 1517–26. [101, 368]

Talagrand, O., and P. Courtier (1987). Variational assimilation of meteorological observations with the adjoint vorticity equation. I: Theory. *Q. J. Roy. Meteorol. Soc., 113*, 1311–28. [318]

Tappert, F. D. (1977). The parabolic approximation method. In *Wave Propagation and Underwater Acoustics*, ed. J. B. Keller, pp. 224–84. Berlin: Springer-Verlag. [159, 334]

Tarantola, A. (1987). *Inverse Problem Theory. Methods for Data Fitting and Model Parameter Estimation*. Amsterdam: Elsevier. [224]

Tarantola, A., and B. Valette (1982). Generalized nonlinear inverse problems solved using the least squares criterion. *Rev. Geophys. Space Phys.*, *20*, 219–32. [280, 281, 282]

Taroudakis, M., and J. S. Papadakis (1993). A modal inversion scheme for ocean acoustic tomography. *J. Computational Acoust.*, *1*, 395–421. [159]

Thacker, W. C. (1986). Relationships between statistical and deterministic methods of data assimilation. In *Variational Methods in Geosciences*, ed. Y. K. Sasaki, pp. 173–9. Amsterdam: Elsevier. [319]

Thacker, W. C., and R. B. Long (1988). Fitting dynamics to data. *J. Geophys. Res.*, *93*, 1227–40. [318, 319]

THETIS Group (1994). Open-ocean deep convection explored in the Mediterranean. *EOS, Trans. Am. Geophys. Union*, *75*, 219–21. (Members of the THETIS Group are F. Schott, U. Send, G. Krahmann, C. Mertens, M. Rhein, M. Visbeck, Y. Desaubies, F. Gaillard, T. Terre, J. Papadakis, M. Taroudakis, G. Athanassoulis and E. Skarsoulis.) [348, 369]

Thompson, R. J. (1972). Ray theory for an inhomogeneous moving medium. *J. Acoust. Soc. Am.*, *51*, 1675-82. [117]

Tolstoy, I., and C. S. Clay (1966). *Ocean Acoustics*. New York: McGraw-Hill. [61]

Tolstoy, A., O. Diachok, and L. N. Frazer (1991). Acoustic tomography via matched field processing. *J. Acoust. Soc. Am.*, *89*, 1119–27. [302]

Turin, G. L. (1960). An introduction to matched filters. *I.R.E. Trans. Inform. Theory*, *IT-6*, 311–29. [188, 189, 190]

Turner, J. S. (1973). *Buoyancy Effects in Fluids*. Cambridge University Press. [3]

Tziperman, E., and W. C. Thacker (1989). An optimal control/adjoint equations approach to studying the oceanic general circulation. *J. Phys. Oceanogr.*, *19*, 1471–85. [318]

Uginčius, P. (1965). Acoustic-ray equations for a moving, inhomogeneous medium. *J. Acoust. Soc. Am.*, *37*, 476–9. [118]

Uginčius, P. (1972). Ray acoustics and Fermat's principle in a moving inhomogeneous medium. *J. Acoust. Soc. Am.*, *51*, 1759-63. [118, 119]

Urick, R. J. (1983). *Principles of Underwater Sound*, 3rd ed. New York: McGraw-Hill. [33, 176, 179, 180]

Van Huffel, S., and J. Vandewalle (1991). *The Total Least Squares Problem. Computational Aspects and Analysis*. Philadelphia: SIAM. [263, 281]

Wagner, H. M. (1969). *Principles of Operations Research. With Applications to Managerial Decisions*. Englewood Cliffs, NJ: Prentice-Hall. [277]

Wahba, G. (1990). *Spline Models for Observational Data*. Philadelphia: Society for Industrial and Applied Mathematics. [275]

Wales, S. C., and O. I. Diachok (1981). Ambient noise vertical directionality in the northwest Atlantic. *J. Acoust. Soc. Am.*, *70*, 577–82. [181]

Westreich, E. L., C.-S. Chiu, J. H. Miller, J. F. Lynch, and M. D. Collins (in press). Modeling pulse transmission in the Monterey Bay using parabolic equation methods. *J. Acoust. Soc. Am.*. [368]

Widfeldt, J. A., and M. J. Jacobson (1976). Acoustic phase and amplitude of a signal transmitted through a uniform flow in the deep ocean. *J. Acoust. Soc. Am.*, *59*, 852–60. [119]

Wiggins, R. A. (1972). The general linear inverse problem: implication of surface waves and free oscillations for earth structure. *Rev. Geophys. Space Phys.*, *10*, 251–85. [249]

Wood, A. B. (1965). From the Board of Invention and Research to the Royal Naval Scientific Service. *J. Roy. Nav. Sci. Serv., 20*, 16. [357]

Worcester, P. F. (1977*a*). Reciprocal acoustic transmission in a mid-ocean environment. Ph.D. thesis, Scripps Institution of Oceanography, University of California, San Diego, La Jolla, CA. [121, 127, 129, 360, 361, 367]

Worcester, P. F. (1977*b*). Reciprocal acoustic transmission in a midocean environment. *J. Acoust. Soc. Am., 62*, 895–905. [121, 126, 127, 129, 130, 193, 210, 360, 367]

Worcester, P. F. (1979). Reciprocal acoustic transmission in a midocean environment: fluctuations. *J. Acoust. Soc. Am., 66*, 1173–81. [367]

Worcester, P. F. (1981). An example of ocean acoustic multipath identification at long range using both travel time and vertical arrival angle. *J. Acoust. Soc. Am., 70*, 1743–7. [51, 101, 202, 216, 363]

Worcester, P. F. (1989). Remote sensing of the ocean using acoustic tomography. In *RSRM '87: Advances in Remote Sensing Retrieval Methods*, ed. A. Deepak, H. E. Fleming, and J. S. Theon, pp. 1–11. Hampton, VA: A. Deepak Publishing. [366]

Worcester, P. F., and B. D. Cornuelle (1982). Ocean acoustic tomography: currents. In *Proceedings of the IEEE Second Working Conference on Current Measurements* (19–21 Jan. 1982, Hilton Head, SC), pp. 131–5. New York: IEEE. [367]

Worcester, P. F., B. D. Cornuelle, J. A. Hildebrand, W. S. Hodgkiss, Jr., T. F. Duda, J. Boyd, B. M. Howe, J. A. Mercer, and R. C. Spindel (1994). A comparison of measured and predicted broadband acoustic arrival patterns in travel time–depth coordinates at 1000-km range. *J. Acoust. Soc. Am., 95*, 3118–28. [10, 99, 368, 375, 376]

Worcester, P. F., B. D. Cornuelle, and R. C. Spindel (1991*a*). A review of ocean acoustic tomography: 1987–1990. In *Reviews of Geophysics, Supplement, U.S. National Report to the International Union of Geodesy and Geophysics, 1987–1990*, pp. 557–70. [366]

Worcester, P. F., B. D. Dushaw, and B. M. Howe (1991*b*). Gyre-scale reciprocal acoustic transmissions. In *Ocean Variability and Acoustic Propagation*, ed. J. Potter and A. Warn-Varnas, pp. 119–34. Dordrecht: Kluwer. [130, 131, 332, 333, 350, 351, 368]

Worcester, P. F., B. D. Dushaw, and B. M. Howe (1990). Gyre-scale current measurements using reciprocal acoustic transmissions. In *Proceedings of the IEEE Fourth Working Conference on Current Measurement* (3–5 Apr. 1990, Clinton, MD), pp. 65–70. New York: IEEE. [350, 368]

Worcester, P. F., J. F. Lynch, W. M. L. Morawitz, R. Pawlowicz, P. J. Sutton, B. D. Cornuelle, O. M. Johannessen, W. H. Munk, W. B. Owens, R. Shuchman, and R. C. Spindel (1993). Evolution of the large-scale temperature field in the Greenland Sea during 1988–1989 from tomographic measurements. *Geophys. Res. Lett., 20*, 2211–14. [78, 101, 216, 347, 368]

Worcester, P. F., and B. Ma (in press). Ocean acoustic propagation atlas. Scripps Institution of Oceanography Reference Series. [382]

Worcester, P. F., D. A. Peckham, K. R. Hardy, and F. O. Dormer (1985*a*). AVATAR: second-generation transceiver electronics for ocean acoustic tomography. In *Oceans '85: Ocean Engineering and the Environment* (12–14 Nov. 1985, San Diego, CA), pp. 654–62. New York: IEEE. [208, 209, 368, 373]

Worcester, P. F., R. Spindel, and B. Howe (1985*b*). Reciprocal acoustic transmissions: instrumentation for mesoscale monitoring of ocean currents. *IEEE J. Ocean. Eng., OE-10*, 123–37. [11, 130, 202, 203, 208, 210, 212, 217, 367, 371]

Worcester, P. F., G. O. Williams, and S. M. Flatté (1981). Fluctuations of resolved acoustic multipaths at short range in the ocean. *J. Acoust Soc. Am., 70,* 825–40. [129]

Wunsch, C. (1975). Internal tides in the ocean. *Rev. Geophys. Space Phys., 13,* 167–82. [131, 352]

Wunsch, C. (1978). The North Atlantic general circulation west of 50 W determined by inverse methods. *Rev. Geophys. Space Phys., 16,* 583–620. [249]

Wunsch, C. (1984). An eclectic Atlantic Ocean circulation model. Part I: The meridional flux of heat. *J. Phys. Oceanogr., 14,* 1712–33. [276]

Wunsch, C. (1987). Acoustic tomography by Hamiltonian methods including the adiabatic approximation. *Rev. Geophys. 25,* 41–53. [44, 56, 137]

Wunsch, C. (1988). Transient tracers as a problem in control theory. *J. Geophys. Res., 93,* 8099–110. [318, 319]

Wunsch, C. (1989). Comments on oceanographic instrumentation development. *J. Oceanogr. Soc., 2,* 26–7, 64. [346]

Wunsch, C. (in press). *The Ocean Circulation Inverse Problem.* Cambridge University Press. [249, 301, 319, 356]

Wunsch, C., and J.-F. Minster (1982). Methods for box models and ocean circulation tracers: mathematical programming and non-linear inverse theory. *J. Geophys. Res., 87,* 5647–62. [276]

INDEX OF AUTHORS & SUBJECTS